AN EIGHT PEAK INDEX OF MASS SPECTRA OF COMPOUNDS OF FORENSIC INTEREST

An Eight Peak Index of
MASS SPECTRA
OF COMPOUNDS
of Forensic Interest

R. E. ARDREY B.Sc., Ph.D.

C. BROWN B.Sc., A.M.B.C.S., M.I.D.P.M., M.I.Inf.Sc.

A. R. ALLAN B.Sc., Ph.D., C.Chem., M.R.S.C.

T. S. BAL B.Sc., M.Sc., Ph.D.

A. C. MOFFAT B.Pharm., Ph.D., F.P.S., C.Chem., F.R.S.C.

Published for

THE FORENSIC SCIENCE SOCIETY

by

The Scottish Academic Press

Edinburgh and London

Published for

THE FORENSIC SCIENCE SOCIETY

by

Scottish Academic Press
33 Montgomery Street
Edinburgh EH7 5JX

ISBN 0 7073 0341 9

British Library Cataloguing in Publication Data

An eight peak index of mass spectra of compounds
of forensic interest
    1. Mass spectrometry   2. Chemistry, organic
    I. Ardrey, R. E.
        547.3′0873′024618   QC463

Printed in Great Britain by
Clark Constable (1982) Ltd, Edinburgh

CONTENTS

INTRODUCTION

The central collection of mass spectral data for the Home Office Forensic Science Service was begun at the Central Research Establishment (CRE) at Aldermaston in 1974. The collection was intended to provide a small, but highly relevant collection, which would aid the identification of the majority of unknown compounds encountered during the mass spectral analysis of forensic samples.

The eight peak abbreviation was chosen to facilitate the exchange of data between laboratories, especially those who did not possess data systems. Methods of identifying compounds using various types of spectral abbreviation had been compared (1) and reviewed (2) and the method chosen for this collection had been shown to compare favourably with others. With data systems now being routinely used with mass spectrometers, most laboratories have the ability to rapidly search large libraries of full spectra. Work in our own laboratory has shown that, with minor modifications, these eight-peak abbreviated spectra may be combined with collections of full spectra and the identification of unknown compounds successfully achieved.

The original eight-peak data was extracted from spectra of some 200 drugs and metabolites run at CRE. This original data base has been expanded to include data from further samples run at CRE, together with data submitted by colleagues worldwide and especially from the Home Office and Metropolitan Police Forensic Science Laboratories in the UK and also from spectra appearing in the scientific literature. A list of laboratories which have submitted data is to be found in Appendix 1.

No attempt has been made to rationalise the data submitted in terms of each particular structure. Contributors have been asked to submit data only from accredited spectra of single compounds. Molecular weights have been checked and duplicate data have been rejected. A duplicate has been defined as a spectrum with five or more ions in common with another spectrum from the same compound unless the base peak or second most intense peak in either spectrum differs. Several major revisions of the data base have taken place since 1974 and the collection now comprises 3687 spectra from 2787 compounds. No attempt has been made to categorise the type of mass spectrometer on which the data have been acquired. The majority have been contributed by Forensic Science Laboratories in the UK and have been obtained using magnetic sector instruments with 70eV electron energy. This should be borne in mind, especially when attempting to identify spectra acquired on a quadrupole mass spectrometer.

Three listings are provided. The first, Appendix 2, is arranged in alphabetical order of compound name. It should be noted that the positions of substitution, e.g. *o-*, *p-*, *N-*, etc. and stereochemistry, e.g. *l-*, *d-*, etc. appear after the name to facilitate computer sorting.

The compounds whose spectra appear in the collection are listed in order of ascending molecular weight in Appendix 3. Certain decomposition products whose structures, and therefore molecular weights, are not known but whose origin is not in dispute, e.g. probe decomposition products, are included and these have been allocated a molecular weight of 999.

Appendix 4 is the major listing and shows the eight most intense peaks in each spectrum in descending order of intensity, indexed under both the base peak (indicated by an asterisk) and the second most intense peak. In the cases in which the asterisked peak appears second the true spectrum is obtained by interchanging the first and second peaks.

Searching of these indexes, either manually or using a computer, should allow the tentative identification of an unknown from the most intense ions in the mass spectrum. Since intensity information is not included in this collection, confirmation of identity should be made by comparison of the spectrum of the unknown with that from a control sample preferably run under identical experimental conditions. If this is not possible spectra obtained under controlled conditions may be used. To this end some 1000 full spectra of compounds contained in this index have been obtained at CRE and are to be published elsewhere.

If no identification is possible using this index the search may be widened by consulting the larger collections available. Two large, hard-copy, indexes are available. The first, The MSDC 8 Peak-Index (3) contains data from some 50,000 spectra. The second, Compilation of Mass Spectral Data (4), lists data from some 10,000 spectra. Both are indexed in terms of molecular weight, molecular formula and, for searching, the three most intense mass spectral ions. A further large, general collection of some 34,000 spectra is available through the National Bureau of Standards in the USA (5). This last collection is also produced as bar charts of the complete spectra arranged in molecular weight order.

The combination of libraries outlined above allows over 90% of the spectra encountered during routine forensic mass spectrometric examination to be identified. As the data bases, especially the smaller in-house collections, are updated the percentage will increase. On the rare occasions that it is not possible to identify an unknown by library matching it is necessary to extract as much information from the mass spectrum and attempt a first order interpretation (6).

REFERENCES

1. B. A. Knock, I. C. Smith, D. E. Wright, R. G. Ridley and W. Kelley, *Anal. Chem.*, 1970; *42*: 1516-1520.
2. R. G. Ridley. In Biochemical Applications of Mass Spectrometry, ed. G. R. Waller, New York: John Wiley; 1972, p. 177.
3. The Eight Peak Index of Mass Spectra, Third edition, Royal Society of Chemistry; 1983.
4. A. Cornu and R. Massot, Compilation of Mass Spectral Data, Second edition, London: Heyden; 1975.
5. EPA/NIH Mass Spectral Data Base, NSRDS-NBS 63. Washington: US Department of Commerce; 1978. First Supplement 1980.
6. F. W. McLafferty, Interpretation of Mass Spectra, Third edition, Mill Valley, California: University Science Books; 1980.

APPENDIX 1

LABORATORIES CONTRIBUTING DATA TO THIS COLLECTION

Abteilung fur Toxikologische Chemie, Karl Marx University, Leipzig, East Germany

Bundeskriminalamt, Wiesbaden, West Germany

Center for Human Toxicology, Salt Lake City, Utah, USA

Department of Forensic Medicine, University of Glasgow

Drug Control and Teaching Centre, University of London

Drug Enforcement Administration, Washington DC, USA

Fachbereich Klinische Medizin, University des Saarlandes, West Germany

Gerechtelijk Laboratorium Von Het Ministerie Van Justitie, Rijswijk, Netherlands

Home Office Forensic Science Laboratory—Aldermaston

Home Office Forensic Science Laboratory—Birmingham

Home Office Forensic Science Laboratory—Chorley

Home Office Forensic Science Laboratory—Wetherby

Institut fur Rechtsmedizin, Giessen, West Germany

Israel Police Research and Development Division, Jerusalem, Israel

Mass Spectrometry Data Centre

Metropolitan Police Forensic Science Laboratory, London

NIH/MIT Literature Collection

Royal Hong Kong Jockey Club

Statens Kriminaltekniska Laboratorium, Linköping, Sweden

Utah State Division of Health, Salt Lake City, Utah, USA

APPENDIX 2

ALPHABETICAL LISTING
Eight peak mass spectra in alphabetical order of
compound name

An asterisk after the compound name indicates that this spectrum forms
part of the collection of full spectra available at CRE to be published
separately

COMPOUND NAME	PEAKS								MWT
Abietic Acid	302	43	41	135	287	91	105	121	302
Absisic Acid	190	125	134	162	91	41	39	135	278
Acebutolol*	72	43	30	56	151	221	41	98	336
Acenocoumarin	310	121	92	43	120	311	63	65	353
Aceperone	245	123	258	165	186	95	246	0	396
Acephate	136	42	94	43	95	96	47	79	183
Acepifylline*	44	194	109	67	86	238	56	85	562
Acepromazine	58	197	86	42	43	44	85	196	326
Acepromazine	58	86	85	326	197	43	280	241	326
Acepromazine Maleate*	100	72	240	340	44	197	254	43	442
Acetaldehyde	29	44	43	42	26	41	27	25	44
7-Acetamidoflunitrazepam	325	297	324	43	256	306	296	326	325
3-Acetamido-2,4,6-triiodophenoxyacetic Acid	460	43	401	359	487	545	402	461	587
3-Acetamido-2,4,6-triiodophenoxyacetic Acid Methyl Ester	474	43	401	359	432	487	418	475	601
Acetaminophen	109	80	151	108	43	81	53	52	151
Acetaminosalol (Phenetsal, Salophen)	121	109	151	65	93	43	39	271	271
Acetanilide	93	135	94	66	43	65	39	0	135
Acetanilide	93	135	65	43	39	66	63	92	135
Acetazolamide*	43	180	42	45	44	64	100	222	222
Acetic Acid	45	29	43	60	42	44	41	31	60
7-Acetoamidoclonazepam	43	327	299	298	292	329	328	256	327
7-Acetoamidonitrazepam	293	265	264	292	43	222	223	294	293
7-Acetoamidonitrazepam	293	264	43	292	263	213	212	294	293
Acetohexamide*	210	56	43	184	211	75	99	76	324
5-Aceto-8-hydroxyquinoline Sulphate	94	77	138	95	51	66	172	45	472
Acetone	43	58	42	27	40	29	26	39	58
Acetonide	185	200	186	89	57	42	103	131	200
Acetonitrile	41	40	39	38	32	29	0	0	41
Acetophenazine	254	143	411	70	255	42	380	157	411
Acetophenazine	72	45	55	54	44	254	43	53	411
Acetophenazine	254	42	70	43	56	143	222	44	411
8-Acetoxychlorpromazine	58	86	289	43	291	85	248	288	376
3-Acetoxychlorpromazine	58	86	376	43	378	42	87	59	376
3-Acetoxydiazepam	271	61	43	45	70	300	256	273	342
3-Acetoxyprazepam	257	43	55	311	340	259	77	313	382
Acetoxytetrahydrocannabinol (8-alpha-Acetoxy-delta 9-)	149	312	270	354	297	269	167	256	414
4-Acetylaminoantipyrine	56	84	245	57	203	83	43	42	245
4-Acetylaminophenol(Mono-TMS-)	181	166	223	73	43	45	208	75	223
Acetylcarbromal	41	43	69	44	39	55	53	70	278
Acetylcarbromal	43	41	129	69	55	39	149	42	278
Acetylcarbromal*	43	129	69	41	86	97	55	44	278
Acetylcholine	58	43	57	149	71	42	41	55	163
Acetylcodeine	58	43	73	341	282	229	342	299	341
Acetylcodeine	341	282	43	229	42	342	204	162	341
6-Acetylcodeine	341	282	229	204	81	70	149	124	341
Acetylcodeine	341	282	42	43	229	59	204	342	341
6-Acetylcodeine*	341	282	229	42	43	59	342	204	341
Acetyldapsone	290	248	108	140	43	65	93	92	290
Acetyldihydrocodeine	343	300	184	43	70	59	344	326	343
Acetyldihydrocodeine	343	184	300	70	43	59	334	226	343
Acetyldihydrocodeine*	343	43	70	284	59	300	344	42	343
2-Acetylfuran	95	110	43	39	96	68	0	0	110
1-Acetyllysergic Acid Diethylamide	365	263	221	264	223	0	0	0	365
Acetylmorphine	327	268	215	328	269	267	146	124	327

COMPOUND NAME	PEAKS								MWT
Acetylmorphine	327	268	43	42	215	44	328	146	327
3-Acetylmorphine	162	285	327	215	124	81	115	267	327
6-Acetylmorphine	327	268	215	59	204	70	146	81	327
6-Acetylmorphine	327	43	268	42	215	81	146	59	327
6-Acetylmorphine*	327	268	328	42	43	215	44	59	327
3-Acetylmorphine 6-Heptafluorobutyrate Derivative	481	43	44	53	268	310	523	69	523
6-Acetylmorphine (3-Heptafluorobutyrate Derivative)	464	43	81	523	204	69	70	411	523
Acetylmorphine TMS Ether	399	340	73	287	43	400	342	341	399
Acetylnorpropoxyphene (N-)	44	220	86	57	74	91	43	115	367
Acetylnorpropoxyphene (N-)	44	220	86	74	58	91	43	115	367
Acetylprocainamide (N-Acetyl)	86	58	99	56	162	132	149	205	277
Acetylprocaine Acetate	86	58	162	120	99	43	71	278	278
Acetylprocaine Hydrochloride	86	99	30	58	87	120	162	206	278
Acetylsalicylic Acid	120	43	138	92	42	121	39	45	180
Acetylsalicylic Acid	120	138	43	92	64	65	63	42	180
Acetylsalicylic Acid	120	138	43	92	121	39	63	64	180
Acetylsalicylic Acid*	120	43	138	92	121	39	64	63	180
Acetylsalicylic Acid TMS Ester	195	120	43	210	135	75	73	92	252
Acetylserotonin (N-)	159	146	218	160	147	219	0	0	218
Acetylsulphamethoxazole	43	198	134	65	108	295	92	135	295
Acetyl Tri-n-butylcitrate	185	57	43	129	29	41	259	157	402
Acetyl Tri-n-butyl Citrate	185	43	129	41	57	259	157	112	402
Acetylurea(N-)	43	28	44	59	42	102	41	74	102
Acrolein	27	56	26	55	29	25	37	38	56
Acrylonitrile	53	26	52	43	51	27	41	31	53
Adenine	135	108	54	53	81	43	136	66	135
Adiphenine	42	44	58	86	56	167	165	152	311
Adiphenine	86	165	167	99	152	166	58	56	311
Adiphenine*	86	30	167	99	87	58	165	29	311
Adiponitrile(1,4-Dicyanobutane)	41	68	54	55	39	27	40	29	108
Adrenaline	44	93	65	43	166	137	111	139	183
Adrenaline*	44	42	124	165	163	123	93	65	183
Adriamycin Metabolite	338	339	77	75	296	310	78	76	398
Ajmalicine	156	352	351	184	169	209	353	129	352
Alachlor	45	160	188	237	162	224	146	161	269
Alclofenac*	41	141	226	143	181	0	0	0	226
Alclofenac*	41	226	77	143	181	141	39	145	226
Alclofenac Metabolite	141	143	77	51	105	0	0	0	186
Alclofenac Metabolite	141	186	77	143	105	0	0	0	260
Aldrin	66	263	79	91	265	101	261	65	362
Aletamine	70	120	43	91	41	65	77	98	161
Aletamine	71	120	43	91	41	0	0	0	161
Allantoin Pentamethyl Derivative	42	156	127	228	142	140	83	72	228
Allobarbitone	167	41	124	80	39	32	166	53	208
Allobarbitone*	41	167	124	80	53	68	141	208	208
Allopurinol	136	135	52	28	137	109	29	18	136
Alloxan	43	44	142	86	42	114	71	70	142
Alloxantin	44	45	101	58	42	86	142	128	286
Allyl Acetate	43	58	41	39	57	29	27	0	100
Allyl Cyanide	41	67	39	27	40	38	66	37	67
Allylcyclopentenylbarbitone Dimethyl Derivative	221	67	196	181	41	164	111	107	262
Allyl Isothiocyanate	99	41	39	72	45	98	71	38	99
Allyl Isothiocyanate	41	99	39	72	45	38	27	40	99
2-(Allyloxyphenyl) N-Methylcarbamate	109	150	78	41	110	81	58	39	207

Compound Name									MWT
Allyl-3-piperidylbenzilate (N-Allyl)	123	105	77	96	183	110	165	351	351
Allylprodine*	172	214	42	110	57	173	91	44	287
Aloxiprin*	43	45	60	120	42	92	138	44	999
Alphacetylmethadol(d)	72	43	73	36	46	91	71	42	353
Alphacetylmethadol(1)	72	73	43	91	224	56	71	129	353
Alphacetylmethadol(1)	72	43	73	36	91	56	42	225	353
Alphadolone Acetate*	317	43	271	147	289	390	95	81	390
Alphameprodine	97	172	91	105	57	42	77	201	275
Alphameprodine	172	42	98	57	44	201	91	36	275
Alphameprodine	172	98	201	91	202	275	96	200	275
Alphamethadol	72	91	73	193	165	58	115	178	311
Alphamethadol	72	56	91	129	105	130	73	57	311
Alphaprodine	172	187	84	57	42	188	44	43	261
Alphaprodine	172	187	144	84	42	57	188	44	261
Alphaprodine	172	187	84	42	57	171	144	186	261
Alphaprodine (alpha-Prodine)	172	187	42	84	129	144	44	91	261
Alphenal	215	41	104	77	132	39	128	244	244
Alphenal Dimethyl Derivative	118	243	272	104	77	129	128	89	272
Alprenolol*	72	30	56	73	249	98	234	102	249
Alverine	72	176	91	58	177	65	281	42	281
Alverine*	72	176	91	58	177	41	42	30	281
Amantadine	94	41	39	42	77	151	95	93	151
Amantadine*	94	151	57	95	40	41	58	108	151
Ambucetamide*	248	44	249	121	136	29	164	192	292
Ametazole	82	36	81	38	55	83	54	42	111
Ametazole	82	81	55	83	54	42	41	39	111
Ametazole*	82	30	81	83	55	54	42	27	111
Amethocaine	58	71	36	176	150	72	193	59	264
Amethocaine*	58	71	150	176	72	193	105	59	264
Amidephrine*	44	42	147	120	65	43	45	39	244
Amidopyrine	56	97	231	42	111	77	71	112	231
Amidopyrine	56	231	97	204	77	111	112	0	231
Amidopyrine*	56	231	97	111	112	42	77	71	231
Amiloride*	43	187	42	171	229	86	144	170	229
4-Aminoantipyrine	56	84	57	203	42	83	77	93	203
Aminoantipyrine	56	57	84	83	203	77	0	0	203
3-Aminobenzoic Acid	137	92	120	65	39	52	66	138	137
2-Aminobenzoic Acid	119	137	92	65	39	64	120	91	137
4-Aminobenzoic Acid*	137	120	92	65	39	138	121	63	137
2-(2-Amino-5-bromobenzoyl)pyridine(Bromazepam Benzophenone analogue)	45	62	43	61	44	63	249	247	276
2-Amino-5-chlorobenzophenone	77	230	231	105	154	232	126	51	231
2-Amino-5-chlorobenzophenone*	230	231	77	232	105	154	233	126	231
2-Amino-5-Chloro-2'-Fluorobenzophenone	249	248	123	154	250	95	251	230	249
2-Amino-2'-chloro-5-nitrobenzophenone*	241	276	139	165	195	111	278	242	276
7-Aminoclonazepam	285	256	43	110	257	84	287	111	285
7-Aminodesmethylflunitrazepam	269	240	241	268	270	107	121	213	269
2-Amino-5,2'-Dichlorobenzophenone	139	156	111	44	75	141	230	50	265
2-Amino-4,6-di-t-butylphenol	206	221	57	41	207	150	222	68	221
4-(2-Aminoethyl)pyrocatechol	124	30	123	153	77	51	125	78	153
7-Aminoflunitrazepam	283	44	255	282	254	284	264	256	283
Aminoglutethimide*	203	232	132	175	204	233	160	118	232
Aminoheptane	44	30	41	42	100	43	45	55	100
2-Aminohexane	44	58	41	86	42	45	43	0	101
2-Amino-3-Hydroxy-5-Chlorbenzophenone	77	246	247	105	78	45	51	248	247

Compound Name	Peaks								MWT
7-Aminonitrazepam	251	222	223	250	252	195	110	97	251
7-Aminonitrazepam	251	44	223	222	43	84	98	111	251
2-Amino-5-nitrobenzophenone	241	77	242	105	44	43	195	57	242
2-Amino-5-Nitro-2'-Chlorobenzophenone	241	139	276	195	111	44	165	119	276
2-Amino-5-Nitro-2'-Fluorobenzophenone	260	43	123	259	95	77	57	165	260
Aminoparathion	261	125	109	108	97	80	233	205	261
2-Aminophenol	109	80	53	52	39	64	63	81	109
3-Aminophenol	109	80	81	110	41	39	53	74	109
4-Aminophenol	109	80	53	81	108	52	54	110	109
2-Amino-1-phenylpropanone	44	40	51	77	42	105	50	45	149
Aminophylline*	180	95	181	68	41	123	53	96	420
Aminopromazine	198	70	115	199	71	269	72	41	327
Aminopyrine	56	97	77	231	42	91	51	55	231
Aminosalicylic Acid*	109	80	44	81	53	39	52	54	153
Aminosalicylic Acid(p-)	79	135	153	107	52	53	80	51	153
Aminozide	59	60	118	100	44	45	42	101	160
Amiodarone*	86	36	87	84	58	56	44	38	645
Amiodarone Probe Product*	546	86	420	517	547	391	263	250	546
Amiperone	279	123	412	165	206	281	292	95	430
Amiphenazole*	191	121	77	104	122	43	51	192	191
Amitraz	162	121	132	147	293	344	120	106	293
Amitriptyline	58	59	202	91	203	215	218	217	277
Amitriptyline	58	42	59	57	202	203	43	91	277
Amitriptyline	58	59	42	30	202	91	203	115	277
Amitriptyline	58	59	42	202	215	203	91	189	277
Amitriptyline*	58	59	202	42	203	214	217	0	277
Amitriptyline Metabolite	58	215	59	229	228	227	226	218	275
Amitriptyline TMS Derivative	58	59	73	75	215	202	203	217	365
Amitrol	26	84	31	57	43	42	58	40	84
Ammonium Nitrate	30	44	46	0	0	0	0	0	80
Amobarbitone	156	158	141	143	157	159	142	0	226
Amobarbitone (Amytal)	156	141	157	41	43	142	197	55	226
Amodiaquine*	58	282	30	284	355	73	283	44	355
Amolanone*	86	30	87	294	58	42	57	309	309
Amotriphene*	58	359	358	403	360	343	388	227	403
Amphetamine	44	91	65	42	45	120	40	92	135
Amphetamine	44	103	91	65	42	104	45	102	135
Amphetamine	44	91	65	51	63	77	0	0	135
Amphetamine	44	91	65	58	39	42	51	63	135
Amphetamine*	44	91	40	42	65	45	42	39	135
Amphetamine-NCS Derivative	91	86	177	65	0	0	0	0	177
Amphetamine TFA Derivative	140	118	91	29	46	69	65	39	231
Amphetamine Trifluoroacetate	140	118	91	69	45	0	0	0	231
Amphetaminil	132	100	91	77	65	89	79	0	250
Ampicillin	44	100	34	55	115	147	41	0	349
Amprotropine*	86	87	58	30	103	121	72	56	307
Amydricaine	58	112	105	96	42	111	77	51	278
Amygdalin	43	31	57	73	60	29	55	44	457
Amyl Acetate	43	70	55	42	41	73	61	87	130
Amyl Alcohol	55	41	42	43	70	57	29	31	88
Amyl Nitrite	70	43	55	41	57	42	71	0	117
Amyl Nitrite	41	57	29	60	43	30	71	39	117
Amylobarbitone	156	141	55	157	142	197	69	98	226
Amylobarbitone	141	156	157	40	69	98	142	70	226

Compound Name									MWT
Amylobarbitone*	156	141	157	41	55	142	98	39	226
Amylocaine	58	105	77	42	59	113	51	30	271
Amylocaine*	58	105	77	98	122	42	113	51	271
Androsterone	290	67	108	107	79	55	41	93	290
Anileridine	246	247	42	120	218	172	106	91	352
Anileridine	246	42	247	120	172	80	218	106	352
Anisidine(o-)	123	108	80	53	65	124	109	52	123
Anisole(Methyl Phenyl Ether)	108	78	65	39	77	93	79	51	108
Anisotropine	124	82	83	57	41	94	42	43	267
Antazoline	84	91	55	77	85	182	83	65	265
Antazoline	84	91	77	65	104	85	83	182	265
Antazoline	84	91	36	51	55	0	0	0	265
Antazoline*	84	91	77	55	182	85	65	104	265
Antimony Trichloride	193	191	121	195	123	158	156	228	226
Antipyrine	77	96	188	105	38	82	93	106	188
Apiol	222	207	149	177	223	221	195	191	222
Apomorphine	266	267	42	220	152	224	165	178	267
Apomorphine*	266	267	224	220	268	44	250	248	267
Apomorphine Dimethyl Ether	294	295	221	252	264	280	165	237	295
Aprobarbitone	167	41	168	124	43	97	169	96	210
Aprobarbitone*	167	41	124	168	97	39	169	45	210
Aprobarbitone (Allypropymal, Aprozal, Alurate)	167	41	39	124	168	29	43	169	210
Arecoline	155	96	140	43	42	81	94	53	155
Aromatised 2-oxo-LSD	235	237	337	236	221	338	209	238	337
Ascorbic Acid(Tetra TMS)	73	147	332	117	205	45	74	133	464
Ascorbic Acid (Vitamin C)	29	41	39	42	69	116	167	168	176
Atenolol	72	30	107	56	43	73	222	57	266
Atenolol*	72	30	56	98	43	107	41	73	266
Atrazine	200	215	149	173	202	58	217	92	215
Atrazine	200	215	58	202	43	68	173	69	215
Atropine	124	83	82	94	289	42	96	125	289
Atropine	124	42	82	94	289	67	83	96	289
Atropine*	124	82	83	42	96	103	67		289
Atropine (dl-Hyoscyamine)	124	82	42	83	94	67	96	39	289
Atropine TMS Ether	124	83	82	73	94	96	125	42	361
Azacyclonol	84	105	85	77	55	56	183	42	267
Azacyclonol	85	84	183	107	77	55	56	184	267
Azacyclonol*	85	84	183	105	56	77	55	30	267
Azaperone	107	165	123	95	121	0	0	0	327
Azapetine	165	41	234	193	179	42	39	178	235
Azapetine*	165	234	194	235	193	196	166	179	235
Azapropazone*	160	300	189	145	188	301	161	42	300
Azar	220	57	41	221	206	58	29	55	277
Azathioprine	42	231	119	152	247	0	0	0	277
Azathioprine*	231	42	119	232	92	65	67	74	277
Azathioprine Metabolite	43	44	87	111	129	167	209	159	288
Azinphos Ethyl	132	77	160	104	105	76	97	129	345
Azinphos Methyl	160	132	77	93	125	104	147	105	317
Azopyridine	213	108	81	36	44	136	54	77	213
Baclofen*	30	138	195	140	103	197	77	196	213
Bamethan*	86	30	57	108	44	29	84	41	209
Bamethan Probe Product*	30	73	107	27	108	41	57	39	999
Bamifylline*	102	84	58	56	42	30	103	91	385
Banol	156	121	158	93	58	65	141	51	213

Compound Name									MWT
Barban	222	51	87	143	153	104	224	69	257
Barbitone	156	141	98	155	55	112	41	83	184
Barbitone	141	156	41	39	55	98	112	44	184
Barbitone*	156	141	55	155	98	39	82	43	184
Barbituric Acid	42	128	85	43	44	41	69	70	128
Beclamide*	91	106	197	162	107	148	27	63	197
Bemegride	55	83	82	113	70	69	97	155	155
Bemegride	55	82	39	83	41	113	70	127	155
Bemegride*	55	83	82	41	113	70	29	69	155
Bemegride Methyl Derivative	55	169	83	70	127	140	41	97	169
Benactyzine	86	183	182	105	77	165	116	312	327
Benactyzine*	86	105	77	87	182	99	183	58	327
Benactyzine Probe Product*	105	77	182	44	86	165	51	30	999
Benazolin	170	134	199	243	172	198	201	200	243
Bendiocarb	151	126	166	31	51	43	58	223	223
Bendrofluazide*	330	118	91	319	219	64	92	421	421
Benefin	294	266	42	43	278	58	295	207	335
Benorylate*	121	163	151	109	43	122	108	65	313
Benoxinate	86	99	71	58	87	140	84	100	308
Benoxinate	86	99	30	58	71	41	136	56	308
Benperidol*	230	109	82	187	243	363	42	123	381
Benserazide*	60	42	88	30	31	43	29	70	257
Benzaldehyde	77	106	105	51	50	78	52	39	106
Benzamine*	110	126	44	77	41	58	42	105	247
Benzathine	91	120	121	92	65	106	30	135	240
Benzene	78	52	51	77	50	39	79	43	78
Benzene	78	77	52	51	50	39	79	74	78
Benzene-d6	84	56	54	42	52	82	85	40	84
Benzestrol	121	107	135	163	177	298	77	136	298
Benzethidine	246	91	42	247	233	57	218	56	367
Benzethidine	246	91	42	233	162	57	56	149	367
Benzhexol	98	99	218	55	41	77	96	219	301
Benzhexol	98	218	99	219	55	41	42	85	301
Benzhexol*	98	105	55	41	99	77	218	84	301
Benziodarone*	518	173	264	519	373	376	520	249	518
Benzitramide*	286	300	96	42	244	230	301	82	492
Benzocaine	120	92	165	65	137	0	0	0	165
Benzocaine*	120	165	92	65	137	39	121	93	165
Benzoctamine*	218	44	191	221	219	178	42	180	249
Benzoic Acid	105	122	77	51	50	39	74	78	122
Benzonitrile	103	76	50	104	51	75	77	52	103
5,6-Benzoquinoline	179	178	151	180	152	177	150	90	179
Benzoylecgonine	82	105	122	77	51	42	94	81	289
Benzoylecgonine	82	124	105	77	168	122	42	83	289
Benzoylecgonine	105	77	122	51	50	44	78	52	289
Benzoylecgonine*	124	82	168	77	105	42	94	83	289
Benzphetamine*	91	148	149	65	92	42	56	39	239
Benzquinamide*	205	244	191	345	206	72	272	246	404
Benzthiazide*	91	121	65	122	309	64	230	123	431
Benztropine	140	83	82	124	96	97	167	125	307
Benztropine	83	82	42	140	124	96	77	67	307
Benztropine*	83	140	82	124	96	97	42	125	307
Benzydamine*	85	58	86	91	84	70	42	225	309
Benzyl Alcohol	79	77	108	101	51	50	39	40	108

Compound Name									MWT
Benzyl Alcohol	79	108	107	77	51	91	50	78	108
2-Benzylaziridine	91	42	132	117	43	41	104	57	133
Benzyl Cyanide	117	90	116	89	63	51	50	39	117
5-Benzylidenebarbituric Acid	215	216	172	102	129	118	128	117	216
5-Benzylidene-1-phenylbarbituric Acid	292	291	119	248	193	249	220	221	292
Benzyl Methyl Ketone	91	134	92	43	65	77	89	51	134
Benzyl Methyl Ketoxime	91	116	131	65	149	92	150	90	149
Benzyl Methyl Ketoxime	91	75	92	65	42	39	40	116	149
Benzyl Methyl Ketoxime	91	41	92	65	39	42	40	116	149
2-Benzyl-2-methyl-5-phenyl-2,3-dihydropyrimid-4-one	186	91	150	187	143	65	115	116	277
Benzyl Morphine	284	91	375	81	42	36	285	175	375
Benzyl-N-methylphenethylamine (alpha-)	134	91	42	119	65	135	58	86	225
Benzylphenethylemine(alpha)	120	91	103	121	77	65	42	39	211
4-Benzylpyrimidine	169	170	91	115	142	65	116	143	170
Berberine	321	278	320	292	306	191	322	304	353
Betacaine	109	44	77	125	58	105	51	0	247
Betacetylmethadol	72	43	44	91	278	0	0	0	353
Betahistine	79	105	104	51	78	52	50	77	136
Betameprodine	97	172	91	105	57	42	77	201	275
Betamethadol	72	73	91	58	44	115	105	165	311
Betamethadol	72	73	91	58	253	193	42	115	311
Betamethasone	122	121	91	312	342	0	0	0	392
Betamethasone	121	43	122	223	147	91	41	135	392
Betaprodine	172	187	144	84	42	57	188	44	261
Betaprodine*	172	187	84	42	44	57	29	43	261
Bethanechol	43	58	42	143	85	171	157	102	196
Bethanidine	177	106	0	0	0	0	0	0	177
Bethanidine	71	91	57	106	65	72	55	77	177
Bethanidine*	71	91	106	177	57	72	65	30	177
BHC(A-)	181	183	219	217	109	111	221	51	288
Bialamicol*	58	30	363	44	72	29	27	364	436
Bibenzyl	91	65	182	92	39	51	63	77	182
Bilirubin	43	286	299	41	44	227	300	211	584
Bioresmethrin	123	171	143	81	128	91	172	41	338
Biperiden	98	218	99	55	130	41	42	85	311
Biperiden*	98	218	99	55	41	42	77	84	311
Bisacodyl	361	276	277	199	319	43	318	362	361
Bisacodyl*	361	277	319	276	199	318	362	43	361
Bis(2-chloroethyl) Ether	93	63	27	95	65	31	94	36	142
Bis-(2-ethylhexyl)-4-hydroxyphthalate	165	183	70	59	71	43	295	112	406
Bis(2-ethylhexyl)-3-hydroxyphthalate	165	112	71	185	113	164	182	149	406
Bisnortilidine	69	68	70	77	103	56	54	51	245
Boldenone	122	121	91	123	147	108	77	286	286
Boldine Dimethyl Ether	354	355	340	324	281	162	297	312	355
Boric Acid	45	62	44	61	43	63	0	0	62
Bornanone-2	95	81	108	152	41	83	109	69	152
Bornanone-3	69	81	152	41	109	95	80	68	152
Brallobarbitone*	207	41	39	124	91	165	122	44	286
Brallobarbitone Methylation Artifact*	39	138	195	136	119	53	91	120	234
Bromacil	205	207	161	163	110	112	260	262	260
Bromacil	205	207	40	39	70	206	190	162	260
Bromacil	205	207	42	162	70	164	231	188	260
Bromazepam	236	315	317	44	77	91	287	316	315
Bromazepam	236	315	317	286	288	316	208	179	315

Compound Name	Peaks								MWT
Bromazepam*	236	317	315	288	316	286	208	78	315
Bromazepam Benzophenone	249	247	248	250	276	278	198	200	276
Bromazepam-N(Py)-Oxide	79	80	225	182	107	63	78	91	331
Bromhexine	70	293	264	112	305	291	262	295	374
Bromhexine*	70	112	293	264	44	42	305	41	374
5-(2-Bromoallyl)-barbituric Acid	168	167	43	97	169	41	124	153	246
1-Bromobutane	57	41	29	56	27	39	55	43	136
Bromocamphor	123	83	55	41	151	81	69	39	230
1-Bromo-2-chlorobenzene	192	190	111	75	194	50	113	74	190
Bromochloro-1,1-difluoroethylene	178	176	97	128	31	180	126	47	176
Bromochloro-1,2-difluoroethylene	178	176	97	61	128	180	31	177	176
Bromochloromethane	49	130	128	51	93	132	95	47	128
Bromocriptine*	70	43	154	71	41	209	86	195	653
Bromocyclohexane	83	55	41	67	54	39	82	27	162
Bromodan	359	357	361	237	355	272	251	239	390
Bromodichloromethane	83	85	129	47	127	87	48	81	162
Bromodiethylaniline(p-Br,N,N-diEt)	212	214	227	229	184	186	105	118	227
4-Bromo-2,5-dimethoxyamphetamine Impurity	56	254	256	229	231	199	201	0	285
4-Bromo-2,5-dimethoxyphenethylamine	30	230	232	44	215	217	77	156	259
Bromodiphenhydramine	58	73	45	57	43	44	167	165	333
Bromodiphenhydramine	58	165	41	73	57	166	44	59	333
Bromodiphenhydramine*	58	73	45	165	59	42	166	44	333
Bromodiphenhydramine*	58	73	45	165	59	42	166	149	333
5-(2-Bromoethyl)-5-butylbarbituric Acid	124	141	154	167	155	184	142	98	291
2-Bromo-2-ethylbutyramide(Carbromal Metabolite)	69	43	41	44	71	167	165	55	193
5-(2-Bromoethyl)-5-ethylbarbituric Acid	156	141	124	139	98	157	155	112	263
2-Bromo-2-ethyl-3-hydroxybutyramide(Carbromal Metabolite)	150	152	165	41	167	43	44	130	209
5-(2-Bromoethyl)-5-isopentylbarbituric Acid	124	154	155	141	142	181	125	198	305
5-(2-Bromoethyl)-5-pentylbarbituric Acid	124	155	154	141	142	181	153	98	305
Bromofluoromethane	112	114	33	93	95	91	81	79	112
Bromoform	173	171	175	91	93	92	94	79	256
2-Bromoheptane	57	41	56	55	43	69	29	42	178
3-Bromohexane	43	85	55	41	42	56	84	29	164
1-Bromo-2-iodobenzene	282	284	155	157	76	75	50	74	282
Bromophenoxim	277	184	88	279	275	63	278	53	459
Bromophos	331	329	125	333	93	109	332	62	364
Bromophos	331	329	125	333	79	109	47	93	364
Bromophos Ethyl	97	359	242	303	357	331	240	301	392
Bromophos Ethyl	303	359	97	357	331	242	125	109	392
2-Bromopropane	43	41	27	42	124	122	45	44	122
5-(3-Bromopropyl)-5-isopropylbarbituric Acid	138	124	141	169	168	247	249	98	318
Bromo STP	44	230	232	273	275	77	42	45	273
Bromothen	58	72	71	177	175	96	42	78	339
Bromotrichloromethane	117	119	163	82	161	121	47	165	196
Bromotrifluoromethane	69	148	150	129	131	79	81	50	148
Bromoxynil	88	62	61	53	37	277	63	89	275
Brompheniramine	249	247	58	72	167	248	168	250	318
Brompheniramine	58	42	167	249	247	168	72	248	318
Brompheniramine*	247	249	58	72	248	167	250	168	318
Brompheniramine (dex) (d-Parabromdylamine)	58	249	247	72	167	248	168	42	318
Bromvaletone	44	83	180	182	137	143	139	41	222
Broxyquinoline*	303	301	305	115	194	196	114	87	301
Brucine	394	379	395	197	107	120	203	55	394
Brucine	394	395	120	107	146	91	134	79	394

Compound Name	Peaks								MWT
Brucine*	394	395	379	392	120	197	203	393	394
BTS 29101	121	252	120	132	106	122	105	383	383
Buclizine	147	165	201	167	166	105	203	117	432
Buclizine	36	231	147	38	285	132	201	165	432
Buclizine	147	231	165	285	132	166	201	117	432
Buclizine*	231	147	285	232	201	132	165	166	432
Buclosamide*	155	157	227	154	185	44	30	156	227
Budralazine	225	149	185	240	103	131	89	129	240
Bufexamac*	107	163	223	108	29	164	41	57	223
Buformin*	43	114	85	30	101	86	44	72	157
Bufotenine	58	204	146	59	42	160	43	159	204
Bufylline*	180	95	68	41	58	53	123	96	269
Bumetanide*	321	364	304	240	168	91	322	365	364
Buphenine	91	176	148	133	44	174	65	107	299
Buphenine*	91	133	176	174	44	107	92	177	299
Bupirimate	273	208	166	193	150	316	108	96	316
Bupivacaine	140	141	84	98	56	138	41	96	288
Bupivacaine*	140	141	84	41	29	96	56	55	288
Butabarbitone	156	141	41	57	157	39	98	55	212
Butacaine	120	41	100	142	178	42	92	44	306
Butacaine*	120	263	142	178	100	264	41	29	306
Butalamine*	142	143	100	155	44	112	57	29	316
Butalbital	168	41	124	181	167	141	97	98	224
Butalbital	168	167	41	181	124	97	169	141	224
Butalbital	168	167	41	39	124	97	43	141	224
Butalbital*	41	167	168	39	124	97	141	181	224
Butanal(n-)	27	29	44	43	41	72	39	57	72
Butane(n-)	43	29	41	27	58	42	39	44	58
Butanilacaine*	86	30	72	44	141	42	29	57	254
2-Butanol	43	31	42	41	33	29	39	0	74
2-Butanol	45	59	31	41	27	43	44	29	74
1-Butanol	56	31	41	43	27	42	29	0	74
1-Butanol	31	56	41	43	27	42	29	39	74
Butanol (tert-)	59	31	41	57	43	29	39	0	74
2-Butanone	43	29	27	72	42	57	44	39	72
Butaperazine	70	113	43	42	71	141	56	72	409
Butethal	141	156	55	41	29	142	39	98	212
Butethamate*	86	99	91	191	87	119	58	248	263
Butethamate Prode Product*	44	43	39	40	68	38	27	37	999
Butethamine	120	86	58	99	56	44	57	36	236
Butethamine	120	65	86	92	56	99	41	44	236
Butethamine	120	85	58	99	56	44	57	149	236
Butethamine*	120	99	30	86	56	44	57	193	236
Butobarbitone	156	141	184	55	155	142	170	98	212
Butobarbitone	41	141	39	156	57	55	53	44	212
Butobarbitone	141	156	41	55	98	142	184	40	212
Butobarbitone*	141	156	41	55	98	39	142	155	212
Butriptyline	58	45	59	42	36	115	178	91	293
Butriptyline*	58	293	45	59	193	100	178	294	293
Butropipazone	175	188	132	123	326	95	165	311	326
Butyl Acetate(n-)	43	56	73	41	61	55	27	29	116
Butylamine(n-)	30	73	41	27	42	39	31	56	73
Butylamine(N-ethyl(n-))	58	30	29	27	28	44	101	41	101
Butylamine(t-)	58	41	30	42	57	39	27	59	73

Compound Name	Peaks								MWT
Butylaminobenzoate	120	137	193	92	65	121	138	41	193
2-Butylamino-4-ethylamino-6-methoxy-s-triazine (s-Butyl)	196	169	225	210	43	94	197	44	225
Butylcarbobutoxymethyl Phthalate	149	150	41	57	56	76	104	205	336
5-Butyl-5-(2-chloroethyl)-barbituric Acid	124	141	167	154	125	112	98	190	246
5-Butyl-5-(2-dimethylaminoethyl)-barbituric Acid	58	124	71	199	125	167	154	141	255
Butyl Formate(n-)	56	41	31	43	29	27	55	28	102
Butyl Hippurate (n-)	105	134	77	106	51	135	235	160	235
5-Butyl-5-(2-hydroxyethyl)barbituric Acid	124	99	142	141	154	98	129	125	228
Butylmethoxyphenol(3-t-) Methyl Ether	179	194	164	151	91	77	121	149	194
Butyl Methyl Ketone(t-)	57	41	43	29	39	100	27	0	100
Butyl Nitrate	43	46	76	41	29	57	30	0	119
Butylphenol(p-t-)	135	107	41	150	95	136	77	91	150
1-Butylpyrrole	81	80	123	39	53	41	67	40	123
5-Butyl-5-(2-thiocyanatoethyl)-barbituric Acid	124	154	141	181	198	142	153	155	269
Butylvinal	154	83	71	55	155	43	70	67	224
Butyl Vinyl Ether	41	29	56	57	28	43	44	27	100
3-Butynyl N-(3-chlorophenyl)carbamate	53	223	153	127	171	181	225	155	223
(3-Butynyl)-N'-(4-chlorophenyl) N-Methylcarbamate (N-)	53	236	110	127	153	68	238	58	236
Butyronitrile(n-)	41	29	27	39	40	0	0	0	69
Butyronitrile(t-)	42	41	68	67	39	27	40	29	69
C 10015	122	121	56	58	194	55	45	73	251
Cadaverine	56	55	41	43	45	85	42	84	102
Cadaverine	30	56	28	85	43	45	41	27	102
Caffeine	194	109	55	67	82	42	137	41	194
Caffeine	194	109	67	82	55	193	195	81	194
Caffeine	194	109	55	67	82	193	42	40	194
Caffeine*	194	109	55	67	82	195	42	110	194
Camazepam	58	72	43	78	271	57	44	77	371
Cambendazole	260	216	302	215	242	0	0	0	302
Cambendazole	260	302	216	215	243	189	242	188	302
Cambendazole GC Decomposition Product	242	215	243	214	111	121	216	0	242
Cambendazole Metabolite	205	219	231	318	0	0	0	0	318
Camphene	93	121	79	67	107	95	94	68	136
Camphor	95	81	41	69	55	83	67	137	152
Camphor	95	81	69	41	55	67	108	109	152
Camphor	95	41	81	69	83	108	109	67	152
Camphor	95	81	108	55	109	152	67	83	152
Camphor	95	81	108	152	69	109	83	110	152
Cannabichromanon (C3)	289	179	43	150	231	304	219	55	304
Cannabichromene	231	41	43	232	69	55	174	57	314
Cannabidiol	231	246	314	232	121	193	74	174	314
Cannabielsoin (C3)	177	43	176	219	147	302	69	111	302
Cannabigerol	193	231	247	123	194	233	136	316	316
Cannabinol	295	296	310	238	31	59	311	297	310
Cannabinol	295	296	43	310	238	299	58	41	310
Cannabinol	295	296	310	238	251	223	165	119	310
Captan (Orthocide)	79	80	151	149	77	114	117	107	299
Captodiamine	58	165	255	359	166	73	199	45	359
Caramiphen	86	99	91	144	58	56	41	87	289
Caramiphen	86	91	99	145	115	58	87	56	289
Caran-trans-3-ol(cis-)	93	121	136	81	43	55	41	107	154
Caran-trans-2-ol((-)cis-)	93	121	136	111	43	41	55	79	154
Caran-trans-2-ol(trans-)	111	41	43	93	55	121	69	95	154
Carbachol	43	58	42	44	30	129	36	143	182

Compound Name									MWT
Carbamazepine	193	192	165	191	194	167	190	63	236
Carbamazepine*	193	192	236	191	194	165	190	237	236
Carbamazepine Metabolite	180	179	223	252	0	0	0	0	252
Carbetapentane	86	91	145	84	144	85	58	87	333
Carbetapentane*	86	91	87	145	58	144	30	44	333
Carbethyl Salicylate	193	165	121	120	194	93	65	92	358
Carbidopa*	123	57	42	103	44	85	124	51	226
Carbimazole*	186	114	29	72	42	113	27	109	186
Carbinoxamine	167	71	166	139	41	78	140	168	290
Carbinoxamine	58	71	167	72	42	59	202	45	290
Carbinoxamine	58	71	42	44	167	72	43	57	290
Carbinoxamine*	58	71	26	54	167	72	42	44	290
Carbinoxamine(1-)	58	71	167	72	42	202	203	45	290
Carbocaine	98	70	41	77	91	99	96	120	264
Carbofuran	164	149	57	122	123	131	165	121	221
2-Carbomethoxyquinoline	129	157	128	130	187	158	156	188	187
Carbon Disulphide	76	78	38	44	77	39	64	46	76
Carbon Tetrachloride	117	119	121	82	47	84	35	49	152
Carbon Tetrachloride	117	119	35	47	121	82	84	49	152
Carbonyl Sulphide	60	30	62	44	34	33	46	31	60
Carbromal	44	69	41	55	43	39	71	53	236
Carbromal	44	69	41	55	43	29	208	71	236
Carbromal*	44	69	41	208	210	55	71	43	236
Car-3-ene	93	77	79	91	80	41	92	136	136
Car-3-ene	93	91	92	79	80	77	43	121	136
Carisoprodol	58	55	97	43	104	158	62	56	260
Carisoprodol	55	58	43	104	56	41	97	62	260
Carisoprodol	104	55	43	56	44	58	41	62	260
Carisoprodol*	55	57	43	97	41	56	158	44	260
Carphenazine	268	425	143	269	197	394	157	171	425
Carphenazine*	268	143	425	70	269	42	394	157	425
Carphenazine (Proketazine)	268	143	425	55	70	41	40	269	425
Cartap	104	71	147	42	44	70	56	102	237
Carveol(cis-)	118	133	93	88	91	42	27	0	152
Carvone	82	54	108	93	58	107	106	39	150
Catechol	110	63	64	81	53	55	92	111	110
Catechol	110	64	63	81	39	92	55	111	110
Cedrene(alpha-)	119	93	41	69	105	161	55	91	204
Cedrene(alpha-)	119	41	93	204	69	105	91	55	204
Cephaeline*	178	192	272	466	244	288	191	273	466
Chelidonine	332	333	304	335	176	303	334	162	353
Chlophedianol	58	139	254	77	178	105	179	111	289
Chlophedianol*	58	254	45	44	77	42	59	72	289
Chloral	82	84	111	83	29	47	113	85	146
Chloral Betaine	81	83	46	110	82	112	84	117	281
Chloral Hydrate	82	84	83	111	47	29	85	113	164
Chloral Hydrate	81	83	46	110	112	0	0	0	164
Chloral Hydrate*	82	47	84	29	111	83	113	85	164
Chloralose (alpha)	71	73	85	61	113	36	43	60	308
Chloralose (alpha) TMS Ether	73	117	75	103	205	147	145	129	596
Chlorambucil*	254	256	118	255	303	305	63	45	303
Chloramphenicol	70	150	151	153	77	51	118	60	322
Chloramphenicol	70	153	36	118	117	60	155	150	322
Chloramphenicol	153	170	155	152	172	136	118	60	322

Compound Name				Peaks					MWT
Chloramphenicol TMS Ether	225	242	244	364	362	208	451	453	466
Chlorazepate	242	43	270	269	241	103	243	76	332
Chlorbromuron	61	46	294	206	292	60	45	63	292
Chlorcyclizine	99	56	72	165	300	228	229	242	300
Chlorcyclizine	99	56	165	43	228	229	241	242	300
Chlordiazepoxide	282	283	284	299	241	247	77	253	299
Chlordiazepoxide	44	282	283	77	284	56	91	57	299
Chlordiazepoxide	282	283	77	284	89	91	285	163	299
Chlordiazepoxide	282	299	218	284	283	219	241	214	299
Chlordiazepoxide	282	299	284	283	241	77	55	91	299
Chlordiazepoxide*	282	299	284	283	241	56	301	253	299
Chlordiazepoxide Benzophenone	230	231	77	232	105	154	233	126	231
Chlorfenethol	139	43	251	253	178	111	141	248	266
Chlorfenson	111	175	75	99	113	177	73	63	302
Chlorfenvinphos (alpha)	267	323	269	325	81	109	295	170	358
Chlorfenvinphos (beta)	267	269	323	81	325	109	295	170	358
Chlorimipramine	58	268	269	85	193	229	228	192	314
Chlormequat	50	58	52	36	42	44	38	49	157
Chlormethiazole	112	45	85	161	163	0	0	0	161
Chlormethiazole*	112	161	85	45	163	113	114	59	161
Chlormethiazole Metab (5-Acetyl-4-methylthiazole)	126	141	43	45	71	98	69	72	141
Chlormethiazole Metabolite	112	85	45	141	113	59	71	69	141
Chlormethiazole Metabolite (5-(1-hydroxyethyl)-4-methylthiazole)	128	45	100	73	43	71	97	125	143
Chlormezanone	152	153	154	42	155	125	111	56	273
Chlormezanone*	98	152	154	42	69	174	208	153	273
Chlormezanone (Trancopal)	152	42	98	154	153	174	69	56	273
4-Chloroaniline	127	129	65	92	128	100	99	91	127
2-Chloroaniline	127	129	65	92	64	39	91	63	127
2-Chlorobenzanilide	139	231	141	111	233	75	140	113	231
Chlorobenzene	112	77	114	51	50	113	75	76	112
4-Chlorobenzoic Acid*	139	156	111	75	141	50	158	113	156
4-Chlorobenzoic Acid Methyl Ester	139	111	141	75	170	50	113	172	170
Chlorobenzylidenemalononitrile (o-Chloro - CS gas)	153	188	126	190	161	154	137	189	188
2-Chlorobiphenyl	188	152	190	153	76	189	151	150	188
1-Chlorobutane	56	41	43	27	29	55	57	42	92
3-Chloro-6-cyano-2-norbornanon-o-(methylcarbamoyl) Oxime	184	148	58	118	186	154	149	104	255
Chlorocyclohexane	67	82	54	41	55	83	39	36	118
Chlorodibromoethane	129	127	131	48	47	91	93	81	206
Chlorodifluoroacetonitrile	76	85	50	31	92	87	77	57	111
Chlorodifluoromethane(Freon 22)	51	31	67	35	50	69	37	47	86
Chlorodinitronaphthalene	252	160	206	254	125	162	217	148	252
4-Chlorodiphenoxylic Acid	232	411	472	91	413	474	0	0	472
2-Chloroethanol	31	27	43	29	49	51	80	26	80
5-(2-Chloroethyl)-5-isopentyl-barbituric Acid	190	141	154	124	191	155	142	181	260
Chlorofluoromethane	68	70	33	49	47	51	48	50	68
Chloroform	83	85	47	87	48	49	35	82	118
3-Chloroheptane	56	41	69	55	43	57	70	29	134
1-Chlorohexane	91	55	43	56	41	42	29	93	120
2-Chloro-4'-hydroxybenzanilide	139	247	141	111	75	249	140	248	247
1,Chloro-3-hydroxypropane	58	57	30	36	27	26	76	29	94
1-Chloro-2-iodobenzene	238	111	240	75	113	50	74	127	238
1-Chloro-3-methyl-2-butene	68	67	53	41	39	40	27	69	104
4-Chloro-2-methylphenoxybutyric Acid (MCPB)	142	228	141	0	0	0	0	0	228
4-Chloro-2-methylphenoxypropionic Acid (CMPP)	142	214	141	169	107	77	0	0	214

Compound Name				Peaks					MWT
Chloromorphide (beta)	268	303	269	305	304	0	0	0	303
1-Chloro-4-nitrobenzene	111	75	157	113	50	127	99	159	157
Chloropentafluoroethane	85	119	69	31	87	135	29	50	154
3-Chlorophenol	128	130	65	64	39	26	63	129	128
4-Chlorophenol	128	130	65	39	64	63	129	99	128
2-Chlorophenol	128	64	130	63	65	92	39	129	128
2-Chlorophenothiazine	233	198	201	199	116	166	171	197	233
Chlorophenoxyacetamide (p-Chloro) (Iproclozide Metab)	185	58	141	111	128	185	113	75	185
3-(4-Chlorophenyl)-1-methyl-1-methoxyurea	61	214	185	126	153	140	93	127	214
Chloroprocaine*	86	99	154	30	87	58	29	156	270
4-Chloropyrazole	102	104	75	48	77	50	38	47	102
Chloropyrilene (Chlorothen, Tagathen)	58	131	72	71	79	42	30	78	295
Chloroquine	86	58	73	87	319	41	99	245	319
Chloroquine	86	81	98	58	41	57	43	319	319
Chloroquine	86	84	169	170	41	205	99	155	319
Chloroquine*	86	58	319	87	73	247	245	112	319
Chlorothen	58	131	71	72	42	79	133	78	295
3-Chlorotoluene	91	126	128	125	65	127	92	89	126
2-Chlorotoluene	91	126	128	89	125	90	65	63	126
4-Chlorotoluene	91	126	125	128	65	127	92	89	126
Chlorotoluron	72	44	212	45	77	42	132	214	212
Chlorotrianisene	380	223	382	238	152	345	215	113	380
Chlorotrianisene	380	382	345	381	190	223	238	113	380
Chlorotrifluoroethylene	116	31	66	85	118	97	47	81	116
Chlorotrifluoromethane	69	85	50	87	35	31	37	33	104
Chloroxuron	72	245	44	290	45	40	75	247	290
4-Chloro-3,5-xylenol	156	121	158	91	77	65	157	122	156
Chlorphenesin*	128	130	202	31	29	65	43	111	202
Chlorphenesin Carbamate	128	118	130	75	57	43	44	61	245
Chlorphenesin Carbamate	118	128	75	130	44	57	62	61	245
Chlorphenesin Carbamate*	128	130	202	43	129	111	75	204	245
Chlorphenethiazine	58	304	214	42	59	306	232	233	304
Chlorpheniramine	203	58	205	204	72	167	202	168	274
Chlorpheniramine*	58	203	167	72	205	202	168	204	274
Chlorpheniramine*	203	58	44	205	54	204	72	202	274
Chlorpheniramine(dex)	203	58	43	57	205	71	72	32	274
Chlorphenoxamine*	58	59	179	42	178	72	77	30	303
Chlorphentermine	58	125	42	168	60	41	89	127	183
Chlorphentermine	58	42	41	125	59	168	89	63	183
Chlorprocaine	86	99	154	58	56	87	42	84	266
Chlorproguanil*	127	43	229	44	161	231	85	186	287
Chlorpromazine	58	318	86	272	85	320	232	42	318
Chlorpromazine	58	319	86	321	85	273	36	274	318
Chlorpromazine	58	86	42	85	44	57	59	43	318
Chlorpromazine*	58	86	318	85	320	272	319	273	318
Chlorpromazine Metabolite	72	71	233	232	0	0	0	0	304
Chlorpromazine Metabolite	58	246	233	318	86	272	248	232	334
Chlorpropamide*	111	175	75	85	30	276	127	113	276
Chlorpropham	43	127	213	171	129	41	154	153	213
Chlorprothixene	58	42	221	59	57	43	222	189	315
Chlorprothixene*	58	59	221	30	42	222	255	43	315
Chlorprothixene Metabolite	98	231	232	197	0	0	0	0	232
Chlorpyrifos	97	197	199	314	316	258	286	125	349
Chlorpyriphos	197	199	97	314	248	316	286	250	349

COMPOUND NAME				PEAKS					MWT
Chlorthalidone	239	76	240	104	285	241	177	102	338
Chlorthalidone	148	76	130	75	102	104	239	50	338
Chlorthalidone*	76	239	104	240	285	241	50	75	338
Chlorthalidone Tetramethyl Derivative	363	176	286	255	365	288	192	220	394
Chlorthalidone Trimethyl Derivative	287	363	176	255	365	289	351	220	380
Chlorthal Methyl	301	299	303	332	45	330	221	334	330
Chlorthiamid	170	205	172	171	207	173	60	136	205
Chlorthiazide	295	268	297	97	57	270	62	64	295
Chlorthion	109	125	79	297	128	47	63	93	297
Chlorzoxazone	169	78	113	171	51	63	44	50	169
Chlorzoxazone	169	113	78	171	115	76	63	51	169
Chlorzoxazone*	169	78	171	113	115	63	51	170	169
3,5-Cholestadiene	43	57	41	55	44	0	0	0	368
3,5-Cholestadiene-7-one	174	382	161	187	159	175	383	41	382
Cholesterol	43	57	41	55	44	81	93	105	386
Cholesterol	81	107	105	91	95	79	93	67	386
Cholesterol	368	386	275	149	353	147	145	247	386
Cholesterol	315	81	43	·55	368	41	107	95	386
Cholesterol*	386	368	43	55	275	81	57	95	386
Chromonar*	86	87	58	30	29	84	56	42	361
Cimetidine	53	94	57	30	82	99	67	111	252
Cimetidine*	30	57	82	116	99	53	55	42	252
Cinchocaine	86	58	87	29	30	116	41	56	379
Cinchocaine*	86	87	58	149	111	99	57	41	379
Cinchocaine	86	87	58	228	326	116	57	113	343
Cinchocaine	86	116	41	58	73	56	57	228	343
Cinchomeronic Acid	50	43	105	44	123	77	78	51	167
Cinchonidine	136	137	81	79	42	41	55	159	294
Cinchonidine	136	42	41	81	55	79	130	77	294
Cinchonidine*	136	81	137	42	41	55	130	128	294
Cinchonine	136	42	58	55	81	294	41	159	294
Cinchonine	136	42	41	55	81	79	130	77	294
Cinchonine*	136	294	81	159	55	42	41	143	294
Cinchophen Methyl Ester	205	263	204	75	102	101	51	88	263
Cinepazide	221	82	80	98	81	97	319	417	417
Cinnamoylcocaine	82	182	83	96	103	147	148	131	329
Cinnamoylcocaine	96	83	82	182	93	36	42	131	329
Cinnamoylcocaine	152	83	42	82	85	181	57	122	329
5-Cinnamylidenebarbituric Acid	242	128	127	155	171	143	154	156	242
Cinnamyl-3-piperidylbenzilate (N-Cinnamyl)	117	199	105	77	183	427	91	336	427
Cinnarizine*	201	117	167	202	251	165	118	115	368
Citral	41	69	39	84	94	27	29	67	152
Citric Acid	42	43	84	56	60	27	44	102	192
Citric Acid	42	84	39	56	43	40	60	102	192
Citronellal	41	69	29	95	55	39	43	27	154
Citronellal	69	41	55	95	67	56	81	71	154
Citronellol	41	69	55	67	81	82	68	43	156
Clamoxyquin*	86	72	191	58	30	128	193	163	321
Clefamide*	228	182	363	88	229	76	276	257	398
Clemastine	84	128	42	43	77	139	82	179	343
Clemastine*	84	128	179	42	85	178	214	98	343
Clemizole	255	131	254	125	257	256	57	58	325
Clemizole	255	131	60	73	255	43	125	258	325
Clidinium Bromide	105	77	96	183	51	42	182	94	431

Compound Name									MWT
Cliquinol*	305	150	307	115	152	114	306	123	305
Clobazam	300	77	258	51	255	259	256	283	300
Clobazam*	300	258	77	259	283	302	231	256	300
Clobetasone*	331	43	71	332	121	147	131	41	408
Clofazimine*	455	457	472	474	459	456	458	473	472
Clofibrate*	128	130	169	87	41	129	242	171	242
Clofibrate Derivative	128	69	143	73	169	75	159	41	324
Clofibrate Derivative	128	130	169	41	129	228	75	69	228
Clofibrate Glucuronide Methyl Ester TMS Ether	73	317	169	171	41	318	217	75	620
Clomiphene	86	87	100	30	58	44	56	42	405
Clomipramine	58	269	85	268	270	271	314	242	314
Clomipramine	58	85	269	268	270	271	229	227	314
Clomipramine*	58	85	269	268	270	271	314	242	314
Clonazepam	280	314	315	234	289	240	75	76	315
Clonazepam*	280	314	315	286	234	288	316	240	315
Clonazepam Benzophenone	241	276	139	165	195	111	242	277	276
Clonidine	229	231	30	172	194	174	200	230	229
Clonidine	229	231	172	194	174	230	36	228	229
Clonidine	30	229	172	42	231	194	174	43	229
Clonidine*	229	30	231	172	194	174	200	230	229
Clonitazene	86	57	43	71	55	69	58	41	386
Clopamide*	111	127	55	83	59	41	112	42	345
Clopenthixol*	143	70	100	144	42	56	98	221	400
Clorexolone*	247	285	328	249	41	287	330	55	328
Clorindione	256	165	258	193	257	76	89	104	256
Clorprenaline*	72	30	43	77	73	51	41	27	213
Clothiapine	83	70	42	244	71	43	273	56	343
Clothiapine*	83	70	273	244	209	42	71	43	343
Clotiazepam	289	318	291	320	275	290	319	317	334
Clozapine*	243	256	70	245	192	227	258	326	326
Cocaine*	82	182	83	105	303	77	94	96	303
Cocaine(beta)	182	82	83	105	94	77	96	303	303
Codeine	299	162	229	124	300	214	298	42	299
Codeine	299	81	42	229	162	124	300	44	299
Codeine	299	162	229	298	124	214	297	282	299
Codeine	299	42	162	229	44	124	59	300	299
Codeine	299	42	124	162	300	297	44	59	299
Codeine	299	162	298	229	115	42	214	124	299
Codeine*	299	42	162	124	229	59	300	69	299
Codeine Heptafluorobutyrate Derivative	43	282	45	59	58	169	115	495	495
Codeine N-oxide	299	229	162	297	240	241	242	298	315
Colchicine	312	371	43	297	399	281	298	313	399
Colchicine	43	312	297	281	254	298	152	139	399
Colchicine*	312	43	399	297	356	281	371	311	399
Conessine*	84	71	85	82	80	341	70	356	356
Coniine	84	56	43	55	85	41	70	42	127
Coniine*	84	82	80	56	43	28	30	41	127
Cortolone(Tetra TMS)	73	449	359	450	147	269	75	243	654
Cotinine	98	42	65	78	176	0	0	0	176
Cotinine	98	176	42	118	41	119	51	39	176
Coumaphos	362	109	97	226	210	125	364	29	362
Cresol(m-)	108	107	79	77	39	90	109	51	108
Cresol(o-)	108	107	77	79	90	39	51	109	108
Cresol(p-)	108	107	77	79	109	90	39	53	108

Cresol(p-)	107	108	77	51	79	39	53	50	108
Crimidine	104	71	147	42	44	70	56	102	171
Crotonaldehyde	41	70	39	69	42	27	38	29	70
Crotononitrile(cis-)	41	67	39	40	66	38	27	37	67
Crotononitrile(trans-)	41	67	39	40	66	38	27	37	67
Crotoxyphos	127	105	193	104	121	43	179	166	314
Crotylbarbitone	181	156	55	141	180	138	155	39	210
Cruformate	256	182	276	108	169	184	291	278	291
Cruformate Metabolite	183	185	155	143	77	89	242	148	242
Cruformate Metabolite	336	109	227	249	115	338	0	0	336
Cruformate Metabolite	108	246	151	229	95	152	153	92	246
Cruformate Metabolite	182	276	108	169	184	278	171	305	305
Cruformate Metabolite	42	94	77	250	249	291	198	51	291
Cruformate Metabolite	263	42	265	164	278	319	304	264	319
Cyanazine	225	68	44	173	240	198	172	43	240
6-Cyano-6-desmethyl-LSD	335	193	192	207	234	233	167	180	335
3-Cyano-3,3-Diphenylpropionic Acid(Difenoxin metab.)	192	165	251	193	190	77	51	166	251
Cyanthoate	111	138	81	109	82	93	97	68	294
Cyclandelate	107	69	125	83	79	55	41	77	276
Cyclazocine*	230	271	55	256	164	270	71	124	271
Cyclizine	56	99	165	167	42	0	0	0	266
Cyclizine	99	56	167	207	194	164	195	208	266
Cyclizine	99	56	42	43	167	70	44	207	266
Cyclizine	99	165	56	194	167	207	152	208	266
Cyclizine*	99	56	167	207	194	266	195	165	266
Cycloate	72	44	41	45	89	42	43	55	198
Cyclobarbitone	207	67	79	81	141	77	55	91	236
Cyclobarbitone	207	141	40	81	79	44	77	41	236
Cyclobarbitone*	207	141	81	79	67	80	41	77	236
Cyclobarbitone Dimethyl Derivative	235	169	79	236	77	91	81	112	264
Cyclobenzaprine	58	59	42	215	202	57	189	43	275
Cyclobenzaprine N-oxide	229	230	215	228	101	202	226	227	291
Cyclofuramid	123	139	221	122	43	140	81	124	221
Cyclohexane	56	41	84	54	55	42	69	0	84
Cyclohexane	56	84	41	55	42	69	39	27	84
Cyclohexanol	57	82	44	67	41	71	29	56	100
Cyclomethycaine*	112	344	121	41	67	55	54	345	359
Cyclomethycaine	112	121	344	55	110	41	179	96	359
Cyclopal	193	169	67	66	57	233	65	205	234
Cyclopentamine	58	36	41	59	126	38	44	56	141
Cyclopentamine	58	59	45	41	40	74	42	59	141
Cyclopentamine*	58	41	30	126	59	69	56	44	141
Cyclopentene	67	68	39	53	41	65	42	27	68
Cyclopenthiazide*	296	41	44	110	298	285	55	268	379
Cyclopentobarbitone*	67	193	66	41	169	39	65	77	234
Cyclopentolate	58	55	89	90	118	42	71	41	291
Cyclopentolate*	58	71	72	207	42	91	59	118	291
Cyclopentolate	58	42	91	71	65	30	57	56	291
Cyclopentylmethanol	41	68	69	67	39	27	31	29	100
2-Cyclopropylmethylamino-5-Chlorobenzophenone	55	285	77	270	91	105	166	56	285
Cycloserine*	43	29	28	59	30	42	74	102	102
Cyclothiazide*	66	120	39	65	269	205	118	77	389
Cycrimine	98	99	218	85	131	219	84	69	287
Cycrimine	98	58	79	129	91	77	52	172	287

Compound Name									MWT
Cycrimine	98	41	42	99	55	69	77	105	287
Cypenamine	56	161	57	43	144	117	91	129	161
Cyprenorphine*	423	55	364	255	84	59	424	121	423
Cyproheptadine	287	96	215	286	229	213	228	243	287
Cyproheptadine	215	287	42	96	229	189	286	216	287
Cyproheptadine*	287	96	286	215	70	44	58	42	287
Cyproheptadine Metabolite	303	96	302	231	202	245	259	288	303
Cysteine(L-)*	74	76	28	75	59	43	42	47	121
2,3-D	162	164	220	175	185	111	147	222	220
2,4-D	162	164	220	161	63	133	111	222	220
3,4-D	220	175	162	222	177	145	147	164	220
Dalapon	62	97	99	64	61	36	107	45	142
Danthron	240	184	241	212	92	138	63	223	240
Danthron*	240	212	241	184	138	92	128	63	240
Dapsone	108	248	140	65	92	141	109	80	248
Daunomycin Metabolite	339	384	45	43	323	321	322	366	384
Daunomycin Metabolite	43	41	339	42	45	44	382	364	382
Dazomet	42	44	43	57	89	72	41	45	162
DDE	246	75	318	248	316	73	176	55	316
DDE (p,p-)	246	318	316	248	320	176	210	0	316
DDE (p,p'-)	318	246	316	248	320	176	105	247	316
DDT	235	237	165	212	246	75	176	36	352
DDT (o,p-)	235	237	165	75	199	246	352	0	352
DDT (p,p'-)	235	237	165	75	50	51	352	0	352
DDT(p,p'-)	235	237	165	212	176	282	236	284	352
Dealkylated Ethotoin	104	105	176	133	0	0	0	0	176
Debrisoquine	132	104	78	130	77	103	43	48	175
Debrisoquine*	132	104	44	175	130	117	103	43	175
Decafentin	154	309	197	77	155	195	307	353	864
Dehydrated Norpropoxyphene Amide	44	220	100	57	205	129	91	307	307
Dehydroemetine*	192	193	287	176	191	286	270	285	478
Demecoline*	207	371	312	342	208	372	42	328	371
Demeton	88	60	89	126	61	115	114	170	258
Demeton Methyl	110	109	79	156	80	47	125	126	230
Demeton O (technical)	88	89	60	61	171	97	115	59	258
Demeton PO	88	101	73	60	81	93	109	111	258
Demeton PS	88	89	60	61	171	97	115	59	258
Demeton S-Methyl Sulphone	169	109	125	168	79	110	142	170	262
Demoxepam*	285	286	269	287	241	242	77	270	286
6-Deoxy-6-azidodihydroisocodeine	123	326	59	298	70	198	185	115	326
6-Deoxy-6-azidodihydroisomorphine	123	312	59	58	70	115	171	270	312
6-Deoxy-6-azido-14-hydroxydihydroisocodeine	342	70	230	201	314	115	58	300	342
6-Deoxy-6-azido-14-hydroxydihydroisomorphine	328	70	216	58	187	115	286	285	328
Deoxycholic Acid	91	282	79	105	77	81	67	338	392
Deoxycortone Acetate	43	55	299	91	147	79	253	271	372
Deptropine*	83	140	82	124	43	96	97	42	333
Desalkylclomipramine(N-)	229	228	231	194	230	193	214	232	229
Desalkylflurazepam(N1-)	259	260	288	287	261	289	262	290	288
Desalkylflurazepam(N1-)	260	288	287	259	261	289	102	262	288
Desalkylflurazepam (N1-Desalkyl)	259	260	288	287	261	289	262	290	288
Desalkyl-3-hydroxyflurazepam Dehydration Product (N1-Desalkyl)	258	223	286	257	259	75	122	251	286
Desalkyl-3-hydroxyflurazepam (N1-Desalkyl)	275	277	258	257	259	75	223	122	304
Deserpidine*	578	195	577	367	351	579	366	365	578
Deserpidine	195	169	351	184	156	221	170	365	578

Compound Name									MWT
Deserpidine (Canescine, Harmonyl)	365	195	221	366	31	29	212	197	578
Desipramine	195	235	234	193	208	266	194	84	266
Desipramine	71	85	193	195	42	130	70	194	266
Desipramine	44	193	195	194	208	234	71	235	266
Desipramine*	235	195	208	44	234	193	194	71	266
2-Desmethylchlordiazepoxide	285	268	284	77	42	286	233	287	285
Desmethylclobazam	286	244	77	218	51	217	288	215	286
Desmethylclomipramine	269	229	268	232	71	44	242	211	300
Desmethylcyclobenzaprine	44	218	217	58	57	261	216	219	261
Desmethyldiazepam	242	270	269	241	243	271	244	103	270
Desmethyldiazepam	241	242	269	77	243	103	270	51	270
Desmethyldiazepam*	242	269	270	241	243	271	244	272	270
Desmethyldiazepam TMS Derivative	73	341	342	343	45	344	327	91	342
Desmethyldoxepin N-Acetyl	234	86	221	219	233	235	178	217	307
Desmethylflunitrazepam	298	271	299	224	272	270	252	280	299
Desmethylmedazepam	193	255	228	256	257	165	230	258	256
Desmethylmedazepam (N-)	193	228	89	255	77	165	51	110	256
Desmethyloxetorone	44	262	215	202	218	189	231	203	305
3-Desmethylprodine	96	44	36	173	172	56	70	129	247
Desmethylpromethazine	58	213	198	180	214	57	212	270	270
Desmethyltetrazepam	239	274	273	275	240	276	245	211	274
Desmethyltrimeprazine	314	229	242	44	269	86	283	71	314
Desomorphine	271	214	44	270	32	228	272	42	271
Deuterobenzene	84	56	54	52	42	82	40	85	84
Dexamethason	121	122	315	43	147	223	135	41	392
Dextromethorphan*	59	271	150	270	31	214	42	171	271
Dextromethorphan	59	150	271	270	214	171	128	212	271
Dextromethorphan (Racemethorphan)	59	42	150	271	44	171	115	128	271
Dextromoramide	100	128	265	55	56	41	42	40	392
Dextromoramide*	100	265	128	266	44	98	56	101	392
Dextropropoxyphene	58	57	59	91	42	77	105	115	339
Dextropropoxyphene	58	57	59	285	286	91	115	208	339
Dextropropoxyphene*	58	117	208	115	193	91	179	130	339
Dextropropoxyphene	58	205	42	191	178	128	91	129	339
Dextropropoxyphene	58	91	57	115	208	117	59	42	339
Dextropropoxyphene Carbinol	58	45	91	77	105	192	59	44	283
Dextropropoxyphene Carbinol	58	44	91	77	105	59	192	43	283
Dextropropoxyphene Metabolite 1	44	220	100	57	205	91	129	221	325
Dextropropoxyphene Metabolite	44	91	129	0	0	0	0	0	251
Dextropropoxyphene Metabolite	58	91	59	42	129	0	0	0	265
Dextropropoxyphene Metabolite	208	115	117	91	193	0	0	0	208
Dextropropoxyphene Metabolite	44	220	100	205	57	0	0	0	307
Dextropropoxyphene Metabolite	44	57	105	91	88	0	0	0	325
Dextropropoxyphene Metabolite	44	91	129	115	178	205	220	251	251
Dextropropoxyphene Metabolite 1	44	220	100	59	205	57	91	129	251
Diacetimide	42	59	73	42	44	101	41	0	101
Diacetone Alcohol	43	59	58	101	29	0	0	0	116
3,5-Diacetoxybenzoic Acid	154	43	196	137	155	69	238	42	238
3,5-Diacetoxybenzoyl Chloride	43	135	170	212	219	69	172	15	256
Dialifor	208	210	94	40	77	44	76	209	393
Diallyl Phthalate	189	41	190	104	149	103	39	115	246
2,5-Diaminobenzophenone	211	212	107	77	43	57	106	139	212
2,5-Diamino-2'-Chlorobenzophenone	246	211	107	43	245	57	248	80	246
2,5-Diamino-2'-Fluorobenzophenone	230	229	107	86	43	211	95	123	230

Compound Name									MWT
1,5-Diaminopentane	30	56	28	85	43	45	41	27	102
9,10-Diaminophenanthrene	208	180	205	207	206	209	104	77	208
Diamorphine	42	81	204	215	146	115	162	94	369
Diamorphine	327	369	43	268	310	42	215	128	369
Diamorphine*	327	43	369	268	310	42	215	204	369
Diampromide	162	105	44	163	57	190	106	58	324
Diampromide*	162	105	190	163	106	29	134	77	324
Diazepam	256	77	283	221	255	257	89	165	284
Diazepam*	256	283	284	285	257	255	258	286	284
Diazepam Benzophenone	245	77	244	105	228	246	51	247	245
Diazepam Metabolite	271	77	273	300	0	0	0	0	300
Diazepam-N-Oxide	299	300	301	283	77	43	256	302	300
Diazinon	179	137	152	304	93	153	97	135	304
Diazoxide*	125	189	230	127	191	63	232	90	230
Dibenzepin*	58	224	209	71	225	72	210	180	295
Dibenzo(a,d)cycloheptene-10,11-epoxide (5H-Dibenzo)	179	178	180	208	152	177	176	206	208
Dibenzo(a,d)cycloheptene (5H-Dibenzo)	192	191	189	165	152	190	193	139	192
Dibenzosuberone	208	89	179	76	165	77	209	88	208
Dibenzyl Ketone	91	65	39	119	92	63	89	51	210
1,2-Dibromobenzene	236	234	238	155	157	75	50	76	236
1,4-Dibromo-2-butene	53	133	135	54	39	27	89	51	212
Dibromochloromethane	129	127	131	48	47	93	91	79	206
Dibromochloropropane	57	157	75	155	39	49	77	41	234
1,2-Dibromocyclohexane(trans-)	81	79	80	41	82	163	161	77	240
Dibromodichloromethane	163	161	165	79	81	82	47	36	240
1,2-Dibromoethane	27	107	109	26	45	108	105	95	186
Dibromofluoromethane	111	113	192	41	43	190	194	79	190
Dibromomethane	93	174	95	59	172	176	74	43	172
1,2-Dibromopentane	69	41	43	42	27	149	29	151	228
1,5-Dibromopentane	69	41	27	55	43	42	39	29	228
1,3-Dibromopropane	41	121	123	39	202	27	204	200	200
1,2-Dibromopropane	41	121	123	39	27	42	40	38	200
Dibutylglycol Phthalate(n-)	57	45	56	29	101	149	85	41	366
Dibutyl Phthalate	149	41	29	57	56	104	32	65	278
Dibutyl Phthalate(n-)	149	150	41	57	223	205	56	104	278
Dibutyl Terephthalate(n-)	56	205	223	167	149	41	57	65	278
Dicamba Methyl Ester	203	205	234	236	188	201	204	190	234
Dichlobenil	171	100	173	75	74	136	99	50	171
Dichlofenthion	279	97	223	88	251	162	281	164	314
Dichlofluanid	123	77	167	92	44	91	224	42	332
Dichloralphenazone*	188	47	82	96	29	77	84	56	516
Dichloralphenazone Probe Artifact*	217	218	215	235	108	36	363	247	999
Dichloroacetic Acid	36	28	38	44	48	49	29	35	128
Dichloroacetylene	94	96	44	47	59	35	24	98	94
3,4-Dichloroaniline	161	163	99	90	165	126	134	73	161
2,6-Dichloroaniline	161	163	90	99	63	165	162	126	161
1,2-Dichlorobenzene	146	148	111	75	113	74	50	150	146
1,3-Dichlorobenzene	146	148	111	75	50	113	150	147	146
1,4-Dichlorobenzene	146	148	111	75	50	150	113	147	146
4,4'-Dichlorobiphenyl	222	152	224	151	223	150	75	153	222
1,3-Dichlorobutane	55	27	63	90	41	39	29	54	126
1,4-Dichloro-2-butene	75	53	89	77	27	39	62	54	124
Dichlorodifluoromethane(Freon 12)	85	87	50	101	103	31	35	66	120
1,4-Dichloro-2,5-dimethoxybenzene (Chloroneb)	191	206	193	208	141	195	163	143	206

Compound Name				Peaks					MWT
1,2-Dichloroethane	62	49	64	63	51	61	0	0	98
1,2-Dichloroethane	62	27	49	64	63	98	51	61	98
1,1-Dichloroethane	63	27	65	83	26	85	61	98	98
1,2-Dichloroethane	27	62	49	64	63	98	51	61	98
1,2-Dichloroethylene	61	96	98	63	26	60	62	100	96
2,2-Dichloroethyl Methyl Ether	45	29	61	43	27	49	46	36	128
Dichlorofluoromethane	67	69	35	47	31	48	83	49	102
1,6-Dichlorohexane	55	41	69	56	82	42	67	43	154
1,3-Dichloro-2-hydroxypropane	79	43	81	49	29	27	36	57	128
Dichloromethane	49	84	86	51	47	48	88	50	84
Dichloromonofluoromethane	67	69	35	47	31	32	48	83	102
2,3-Dichloro-1,4-naphthoquinone (Dichlone)	191	226	228	163	193	192	165	99	226
2,6-Dichloro-4-nitroaniline (DCNA)	206	124	176	208	160	178	162	126	206
2,4-Dichloro-6-(o-chloroaniline)-s-triazine (Dyrene)	239	241	178	143	274	276	240	242	274
1,5-Dichloropentane	55	42	41	68	27	29	69	39	140
Dichlorophen*	128	141	268	130	270	77	143	233	268
2,4-Dichlorophenol	162	63	164	98	99	49	62	73	162
2,4-Dichlorophenol	162	164	63	98	73	166	100	99	162
2,6-Dichlorophenol	162	164	63	98	126	166	73	99	162
2,4-Dichlorophenoxyacetic Acid	162	220	161	175	0	0	0	0	220
2,4-Dichlorophenoxymethyl Acetate	175	234	45	199	177	236	73	161	234
2,4-Dichlorophenoxypropionic Acid (2,4-DP)	162	234	189	161	0	0	0	0	234
(3,4-Dichlorophenyl)methylacrylamide (N-)	69	229	231	161	230	133	162	233	229
(3,4-Dichlorophenyl)-2'-methyl-2',3'-dihydroxypropionamide (N-)	75	161	163	263	265	127	126	162	263
2,4-Dichlorophenyl Methyl Ether	176	161	178	163	133	135	63	180	176
(3,4-Dichlorophenyl)-2-methyl-3'-hydroxyvaleramide (N-)	161	163	275	277	217	165	219	165	275
(3,4-Dichlorophenyl)-2'-methylvaleramide (N-)	161	163	259	261	217	165	219	162	259
2,4-Dichloroprop Methyl Ester	162	164	189	59	191	55	248	87	248
Dichlorphenamide*	304	306	64	74	109	177	176	48	304
Dichlorvos	109	185	79	202	145	187	200	199	220
Dichlorvos	109	96	15	79	185	47	45	95	220
Dichlorvos	109	185	79	145	187	47	220	110	220
Dichlozoline	186	41	201	188	43	152	273	203	273
Diclofenac Methyl Ester	214	242	216	309	215	311	179	151	309
Dicophane*	235	237	165	236	199	239	238	75	352
Dicoumarol	121	92	120	65	162	63	93	64	336
Dicoumarol*	336	121	120	215	162	92	337	187	336
Dicrotophos	127	67	72	193	237	44	109	111	237
1,4-Dicyano-1-butene	66	39	106	79	52	53	51	40	106
Dicyclohexylamine	138	56	55	41	44	30	28	82	181
Dicyclohexyl Phthalate	104	76	50	67	54	82	41	148	320
Dicyclomine	86	99	55	100	87	41	44	165	309
Dicyclomine*	86	71	99	58	55	56	100	87	309
Dicyclomine	86	55	99	41	44	100	83	56	309
Didesethylflurazepam	246	30	302	36	274	211	273	248	331
Didesethylflurazepam	314	255	302	246	316	211	315	273	331
Didesethylflurazepam Dehydration Product	313	315	314	312	137	250	183	273	313
Dieldrin	79	108	263	277	279	345	378	0	378
Dieldrin	79	82	81	263	77	108	277	80	378
Dienochlor	237	239	235	241	332	334	404	402	470
Diethazine	86	298	83	58	87	30	85	180	298
Diethazine	86	180	58	87	198	298	41	212	298
Diethazine*	86	298	87	30	58	299	212	180	298
2-Diethylaminoethylamino-5-Chloro-2'-Fluorobenzophenone	86	87	58	78	56	109	95	123	348

Compound									MWT
Diethylaminoethyltheophylline	86	87	99	58	71	207	84	56	279
2,4-Diethylamino-s-triazine	43	44	167	57	55	139	71	152	167
Diethylamyl Phosphate	99	155	127	81	109	0	0	0	224
Diethylcarbamazine	71	28	100	72	58	199	129	83	199
Diethylcarbamazine*	71	72	58	100	83	56	70	44	199
Diethylcarbamazine Probe Product*	44	43	39	40	71	58	68	72	999
2,2-Diethylcyclohexanone	55	41	126	68	97	110	81	43	154
2,6-Diethylcyclohexanone	55	41	83	42	126	82	56	97	154
Di-2-Ethyl-1,5-Dimethyl-3,3-Diphenyl-1-Pyrrolinium(Methadone metab.)	277	276	262	42	115	105	91	56	277
1,3-Diethyl-5,5-diphenylhydantoin	180	308	279	208	77	251	165	104	308
2,3-Diethyl-5,5-diphenylhydantoin	279	208	77	308	280	104	149	180	308
Diethyl Ether	31	29	59	45	74	43	0	0	74
Diethyl Ether	31	59	29	45	74	27	41	43	74
Di-2-ethylhexyl Phthalate	149	167	57	71	70	43	150	41	390
1,3-Diethylhydroxyphenobarbitone Ethyl Ether	332	303	275	304	190	276	134	333	332
1,3-Diethylhydroxyphenobarbitone TMS Ether	376	347	319	348	377	361	73	192	376
1,3-Diethyl-9-methylxanthine	123	42	222	207	179	149	150	166	222
1,7-Diethyl-3-methylxanthine	222	95	194	166	207	179	123	67	222
1,3-Diethyl-7-methylxanthine	222	136	166	194	150	123	207	67	222
3,7-Diethyl-1-methylxanthine	222	150	194	179	166	207	109	43	222
Diethyl Nitrosamine	102	44	42	29	56	57	27	30	102
Diethyl-O-(2,5-dichlorophenyl) Phosphorothioate (O,O-Diethyl)	223	279	97	162	251	225	164	281	314
Diethyl-O-(3,6-dichloro-2-pyridyl) Phosphorothioate	163	165	280	97	224	128	252	174	315
1,3-Diethylphenobarbitone	146	260	118	117	103	261	91	232	288
Diethyl Phthalate	149	177	150	65	76	105	176	104	222
Diethyl Phthalate	149	178	65	177	150	76	121	105	222
Diethyl-p-nitrophenyl Phosphate	109	81	149	81	275	139	99	65	275
Diethylpropion	100	77	44	42	72	105	101	51	205
Diethylpropion*	100	44	72	101	77	56	42	105	205
1,3-Diethylsecobarbitone	224	223	41	225	125	209	43	109	294
Diethylstilbestrol	107	145	268	238	121	133	159	224	268
Diethyl Terephthalate	177	149	166	65	194	178	104	121	222
Diethylthiambutene	276	111	219	277	42	97	100	135	291
Diethylthiambutene	276	219	111	277	278	42	100	97	291
Diethyltryptamine	86	30	58	130	29	77	87	42	216
Diethyltryptamine (N,N-)	86	58	30	87	42	130	77	56	216
Difenoxin	218	42	219	91	155	165	115	56	424
Diflunisal*	232	250	175	204	176	233	102	78	250
Difluorobromomethane	51	31	130	132	79	81	111	113	130
Difluorochlorobromomethane	85	87	129	131	147	79	81	50	163
1,1-Difluoro-1-chloroethane	65	45	85	64	44	31	61	81	100
Difluorodibromomethane	129	131	191	81	79	50	31	189	208
1,2-Difluoroethane	51	65	27	45	62	47	26	64	66
Difluoromethane	33	51	31	52	50	34	20	0	52
Diftalone	132	264	90	133	89	105	118	265	264
Digoxin*	68	39	29	43	94	44	336	95	780
Digoxin*	73	58	57	43	41	39	29	45	780
Dihexyverine	98	111	55	41	112	99	96	97	321
Dihydrocodeine	301	300	302	164	244	242	284	286	301
Dihydrocodeine*	301	44	42	59	164	70	302	242	301
Dihydrocodeinone	299	242	243	59	57	214	185	300	299
Dihydrocodienone	299	185	242	115	42	214	128	243	299
Dihydroergocornine	43	70	41	71	269	154	195	167	563
Dihydroergocornine*	70	71	269	154	195	55	59	57	563

Compound Name									MWT
Dihydroergocristine*	125	70	91	153	41	244	43	71	611
Dihydroergocristine Mesilate	125	70	91	153	41	43	244	349	611
Dihydroergocryptine	154	246	70	349	43	223	225	167	577
Dihydroergocryptine*	154	70	155	167	223	225	349	153	577
Dihydroergotamine	43	70	40	269	125	91	153	131	583
Dihydroergotamine*	70	125	91	153	43	41	44	244	583
2,3-Dihydro-LSD	325	223	225	224	326	226	182	167	325
Dihydromorphine	287	70	44	42	163	59	288	286	287
Dihydromorphine	287	70	164	44	42	286	285	288	287
Dihydromorphine*	287	70	44	164	42	288	59	230	287
Dihydromorphinone	285	36	229	42	96	228	44	286	285
2,5-Dihydrothiophene(2-oxy-)	100	72	39	71	55	45	46	101	100
3,5-Dihydroxybenzoic Acid	154	137	109	69	81	53	155	51	154
1,4-Dihydroxybenzoic Acid	110	136	44	108	154	81	69	80	154
2,4-Dihydroxybenzoic Acid*	136	154	108	80	52	95	137	69	154
5-(3,4-Dihydroxycyclohexa-1,5-dienyl)-3-me-5-phenylhydantoin TMS Deriv	73	191	75	45	74	104	147	167	516
7,14-Dihydroxydiftalone	133	146	89	278	105	77	132	118	296
Dihydroxyphencyclidine	91	115	128	117	129	219	157	0	275
3,4-Dihydroxyproline	102	87	69	41	59	0	0	0	101
Dihydroxyprotriptyline	44	70	179	178	207	280	250	236	297
2,3-Dihydroxyquinalbarbitone	171	43	143	41	128	55	141	159	270
Dihydroxytrazodone (Trazodone Metabolite)	239	70	209	210	166	139	138	211	405
Dihydroxyzolimidine (Metab 1)	277	306	288	278	209	198	222	237	306
Dihydroxyzolimidine (Metab 2)	249	306	250	170	307	251	171	277	306
1,2-Diiodobenzene	203	76	50	330	74	75	127	165	330
Diiodohydroxyquinoline*	397	115	242	398	88	143	271	62	397
Diisobutyl Phthalate	149	57	41	150	56	223	104	76	278
Di-iso-octyl Phthalate	57	149	71	43	41	69	55	113	390
2,4-Diisopropylamino-s-triazine	58	138	195	180	43	41	152	69	195
Di-isopropyl Ether	45	43	87	41	59	27	69	39	102
Di-(1-isopropylphenyl)amine	91	44	162	119	41	65	43	42	253
Di-(1-isopropylphenyl)amine*	162	91	44	119	163	41	70	65	253
Di-(1-isopropylphenyl)methylamine	91	58	176	90	41	42	119	65	267
Diisopropyl Phthalate	149	104	76	50	150	41	43	42	250
Diisopropyltryptamine (N,N-)	144	72	30	115	43	130	144	56	244
Diloxanide Furoate*	95	327	39	329	96	122	244	67	327
Dimefline*	279	163	323	58	277	308	280	322	323
Dimenhydrinate	58	73	45	43	57	167	44	165	469
Dimenoxadole	58	105	57	43	71	55	41	167	327
Dimethacrine	86	58	278	42	87	294	193	44	294
Dimethindene*	58	59	72	45	292	218	42	0	292
Dimethindene	58	42	59	57	128	115	218	78	292
Dimethirimol (tech)	166	209	96	167	71	180	93	55	209
Dimethisoquin*	71	58	72	43	159	56	42	201	272
Dimethisoquin	58	71	159	56	72	41	42	115	272
Dimethisterone	67	137	91	138	55	79	41	95	340
Dimethoate (Cygon)	87	93	125	58	47	79	63	229	229
Dimethocaine*	86	120	30	87	58	92	84	56	278
Dimethothiazine	72	73	320	71	70	56	210	198	391
Dimethoxanate	58	202	116	198	59	44	72	42	358
Dimethoxanate*	58	116	198	199	72	59	42	44	358
2,5-Dimethoxyamphetamine	44	152	137	77	65	91	78	121	195
3,5-Dimethoxyamphetamine	44	152	137	195	179	0	0	0	195
3,5-Dimethoxyamphetamine	44	152	195	151	77	78	65	51	195

Compound Name									MWT
2,6-Dimethoxyamphetamine	44	152	91	77	65	153	151	195	195
3,4-Dimethoxyamphetamine	44	152	151	65	51	137	78	77	195
2,3-Dimethoxyamphetamine	44	65	91	152	77	51	164	137	195
2,4-Dimethoxyamphetamine	44	152	151	121	153	77	91	78	195
2,5-Dimethoxyamphetamine	44	152	137	121	153	65	77	91	195
2,3-Dimethoxyamphetamine	44	40	152	91	65	165	42	164	195
2,4-Dimethoxyamphetamine	44	152	40	121	151	91	77	153	195
2,5-Dimethoxyamphetamine	44	152	40	137	91	77	65	121	195
2,6-Dimethoxyamphetamine	44	152	91	77	65	151	121	78	195
3,4-Dimethoxyamphetamine	44	152	151	137	153	107	65	43	195
3,5-Dimethoxyamphetamine	44	152	40	77	42	91	151	78	195
2,5-Dimethoxy-b-methyl-b-nitrostyrene	223	162	161	147	176	91	77	119	223
2,5-Dimethoxy-b-nitrostyrene	209	162	133	148	77	51	147	105	209
2,5-Dimethoxy-4-methylamphetamine(STP)	44	166	151	91	77	79	39	42	209
2,5-Dimethoxy-4-methylphenethylamine	166	151	30	135	195	165	44	167	195
2,5-Dimethoxytoluene	137	152	109	77	78	39	138	15	152
1,2-Dimethylallobarbital (or 1,4-Dimethyl-)	195	41	194	53	138	70	137	79	236
1,3-Dimethylallobarbitone	195	138	41	53	194	80	110	58	236
1,3-Dimethylalphenal	118	243	104	77	129	231	130	128	272
Dimethylaminoacetone(N,N-)	58	42	30	43	57	44	0	0	101
Dimethylaminobenzaldehyde(p-)	148	149	77	42	51	132	105	50	149
5-Dimethylaminobenzylidene-(para)-barbituric Acid	259	258	260	257	216	173	217	172	259
2-Dimethylamino-4,4-diphenyl-5-nonane	72	73	322	251	165	167	337	0	337
Dimethylaminoethyl-3-piperidylbenzilate (N-Dimethyl)	58	324	154	114	105	96	72	77	382
Dimethylaminomethane(bis-)	58	28	42	30	102	45	59	0	102
4-Dimethylamino-3-methylphenyl n-methylcarbamate (Bayer 44646)	151	150	136	208	58	77	45	40	208
3-Dimethylaminopropylamine	58	30	42	85	28	44	56	57	102
((3-Dimethylaminopropyl)-2-hydroxyiminodibenzyl)(N-)(Imipramine Metab)	58	251	250	296	211	85	224	209	296
1,2-Dimethylamobarbital (or 1,4-Dimethyl-)	169	184	140	185	126	226	55	41	254
1,3-Dimethylamobarbitone	169	184	185	170	55	112	183	69	254
Dimethylamobarbitone	169	184	120	112	121	126	225	226	254
Dimethylamphetamine	72	42	91	44	73	56	39	70	163
Dimethylamphetamine (N,N-)	72	44	42	91	56	65	58	162	163
Dimethylamylobarbitone (N,N-Dimethyl)	169	184	170	207	57	55	69	225	254
2,4-Dimethylaniline	121	120	106	77	91	122	93	118	121
1,3-Dimethylaprobarbitone	195	196	41	138	53	111	181	58	238
1,3-Dimethylbarbitone	169	184	183	126	112	83	41	40	212
Dimethylbarbitone (N,N-Dimethyl-)	169	184	59	55	112	183	126	69	212
1,2-Dimethylbarbitone (or 1,4-Dimethyl-)	184	40	169	140	126	183	44	55	212
1,3-Dimethylbutabarbitone	169	184	41	112	185	69	183	55	240
1,3-Dimethylbutalbital	196	195	41	138	181	111	209	169	252
1,2-Dimethylbutalbital (or 1,4-Dimethyl-)	196	41	195	209	138	181	67	43	252
1,1-Dimethylbutanol	59	45	87	43	41	31	73	55	102
2,2-Dimethylbutanol	43	71	70	55	41	73	45	29	102
2,2-Dimethylbutan-2-ol	59	87	41	69	43	31	45	39	102
2,2-Dimethylbutan-3-ol	57	45	56	41	87	69	29	43	102
3,3-Dimethylbutan-1-ol	57	69	41	56	43	29	45	31	102
3,3-Dimethylbutan-2-one	57	41	43	29	100	39	28	56	100
1,3-Dimethylbutethal	184	112	42	55	170	183	212	58	240
Dimethylbutobarbitone (N,N-Dimethyl)	169	184	183	170	112	212	55	185	240
Dimethylbutylamine	58	101	44	42	59	29	30	0	101
Dimethylcyclophosphoramide Mustard	108	199	163	201	92	149	132	170	248
Dimethyldemerol	186	70	83	42	85	71	57	56	275
1,3-Dimethyldihydroxysecobarbitone di-TMS Ether	73	341	43	271	75	147	41	342	444

Compound Name									MWT
1,3-Dimethyldilantin	251	280	72	134	77	265	208	175	280
2,6-Dimethyl-3,5-diphenylpyrimidine	259	260	115	244	116	95	108	215	259
2,4-Dimethyl-3,5-diphenylpyrimidine	259	260	244	115	108	85	91	215	259
Dimethyldisulphide	94	45	79	49	46	47	48	61	94
Dimethylformamide	73	44	42	30	43	72	29	58	73
2,4-Dimethylformanilide	120	149	106	121	77	91	132	150	149
2,5-Dimethylfuran	43	96	95	53	81	91	27	41	96
2,5-Dimethylfuran	96	95	43	53	81	27	51	50	96
Dimethylglycol Phthalate	59	58	31	45	149	104	43	76	282
1,3-Dimethylhydroxyamobarbitone TMS Ether	131	327	73	143	75	132	328	169	342
Dimethyl-3-hydroxyamylobarbitone (N,N'-)	184	169	0	0	0	0	0	0	270
2,3-Dimethyl-3-hydroxy-1-butene	59	43	86	85	40	41	38	42	100
1,3-Dimethylhydroxypentobarbitone Glucuronide Me Ester TMS Derivative	73	253	217	185	317	75	204	69	676
1,3-Dimethylhydroxypentobarbitone TMS Ether	117	73	327	75	143	256	118	69	342
1,3-Dimethylhydroxyphenobarbitone Methyl Ether	261	290	233	148	262	133	176	260	290
1,3-Dimethylhydroxyphenobarbitone TMS Ether	73	319	291	348	206	320	45	333	348
Dimethyl-3'-hydroxyquinalbarbitone	196	195	237	181	138	45	41	69	268
1,3-Dimethylhydroxysecobarbitone Me Ester TMS Ether	73	265	217	75	204	69	41	147	688
Dimethyl Isophthalate	163	194	135	76	164	103	120	77	194
Dimethyl-3'-ketoquinalbarbitone	196	195	43	138	181	237	197	69	280
1,2-Dimethyl-5-methoxyindole-3-acetic Acid Me Ester (Indomethacin Der)	247	188	160	63	215	43	106	216	247
Dimethylmorphine	313	229	138	162	42	0	0	0	313
Dimethylnealbarbitone (N,N-Dimethyl)	169	195	57	209	251	112	210	196	266
3,3-Dimethylpentane	43	71	27	41	29	70	85	39	100
2,2-Dimethylpentane	57	43	41	56	85	29	27	39	100
2,3-Dimethylpentane	43	56	57	41	29	27	71	42	100
2,4-Dimethylpentane	43	57	41	56	42	27	29	39	100
1,3-Dimethylpentobarbitone	169	184	41	43	183	69	112	55	254
Dimethylpentobarbitone (N,N-Dimethyl)	169	184	112	183	69	55	185	58	254
1,2-Dimethylphenobarbital (or 1,4-Dimethyl-)	203	188	232	117	204	40	115	70	260
1,3-Dimethylphenobarbitone	232	146	117	175	118	233	103	188	260
Dimethylphenobarbitone	232	146	118	175	117	120	121	188	260
Dimethylphenobarbitone (N,N-)	232	118	117	146	175	233	188	260	260
Dimethylphenobarbitone (N,N-Dimethyl)	232	118	146	117	103	77	175	115	260
2,4-Dimethylphenol	107	122	121	77	91	79	51	39	122
2,4-Dimethyl-3-phenyl-6-(benzyl)pyrimidine	272	273	258	259	115	243	91	260	273
Dimethyl-2-phenylethylamine (N,N-)	58	42	91	31	59	65	39	51	149
2,4-Dimethylphenylformamidine (N,N'-bis-2,4-Dimethyl-)	121	252	105	77	126	79	237	251	252
2,4-Dimethylphenyl-N'-methylformamidine (N-2,4-Dimethyl-)	162	132	121	120	106	118	147	77	162
Dimethyl Phthalate	163	77	164	135	194	76	92	50	194
2,2-Dimethylpropanoic Acid	57	41	29	39	27	45	59	56	102
Dimethylpropiolactone(beta,beta-)	41	44	39	56	28	27	55	40	100
Dimethylpropion	72	40	42	44	73	105	77	56	177
2,6-Dimethylpyrazine	42	108	39	40	38	27	37	41	108
2,5-Dimethylpyrazine	108	42	39	40	81	38	52	41	108
2,3-Dimethylpyrazine	108	67	42	40	41	26	39	109	108
2,3-Dimethylpyridine	107	106	66	39	65	92	79	108	107
2,5-Dimethylpyridine	107	106	79	77	92	39	65	108	107
2,6-Dimethylpyridine	107	106	66	92	65	39	108	79	107
3,4-Dimethylpyridine	107	106	79	92	77	39	108	65	107
3,5-Dimethylpyridine	107	106	79	92	77	39	108	80	107
2,4-Dimethylpyridine(2,4-Lutidine)	107	106	79	92	39	65	80	77	107
2,4-Dimethylpyrrole	94	95	80	39	41	27	67	53	95
3,5-Dimethylpyrrole-2,4-dicarboxylic Acid Ethyl Ester	239	148	194	164	210	193	182	147	239

Compound Name	Peaks								MWT
Dimethylquinalbarbitone (N,N-Dimethyl)	196	195	181	111	138	55	197	266	266
2,6-Dimethylquinoline	157	156	158	115	142	128	89	154	157
2,4-Dimethylquinoline	157	156	115	158	142	116	128	77	157
Dimethyl-s-(4-oxo-1,2,3-benzotriazin-3(4h)-ylmethyl) Phosphorothiolate	160	132	77	104	105	76	159	109	301
Dimethylsulfoxide	63	78	45	61	46	62	48	47	78
2,4-Dimethyltalbutal (or 4,6-Dimethyl-)	95	115	196	41	67	43	96	209	252
2,5-Dimethyltetrahydrofuran	56	41	43	57	29	45	67	27	100
3,4-Dimethyltetrahydrofuran	55	70	42	100	41	29	39	43	100
2,4-Dimethyltetrahydrofuran	85	41	55	56	43	42	57	70	100
Dimethylthiambutene*	248	97	219	218	111	249	217	263	263
Dimethyltryptamine	58	42	130	77	30	103	115	102	188
Dimethyltryptamine (N,N-)	58	44	188	130	42	143	59	77	188
Dimethyltryptamine (N,N-)	58	59	42	188	72	115	145	104	188
Dimethyltryptamine(N,N-)*	58	188	130	59	42	143	129	115	188
1,7-Dimethylxanthine	55	82	67	180	109	0	0	0	180
Dimetilan	72	240	42	39	40	41	44	29	240
Dimetilan	72	240	169	44	42	73	170	56	240
Dimexan	75	76	47	60	214	107	121	77	214
Diminazine*	30	43	72	102	73	42	118	99	281
Dimoxyline	352	338	367	366	353	336	322	339	367
2,4-Dinitrophenol	184	44	91	63	53	107	92	51	184
Dinitrorhodane Benzene	30	63	69	74	75	79	62	131	225
2,4-Dinitrotoluene	165	119	30	78	90	92	91	79	182
3,4-Dinitrotoluene	30	182	94	66	65	78	92	53	182
2,6-Dinitrotoluene	165	119	30	118	92	91	90	78	182
2,4-Dinitrotoluene	165	89	63	30	39	119	182	78	182
Dinonyl Phthalate	57	149	71	70	293	69	111	43	418
Dinoseb	211	163	147	117	240	77	89	205	240
Dinoterb	225	177	131	41	77	240	38	103	240
Dioctyl Adipate*	129	57	70	71	55	112	43	41	370
Dioctyl Phthalate(n-)	149	57	167	71	70	43	113	41	390
Diosgenin*	139	282	69	55	41	271	91	105	414
Dioxaphetyl Butyrate	100	72	56	165	42	101	91	115	353
Dioxaphetyl butyrate	100	42	56	101	114	353	165	91	353
Dioxaphetyl butyrate	100	114	353	91	165	70	161	178	353
Dioxathion	97	125	65	153	93	45	121	73	456
Dioxepane	71	42	41	102	43	31	55	44	102
Dioxopromethazine	72	73	56	71	70	231	152	180	316
Dioxyparaquat	216	147	158	57	217	131	160	187	216
Dipentene	68	93	67	136	41	121	79	39	136
Dipentyl Phthalate	149	43	150	55	41	42	237	71	306
Diperodon*	98	119	91	99	124	64	41	55	397
Diphenadione*	173	340	168	167	165	341	174	322	340
Diphenazoline*	84	160	167	159	165	152	55	77	266
Diphenhydramine	58	73	165	45	105	160	44	77	255
Diphenhydramine	58	42	73	165	45	77	57	152	255
Diphenhydramine	58	73	45	44	59	165	167	42	255
Diphenhydramine*	58	73	45	167	165	166	44	152	255
Diphenidol	98	105	77	41	42	55	99	84	309
Diphenidol*	98	99	105	77	55	41	127	111	309
Diphenoxylate	246	42	247	377	91	172	47	165	452
Diphenoxylate	246	91	165	377	452	0	0	0	452
Diphenoxylate	246	375	42	247	91	376	156	184	452
Diphenoxylate*	246	42	247	91	103	165	115	56	452

Compound Name				Peaks					MWT
Diphenoxylic Acid	232	438	378	380	246	91	407	423	438
3,3-Diphenyl-1,5-dimethyl-2-pyrrolidone(Methadone metab.)	265	193	115	130	42	208	56	264	265
Diphenyl Ether	170	77	51	141	169	171	142	115	170
1,3-Diphenylguanidine	93	77	119	51	211	0	0	0	211
Diphenylhydantoin	180	77	104	209	223	252	51	181	252
Diphenylhydantoin	180	209	104	223	252	77	181	165	252
Diphenylhydantoin	180	104	77	51	209	223	165	181	252
Diphenyl Mercury	77	51	50	356	78	354	353	279	356
Diphenyl Phthalate	225	77	226	104	76	149	153	43	318
Diphenylpyraline	99	42	98	70	114	96	44	43	281
Diphenylpyraline*	99	114	98	167	70	165	57	43	281
1,2-Diphenyl-3,5-pyrazolidinedione	183	77	252	184	105	91	51	64	252
Diphylline	180	194	223	109	95	42	254	193	254
Dipipanone	112	113	56	57	110	91	55	42	349
Dipipanone	112	113	41	56	69	334	55	91	349
Dipipanone	112	41	113	334	56	44	55	69	349
Dipipanone*	112	264	113	91	179	110	178	115	349
Dipiproverine	174	247	98	175	173	248	112	245	330
Diprophylline*	180	223	194	254	109	95	193	166	254
Dipropylamine(n-)	72	30	43	101	41	27	58	28	101
Dipropyl Ether	43	41	73	29	39	102	31	0	102
Dipyridamole*	504	473	429	505	221	474	84	430	504
Dipyrone	56	83	42	217	123	64	57	119	333
Dipyrone	56	83	230	217	123	97	77	64	333
Diquat	156	82	80	128	81	155	78	79	342
Disopyramide*	195	212	114	30	194	72	44	43	339
Disulfiram	116	88	43	148	60	56	72	117	296
Disulfiram*	116	88	29	44	60	148	56	27	296
Disulfoton	88	89	60	61	97	274	142	186	274
Disulfoton	88	89	29	97	61	60	27	65	274
2,6-Di-t-butyl-4-methylphenol (Antioxidant)	205	220	57	206	41	145	219	55	220
Dithianon	296	76	104	240	50	295	268	297	296
Ditolyl Ether(p-)	198	91	65	199	197	155	107	92	198
Diuron	72	232	234	161	73	45	163	124	232
Dixyrazine	212	42	187	45	70	180	56	98	427
Dodemorph	154	55	41	141	42	155	281	70	281
Dodine	73	43	72	86	100	87	59	128	227
Dopamine TFA Derivative	328	69	329	126	441	315	0	0	441
Dothiepin	58	59	236	202	203	221	42	57	295
Dothiepin*	58	236	40	202	235	203	42	44	295
Dothiepin	58	59	295	57	221	202	204	203	295
Dothiepin (Prothiaden)	58	59	221	204	202	203	293	234	295
Dowco 199	130	148	243	299	209	102	194	76	299
Doxapram	100	113	91	101	87	165	115	56	378
Doxapram	100	378	113	56	101	87	379	194	378
Doxapram*	100	113	56	101	87	378	194	91	378
Doxepin	58	220	59	219	277	179	191	193	279
Doxepin	58	42	59	57	165	43	115	178	279
Doxepin*	58	220	219	59	191	189	42	205	279
Doxepin	58	59	42	178	165	277	219	202	279
Doxylamine	71	58	72	167	182	42	59	0	270
Doxylamine	71	58	167	180	72	42	182	78	270
Doxylamine*	58	71	72	167	182	42	180	59	270
Droperidol	165	123	246	199	95	42	108	214	379

COMPOUND NAME	PEAKS								MWT
Droperidol*	246	165	42	123	199	247	214	108	379
Droperidol Hofmann Reaction Product	123	95	164	69	0	0	0	0	164
Dybenal	141	77	176	111	113	178	75	143	176
Dydrogesterone	43	91	227	268	312	79	55	77	312
Dyphylline	180	194	223	109	95	193	166	42	254
Ecgonine	82	97	42	83	96	57	94	55	185
Ecgonine	82	96	83	124	138	168	185	186	185
Ectylurea	41	43	113	96	44	98	39	69	156
Embramine*	58	59	42	165	103	72	30	0	347
Emepromium Bromide GC Breakdown Product	72	167	73	165	152	253	166	168	253
Emepromium Bromide GC Breakdown Product	86	167	58	87	165	152	263	42	267
Emetine*	192	206	272	480	288	246	205	191	480
Emylcamate	73	43	84	55	69	41	44	85	145
Enallylpropymal*	181	41	182	39	124	53	138	97	224
Encyprate	91	146	65	219	92	41	39	190	219
Endosulphan	195	36	237	41	241	75	239	170	404
Endrin	67	81	263	36	79	82	261	265	378
Ephedrine	56	77	91	105	106	42	0	0	165
Ephedrine	58	77	105	56	57	51	42	79	165
Ephedrine	58	105	77	42	106	56	57	51	165
Ephedrine	58	56	77	108	42	105	146	91	165
Ephedrine*	58	146	56	105	77	42	106	40	165
Ephedrine GC Decomposition	58	30	77	59	56	42	51	79	165
Ephedrine GC Decomposition	58	71	56	77	42	105	146	59	165
Ephedrine GC Decomposition	58	85	70	57	42	148	56	77	165
Ephedrine GC Decomposition	58	85	42	77	56	70	105	57	165
Ephedrine TMS Ether	58	73	59	88	45	75	43	56	237
10,11-Epoxycarbamazepine	180	223	252	179	152	0	0	0	252
10,11-Epoxycyclobenzaprine	58	203	202	232	84	85	101	215	291
10,11-Epoxyprotriptyline	44	70	178	250	179	279	0	0	279
Ergocornine*	43	70	71	54	44	154	267	55	561
Ergocornine Hydrogen Maleate	70	43	154	71	41	267	195	221	561
Ergocristine	57	43	71	85	99	113	243	244	609
Ergocristine*	70	125	71	91	153	267	154	221	609
Ergocryptine*	43	70	71	154	41	209	69	267	575
Ergocryptinine*	43	70	71	154	267	209	221	69	575
Ergometrine	325	221	207	196	181	44	223	42	325
Ergometrine*	221	72	325	54	196	55	207	181	325
Ergometrine Maleate	221	196	222	307	112	181	205	154	325
Ergonovine	221	181	207	196	223	325	222	180	325
Ergosine*	43	70	57	154	71	55	69	85	547
Ergosinine	70	43	40	154	41	44	267	69	547
Ergosinine*	70	154	69	224	210	55	196	209	547
Ergotamine*	125	44	70	91	41	40	244	153	581
Ergotamine Tartrate	125	70	91	153	244	40	43	314	581
Ergothioneine	57	71	43	85	40	55	41	149	229
Ergotoxine	40	43	44	70	41	149	71	55	561
Etafedrine	58	86	87	77	56	42	30	44	193
Etafedrine*	86	58	87	42	56	77	44	43	193
Etamphylline*	86	109	30	151	87	81	99	58	279
Ethacrynic Acid*	247	189	249	191	243	55	29	245	302
Ethacrynic Acid Methyl Ester	261	45	263	243	55	73	245	316	316
Ethambutol*	102	30	72	44	116	55	173	71	204
Ethamivan	151	52	123	108	42	65	51	72	223

Compound Name									MWT
Ethamivan	151	223	222	72	123	152	108	224	223
Ethamivan*	151	72	223	123	222	152	52	29	223
1,2-Ethanediamine	30	42	43	27	44	29	31	41	60
Ethanol	31	29	45	27	26	46	43	30	46
Ethanol	31	45	29	27	46	43	30	42	46
Ethanolamine	42	31	61	43	29	27	44	41	61
Ethchlorvynol (Placidyl)	115	117	89	53	109	51	91	39	144
Ethenzamide*	120	92	105	148	150	121	133	65	165
Ethephon (tech)	82	77	105	51	81	50	122	109	144
Ethiazide*	296	298	205	221	64	63	41	125	325
Ethinamate	91	43	67	81	78	79	106	44	167
Ethinamate*	91	81	106	78	39	95	68	43	167
Ethinamate (Ethynylcyclohexyl Carbamate)	81	91	106	95	79	68	67	78	167
Ethinylestradiol	213	160	159	296	133	145	212	157	300
Ethion	231	97	153	121	125	29	65	93	384
Ethionamide*	166	165	167	138	133	105	60	106	166
Ethirimol	166	96	209	55	167	71	42	69	209
Ethisterone	122	121	91	147	161	43	107	120	300
Ethoheptazine	107	149	57	78	108	79	77	72	261
Ethoheptazine	57	42	186	58	44	70	83	84	261
Ethoheptazine	57	58	44	42	39	70	40	68	261
Ethoheptazine*	57	58	70	42	44	188	84	43	261
Ethomoxane*	86	44	30	265	41	180	29	87	265
Ethopropazine	100	101	72	197	312	84	179	212	312
Ethopropazine	100	44	101	72	42	56	180	198	312
Ethopropazine*	100	101	44	72	198	180	42	29	312
Ethosuximide	113	55	70	42	41	39	85	69	141
Ethosuximide(i)*	55	70	113	42	41	39	44	69	141
Ethosuximide(ii)*	113	70	55	42	41	39	85	69	141
Ethotoin	104	105	204	77	78	133	51	132	204
Ethotoin (Peganone)	204	105	104	133	77	72	205	78	204
3-Ethoxy-6-deoxy-6-azidodihydroisomorphine	123	340	59	283	115	322	199	70	340
7-Ethoxydiftalone	263	133	132	308	264	105	77	279	308
2-Ethoxyethanol	31	59	29	45	27	72	43	0	90
3-Ethoxypropionaldehyde	31	29	45	27	28	58	43	74	102
Ethoxyquin	202	108	174	137	109	80	203	145	217
5-(2-Ethoxythiocarbonylthioethyl)-5-isopentylbarbituric Acid	124	55	139	99	142	141	125	154	346
Ethyl Acetate	43	45	61	70	29	27	42	73	88
Ethyl Acrylate	55	27	29	56	45	28	73	26	100
Ethyl-4-aminobenzoate	120	65	92	39	165	137	63	41	165
2-Ethylamino-1-phenylpropanol	72	44	73	77	105	42	132	160	179
2-Ethylamino-1-phenylpropanone	72	44	134	77	42	73	105	51	177
Ethylamphetamine	72	44	91	73	42	56	65	39	163
2-Ethylaniline	106	121	77	107	79	53	120	78	121
Ethylaniline(N-)	106	121	77	107	120	104	51	79	121
Ethylbenzene	91	106	51	65	77	92	39	78	106
Ethyl Biscoumacetate*	121	318	317	173	120	362	44	31	408
Ethyl Bromopyruvate	29	42	43	120	122	27	93	95	194
2-Ethyl-1-butanol	43	70	71	55	41	29	56	84	102
Ethyl Butylamine(n-)	58	30	101	44	29	28	41	27	101
Ethyl Butyl Ether	59	31	29	57	41	56	27	73	102
2-Ethyl Butyraldehyde	43	72	27	41	29	71	57	39	100
Ethyl Butyrate	71	43	88	60	41	73	45	89	116
Ethylbutyrylurea(Carbromal Metabolite)	45	130	44	71	42	61	115	55	158

Compound Name									MWT
Ethyl Caprate	88	101	43	61	60	41	73	70	200
Ethyl Caprylate	88	57	101	60	61	127	73	43	172
Ethyl 2,4-dichlorophenoxyacetate	175	248	177	250	185	145	69	57	248
2-Ethyl-2-(3,4-dihydroxyphenyl)glutarimide (Glutethimide Metab)	249	220	192	149	164	131	135	147	249
2-Ethyl-1,3-dioxolane	73	45	28	27	29	43	57	0	102
3-Ethyl-5,5-diphenylhydantoin	180	209	280	77	181	104	251	165	280
Ethylene dibromide	109	107	73	108	106	123	43	121	186
Ethylene Glycol	31	33	29	43	27	42	62	44	62
Ethylephedrine	86	58	42	87	77	30	56	51	193
Ethylethosuximide (N-)	141	70	55	41	42	69	112	126	169
Ethyl Heptanoate	88	43	60	113	101	61	73	70	158
2-Ethyl-1-hexanol	57	43	41	70	55	56	83	29	130
3-Ethylhexobarbitone	249	81	183	250	79	185	184	264	264
2-(2-Ethylhexyl)-4-hydroxyphthalate	120	57	164	165	70	69	84	98	294
1-(2-Ethylhexyl)-4-hydroxyphthalate	57	165	70	120	43	41	92	83	294
1-(2-Ethylhexyl)-3-hydroxyphthalate	164	120	92	83	70	112	221	165	294
2-Ethylhexyl Phthalate	149	43	167	57	104	112	279	113	278
5-Ethyl-5-(2-hydroxyethyl)barbituric Acid	124	141	156	139	98	99	101	114	200
2-Ethyl-2-(4-hydroxy-3-methoxyphenyl)glutarimide (Glutethimide Metab)	234	263	206	163	178	131	149	161	263
2-Ethyl-2-(4-hydroxyphenyl)glutarimide (Glutethimide Metab)	204	133	176	233	148	131	188	119	233
Ethyl Isobutyl Ether	59	31	29	27	41	44	39	57	102
Ethyl Isopropyl Ketone	57	43	29	27	41	28	100	71	100
Ethyl Laurate	88	101	43	41	73	61	70	55	228
4-Ethylmephobarbitone	217	218	132	146	117	103	246	118	274
3-Ethylmephobarbitone	246	146	117	118	247	175	103	77	274
Ethyl-2-methylallyl Ether	57	55	85	29	100	72	41	43	100
2-Ethyl-5-Methyl-3,3-Diphenyl-1-Pyrroline	208	193	130	115	91	165	179	207	265
Ethylmethylthiambutene	262	135	111	219	97	263	56	0	277
Ethylmethylthiambutene	262	219	111	97	263	56	218	42	277
Ethylmethylthiambutene*	262	111	219	97	263	86	42	264	277
Ethylmorphine	313	42	162	36	81	124	59	44	313
Ethylmorphine	313	243	162	124	59	42	112	0	313
Ethylmorphine	42	162	313	44	124	59	115	81	313
Ethylmorphine*	313	162	314	124	284	59	42	243	313
Ethyl Myristate	88	101	43	41	73	55	57	70	256
Ethyl-N-demethyl-4-phenylpiperidine-4-carboxylate	57	233	42	56	43	158	131	160	233
Ethyl Nitrate	46	29	76	30	0	0	0	0	91
1-Ethyl-N-methyl-2-phenylethylamine	72	134	91	42	57	31	65	44	163
Ethylnorepinephrine	58	30	41	65	36	39	93	59	179
Ethylnorpethidine (N-)	246	261	260	103	186	91	77	0	261
Ethyl Palmitate	88	101	43	57	55	41	89	73	284
3-Ethylpentane	43	71	70	29	27	41	55	39	100
Ethyl-1-phenylcyclohexylamine (N-)	160	203	91	146	161	117	44	104	203
Ethyl-1-phenylcyclohexylamine (N-)	160	91	129	130	117	115	104	203	203
1-Ethyl-2-phenylethylamine	58	41	91	31	120	65	29	42	149
Ethyl Phenylmalondiamide (Primidone Derivative) di TMS Derivative	235	220	73	75	204	145	236	130	350
2-Ethyl-2-phenylmalondiamide (Primidone Metab)	148	163	91	103	120	115	117	77	163
Ethyl Phenylmethyl Ketone (ortho)	100	133	77	52	148	50	76	0	148
Ethyl-3-piperidyl Benzilate (N-)	111	36	77	105	96	112	183	42	339
Ethyl-3-piperidyl Benzilate (N-)	111	105	96	77	183	165	128	324	339
Ethyl-3-piperidylcyclopentyl Glycolate (N-)	111	96	175	128	157	316	262	331	331
Ethyl-3-piperidyldiphenyl Acetate (N-)	111	96	167	165	152	105	128	139	323
Ethylpropion	72	44	42	70	77	128	51	105	177
Ethyl Propionate	57	29	75	27	101	74	28	45	102

Compound Name	Peaks								MWT
2-Ethylpyridine	106	107	79	78	52	51	65	80	107
3-Ethylpyridine	107	92	106	65	79	39	108	77	107
4-Ethylpyridine	107	106	92	65	79	39	51	108	107
Ethyl Stearate	88	101	43	57	89	55	41	73	312
1-Ethyltheobromine	208	180	109	67	42	137	55	179	208
7-Ethyltheophylline	208	95	193	67	180	123	73	43	208
Ethyltryptamine (N-)	58	131	130	30	77	132	103	59	188
Ethynylestradiol	213	160	296	133	159	145	146	214	296
Etilefrine*	58	30	59	77	29	95	65	57	181
Etonitazene	86	36	162	58	87	107	105	0	396
Etorphine*	44	215	411	324	45	164	42	216	411
Etorphine TMS Ether	44	73	272	45	164	162	396	250	483
Etoxeridine	246	247	42	36	45	91	56	219	321
Eucatropine*	124	276	58	140	56	72	125	41	291
Eugenol	164	149	131	137	103	77	133	165	164
Fencamfamin*	98	215	58	84	91	56	71	186	215
Fencamfamine	98	91	84	41	56	58	39	115	215
Fenchene(alpha-)	93	79	80	121	94	107	81	136	136
Fenchene(alpha-)	93	136	111	78	79	40	94	80	136
Fenchlorvos	285	125	287	109	93	79	47	289	320
Fenchlorvos	287	125	285	79	109	93	289	47	320
Fenchlorvos (Ronnel)	285	287	125	109	289	79	47	93	320
Fendroprox	97	56	32	91	68	98	65	57	202
Fenethylline	250	207	70	91	251	119	148	65	341
Fenethylline*	250	70	207	91	251	119	148	56	341
Fenfluramine*	72	44	159	73	58	42	109	56	231
Fenitrothion	127	109	125	277	260	79	192	93	277
Fenoprofen*	197	241	198	77	242	104	91	103	242
Fenpipramide*	98	112	99	55	42	41	211	84	322
Fenson	77	51	141	99	63	73	50	268	268
Fentanyl	121	152	93	65	39	122	63	153	336
Fentanyl	146	57	245	189	42	91	77	132	336
Fenthion	278	125	109	153	168	169	93	79	278
Flavone	222	120	92	194	64	63	221	97	222
Flavoxate	98	147	111	42	55	115	70	96	391
Flavoxate*	98	111	99	147	55	41	42	96	391
Fluanisone	205	218	123	356	219	162	95	190	356
Flufenamic Acid	263	281	166	92	145	167	235	139	281
Flufenamic Acid Methyl Ester	263	295	264	166	92	235	145	243	295
Flunitrazepam	285	63	75	109	312	183	238	266	313
Flunitrazepam	313	285	312	286	266	238	294	239	313
Flunitrazepam*	285	312	313	286	266	238	294	284	313
Flunixin Methyl Ester	295	263	310	251	294	249	277	181	310
Flunixin TMS Ester	263	353	251	368	73	249	75	277	368
Fluopromazine*	58	352	86	353	85	306	42	266	352
Fluorochlorobromomethane	67	69	31	111	113	79	48	47	146
Fluorocyclohexane	67	41	82	54	56	59	55	39	102
Fluorodichlorobromomethane	101	103	147	145	105	31	149	66	179
Fluoxymesterone	43	71	55	79	91	109	123	336	336
Flupenthixol*	143	70	100	144	42	98	58	56	434
Fluphenazine	42	280	70	143	56	113	72	100	437
Fluphenazine	280	70	143	42	113	406	281	437	437
Fluphenazine	280	42	143	70	56	113	281	265	437
Fluphenazine*	280	143	42	70	437	406	113	56	437

Compound Name				Peaks					MWT
Flurazepam	86	99	58	87	56	85	183	387	387
Flurazepam	86	58	99	87	56	42	71	84	387
Flurazepam*	86	87	99	58	84	387	315	56	387
Flurazepam Benzophenone	248	249	123	250	251	154	95	230	249
Flurazepam Benzophenone	86	87	58	30	109	43	95	123	348
Flurazepam Benzophenone	86	87	109	95	123	348	166	262	348
Flurazepam Benzophenone 2	262	109	166	264	293	95	123	75	293
Flurazepam Benzophenone 4	86	30	58	87	109	123	166	262	348
Flurazepam Chloroethyl Artifact Benzophenone	262	109	166	264	275	311	313	123	311
Flurazepam Hydroxyethyl Metabolite Benzophenone	262	264	293	263	295	265	0	0	293
Flurazepam Metabolite	245	183	89	247	0	0	0	0	245
Flurazepam Metabolite	42	44	259	260	261	0	0	0	288
Flurazepam Metabolite	313	137	315	164	312	0	0	0	313
Flurazepam Metabolite 1	313	137	315	314	0	0	0	0	313
Flurazepam Metabolite	84	70	83	68	288	260	273	289	371
Flurazepam N1-Acetic Acid	346	318	259	345	347	273	348	320	346
Flurazepam N1-Acetic Acid	245	44	247	209	183	246	271	259	346
Flurazepam N1-Acetic Acid Decomposition Product	245	247	183	210	89	105	122	246	245
Flurazepam(N1-Ethanol-)	288	273	331	287	304	290	289	275	331
Flurbiprofen*	199	244	200	178	179	184	183	245	244
Fluspirilene*	244	42	72	475	109	245	85	476	475
Fomocaine	100	131	311	218	0	0	0	0	311
Fomocaine Metabolite	219	131	100	218	0	0	0	0	327
Formetanate	221	163	164	44	149	122	42	36	221
Formic Acid	29	46	45	44	30	47	31	48	46
Formylamphetamine (N-Formyl)	72	44	118	91	65	117	42	39	163
Formyldi-(1-isopropylphenyl)amine* (N-)	91	190	119	72	191	41	44	162	281
Formyl-1,3-diphenylisopropylamine (N-)	120	91	148	103	194	77	65	0	239
Formylmethylamphetamine (N-)	86	58	91	56	65	42	39	118	177
Fructose	73	60	86	71	57	61	103	149	180
Frusemide*	81	53	330	96	82	332	64	63	330
Frusemide (Fursemide)	81	53	330	82	96	332	222	250	330
Furaltadone*	100	56	101	42	185	184	41	128	324
Furan	39	68	38	29	40	37	42	26	68
Furazolidone*	87	79	51	225	42	50	86	80	225
Furcarbanil	123	215	43	81	53	124	216	94	215
Furethidine	246	247	42	71	43	56	232	0	361
Furethidine*	246	42	43	247	71	41	56	91	361
Furfural	39	96	95	29	0	0	0	0	96
Furfural	96	95	39	38	29	37	97	67	96
Furfuryl Alcohol	98	41	81	42	39	97	53	69	98
5-Furfurylidenebarbituric Acid	206	128	107	119	92	135	178	106	206
Furfurylmethylamphetamine	81	138	53	82	91	139	56	42	215
Furosemide	81	53	80	64	52	330	51	63	330
Furosemide	81	53	64	96	251	82	39	48	330
Furosemide Trimethyl Derivative	81	53	96	372	82	374	339	357	372
Gallic Acid	170	153	39	51	79	125	53	126	170
Gentian Violet	253	239	359	373	252	238	237	36	407
Geraniol	69	41	39	65	27	29	53	154	154
Geraniol	69	41	68	67	93	55	39	0	154
Glenbar	301	299	303	221	142	223	317	315	346
Glibornuride*	91	155	197	65	84	39	95	41	366
Glipizide*	150	121	56	93	39	151	66	94	445
Glucose	73	60	57	71	61	74	101	98	180

Compound Name	Peaks								MWT
Glutaraldehyde	44	43	29	27	41	82	0	0	100
Glutethimide	189	117	132	160	91	115	77	103	217
Glutethimide*	189	132	117	160	91	115	103	77	217
Glutethimide (Doriden)	117	189	132	115	91	160	77	39	217
Glutethimide Metabolite	233	146	188	104	103	0	0	0	233
Glutethimide Metabolite	91	161	117	103	115	189	143	55	233
Glutethimide Methyl Derivative	203	132	117	174	115	91	231	103	231
Glycerol guiacolate	124	198	109	77	81	110	125	167	198
Glycoaldehyde	31	29	60	30	42	41	26	56	60
Glymidine*	244	59	77	29	43	31	45	55	309
Griseofulvin	138	352	215	310	214	69	321	354	352
GS 29696	183	58	94	42	182	40	196	73	240
Guaiacol Glyceryl Ether	124	109	81	52	77	51	65	95	198
Guaiphenesin	124	109	198	77	81	43	180	122	198
Guaiphenesin	124	109	198	31	81	77	95	0	198
Guanethidine	126	44	55	58	139	42	43	96	198
Guanethidine	126	139	44	58	43	55	42	41	198
Guanylurea	86	43	102	69	44	42	59	0	102
Guthion	160	132	77	93	105	76	104	51	317
Haloperidol	224	238	226	123	340	0	0	0	375
Haloperidol	224	42	237	123	95	57	226	206	375
Haloperidol	123	95	224	42	237	206	56	84	375
Haloperidol*	224	42	237	226	123	206	239	56	375
Haloperidol Hofmann Reaction Product	207	225	96	111	0	0	0	0	225
Haloperidol Hofmann Reaction Product 1	123	95	164	69	0	0	0	0	164
Halothane	116	197	196	119	67	129	69	98	197
Harmaline	213	214	170	198	169	115	63	143	214
Harmaline	213	214	198	170	199	215	169	172	214
Harmalol	199	200	198	170	172	171	63	42	200
Harman	182	57	43	55	40	69	41	181	182
Harmine	212	197	169	213	211	170	106	168	212
Harmine*	212	169	197	213	106	211	170	168	212
Heptabarbitone	221	141	81	79	222	41	67	93	250
Heptabarbitone	221	141	79	81	77	40	38	67	250
Heptabarbitone*	221	43	78	93	80	41	141	39	250
Heptachlor	100	272	274	270	237	102	65	276	370
Heptachlor Epoxide	81	353	355	351	357	237	386	0	386
Heptacosafluorobutylamine	69	131	100	502	119	414	464	614	614
Heptacosafluorotributylamine	69	219	131	100	264	502	119	414	614
Heptaminol*	44	43	59	56	69	55	41	113	145
Heptane(n-)	43	41	57	71	29	56	42	27	100
2-Heptanone	43	58	71	41	59	27	29	85	114
3-Heptanone	57	29	85	72	41	43	27	114	114
4-Heptanone	43	71	27	41	39	42	29	58	114
Hexachlorobenzene (HCB)	284	286	288	282	249	247	251	214	282
2,4,5,2',3',4'-Hexachlorobiphenyl	360	362	290	358	288	364	145	292	358
Hexachlorobiphenyl(bis-2,4,5-)	360	362	358	290	288	364	292	145	358
Hexachloro-1,3-butadiene	225	227	223	190	260	141	118	188	258
Hexachloroethane	201	117	119	203	199	166	94	47	234
Hexachloroethane	117	119	201	166	164	203	199	168	234
1,5-Hexadiene	67	41	54	39	27	53	81	68	82
Hexafluorenium Bromide	165	58	163	42	166	164	139	63	660
Hexamethyldisiloxane*	147	148	66	73	149	45	59	131	162
Hexamethylene Oxide	42	55	41	100	39	68	29	27	100

Compound									MWT
Hexamine*	42	140	112	41	85	43	71	141	140
Hexanal	44	56	43	41	29	72	82	67	100
2,5-Hexanedione(Acetonyl Acetone)	43	99	71	57	114	27	42	72	114
Hexane(n-)	57	43	41	29	27	56	42	39	86
1-Hexanol	56	43	55	42	41	69	31	29	102
2-Hexanol	45	69	41	43	44	87	27	29	102
3-Hexanol	59	55	73	31	43	41	27	29	102
3-Hexanone	43	57	29	27	71	41	100	39	100
2-Hexanone	43	58	29	27	41	57	39	100	100
Hexapropymate	99	81	55	44	41	43	120	79	181
Hexethal*	156	141	55	41	157	43	98	39	240
Hexetidine	142	57	42	197	185	339	240	226	339
Hexobarbitone	81	221	79	80	39	77	157	41	236
Hexobarbitone*	221	81	157	80	79	155	41	77	236
Hexobendine*	296	195	58	297	253	196	212	84	592
Hexylamine(n-)	44	30	58	41	86	42	27	0	101
Hexylcaine	112	77	105	139	55	41	96	56	261
1-Hexyne	67	41	43	27	39	54	40	53	82
Hippuric Acid	105	135	51	134	77	106	50	78	179
Hippuric Acid	105	77	51	135	134	50	106	117	179
Histamine	82	30	81	54	28	55	83	41	111
Histamine (4-(omega-Aminoethyl)-1,3-diazole)	82	81	44	54	55	83	94	41	111
Homatropine	124	82	83	275	94	96	67	80	275
Homatropine	124	82	94	77	42	44	105	106	275
Homatropine	124	94	82	77	79	123	96	67	275
Homatropine	124	42	82	83	94	67	96	77	275
Homatropine*	124	107	82	83	42	77	79	94	275
Hydrallazine	50	130	76	74	75	63	51	38	160
Hydrallazine*	160	103	89	131	115	76	161	104	160
Hydrallazine Metabolite	200	171	129	117	115	145	183	199	200
Hydrochlorothiazide*	269	205	221	297	271	62	285	124	297
Hydrochlorothiazide	269	64	205	297	271	43	44	31	297
Hydrochlorothiazide Tetramethyl Derivative	42	353	310	138	44	75	288	218	353
Hydrocodone	299	242	256	96	243	58	60	214	299
Hydrocodone	299	44	242	59	96	42	76	243	299
Hydrocodone*	299	242	59	243	42	96	70	214	299
Hydrocotarnine	220	178	221	205	163	177	179	42	221
Hydroflumethiazide*	303	331	239	255	30	158	64	159	331
Hydroflumethiazide Tetramethyl Derivative	42	44	387	172	145	252	236	188	387
Hydrolysed Dextropropoxyphene	58	91	191	129	128	205	178	174	265
Hydromorphinol	303	70	58	44	57	42	216	286	303
Hydromorphone	285	96	229	228	70	214	115	200	285
Hydromorphone	42	285	115	44	96	58	228	229	285
Hydrophenyllactic Acid(para)	107	44	77	40	108	39	51	43	182
Hydroquinidine*	138	326	55	110	189	82	160	139	326
Hydroxyacetanilide(p-)	109	151	43	80	81	0	0	0	151
3'-Hydroxyamobarbitone	141	143	156	158	0	0	0	0	242
Hydroxyamobarbitone Glucuronide Me TMS Derivative (peak 2)	73	217	204	75	147	185	253	45	676
Hydroxyamobarbitone Glucuronide Me TMS Derivative (peak 1)	73	253	75	217	147	185	317	45	676
Hydroxyamphetamine	72	44	57	43	149	41	55	69	151
Hydroxyamphetamine	44	107	77	43	78	51	42	108	151
3'-Hydroxyamylobarbitone	59	157	156	141	43	41	71	69	242
3-Hydroxyamylobarbitone	59	156	157	69	56	43	41	55	242
Hydroxyamylobarbitone (N-Hydroxy)	200	143	185	169	126	201	170	155	242

Compound Name									MWT
Hydroxyaprobarbitone (N-Hydroxy)	41	43	183	167	140	184	124	109	226
4-Hydroxybenzaldehyde	121	122	123	65	39	93	63	66	122
4-Hydroxybenzoic Acid	121	138	93	65	39	63	127	53	138
2-Hydroxybiphenyl	170	169	141	115	171	139	142	39	170
3-Hydroxybromazepam	79	78	52	105	304	314	316	51	331
Hydroxybutobarbitone	156	141	157	45	211	199	181	55	228
3-Hydroxycarbofuran	137	180	147	151	162	58	134	65	237
Hydroxychloroquine	64	48	247	102	58	245	304	231	335
Hydroxychloroquine*	102	245	247	304	305	306	58	126	335
3-Hydroxychlorpromazine	58	86	42	334	220	59	87	44	334
8-Hydroxychlorpromazine	58	86	334	42	85	336	44	243	334
Hydroxycyclobenzaprine	58	202	231	215	289	0	0	0	291
Hydroxycyclobenzaprine N-oxide	246	245	231	215	202	116	217	203	307
4-Hydroxycyclophosphamide	92	56	94	63	84	86	184	93	276
3-Hydroxycyproheptadine	303	96	302	231	202	245	259	304	303
Hydroxydesmethylcyclobenzaprine	44	234	233	58	231	202	215	218	277
3-Hydroxydesmethyldiazepam	73	429	430	45	431	147	432	75	430
3-Hydroxydiazepam	271	77	273	300	255	0	0	0	300
3-Hydroxydiazepam	271	56	43	41	55	42	273	77	300
3-Hydroxydiazepam(N-Methyloxazepam)	271	273	77	272	256	300	255	257	300
3-Hydroxydiazepam TMS Derivative	73	343	257	256	345	372	283	45	372
4'-Hydroxydiazepam TMS Derivative	344	73	372	371	346	345	373	374	372
3-Hydroxydiazepam TMS Derivative	343	372	345	357	257	0	0	0	372
7-Hydroxydiftalone	133	147	280	105	77	132	134	90	280
7-Hydroxydiftalone	132	133	280	262	235	104	77	89	280
6-Hydroxy-3-(2-dimethylaminoethyl)benzo(b)thiophene	58	30	148	115	59	77	91	89	221
6-Hydroxydopamine TFA Derivative	440	69	126	441	0	0	0	0	553
Hydroxyephedrine*	58	30	59	56	77	0	0	0	181
Hydroxyephedrine(p-)	58	56	77	107	160	65	38	57	181
2-Hydroxyethylamino-5-Chloro-2'-Fluorobenzophenone	262	109	166	264	293	123	168	95	293
Hydroxyethylflurazepam	288	273	331	287	304	289	290	253	332
5-(2-Hydroxyethyl)-5-isopentylbarbituric Acid	124	142	154	141	99	172	155	98	242
5-(2-Hydroxyethyl)-5-methylbarbituric Acid	142	100	99	125	143	98	124	156	186
4-Hydroxyglutethimide	146	233	103	133	91	117	115	77	233
Hydroxyharman	198	168	180	140	197	169	179	196	198
Hydroxyhippuric Acid(ortho)	121	120	92	65	195	39	93	149	195
2-Hydroxyiminodibenzyl(Imipramine Metabolite)	211	210	196	212	105	180	167	77	211
5-Hydroxyindole-3-acetic Acid	146	191	147	91	117	63	65	39	191
5-Hydroxyindole-3-acetic Acid	146	191	147	130	57	145	117	89	191
Hydroxylated Ethotoin	121	120	220	148	0	0	0	0	220
13-Hydroxy-LSD	339	237	197	223	238	212	229	340	339
12-Hydroxy-LSD	339	237	197	238	239	223	196	212	339
7-Hydroxymecloqualone	251	286	252	78	288	111	152	271	286
8-Hydroxymecloqualone	251	286	152	252	288	111	154	160	286
5-Hydroxymecloqualone	251	286	252	111	288	271	152	273	286
6-Hydroxymecloqualone	251	286	252	288	111	154	271	152	286
Hydroxymethaqualone	160	266	235	251	77	146	58	247	266
5-(4'-Hydroxy-2'-methoxybenzylidene)-barbituric Acid	260	173	259	244	201	216	128	145	262
2-Hydroxy-4-methyl-5-b-chloroethiazole	128	45	73	100	177	0	0	0	177
4-Hydroxy-3-methyl-2-butanone	43	61	42	31	41	57	84	29	102
3-Hydroxy-3-methyl-2-butanone	59	31	43	41	71	87	39	60	102
2-Hydroxy-2-methyl-3-butanone	59	31	43	41	29	39	42	0	102
7-Hydroxy-4-methylcoumarin	148	176	147	91	39	51	120	63	176
6-Hydroxy-4-methylcoumarin	147	176	148	91	39	65	51	63	176

Compound Name									MWT
2-Hydroxy-4-methyl-5-ethylthiazole	128	100	45	73	43	143	0	0	143
5-(Hydroxymethyl)-2-furfuraldehyde	97	41	126	39	69	53	51	125	126
2-Hydroxymethyl-4-hydroxybenzoic Acid	121	150	65	93	168	0	0	0	168
2-Hydroxymethyl-5-hydroxybenzoic Acid	121	150	93	65	168	0	0	0	168
2-Hydroxymethyl-4-hydroxybenzolactone	121	150	65	93	0	0	0	0	150
2-Hydroxymethyl-5-hydroxybenzolactone	121	150	93	65	0	0	0	0	150
2-Hydroxy-2-methyltetrahydrofuran	71	43	41	27	42	31	44	39	102
Hydroxynaltrexone (alpha)	343	55	110	36	98	302	84	344	343
5-Hydroxyniflumic Acid	298	44	252	253	297	279	280	145	298
4'-Hydroxyniflumic Acid	298	232	204	252	251	279	233	280	298
3-Hydroxy-1-nitrosopyrrolidine	42	44	116	41	30	57	56	68	116
3-Hydroxy-N-methylmorphinan	257	59	150	256	200	157	76	189	257
Hydroxypentobarbitone	156	141	157	69	45	197	195	98	242
Hydroxypethidine	71	140	70	263	262	189	57	42	263
Hydroxypethidine(p-)	71	70	263	57	42	44	262	43	263
Hydroxyphenamate	135	57	91	77	43	119	105	180	209
4-Hydroxyphenobarbitone	219	248	148	220	120	218	133	65	248
Hydroxyphenylacetic Acid(meta)	107	152	77	39	108	51	79	53	152
Hydroxyphenylacetic Acid(ortho)	78	134	106	51	77	39	40	107	152
Hydroxyphenylacetic Acid(para)	107	77	51	39	53	78	50	52	152
Hydroxyphenylethanol (beta-m-Hydroxy)	107	108	77	138	51	53	79	78	138
5-(4-Hydroxyphenyl)-3-methyl-5-phenylhydantoin Glucuronide me Ester TMS	73	317	217	75	147	318	43	79	688
5-(3-Hydroxyphenyl)-3-methyl-5-phenylhydantoin TMS Ether	354	73	104	268	325	282	355	77	354
5-(4-Hydroxyphenyl)-3-methyl-5-phenylhydantoin TMS Ether	354	277	325	104	268	73	269	355	354
5-(4-Hydroxyphenyl)-5-Phenylhydantoin(Para HPPH)	239	196	268	120	197	77	225	104	268
3-(4-Hydroxyphenyl)propionic Acid	107	166	77	39	108	45	65	120	166
3-Hydroxypiperidine	44	30	28	57	29	56	43	101	101
3-Hydroxyprazepam	257	55	311	77	259	313	44	312	340
Hydroxyprotriptyline	70	44	207	178	279	249	0	0	279
Hydroxyquinoline*	145	117	122	89	105	90	63	146	145
Hydroxystilbamidine*	96	44	43	31	45	27	78	264	280
1-Hydroxy-1,2,3,4,-tetrahydronaphthalene	130	120	91	119	129	115	147	148	148
2-Hydroxytetrahydropyran	85	55	41	29	56	28	84	39	102
11-HydroxyTHC (delta 9)	299	330	59	43	300	0	0	0	330
4'-Hydroxywarfarin	281	187	324	121	43	0	0	0	324
Hydroxyzine	201	165	45	166	56	203	42	58	374
Hydroxyzine	201	170	203	188	165	166	202	142	374
Hydroxyzine*	201	203	165	45	299	166	202	56	374
Hyoscine	94	138	42	108	154	303	136	137	303
Hyoscine*	94	138	42	108	136	41	96	97	303
Hyoscine	138	94	108	136	301	154	77	42	303
Hyoscine TMS Derivative	138	94	73	108	42	154	137	136	375
Ibogaine	136	135	310	149	122	225	155	186	310
Ibogaine*	136	310	135	225	149	122	155	311	310
Ibomal*	167	209	43	124	39	41	53	140	288
Ibomal Methylation Artifact*	194	136	193	81	137	109	39	197	236
Ibuprofen	91	163	161	119	107	117	41	118	206
Ibuprofen*	163	161	119	91	206	117	107	164	206
Ibuprofen Methyl Ester	161	117	177	91	119	118	220	121	220
Idobutal*	167	41	168	124	39	97	141	67	224
Imidazole	44	29	27	41	58	57	43	39	68
Iminodibenzyl(Imipramine Metabolite)	195	194	180	96	193	196	83	167	195
Imipramine	235	58	234	85	280	195	193	35	280
Imipramine	58	85	173	193	194	195	234	235	280

Compound Name				Peaks					MWT
Imipramine	58	234	235	85	193	194	195	192	280
Imipramine	58	235	85	234	195	280	193	194	280
Imipramine	58	85	235	234	42	280	208	193	280
Imipramine*	58	235	85	234	236	195	193	208	280
Impramine Degradation Product	194	195	180	193	89	90	77	63	195
Impramine Degradation Product	77	89	248	69	63	102	51	65	317
5-Indanmethylenebarbituric Acid	242	256	228	255	115	129	141	116	256
Indapamide*	147	131	130	132	119	148	365	218	365
Indene	116	115	58	63	89	117	39	62	116
Indole	117	90	89	118	116	63	59	39	117
3-Indoleacetaldehyde	130	159	77	64	131	103	48	51	159
3-Indoleacetamide	130	174	77	131	103	102	129	51	174
3-Indolealdehyde	144	145	116	89	63	146	90	58	145
3-Indolelactic Acid	130	44	205	77	129	131	103	102	205
3-Indolelactic Acid	130	131	205	129	77	43	41	45	205
Indolyl-3-acetic Acid	130	175	77	131	103	102	129	176	175
3-(3-Indolyl)acetic Acid	130	45	131	159	77	62	43	103	175
3-Indolylglyoxyldimethylamide	144	216	116	89	72	63	0	0	216
Indol-3-yl-N,N-diethylglyoxamide	144	72	145	116	89	29	100	244	244
Indol-3-yl-N,N-disopropylglyoxamide	144	43	100	86	116	145	128	89	272
Indomethacin	44	28	29	31	43	60	73	41	357
Indomethacin	139	111	141	138	140	75	158	113	357
Indomethacin*	139	141	357	111	359	140	113	75	357
Indomethacin Derivative (4-Chlorobenzoic Acid TMS Ester)	213	139	75	169	111	215	141	77	228
Indomethacin Derivative TMS Derivative	139	73	141	111	487	370	75	140	487
Indomethacin Derivative TMS Derivative	139	141	429	111	73	140	431	113	429
Indomethacin Derivative TMS Ester	174	291	73	175	159	131	75	158	291
Indomethacin Methyl Ester	139	141	111	371	140	75	113	158	371
Indomethacin TMS Ester	139	141	73	111	312	429	140	75	429
5-(2-Iodoethyl)-5-isopentylbarbituric Acid	124	225	154	155	128	127	125	141	352
Ionol	205	57	220	41	206	145	29	81	220
Ionone(beta-)	177	43	41	135	91	178	93	95	192
Iopronic Acid	43	101	546	428	331	402	275	487	673
Iopronic Acid Methyl Ester	59	560	331	428	528	402	433	373	687
Ioxynil	57	127	41	43	55	88	243	42	371
Iproniazid	123	106	43	58	78	79	51	164	179
Iproniazid	106	162	78	51	79	123	43	41	179
Iproniazid*	123	31	58	106	79	43	78	51	179
Iproniazid Metabolite	78	106	51	137	79	0	0	0	137
Isoamyl Salicylate	120	43	121	138	39	65	41	29	208
Isoborneol	95	110	41	55	96	69	139	136	154
Isoborneol	95	41	27	43	39	93	55	29	154
Isoborneol	95	41	43	110	93	55	136	67	154
Isobutylamine	43	41	57	71	70	55	73	42	73
Isobutylamine	30	73	41	27	55	39	56	31	73
Isobutylaminobenzoate(p-Amino-)*	120	137	193	65	92	39	41	121	193
Isobutyl Formate	43	56	41	31	29	27	60	39	102
Isobutyl Methyl Ketone	43	58	57	100	41	85	29	39	100
Isobutyl Vinyl Ether	57	41	56	29	44	27	100	39	100
Isobutyraldehyde	43	41	72	27	39	29	42	26	72
Isocarboxazid	91	106	127	110	57	92	65	104	231
Isocarboxazid	91	43	65	127	110	39	51	106	231
Isocarboxazid*	91	127	106	110	43	92	65	120	231
Isocinchomeronic Acid	123	43	44	51	78	77	50	105	167

Compound Name									MWT
Isoetharine	100	58	41	65	43	30	192	56	239
Isoetharine*	100	58	41	101	43	56	65	30	239
Isoetharine Mesylate	56	100	98	58	43	123	41	192	239
Isofenchol(endo-)	139	95	109	41	121	43	55	136	154
Isolan	72	41	45	43	42	29	44	211	211
Iso-lysergide*	221	72	207	181	43	44	323	222	323
Isomethadone	58	128	127	59	193	179	178	115	309
Isomethadone	58	43	128	127	59	0	0	0	309
Isoniazid	78	106	51	137	50	79	107	52	137
Isoniazid	78	106	51	104	77	137	50	0	137
Isoniazid*	106	78	51	137	50	79	107	31	137
Isonicotinic Acid	123	51	78	106	50	52	105	39	123
Isooctyl Phthalate	149	57	167	43	71	70	41	55	390
Isophthalonitrile	128	101	50	75	129	76	64	51	128
Isoprenaline*	72	44	43	124	123	30	42	41	211
Isoprenaline TMS-Derivative	73	68	140	98	41	125	74	99	427
Isopropamide Iodide	86	114	100	44	238	115	56	72	480
Isopropanol	45	43	27	41	31	44	29	0	60
Isopropanol or Propan-2-ol	45	43	29	41	31	0	0	0	60
Isopropenyl Acetate	43	58	41	39	42	27	0	0	100
Isopropoxyphenyl N-methylcarbamate (o-Isopropoxy-)	110	152	111	81	58	43	41	52	209
Isopropyl Acetate	43	61	87	41	59	42	27	39	102
Isopropylbenzene	105	120	40	29	44	77	79	51	120
Isopropyl Ether	45	43	87	59	41	27	39	69	102
Isopropyl-N-(3-chlorophenyl)carbamate (CIPC)	43	41	127	129	213	154	105	171	213
Isopropyl-N-phenylcarbamate (IPC)	43	93	41	179	39	137	65	32	179
1-Isopropyl-2-phenylethylamine	72	120	55	91	65	39	77	31	163
3-Isopropylphenyl-N-methylcarbamate (UC-10854, H 8757)	121	136	91	77	58	39	41	193	193
2-Isopropylquinoxaline-1,4-dioxide	171	173	169	188	76	160	170	129	204
Isopulegol	69	81	67	55	71	68	56	41	154
Isoquinoline	129	102	128	51	130	76	75	103	129
Isothipendyl	72	73	181	86	214	42	56	200	285
Isothipendyl*	72	73	214	200	44	285	86	56	285
Isovaleric Acid	60	43	41	27	45	29	39	74	102
Isoxsuprine	77	174	107	135	176	65	281	146	301
Isoxsuprine*	178	44	135	179	77	84	107	41	301
Isoxsuprine (Duvadilan)	178	135	176	107	70	77	84	0	301
Karbutilate	72	180	84	135	44	57	45	41	279
Kepone	272	274	270	237	235	239	218	216	486
Kepone	272	274	237	270	143	276	239	235	486
Ketamine	180	182	209	152	138	211	102	154	237
Ketamine	180	36	102	115	152	138	146	125	237
Ketamine*	180	209	182	152	181	30	211	138	237
Ketamine Impurity	152	180	138	102	154	182	209	0	237
Ketamine Metabolite	166	131	102	138	77	91	195	0	223
Ketamine Metabolite	153	118	138	102	130	0	0	0	221
Ketazolam	256	43	44	283	84	69	284	257	368
Ketazolam*	256	284	283	285	84	257	258	255	368
Ketene	42	41	29	25	40	24	43	21	42
Ketobemidone	70	44	71	57	247	190	119	191	247
3'-Ketohexobarbitone	250	95	39	235	66	207	41	193	250
Ketoprofen	105	254	177	209	77	210	181	255	254
Ketoprofen*	105	177	77	209	254	210	103	181	254
Ketoprofen Methyl Ester	209	105	77	268	191	103	210	51	268

Compound									MWT
3-Ketoquinalbarbitone	43	69	168	41	85	167	86	97	252
Kynurenic Acid	55	43	45	143	57	189	145	73	189
Lactose	73	85	60	103	71	57	61	58	342
Laevulinic Acid	43	56	45	55	73	99	39	44	116
Lauric acid(n-Dodecanoic Acid)	73	60	43	57	41	55	129	71	200
Lavandulol	69	111	41	68	93	123	67	81	154
Lenacil	153	154	77	94	110	53	40	51	234
Leptazol	55	82	41	39	54	42	56	109	138
Leucinocaine*	142	120	143	86	142	92	72	56	292
Levallorphan	41	85	84	157	56	43	57	70	283
Levallorphan	283	44	282	256	43	176	85	57	283
Levallorphan*	283	282	256	176	157	43	41	57	283
Levodopa	123	124	77	74	197	152	105	179	197
Levodopa	123	77	51	124	39	74	53	78	197
Levodopa*	123	124	77	44	51	74	53	39	197
Levomoramide	100	265	128	266	56	129	55	101	392
Levophacetoperane	84	43	126	77	56	91	105	55	283
Levophenacylmorphan	43	58	256	45	51	55	44	56	361
Levorphanol	31	32	29	257	59	256	150	200	257
Levorphanol	59	42	150	157	44	257	256	41	257
Levorphanol*	59	257	150	256	44	31	200	157	257
Lidoflazine*	343	70	344	109	42	113	491	56	491
Lignocaine	86	85	58	87	120	91	77	56	234
Lignocaine	86	58	87	42	56	72	234	120	234
Lignocaine*	86	87	58	44	72	42	120	85	234
Limonene	68	93	67	39	79	53	27	94	136
Limonene	68	93	67	136	94	41	121	53	136
Limonene	68	67	93	94	79	92	121	107	136
Linalool	71	93	47	43	55	69	80	154	154
Linalool	71	43	41	93	55	69	80	67	154
Linalool	93	71	41	69	55	43	80	121	154
Lindane	183	181	109	111	219	217	51	221	288
Lindane(beta)	109	219	181	183	111	193	288	0	288
Lindane(G-BHC)	181	183	109	219	111	217	51	221	288
Linuron	61	46	160	60	248	133	45	124	248
Lobeline	105	77	120	43	51	96	106	97	337
Lobeline*	96	105	77	97	216	42	218	51	337
Loperamide*	238	42	239	224	240	56	72	226	476
Lorazepam	239	274	302	276	304	275	241	75	320
Lorazepam	77	137	101	177	176	204	100	203	320
Lorazepam*	291	239	274	293	75	302	276	138	320
Lorazepam Benzophenone	230	265	139	267	111	232	264	266	265
Lormetazepam	304	44	75	306	51	57	50	305	334
Loxapine	70	83	42	257	193	56	228	164	327
LSD	323	221	44	199	222	223	207	76	323
LSD*	323	221	181	222	207	72	223	324	323
LSD	221	72	207	181	42	44	222	58	323
LSD (d) Tartrate	221	323	181	222	207	72	223	180	323
LSD (iso)	323	221	207	181	222	196	223	72	323
Lsd-TMS Derivative	395	253	396	294	268	74	221	128	395
Lumi-LSD (10-hydroxy-9,10-dihydro-LSD)	170	171	167	142	172	129	341	323	341
Lynestrenol	91	79	67	201	77	105	93	120	284
Lysergamide*	267	221	207	180	223	154	196	268	267
Lysergic Acid*	268	224	154	180	207	223	192	179	268

Compound Name									MWT
Lysergic Acid (d)	268	154	224	180	207	192	221	223	268
Lysergic Acid Monoethylamide	295	221	181	207	196	180	223	167	295
Lysergol*	254	192	193	154	223	255	180	221	254
Mafenide*	106	30	77	185	105	104	89	141	186
Malathion	173	125	127	93	158	99	143	0	330
Malathion	125	93	127	173	158	99	55	79	330
Maleinehydrazide	112	82	55	54	45	84	41	53	112
Mannitol	73	61	103	74	56	60	57	133	182
Maprotiline*	44	70	59	277	71	191	278	203	277
Maprotiline (Ludiomil)	44	59	70	277	71	191	203	178	277
Mazindol	266	88	102	75	115	128	131	176	284
Mazindol*	266	268	267	255	231	102	88	176	284
MCPA	141	200	77	143	155	125	142	202	200
MCPA Methyl Ester	214	141	155	125	216	77	45	143	214
MCPB	142	107	144	87	43	77	45	108	228
MCPB Methyl Ester	101	59	69	41	142	107	77	102	242
Mebendazole*	186	218	77	295	263	105	51	158	295
Mebeverine*	308	165	309	121	55	154	98	56	429
Mebhydroline	60	73	233	91	232	276	275	0	276
Mebhydroline*	91	233	30	232	31	276	275	65	276
Mebutamate	97	72	55	71	62	110	69	158	232
Mebutamate*	97	55	69	72	71	98	43	62	232
Mecamylamine	98	71	84	56	99	124	41	167	167
Mecamylamine*	98	84	71	56	41	42	99	124	167
Mecarbam	131	97	58	159	125	160	329	65	329
Mechlorethamine	106	42	108	44	63	43	49	41	156
Meclizine	105	189	165	42	166	106	79	77	390
Meclofenamic Acid Methyl Ester	242	244	309	311	243	214	178	277	309
Meclofenoxate	58	42	71	36	59	111	30	75	257
Meclofenoxate*	58	111	71	42	75	59	141	113	257
Meclozine	105	189	36	38	201	165	285	166	390
Meclozine*	189	105	201	285	165	166	190	134	390
Meconic Acid	44	36	102	69	156	38	45	71	200
Meconin	165	194	147	77	121	176	118	51	194
Mecoprop Methyl Ester	142	169	228	107	141	144	171	77	228
Medazepam	207	242	244	165	270	243	208	269	270
Medazepam*	242	207	244	270	243	271	269	165	270
Medazepam Benzophenone	245	77	244	228	105	246	193	247	245
Medinoterbacetate	254	239	44	41	77	91	191	296	296
Mefenamic Acid	180	194	77	222	241	179	242	224	241
Mefenamic Acid*	223	241	208	222	194	180	77	224	241
Mefenamic Acid Methyl Ester	223	255	208	77	222	180	194	96	255
Mefenamic Acid TMS Derivative	223	313	208	224	222	180	194	298	313
Mefruside*	85	43	42	86	44	41	75	110	382
Megesterol	43	91	342	281	79	55	256	241	342
Megesterol Acetate	281	43	282	187	107	91	55	105	384
Melatonin	160	173	232	161	174	145	158	0	232
Menazon	156	93	281	125	43	157	55	63	281
Menthol	81	95	71	41	67	55	138	123	156
Menthol	71	81	95	41	55	82	43	69	156
Menthone(o-)	97	111	83	112	55	41	69	56	154
Mepazine	58	42	41	198	96	112	44	180	310
Mepenzolate Bromide	97	105	77	96	42	183	98	82	419
Meperidine	71	42	70	57	44	43	103	91	247

Compound Name	Peaks								MWT
Mephenesin	108	118	91	107	43	57	182	75	182
Mephenesin*	108	107	91	182	109	77	79	31	182
Mephenesin Carbamate	108	118	91	107	57	43	44	75	225
Mephenesin Carbamate*	118	108	182	91	225	107	57	75	225
Mephenoxalone	124	109	77	52	223	95	81	122	223
Mephenoxalone (Trepidone)	124	223	109	77	122	123	52	95	223
Mephentermine	72	39	148	56	41	71	51	73	163
Mephentermine	72	91	73	56	148	41	57	42	163
Mephenytoin	189	104	77	190	51	105	103	132	218
Mephenytoin	104	189	77	51	105	103	190	56	218
Mephobarbitone	218	118	146	117	103	77	219	91	246
Mephobarbitone	218	217	117	118	146	103	115	91	246
Mephobarbitone (Mebaral, Prominal)	218	117	146	118	219	161	39	51	246
Mepivacaine	98	70	99	36	42	96	99	38	246
Mepivacaine*	98	99	70	42	96	55	41	40	246
Meprobamate	83	55	71	96	114	144	62	56	218
Meprobamate	83	55	71	41	44	62	56	45	218
Meprobamate	44	41	55	83	31	43	39	56	218
Meprobamate*	83	84	55	56	43	71	41	62	218
Meprylcaine	100	58	105	77	56	101	41	70	235
Mepyramine	121	58	72	71	214	122	215	78	285
Mescaline	182	30	167	181	51	107	151	211	211
Mescaline	182	181	149	44	167	211	40	41	211
Mescaline	182	181	167	211	151	148	139	136	211
Mescaline*	182	30	181	167	211	183	151	148	211
Mescaline-NCS Derivative	181	253	72	0	0	0	0	0	253
Mescaline Precursor	239	177	192	149	63	77	92	134	239
Mesitylene(1,3,5-Trimethylbenzene)	105	120	119	77	91	39	106	79	120
Mesoridazine	98	77	94	99	126	0	0	0	386
Mesoridazine	98	70	42	94	77	51	99	126	386
Mesoridazine*	98	70	99	42	386	126	55	41	386
Mestranol	227	310	174	284	147	160	173	199	310
Metaraminol	76	77	95	105	70	121	58	44	167
Metaraminol*	44	77	76	29	95	39	58	42	167
Metasystox	169	109	125	79	59	0	0	0	230
Metaxalone	122	77	107	105	91	79	123	30	221
Metazocine	231	216	84	124	59	42	72	174	231
Metformin*	43	44	85	86	42	129	68	30	129
Methabenzthiazuron	164	136	135	163	96	108	165	57	221
Methadone	72	165	42	180	178	179	91	73	309
Methadone	72	73	58	57	223	165	70	0	309
Methadone	72	73	91	223	165	71	294	57	309
Methadone	72	71	73	57	91	165	70	56	309
Methadone	72	57	43	69	71	55	81	41	309
Methadone*	72	73	91	293	223	165	85	71	309
Methadone (6-Dimethylamino-4,4-diphenyl-3-heptanone)	91	223	294	57	42	56	165	44	309
Methadone Intermediate	58	72	192	42	165	73	71	59	278
Methadone Metabolite 1	277	276	262	105	278	220	200	91	277
Methadone Metabolite	40	277	276	44	262	57	43	41	277
Methadone-N-Oxide	207	129	72	91	208	105	174	128	325
Methamidophos	94	95	141	64	47	46	79	45	141
Methamphetamine	58	91	56	65	51	77	134	0	149
Methamphetamine	58	91	56	65	42	39	51	41	149
Methandriol	253	271	91	105	213	145	107	147	304

COMPOUND NAME	PEAKS								MWT
Methanol	31	32	29	30	28	33	34	27	32
Methantheline Bromide	72	181	85	152	42	44	58	43	419
Methantheline Bromide (Banthine)	181	86	72	182	85	152	108	99	419
Methaprylon Metabolite	83	55	98	155	126	0	0	0	199
Methaprylon Metabolite	153	55	183	138	69	0	0	0	199
Methapyrilene	58	97	72	84	71	191	192	261	261
Methapyrilene	97	58	72	191	71	78	79	190	261
Methapyrilene	58	97	72	71	121	224	149	191	261
Methapyrilene*	58	97	72	71	42	191	79	78	261
Methaqualone	235	91	65	132	76	233	90	250	250
Methaqualone*	235	250	91	233	236	65	76	132	250
Methaqualone Metabolite	251	266	249	77	143	252	0	0	266
Methaqualone Metabolite	235	160	266	77	235	251	146	0	266
Methaqualone Metabolite TMS Derivative	73	338	323	247	179	139	235	154	338
Methaqualone Metabolite TMS Derivative	323	338	73	321	154	143	163	249	338
Methaqualone Metabolite TMS Derivative	323	338	73	321	143	154	0	0	338
Methaqualone Metabolite TMS Derivative	323	338	73	75	249	143	0	0	338
Methaqualone Metabolite TMS Derivative	323	73	338	307	75	249	154	143	338
Methaqualone Metabolite TMS Derivative	323	91	154	338	251	266	309	307	338
Methaqualone Metabolite TMS Derivative	323	338	91	73	321	154	132	149	338
Methaqualone Metabolite TMS Derivative	323	338	91	321	75	154	73	149	338
Methaqualone Metabolite TMS Derivative	323	91	132	338	154	0	0	0	338
Methaqualone Metabolite TMS Derivative	323	235	338	73	176	91	75	247	338
Metharbitone*	155	170	112	169	55	82	41	39	198
Metharbitone (Gemonil)	170	155	41	169	55	112	39	83	198
Methazolamide*	43	221	83	236	56	223	222	55	236
Methazole	159	124	260	161	262	126	163	88	260
Methdilazine	97	98	55	82	199	198	180	296	296
Methenamine	42	140	41	85	112	44	43	40	140
Methetoin	189	104	77	190	51	0	0	0	218
Methidathion (Technical)	145	85	93	125	146	69	58	302	302
Methimazole*	114	42	72	113	69	81	54	115	114
Methiocarb	168	153	109	65	58	139	91	39	225
Methisazone*	234	146	179	206	131	91	117	118	234
Methixene	197	58	99	165	309	112	152	198	309
Methixene*	99	197	44	58	112	309	41	42	309
Methocarbamol	118	124	109	44	57	77	81	45	241
Methocarbamol	124	109	118	43	198	122	123	125	241
Methocarbamol	118	109	124	77	81	62	95	75	241
Methocarbamol*	124	118	198	109	43	57	125	77	241
Methohexital	79	53	221	81	41	77	178	93	262
Methohexital Methyl Derivative	79	235	53	178	81	41	195	261	276
Methohexitone*	41	81	53	221	79	39	178	233	262
Methoin*	189	104	190	77	44	105	132	103	218
Methoin Metabolite (Phenylethylhydantoin)	175	104	77	176	105	132	51	204	204
Methorphan	59	271	150	214	270	171	112	213	271
Methotrimeprazine	58	328	100	42	135	228	229	242	328
Methotrimeprazine	58	328	100	269	229	283	242	243	328
Methotrimeprazine	58	42	185	30	100	228	57	328	328
Methotrimeprazine*	58	328	100	228	185	329	242	229	328
Methotrimeprazine Metabolite	58	328	100	43	71	269	229	207	328
Methotrimeprazine Metabolite	314	229	242	269	44	57	71	70	314
Methotrimeprazine Metabolite	58	328	100	229	269	207	242	283	328
Methotrimeprazine Sulphoxide	58	328	229	242	100	269	283	0	344

Methoxamine	168	137	44	139	43	152	124	167	211
Methoxamine	44	166	168	137	151	63	53	95	211
2-Methoxyamphetamine	44	91	122	45	77	66	78	107	165
3-Methoxyamphetamine	44	45	122	91	77	78	121	42	165
Methoxyamphetamine	44	122	121	78	77	42	51	52	165
Methoxyamphetamine (para)	44	122	121	42	78	77	91	107	165
5-Methoxybenzylidene-(para)-barbituric Acid	246	245	202	117	159	132	145	215	246
4-Methoxy-2-buten-1-ol	41	71	45	29	39	27	70	0	102
Methoxychlor	227	228	274	238	152	36	308	153	344
5-Methoxydiisopropyltryptamine	114	72	30	115	160	43	174	145	274
2-Methoxyethanol	45	31	29	47	43	76	27	46	76
2-Methoxyethyl Vinyl Ether	45	58	59	29	31	43	27	0	102
13-Methoxy-LSD	353	251	211	252	237	253	226	196	353
14-Methoxy-LSD	353	251	237	211	249	338	252	351	353
12-Methoxy-LSD	353	251	211	237	253	252	338	354	353
3-Methoxy-4,5-methylenedioxyamphetamine	44	166	165	42	77	39	64	194	209
4'-Methoxy-2-(methylsulphonyl)acetanilide	243	108	123	122	164	136	95	79	243
4'-Methoxy-N-isopropyl-2-(methylsulphonyl)acetanilide	150	134	164	123	136	79	122	285	285
2-Methoxy-4-nitrosophenol	139	99	124	153	78	79	95	107	153
4-Methoxy-N-methylamphetamine	58	44	59	121	91	56	77	78	179
3-Methoxy-N-methylamphetamine	58	56	91	44	77	42	78	121	179
Methoxyphenamine	58	91	57	56	78	65	77	42	179
Methoxyphenamine	58	30	91	59	56	42	65	77	179
Methoxyphenamine*	58	91	59	56	30	42	121	78	179
Methoxyphenamine (ortho-)	58	91	43	45	30	59	56	42	179
1-Methoxyphenothiazine	229	214	186	114	215	154	230	199	229
3-Methoxyphenothiazine	229	214	186	114	215	230	154	199	229
Methoxypromazine*	58	86	314	229	228	185	42	44	314
Methoxypropoxyphene	58	121	134	127	148	238	270	284	369
Methoxypropoxyphene	58	121	134	59	57	127	105	77	369
2-Methoxytetrahydrofuran	71	41	42	43	61	72	101	102	102
Methscopolamine Bromide	103	148	94	96	108	41	77	91	397
Methsuximide	118	117	203	103	77	78	119	91	203
Methychlothiazide (Endurone)	310	64	36	312	42	43	62	63	359
Methyclonazepam	75	63	89	51	248	125	302	328	329
Methyl Acetate	43	74	42	59	29	31	44	45	74
Methyl Acetylsalicylate	120	152	43	121	92	63	65	64	194
Methyl Acetylsalicylate	135	194	179	136	91	137	180	40	194
Methyl Acrylamide(N-)	29	30	28	27	44	71	55	26	101
Methylamine	30	31	29	27	26	32	0	0	31
2-Methylamino-5-amino-2'-Fluorobenzophenone	244	227	55	57	95	243	69	43	244
4-Methylaminoantipyrine	56	217	83	42	57	98	77	218	217
Methylaminobenzoate (ortho)	119	151	120	92	152	91	93	65	151
Methylaminobenzoate (para)	120	151	92	121	152	65	93	122	151
2-Methylamino-5-chlorobenzophenone	77	245	244	105	228	246	247	51	245
2-Methylamino-5-chlorobenzophenone	245	244	77	105	193	246	228	44	245
2-Methylamino-5-chlorobenzophenone	77	245	244	105	193	228	168	246	245
2-Methylamino-5-chlorobenzophenone*	245	77	246	247	228	105	51	193	245
2-Methylamino-2',5-Dichlorobenzophenone	62	244	279	45	229	57	61	111	279
Methylaminomethylheptane	58	41	55	44	43	59	128	56	141
2-Methylamino-5-nitro-2'-fluorobenzophenone	274	273	211	257	123	199	275	95	274
Methylaminophenazone	56	123	217	83	57	77	215	216	217
Methylaminophenol (para)	123	122	94	108	81	29	60	43	123
((3-Methylaminopropyl)-iminodibenzyl)(N-)(Imipramine Metabolite)	235	195	234	208	193	266	194	45	266

Compound Name	Peaks								MWT
Methylamobarbitone (N-methyl)	155	157	170	172	156	171	158	173	240
Methylamphetamine*	58	91	59	134	65	56	42	57	149
Methylamphetamine TFA	154	118	110	69	91	45	0	0	245
Methyl Amylamine(n-)	44	101	58	30	28	41	42	43	101
Methylamylobarbitone (N-Methyl)	170	155	55	156	171	69	212	112	240
Methylated Coumatetralyl*	306	175	291	115	91	202	187	129	306
Methylated THC-9-oic Acid	299	343	358	231	283	290	315	326	358
Methylbarbitone (N-Methyl)	170	155	169	55	112	83	98	69	198
Methylbemegride	55	58	70	83	69	73	97	127	169
2-Methylbenzanilide	119	91	211	65	120	212	92	90	211
Methylbenzylamine (N-Methyl)	120	91	44	121	42	65	51	77	121
Methyl beta-Phenylpropionate	104	164	91	105	133	103	165	78	164
4-Methyl-5-b-hydroxyethiazole	112	143	113	85	45	0	0	0	143
3-Methylbutanal	44	41	43	29	27	39	58	42	86
2-Methylbutanal	57	29	41	58	27	39	86	43	86
2-Methylbutanoic Acid	74	29	57	41	27	28	87	45	102
3-Methyl-2-butanone	43	41	27	86	39	42	71	44	86
Methylbutobarbitone (N-Methyl)	155	170	156	98	169	112	197	55	226
Methyl Butyrate(n-)	74	43	71	59	41	87	27	0	102
4-Methylcatechol	124	123	78	77	51	39	107	110	124
2-Methyl-4-chlorophenoxyacetic Acid (MCPA)	200	141	155	0	0	0	0	0	200
Methylchlorpropamide (N-Methyl)	111	175	205	141	113	75	112	56	290
Methyl Cinnamate	131	162	79	103	161	52	132	163	162
Methyl Crotonate	69	41	39	85	100	28	59	0	100
Methylcyclopal	221	67	196	41	164	181	222	0	248
2-Methylcyclopentanol	57	41	44	82	67	71	56	39	100
3-Methylcyclopentanone	42	55	69	98	41	56	27	39	98
Methyl Decanoate	74	87	43	55	41	75	143	57	186
Methyldesorphine	283	282	160	42	284	268	110	44	283
Methyldihydromorphine	57	43	149	58	55	41	69	71	301
3-Methyldilantin	180	104	266	77	237	57	209	71	266
Methyldilantin	180	77	104	266	237	209	165	0	266
4-Methyl-2,5-dimethoxybenzaldehyde	180	165	137	179	134	77	39	109	180
Methyl 3,4-dimethoxybenzoate	196	165	197	181	166	59	94	121	196
Methyl 3,4-Dimethoxycinnamate	222	191	223	207	79	147	164	190	222
Methyl 3,4-dimethoxyphenylacetate	151	210	152	211	107	195	59	153	210
Methyl 3,4-Dimethoxyphenylpropionate	151	224	164	152	225	149	165	193	224
4-Methyl-1,3-dioxane	43	31	55	72	101	28	45	32	102
4-Methyl-1,3-dioxane	43	31	55	72	101	28	45	32	102
2-Methyl-1,3-dioxane	43	87	31	45	59	28	29	44	102
2-Methyl-1,4-dioxane	43	28	58	102	29	42	45	44	102
Methyldiphenethylamine (N-)	148	105	91	44	77	79	65	56	239
4-Methyl-1,2-diphenyl-3,5-pyrazolinedione	77	183	266	105	51	91	78	65	266
Methyl dipropylacetate	87	57	116	55	115	129	88	127	158
Methyldopa	88	123	44	42	124	41	122	77	211
Methyldopa	88	42	123	77	51	124	43	166	211
Methyldopa	88	42	123	124	89	77	51	40	211
Methylecgonine	82	96	83	97	42	199	94	168	199
Methylecgonine	96	82	83	42	97	55	68	199	199
Methylene Bromide	174	93	95	172	176	91	81	79	172
Methylene Chloride	49	84	86	51	47	88	48	50	84
Methylenechlorobromide	49	130	128	51	93	81	79	95	128
3,4-Methylenedioxyamphetamine	44	137	136	163	179	0	0	0	179
3,4-Methylenedioxyamphetamine	44	136	135	77	51	179	45	78	179

COMPOUND NAME				PEAKS					MWT
3,4-Methylenedioxybenzylamine	150	151	121	65	93	77	135	0	151
1,2-Methylenedioxy-4-(2-nitropropenyl)benzene	103	160	77	207	41	39	150	51	207
1-(3,4-Methylenedioxyphenyl)-2-nitropropene	103	28	160	77	207	41	150	0	207
Methylene Oxalate	30	44	29	28	102	56	0	0	102
Methyl Enol Ether of Chlorpropamide	115	175	111	58	177	290	113	75	290
Methyl Enol Ether of Tolbutamide	91	155	129	41	284	229	163	184	284
Methylephedrine	72	44	42	77	56	73	51	70	179
Methylephedrine*	58	72	30	77	56	44	42	73	179
Methylergometrine*	339	221	196	181	207	223	222	72	339
Methylergometrine Maleate	221	222	196	321	126	205	207	181	339
Methylergometrine Maleate	339	221	223	207	196	181	222	340	339
Methylergonovine Hydrogen Maleate	221	54	339	72	55	196	207	181	339
Methylethane(N-(2'-methylethyl)-1-amino-2-)	44	86	58	28	42	41	43	27	101
Methylethosuximide (N-)	127	55	70	41	42	128	69	112	155
2-Methyl-5-ethylpyrazine	121	122	39	56	27	42	94	54	122
Methyl Fluoroacetate	59	33	61	41	44	92	31	93	92
Methylformamide (N-Methyl)	59	58	60	28	41	42	0	0	59
Methyl Formate	31	29	60	30	33	44	45	61	60
2-Methylfuran	82	53	81	39	27	51	50	54	82
Methyl Heptadecanoate	74	87	43	41	55	75	57	69	284
6-Methyl-3-heptanone	43	57	29	27	72	41	71	99	128
Methyl Hexadecanoate	74	87	43	75	57	55	143	69	270
3-Methylhexane	43	57	71	41	29	70	27	56	100
2-Methylhexane	43	42	41	85	27	57	29	56	100
Methylhexaneamine	44	41	42	43	39	30	56	45	115
5-Methyl-3-hexanone	57	29	27	85	41	43	114	39	114
3-Methylhexobarbitone	235	81	169	171	79	236	170	91	250
2-Methylhexobarbitone (or 4-Methyl-)	235	150	165	236	79	137	250	164	250
Methyl Hippurate	105	77	134	51	106	193	161	50	193
2-Methyl-4'-hydroxybenzanilide	119	91	227	65	120	228	92	63	227
Methyl Hydroxybenzoate	121	152	40	93	65	0	0	0	152
3-Methyl-3-hydroxydilantin Methyl Ether	296	180	267	210	104	77	134	297	296
Methyl 3-Hydroxydipropylacetate	87	116	55	113	57	145	59	143	174
3-Methylhydroxyhexobarbitone TMS Ether	169	73	170	75	249	233	79	171	338
2-Methyl-3-(4-hydroxy-2'-methylphenyl)-4-(3H)-quinazolinone	251	266	249	77	143	76	252	39	266
Methyl 3-Indoleacetate	130	189	131	79	190	52	103	51	189
Methyl-3-indolebutyrate	130	217	143	131	186	218	144	117	217
Methyl Isoamylamine	44	101	28	43	30	41	43	45	101
Methyl Isobutanoate	43	71	41	39	27	87	102	0	102
Methylisobutyramide(N-)	58	43	101	73	27	41	71	86	101
1-Methyl-4-isopropenylbenzene	132	117	133	92	91	115	118	59	132
1-Methyl-4-isopropenylcyclohexene	68	93	67	39	41	53	136	79	136
Methyl Isopropyl Nitrosamine	102	43	42	57	56	41	40	39	102
3-Methylisoquinoline	143	115	142	116	144	89	72	59	143
Methyl Laurate	74	87	41	43	55	75	57	69	214
Methyl Methacrylate	41	40	39	69	29	38	59	100	100
Methyl Methacrylate	41	69	39	100	40	59	99	0	100
Methyl methoxybenzoate(meta)	166	135	107	167	136	77	108	59	166
Methyl methoxybenzoate(ortho)	135	166	133	77	105	137	134	136	166
Methyl methoxybenzoate(para)	135	166	136	167	77	107	92	137	166
Methyl-5-methoxyindoleacetate	160	219	161	220	145	74	69	83	219
2-Methyl-5-methoxyindole-3-acetic Acid Methyl Ester	174	233	175	159	131	130	234	158	233
Methyl Methoxyphenyl-(meta)-acetate	180	121	181	122	59	91	148	182	180
Methyl Methoxyphenyl-(ortho)-acetate	121	180	91	181	122	148	107	93	180

Compound Name				Peaks					MWT
Methyl Methoxyphenyl-(para)-acetate	121	180	122	181	0	0	0	0	180
Methyl Methoxyphenyl-(para)-propionate	121	194	134	163	122	195	135	108	194
2-Methyl-3-(2'-methyl-4'-hydroxy-5'-methoxyphenyl)-quinazoline	296	281	143	279	76	297	39	77	296
Methylmorpholine (N-)	43	42	101	71	27	44	29	0	101
Methyl Myristate	74	87	43	41	55	57	75	69	242
Methylnicotinamide (N-)	78	106	135	136	51	50	79	52	137
Methylnicotinate*	78	106	51	137	50	136	107	79	137
Methyl Nitrate	46	30	29	28	15	31	0	0	77
Methylnitrazepam	267	77	51	91	63	117	294	248	295
2-Methyl-4-nitroaniline	152	106	77	79	122	59	0	0	152
4-Methyl-2-nitroaniline	152	106	77	79	91	104	78	105	152
Methylnitroaniline(N-Me,o-)	77	152	79	105	78	104	118	51	152
2-Methyl-4-nitrosophenol	137	121	120	94	77	80	107	92	137
3-Methyl-4-nitrosophenol	137	92	80	77	123	109	122	107	137
Methylnorfenfluramine	58	30	42	159	56	59	31	202	231
Methyl-6-n-pentyl-4-hydroxy-2-oxocyclohex-3-ene-1-carboxylate	169	137	125	95	55	41	84	43	240
Methyl Octadecadienoate	67	81	54	55	95	68	82	69	294
Methyl Octanoate	74	87	43	57	41	55	59	127	158
Methyloxazepam TMS Derivative	343	257	256	372	357	283	371	0	372
Methyl 3-Oxodipropylacetate	57	87	130	55	116	101	143	141	172
Methyl Palmitate	74	87	270	75	143	43	55	41	270
Methyl Palmitate	74	87	43	41	55	75	57	69	270
Methylparaben	121	152	93	65	39	0	0	0	152
Methylparaben Dimethyl Ether	135	66	77	107	92	136	63	64	166
Methyl Parathion	109	125	263	79	93	63	47	264	263
Methyl-p-chlorobenzenesulphonamide (N-Methyl)	111	75	141	205	171	112	50	113	205
Methyl-p-chlorobenzenesulphonamide (N-Methyl)	111	175	205	75	113	207	84	177	205
2-Methylpentan-1-al	43	58	41	29	27	71	39	57	100
3-Methylpentane	57	56	41	0	0	0	0	0	86
2-Methylpentane	43	42	71	41	29	57	39	0	86
4-Methyl-2-pentanol	45	43	69	41	44	87	57	56	102
2-Methyl-2-pentanol	59	45	87	43	41	31	73	27	102
2-Methyl-1-pentanol	43	71	70	41	55	69	56	84	102
3-Methyl-1-pentanol	56	69	55	41	43	29	57	31	102
3-Methyl-3-Pentanol	73	55	43	45	87	41	69	29	102
4-Methylpentan-2-one	43	58	41	57	27	39	29	100	100
2-Methylpentan-3-one	57	43	29	100	71	27	41	56	100
3-Methylpentan-2-one	43	57	29	41	72	56	27	100	100
4-Methyl-2-pentanone	43	58	57	41	85	100	29	39	100
3-Methyl-2-pentanone	43	29	57	41	72	27	56	39	100
4-Methyl-3-pentene-2-one	55	83	43	98	29	27	39	53	98
Methylpentynol Carbamate*	69	83	43	79	41	81	53	80	141
Methylphenidate	84	85	56	91	36	55	30	118	233
Methylphenidate	84	91	56	55	121	118	117	85	233
Methylphenidate	84	91	55	150	146	56	85	83	233
Methylphenidate	84	91	56	90	65	89	77	115	233
Methylphenidate	84	56	91	85	55	30	41	42	233
Methylphenidate	84	91	85	56	55	150	41	118	233
Methylphenobarbitone*	218	117	118	146	103	77	91	115	246
Methyl Phenyl Acetate (Methyl Phenidate Decomposition Prod)	91	150	65	92	89	59	0	0	150
Methylphenylaziridine	91	132	42	105	92	43	41	77	133
2-Methyl-3-phenylaziridine	41	91	132	105	42	92	43	105	132
4-Methyl-5-phenyl-2-(benzyl)pyridine	258	259	243	260	115	91	116	108	259
2-Methyl-3-phenyl-6-(benzyl)pyrimidine	258	259	244	260	91	115	65	72	259

Compound Name									MWT
1-Methyl-2-phenylindole	207	206	208	204	103	205	177	102	207
5-Methylphenyl-5-phenylhydantoin(p-5-phenylhydantoin)(Metabolite)	180	77	118	91	104	65	89	51	266
4-Methyl-5-phenylpyrimidine	170	102	169	115	77	51	60	61	170
4-Methyl-5-phenylpyrimidine	170	169	102	115	116	171	51	76	170
4-Methyl-5-phenylpyrimidine*	40	43	170	44	169	102	69	91	170
3-Methyl-p-hydroxydilantin Methyl Ether	296	267	210	219	134	180	77	297	296
2-Methylpiperazine	44	28	85	30	42	56	57	43	100
Methylpiperazine(N-)	58	43	100	42	56	44	29	28	100
Methyl-3-piperidylbenzilate (N-)	97	105	77	183	84	36	42	51	325
Methyl-3-piperidylbenzilate(N-)	97	105	77	96	84	98	183	70	325
Methyl-3-piperidylbenzilate (N-Methyl)	97	105	77	183	165	114	167	152	325
Methyl-4-piperidylbenzilate (N-Methyl)	98	105	183	77	96	55	325	114	325
Methyl-3-piperidylphenylcyclohexyl Glycolate (N-Methyl)	97	107	105	189	128	249	248	331	331
2-Methylpropanal	43	27	41	29	72	39	42	26	72
1-Methylpropyl Formate	43	56	41	29	31	27	60	39	102
Methyl 2-Propylglutarate	142	128	121	120	171	83	87	143	202
Methyl Propyl Ketone	43	86	41	71	58	57	0	0	86
4-(Methyl-2'-propynylamino)-3,5-dimethylphenyl N-Methylcarbamate	150	189	246	57	151	174	188	190	246
2-(Methyl-2'-propynylamino)phenyl N-Methylcarbamate	122	161	94	160	57	120	123	162	218
Methyl-p-tolylsulphonamide (N-Methyl)	91	92	65	155	185	58	77	121	185
Methyl-p-tolylsulphonamide (N-Methyl)	91	155	185	65	92	0	0	0	185
Methylpyrazine	94	67	26	39	40	53	38	42	94
2-Methylpyridine	93	66	92	78	65	39	51	94	93
3-Methylpyridine	93	66	92	65	39	94	67	59	93
4-Methylpyridine	93	66	92	65	39	94	54	67	93
Methyl Pyridine-3-carboxylate	106	137	78	136	79	105	107	138	137
4-Methylpyridine(4-Picoline)	93	66	92	39	65	67	51	54	93
1-Methylpyrrole	81	80	39	42	53	55	54	38	81
Methyl Pyruvate	43	102	42	28	29	59	44	41	102
Methylquinalbarbitone (N-Methyl)	182	181	97	209	167	53	55	124	252
2-Methylquinoline	143	128	115	144	142	101	51	77	143
4-Methylquinoline	143	115	142	144	116	89	77	39	143
6-Methylquinoline	143	142	115	144	141	89	116	63	143
7-Methylquinoline	143	142	115	144	141	116	89	39	143
8-Methylquinoline	143	142	115	144	141	89	116	39	143
Methylrubazonic Acid	56	77	91	84	42	40	373	203	373
Methylsaccharin	133	104	76	105	197	0	0	0	197
Methyl Salicylate	120	92	152	121	65	64	93	63	152
Methyl Salicylate	92	120	39	65	152	121	63	64	152
Methyl Stearate	74	87	43	55	298	143	75	57	298
3-Methyl-s-triazolopthalazine (Hydralazine Metab)	184	115	88	62	185	114	51	50	184
Methylstyrene(alpha-)	118	117	103	78	77	115	51	91	118
Methylstyrene(p-)	117	118	115	91	58	39	103	65	118
2-Methylsulphonylacetanilide	93	213	79	77	106	121	119	104	213
Methylsuximide(N-)	127	55	70	41	42	0	0	0	155
3-Methylsydnone	100	42	28	41	70	45	40	30	100
Methyltestosterone	302	124	43	91	79	121	105	122	302
Methyl Tetradecanoate	74	87	43	75	55	57	41	143	242
1-Methyl-1,2,3,4-tetrahydro-beta-carboline	171	186	156	155	185	130	116	144	186
2-Methyl-3-tetrahydrofuranone	43	28	72	29	27	100	44	45	100
3-Methyltetrahydropyran	42	68	55	45	100	41	70	39	100
3-Methyltetrahydrothiophene	60	102	41	45	74	87	56	39	102
2-Methyltetrahydrothiophene	87	102	45	41	59	39	60	74	102
5-Methyl-1,2,3-thiadiazole	71	45	72	28	31	39	29	27	100

4-Methyl-5-thiazoleacetic Acid	112	157	85	45	158	0	0	0	157
(4-(Methylthio)-3,5-dimethylphenyl N-methylcarbamate (Bayer 37344)	168	153	225	91	58	45	39	77	225
(4-(Methylthio)-3,5-dimethylphenyl N-methylcarbamate (Methiocarb)	168	153	109	91	225	45	169	154	225
Methyltolbutamide (N-Methyl)	91	155	185	65	41	92	121	56	284
Methyltoluidine(N-Me,p-)	120	121	91	65	77	119	122	118	121
3-Methyltriazolophthalazine	184	115	185	114	183	88	129	155	184
Methyl 3,4,5-trimethoxybenzoate	226	211	59	155	66	53	195	183	226
Methyl 3,4,5-trimethoxycinnamate	252	237	253	221	238	177	209	149	252
Methyl Trithion	157	125	159	314	93	45	316	171	314
Methyl Undecanoate(n-)	74	87	169	55	157	143	75	43	200
2-Methylvaleraldehyde	43	58	41	29	27	71	57	39	100
Methyl Vinyl Ketone	55	43	27	70	97	42	41	26	70
Methylwarfarin	279	91	280	322	247	121	201	189	322
1-Methylxanthine	166	54	109	53	81	136	137	0	166
3-Methylxanthine	166	68	95	41	53	123	0	0	166
7-Methylxanthine	166	68	123	53	42	41	95	0	166
Methyprylone	155	140	83	98	55	168	183	0	183
Methyprylone	140	155	83	38	69	98	55	183	183
Methyprylone*	155	140	83	98	55	41	84	69	183
Methyprylone Methyl Derivative	154	169	83	55	41	42	98	69	197
Methyprylone Probe Product (Dimer)*	98	321	83	55	140	41	155	182	999
Methyprylon Metabolite	83	55	98	153	166	41	155	152	199
Methyprylon Metabolite (6-Hydroxy)	83	98	55	166	153	41	84	155	199
Methysergide	235	210	195	221	0	0	0	0	353
Methysergide	55	43	195	41	235	353	221	72	353
Methysergide*	353	210	235	336	72	54	236	195	353
Methysergide Hydrogen Maleate	353	235	210	72	54	45	221	195	353
Metobromuron	61	46	258	91	260	172	170	63	258
Metoclopramide*	86	99	184	58	30	87	201	186	299
Metolazone*	350	91	259	352	348	65	351	107	365
Metopimazine*	141	445	169	123	155	96	42	317	445
Metopon	299	96	242	243	185	228	214	300	299
Metoprolol	72	30	56	107	223	98	73	43	267
Metoprolol*	72	30	107	56	45	41	44	43	267
Metoxuron	72	228	183	230	44	229	168	45	228
Metribuzin	200	42	58	61	43	47	75	104	214
Metronidazole*	81	124	54	53	125	171	45	42	171
Metyrapone*	120	106	78	92	51	226	41	39	226
Mevinphos	127	192	109	67	43	193	39	79	224
Mevinphos GC peak 2	127	192	109	67	43	193	79	39	224
Mexiletine	58	44	41	83	69	77	85	43	179
Mexiletine	58	41	77	42	43	91	47	122	179
Mianserin*	193	264	43	72	71	220	192	194	264
Mianserin (Bolvidon)	193	264	194	43	71	72	220	109	264
Michler's Ketone	148	268	267	224	269	251	120	118	268
Miconazole*	159	161	81	335	333	163	337	205	414
Mirex	272	274	237	239	270	276	235	241	540
MK-251 (Dimethyl-4-(a,a,b,b-tetrafluorophenethyl)benzylamine)	296	169	297	127	310	311	0	0	311
MK-251 Metabolite	297	127	185	170	298	312	311	293	312
MK-251 Metabolite	167	127	168	294	295	285	0	0	294
MK-251 Metabolite	311	127	169	312	295	153	199	184	326
MK-251 Metabolite	127	296	169	281	183	154	141	128	296
MK-251 Metabolite	295	127	168	153	167	296	276	311	341
Modaline	84	94	177	148	42	93	134	41	177

Compound Name									MWT
Molindone	100	56	42	176	98	120	70	189	276
Monalide	85	43	127	41	197	57	239	129	239
Monodesethylflurazepam	341	245	246	274	343	313	58	299	359
Monodesethylflurazepam	58	246	302	71	289	273	274	341	359
Monodesethylflurazepam	246	302	58	71	289	341	56	274	359
Monolinuron	61	46	126	99	214	60	63	45	214
Monopentafluorobenzylmorphine	284	465	81	285	42	175	181	162	465
Monuron	72	198	44	200	73	42	99	199	198
Moperone	204	217	123	186	205	0	0	0	355
Moramide	100	128	265	266	129	246	236	306	392
Moramide Intermediate	100	36	56	101	42	38	165	91	339
Moramide Intermediate	100	56	101	42	165	91	115	70	339
Morazone	56	201	70	176	77	202	71	55	377
Morazone*	201	176	202	56	258	70	71	42	377
Morfamquat	170	114	156	155	113	197	83	42	538
Morpheridine	246	42	36	247	100	218	56	232	346
Morpheridine	246	36	100	42	38	82	91	56	346
Morpheridine	246	247	36	100	42	84	232	56	346
Morpheridine	246	100	42	82	91	56	232	41	346
Morpheridine*	246	247	42	100	232	56	218	172	346
Morphine	285	162	42	44	31	215	70	200	285
Morphine	285	215	162	286	124	284	268	174	285
Morphine	285	42	70	162	44	284	124	59	285
Morphine	115	285	131	128	162	127	77	152	285
Morphine	285	42	215	44	162	124	70	115	285
Morphine*	285	162	42	215	286	124	44	284	285
Morphine bis TMS Ether	73	429	236	196	146	414	430	287	429
Morphine Diheptafluorobutyrate Derivative	69	43	169	45	464	70	58	677	677
Morphine Dimethyl Derivative	313	138	42	282	229	176	115	146	313
Morphine Methobromide	45	58	73	285	72	80	82	42	379
Morphine Methoiodide	285	226	72	58	45	42	44	162	427
Morphine-N-oxide	58	285	72	42	71	59	186	44	301
Morphine N-oxide	285	162	284	215	286	124	268	226	301
Mydocalm (Tolperisone)	98	119	160	84	85	91	245	230	245
Myrcene	93	41	69	77	91	53	67	94	136
Myrcene	41	69	93	39	67	79	27	91	136
Myrcene	41	93	69	79	91	77	67	53	136
Myrophine	585	494	91	43	57	41	73	55	585
Myrophine	91	43	57	41	73	55	60	81	585
Naled	109	145	79	185	147	35	187	47	378
Naled	109	145	15	79	96	185	73	47	378
Nalidixic Acid*	188	189	159	132	53	173	131	145	232
Nalorphine	311	43	60	41	312	45	241	188	311
Nalorphine*	311	312	41	188	80	82	81	241	311
Nalorphine (N-Allylnormorphine)	271	44	270	214	272	42	43	70	311
Nalorphone	41	311	39	188	81	115	42	77	311
Naloxone	327	242	328	286	229	96	41	201	327
Naloxone	41	55	42	115	70	39	96	68	327
Naloxone*	327	328	41	242	286	96	229	70	327
Naltrexone	341	55	36	300	342	110	243	256	341
Naphazoline	209	115	141	210	152	153	139	208	210
Naphazoline*	209	210	141	115	153	208	46	181	210
Naphthylethyldiethyl Phosphate (beta-)	280	144	115	224	154	252	29	281	280
1-Naphthyl-N-methylcarbamate (Carbaryl, Sevin)	144	115	116	58	55	63	201	40	201

Compound Name	Peaks								MWT
1-Naphthyl-N,N-dimethyl carbamate	74	215	115	42	39	127	0	0	215
Naproxen	185	115	141	230	170	45	153	142	230
Naproxen*	185	230	141	186	184	115	170	153	230
Naproxen Methyl Ester	185	244	141	115	170	186	153	154	244
1-Napthanol	144	115	116	145	72	89	63	58	144
Narceine	58	234	44	59	41	42	36	427	445
Narceine	58	42	59	234	77	427	178	133	445
Narceine*	58	427	234	59	50	42	428	91	445
Narcotine	220	221	205	206	218	222	118	148	413
Narcotine	220	42	77	205	51	147	44	53	413
Nealbarbitone	57	141	167	41	181	182	83	223	238
Nealbarbitone*	57	41	141	167	39	83	55	182	238
Nealbarbitone Dimethyl Derivative	169	195	209	57	41	112	138	251	266
Nebol	69	41	39	93	68	29	27	154	154
Nefopam	179	180	178	165	225	89	195	210	253
Nefopam*	58	179	180	225	178	165	42	210	253
Neoisothujyl Alcohol	93	121	136	95	43	81	110	107	154
Neomenthol	71	95	81	41	55	43	69	82	156
Neopine	299	162	229	123	59	42	44	300	299
Neostigmine	72	42	208	108	65	73	66	39	302
Nerol	69	41	93	68	80	121	67	111	154
Nerol	41	69	68	93	39	29	67	43	154
Nialamide	91	106	44	78	51	177	79	107	298
Nialamide	91	106	78	51	58	44	79	149	298
Nialamide*	91	177	44	106	45	78	123	51	298
Nicametate*	86	30	58	123	78	51	29	42	222
Nicocodeine	282	106	229	267	78	42	124	81	404
Nicocodeine	404	282	106	78	229	267	42	124	404
Nicodicodeine	106	78	70	59	42	284	300	44	406
Nicofuranose*	123	106	78	51	105	50	77	52	600
Nicomorphine	36	172	106	42	57	78	44	214	495
Nicotinamide	122	104	78	106	51	0	0	0	122
Nicotinamide*	122	78	106	51	50	52	44	123	122
Nicotine	84	42	133	162	161	51	119	65	162
Nicotine	84	133	42	162	161	39	51	41	162
Nicotine	84	44	133	42	162	161	44	85	162
Nicotine	84	133	76	42	162	161	58	119	162
Nicotine*	84	133	42	162	161	105	77	119	162
Nicotine-N-oxide	84	178	161	133	118	0	0	0	178
Nicotinic Acid*	123	105	78	51	106	77	124	50	123
Nicotinyl Alcohol	109	108	80	57	43	55	53	71	109
Nicotinyl Alcohol*	109	108	80	53	51	39	91	27	109
Nicoumalone*	310	121	353	311	43	120	92	296	353
Nifedipine*	329	284	224	268	330	285	225	270	346
Niflumic Acid	282	236	237	281	263	145	44	93	282
Niflumic Acid Methyl Ester	236	295	296	263	145	235	0	0	296
Nikethamide	106	78	176	51	148	162	56	0	178
Nikethamide	106	78	177	51	178	107	149	163	178
Nikethamide	106	177	78	178	51	105	107	149	178
Nimorazole*	100	56	42	101	55	54	41	30	226
Niridazole*	158	214	57	145	124	45	70	96	214
Nirosodimethylamine(N-)	42	74	30	43	44	41	40	45	74
Nitrazepam*	280	253	281	206	234	252	254	264	281
Nitrazepam Benzophenone	241	242	77	105	195	211	212	51	242

Compound Name									MWT
2-Nitroaniline	138	65	92	77	104	39	90	64	138
3-Nitroaniline	92	128	65	39	66	122	80	93	138
4-Nitroaniline	138	65	92	39	108	106	66	52	138
5-Nitrobarbituric Acid	44	43	42	30	70	54	96	141	173
Nitrobenzene	77	51	123	50	65	93	30	78	123
Nitrobenzene	77	123	51	65	78	93	50	30	123
Nitroethane	29	27	30	26	28	43	46	0	75
Nitrofen	283	285	202	139	50	75	63	76	283
Nitroglycerin	46	30	76	31	45	43	42	47	227
Nitroglycerin	46	30	29	76	28	43	31	44	227
1-Nitronaphthalene	127	173	115	126	77	101	128	143	173
Nitrophenol(para)	139	65	39	109	93	53	81	63	139
6-Nitroquinoline	174	128	101	116	77	102	75	51	174
8-Nitroquinoline	174	128	116	101	77	89	102	75	174
4-Nitrosophenol	123	107	80	78	95	79	93	81	123
Nitrosopyrrolidine	41	100	42	43	30	69	39	68	100
Nitrostyrene	115	91	77	116	117	129	163	0	163
Nitrostyrene	115	91	105	116	40	39	51	77	163
2-Nitrotartranilic Acid	138	44	92	43	195	134	90	106	270
2-Nitrotoluene	120	65	91	92	39	77	121	89	137
Nitrotoluene (meta)	137	91	107	79	65	92	138	77	137
Nitrotoluene (meta)	91	65	137	41	57	39	30	0	137
Nitrotoluene (ortho)	120	92	91	65	121	77	93	137	137
Nitrotoluene (ortho)	65	57	41	91	120	29	43	39	137
Nitrotoluene(p-)	91	137	65	39	107	63	89	77	137
Nitrotoluene (para)	91	65	137	107	39	63	77	79	137
Nitrous oxide	44	30	45	46	31	29	0	0	44
2-Nitroxylene(m-)	134	79	77	106	105	103	151	78	151
4-Nitroxylene(o-)	151	105	77	79	103	121	78	39	151
Nomifensine*	194	195	238	193	72	178	45	196	238
Nonachlorokepone	272	274	270	237	276	251	203	182	452
2-Nonanone	43	58	41	71	59	57	27	29	142
Noracymethadol	58	36	134	43	99	59	91	44	339
Noracymethadol*	58	43	134	222	91	44	59	56	339
Noradrenaline	31	139	93	153	151	77	137	123	169
Noradrenaline	30	65	93	139	39	111	53	51	169
Noradrenaline*	44	45	58	76	60	43	42	46	169
Noradrenaline TFA Derivative	440	126	69	427	0	0	0	0	553
Noraminophenazone	56	123	215	217	83	119	77	57	217
Norantipyrine	174	77	91	105	132	0	0	0	174
Norbormide*	91	58	86	106	231	45	77	230	511
Norcodeine	285	215	81	148	110	164	286	132	285
Norcodeine*	285	81	215	148	286	164	110	115	285
Norephedrine	44	77	79	51	45	42	78	105	151
Norethandolone	302	57	85	231	91	110	79	215	302
Norethindrone	91	110	98	231	272	298	41	77	298
Norethisterone Acetate	43	340	298	325	91	41	231	280	340
Norethynodrel	91	215	79	105	77	55	41	298	298
Norfenfluramine	44	42	159	43	45	184	41	109	217
Norharman	168	140	169	64	63	167	114	141	168
Norlevorphanol	45	243	136	200	159	157	198	242	243
Nor LSD (6-desmethyl-LSD)	207	309	208	182	181	209	180	167	309
Normeperidine Ethylcarbamate	232	202	129	56	42	158	217	305	305
Normethadone	58	72	71	224	59	42	152	57	295

Normethadone	58	72	71	42	59	163	91	179	295
Normorphine	271	81	150	148	45	110	42	82	271
Norpethidine	57	42	56	43	233	158	91	103	233
Norpethidine	57	233	56	158	103	91	160	77	233
Norpethidinic Acid	57	56	205	103	91	158	77	115	205
Norpethidinic Acid Acid Ethyl Ester N-TMS Derivative ...	73	129	305	128	304	276	103	232	305
Norpethidinic Acid Methyl Ester	57	56	42	219	43	103	91	158	219
Norpethidinic Acid TMS Ester	57	42	73	277	103	56	187	43	277
Norpethidinic Acid TMS Ester N-TMS Derivative	73	129	128	114	334	349	115	130	349
Norpipanone*	98	111	99	42	55	41	29	112	335
Norpropoxyphene	44	208	58	117	57	193	130	115	325
Norpropoxyphene Amide	44	43	41	69	57	55	81	75	325
Norpropoxyphene Amide	44	100	234	88	57	91	105	129	325
Norpropoxyphene Carbinol	44	105	91	77	178	208	65	130	269
Norpseudoephedrine	44	77	79	51	42	45	43	105	151
Norpseudoephedrine	44	77	105	106	51	91	79	132	151
Norpseudoephedrine	44	57	43	41	55	69	77	79	151
Norpseudoephedrine	44	77	79	51	105	0	0	0	151
19-Nortesterone	274	110	91	79	105	147	215	256	274
Nortilidine	83	68	82	72	84	77	103	115	259
Nortriptyline	44	57	43	71	41	55	45	56	263
Nortriptyline	44	203	202	45	220	219	215	204	263
Nortriptyline	44	202	259	203	218	215	217	42	263
Nortriptyline	44	45	202	203	91	215	189	115	263
Nortriptyline	44	202	45	215	203	42	220	204	263
Nortriptyline	44	45	202	203	91	99	189	165	263
Nortriptyline*	44	202	45	220	218	215	91	0	263
Nortriptyline Metabolite	44	215	202	216	45	213	91	189	261
Nortriptyline N-Trifluoroacetate	232	219	217	204	233	203	91	202	359
Nortriptyline TMS Derivative	44	218	203	73	217	202	215	219	351
Noscapine	220	205	221	28	147	77	178	42	413
Noscapine	205	220	42	77	147	221	119	118	413
Noscapine*	220	221	205	147	42	193	77	118	413
Noxiptilin	207	206	91	179	180	178	89	208	294
Noxiptyline*	58	71	208	72	59	42	89	57	294
Nylidrin	91	160	65	174	175	279	146	124	299
Ocimene	93	92	91	79	77	80	121	105	136
Ocimene	93	41	27	39	79	80	77	43	136
Ocimene(allo-)	121	136	105	93	79	91	122	107	136
Ocimene-X(beta-)	93	92	91	79	77	80	121	105	136
Ocimene-Y(beta-)	93	80	92	91	79	121	105	77	136
Octachlorokepone	238	240	236	203	201	242	205	182	418
3-Octanone	43	29	57	27	72	71	99	41	128
4-Octanone	43	57	71	27	41	85	58	39	128
Octaverine*	397	368	382	398	354	29	340	312	397
1-Octene	43	41	55	56	42	70	29	39	112
Oleic Acid	41	55	43	69	83	57	67	54	282
Omethoate	156	110	58	109	79	80	126	47	213
Opipramol*	363	206	143	42	70	207	218	113	363
Orcinol	124	123	39	95	51	67	55	69	124
Orciprenaline	72	43	65	41	111	93	39	56	211
Orciprenaline*	72	43	73	41	70	65	40	39	211
Orciprenaline TMS Derivative	68	73	140	98	125	350	405	412	427
Orphenadrine	58	73	165	45	166	42	181	59	269

Compound Name									MWT
Orphenadrine	58	73	45	42	165	59	46	44	269
Orphenadrine*	58	73	44	45	165	42	40	181	269
Orthocaine*	136	167	108	80	53	52	137	51	167
Oxacillin	44	41	45	58	57	144	241	213	401
Oxanamide (Quinactin)	36	57	41	40	44	38	55	39	157
Oxazepam	257	259	77	228	286	0	0	0	286
Oxazepam	268	257	267	239	233	77	205	269	286
Oxazepam	77	75	205	239	104	233	76	177	286
Oxazepam	77	205	104	51	177	75	229	151	286
Oxazepam	77	257	268	205	239	267	233	241	286
Oxazepam*	257	77	268	239	205	267	233	259	286
Oxazepam Benzophenone	230	231	77	232	105	154	51	233	231
Oxazepam Glucuronide TMS Derivative	217	269	375	241	204	341	342	329	822
Oxeladin*	86	91	105	87	144	58	100	56	335
Oxethazaine	145	91	74	72	56	86	277	43	467
Oxethazaine*	72	91	114	145	75	160	117	92	467
Oxethazaine Probe Product 1*	91	120	121	92	65	167	119	106	999
Oxethazaine Probe Product 2*	91	92	425	72	160	44	114	65	999
Oxetorone	58	98	202	231	189	215	218	205	319
Oxetorone N-oxide	274	60	273	259	181	202	218	215	335
2-Oxo-LSD	237	239	339	238	209	240	196	167	339
Oxo-2-phthalazinylmethyl Benzoic Acid (2-(1-(2H))) (Diftalone Metab)	132	280	133	104	235	262	77	89	280
4-Oxo-4,5,6,7-tetrahydrocoumarone-3-carboxylic Acid	152	136	96	124	180	51	39	52	180
Oxprenolol	72	30	221	41	56	73	150	43	265
Oxprenolol*	72	41	56	43	221	73	57	45	265
Oxybuprocaine*	86	99	29	30	100	71	87	192	308
Oxycodone	315	230	314	316	258	201	70	229	315
Oxycodone*	315	230	316	70	44	42	258	140	315
Oxyethyltheophylline	180	95	224	123	68	109	193	194	224
Oxymetazoline*	245	260	44	217	218	246	261	259	260
Oxymorphone	44	42	301	115	70	216	91	43	301
Oxymorphone*	301	216	44	42	70	302	203	57	301
Oxypertine*	175	70	176	132	379	204	56	217	379
Oxyphenbutazone	93	109	199	77	119	162	55	324	324
Oxyphenbutazone	199	92	135	76	324	119	93	134	324
Oxyphenbutazone*	199	324	93	77	65	55	121	135	324
Oxyphencyclimine	129	112	55	41	42	44	262	105	344
Oxyphencyclimine	189	105	107	77	55	91	79	190	344
Oxyphencyclimine*	105	129	112	77	42	313	41	55	344
Oxyphenisatin Acetate*	317	288	359	401	43	318	289	196	401
Palmitic Acid	73	71	83	129	256	98	85	97	256
Papaverine	324	338	339	308	293	325	220	340	339
Papaverine	154	324	89	338	77	51	339	102	339
Papaverine*	339	324	338	325	340	308	154	292	339
Paracetamol	109	151	43	79	80	53	108	110	151
Paracetamol*	109	151	43	80	108	81	53	52	151
Paracetamol Cysteine Conjugate	141	43	183	44	140	108	80	52	254
Paracetamol Glucuronide Methyl Ester TMS Ether	317	223	75	147	217	43	318	73	557
Paracetamol Methyl Ester	108	123	165	43	52	80	53	122	165
Paracetamol TMS Ester	181	223	166	73	43	45	208	75	223
Paraethoxycaine	86	121	99	30	58	65	42	56	265
Paraldehyde	43	44	87	31	45	42	131	71	132
Paraldehyde	45	43	60	44	89	42	87	117	132
Paramethadione	43	129	57	56	41	72	39	58	157

```
Paraoxone ................................................ 109  81 275 149  99 139 127 247   275
Paraquat Dichloride ...................................... 156  50 155  52  51 128 157  76   256
Parathion ................................................ 109  97 291 139 125 137 155 123   291
Parathion ................................................  97 291 109 137 139 155 125  65   291
Parathion ................................................ 291 109  97 137  29 139  78 292   291
Parathion ................................................  97 109 291 139 125 137 155 123   291
Parathion-ethyl .......................................... 291 109  97 137 139 155 125 123   291
Parathion-methyl ......................................... 263 125 109  79  93  47  63 264   263
Parathion-methyl ......................................... 109 263 125  79  63  93 264  64   263
Paraxanthine ............................................. 180  68 123  53  42  95 150 151   180
Pargyline ................................................  91  82  42  39  65 159 158  68   159
Pargyline ................................................  82  91  68  42  92 159 158  65   159
Pargyline* ...............................................  82  68  91 159  42 158  92  65   159
Pecazine (Mepazine) ...................................... 310 111 112  58 199 212  96  41   310
Pemoline ................................................. 176 107  90  89  77 105  79  70   176
Pemoline .................................................  77 107  51  89  42 105  90  79   176
Pemoline* ................................................ 107 176  90  77  70 105  42  79   176
Pemoline Metabolite 2 .................................... 107  79  77  51 152 105  50  78   152
Penfluridol Hofmann Reaction Product 1 ................... 275 293  96 179   0   0   0   0   293
Penicillamine* ...........................................  75  41  57  70  43  59  56  47   149
Pentachlorobenzene ....................................... 250 252 248 108 215  73 213  85   248
Pentachlorobenzene ....................................... 250 248 252 254 215 108 213 178   248
2,3,4,5,6-Pentachlorobenzyl Alcohol (Blastin) ............ 245 243 247 278 262 179 244 181   278
Pentachlorobiphenyl ...................................... 326 324 328 330 254 256 184 258   324
Pentachloroethane ........................................ 167 165 117 119  83 169 130 132   200
2,3,4,5,6-Pentachloromandelonitrile (Oryzon) ............. 277 279 275 278 280 276 281 249   303
Pentachloronitrobenzene (PCNB) ........................... 295 249 237 293 297 214 212 142   293
Pentachlorophenol ........................................ 266 268 264 165 167 270 202 130   264
Pentachlorophenol (PCP) .................................. 266 264 268 165 167 130 202 200   264
Pentaerythritoltetranitrate ..............................  46  76  57  55  56  60  47  97   316
Pentaerythritoltetranitrate ..............................  46  76 240 194 316  58 169  84   316
Pentaerythritoltetranitrate ..............................  76  46  57  55  56  60  47  43   316
2,3-Pentanedione .........................................  43  29  57  27 100  26  42  44   100
2,4-Pentanedione(Acetylacetone) ..........................  43  85 100  27  29  42  41   0   100
Pentane(n-) ..............................................  43  42  41  27  29  39  57  72    72
Pentanoic Acid ...........................................  60  73  27  29  41  43  45  28   102
Pentanolactone(delta-) ...................................  42  41  56 100  55  71   0   0   100
Pentanol(n-) .............................................  42  55  41  31  29  70  43  57    88
3-Pentanone ..............................................  57  29  86  27  56  58  26  43    86
2-Pentanone ..............................................  43  86  41  58  27  71  39  42    86
2-Pentanone ..............................................  43  27  29  57  86  41  39  58    86
Pentaquine* .............................................. 187 241 242 174  30 126 175 301   301
Pentazocine .............................................. 217  45  41  70  69 110 202  72   285
Pentazocine .............................................. 217  70  69 110 202 270 285 284   285
Pentazocine ..............................................  41  69  70  45 110 217 159 145   285
Pentazocine ..............................................  45  41 217  69  70 110 285 202   285
Pentazocine* .............................................  45 217  70  41  69 110 285 202   285
Pentifylline* ............................................ 180 264 193 109 194 181  67  42   264
Pentobarbitone ........................................... 156 141 157  55  69 155  98 197   226
Pentobarbitone ........................................... 156 141  98 112 197 226   0   0   226
Pentobarbitone* .......................................... 141 156  43  41 157  55  39  98   226
3-Pentylcannabidiol ...................................... 231  41  67  43 232  68  91  55   314
Pentylenetetrazol ........................................  55  41  39  82  54  53 109  42   138
Pentyl Nitrate ...........................................  29  41  57  46  76  43  27  28   133
```

Compound Name									MWT
5-Pentyl-5-(2-thiocyanatoethyl)-barbituric Acid	124	154	181	125	141	153	142	112	283
Perazine*	113	339	44	70	141	43	60	340	339
Perhexiline*	84	194	55	85	56	41	30	99	277
Pericyazine	114	44	42	41	142	43	223	56	365
Pericyazine*	114	44	142	365	42	223	115	205	365
Permethrin	183	163	184	165	91	77	127	89	390
Permethylated Allobarbitone*	195	138	41	80	53	194	39	110	236
Permethylated Amylobarbitone*	169	184	170	41	185	112	55	183	254
Permethylated Aprobarbitone*	195	196	138	41	111	181	110	53	238
Permethylated Barbitone*	169	184	126	183	41	112	83	55	212
Permethylated Brallobarbitone*	235	193	236	136	194	39	138	121	314
Permethylated Butalbital*	196	195	41	138	169	111	112	181	252
Permethylated Butobarbitone*	169	184	55	112	41	170	58	183	240
Permethylated Cyclobarbitone*	235	170	40	236	180	81	73	121	264
Permethylated Cyclopentobarbitone*	67	221	196	41	164	111	39	181	262
Permethylated Enallylpropymal*	195	196	41	138	111	58	181	53	252
Permethylated Heptabarbitone*	249	169	250	41	133	67	79	93	278
Permethylated Hexethal*	169	184	55	112	41	185	170	183	268
Permethylated Hexobarbitone*	235	81	79	169	171	170	41	80	250
Permethylated Ibomal*	195	237	138	238	196	43	39	110	316
Permethylated Idobutal*	195	196	41	138	181	169	111	39	252
Permethylated Metharbitone*	169	184	126	183	112	83	55	170	212
Permethylated Methohexitone*	81	41	53	235	178	79	39	195	276
Permethylated Methylphenobarbitone*	232	118	117	146	175	103	233	77	260
Permethylated Nealbarbitone*	169	195	57	209	41	112	250	138	264
Permethylated Pentobarbitone*	169	184	41	43	112	183	58	55	254
Permethylated Phenobarbitone*	232	118	117	146	175	103	77	91	260
Permethylated Phenylmethylbarbituric Acid*	132	104	246	103	78	77	51	133	246
Permethylated Probarbital*	169	184	183	112	197	41	126	170	226
Permethylated Quinalbarbitone*	196	195	41	138	181	111	40	43	266
Permethylated Secbutobarbitone*	169	184	41	58	112	57	128	183	240
Permethylated Sigmodal*	195	265	43	196	41	44	39	110	344
Permethylated Talbutal*	195	196	41	138	181	111	58	110	252
Permethylated Thialbarbitone*	81	47	251	41	79	42	178	53	292
Permethylated Thialbarbitone Oxygen Analogue*	235	81	169	41	79	54	197	196	276
Permethylated Thiopentone*	200	120	185	121	40	41	69	44	270
Permethylated Vinbarbitone*	223	41	166	224	169	67	138	39	252
Perphenazine	42	246	232	70	56	143	214	43	403
Perphenazine	246	42	70	143	403	56	248	112	403
Perphenazine	42	56	246	70	232	233	198	98	403
Perphenazine*	246	143	403	70	404	42	248	113	403
Perthane	223	224	117	167	179	178	165	193	306
Pethidine	71	70	42	57	43	172	247	91	247
Pethidine	71	247	172	218	174	103	96	91	247
Pethidine	71	70	247	246	57	172	42	218	247
Pethidine	71	247	246	70	218	57	174	248	247
Pethidine*	71	70	44	57	42	247	43	246	247
Pethidine (Demerol)	71	70	247	57	42	246	91	103	247
Pethidine Intermediate A	57	43	70	71	42	200	199	91	200
Pethidine Intermediate A	57	42	43	70	71	200	199	44	200
Pethidine Intermediate C	42	43	44	57	70	71	103	91	219
Pethidine Metabolite	57	233	42	56	158	0	0	0	233
Pethidine (N-Methyl-4-phenyl-4-carboxypiperidine Ethyl Ester)	71	70	42	247	57	44	96	43	247
Pethidinic acid*	71	70	57	43	219	42	218	44	219

Compound Name										MWT
Pethidinic Acid Methyl Ester	...	71	70	42	57	233	232	43	44	233
Pethidinic Acid TMS Ester	...	71	70	42	103	73	291	57	44	291
Phellandrene(alpha-)	...	93	91	77	119	92	136	41	39	136
Phellandrene(beta-)	...	93	136	77	94	79	91	80	92	136
Phenacemide	...	118	91	44	92	178	65	0	0	178
Phenacemide	...	91	92	65	44	117	39	90	89	178
Phenacemide*	...	91	92	118	44	43	135	65	178	178
Phenacetin*	...	108	109	179	137	43	81	80	110	179
Phenacetin Metabolite	...	197	155	140	0	0	0	0	0	197
Phenadoxone	...	162	105	57	163	114	190	44	77	351
Phenadoxone	...	114	115	70	57	56	336	91	223	351
Phenaglycodol	...	59	43	156	121	155	157	158	139	214
Phenaglycodol	...	43	59	31	155	121	156	77	41	214
Phenampromide	...	98	125	42	41	29	99	120	190	274
Phenampromide	...	98	125	99	120	41	42	57	77	274
Phenazocine	...	230	58	231	44	173	105	42	159	321
Phenazocine*	...	230	231	58	44	42	105	173	159	321
Phenazone*	...	188	96	77	56	105	189	55	51	188
Phenazopyridine	...	81	54	108	77	51	213	39	52	213
Phenazopyridine	...	108	81	213	54	77	136	51	43	213
Phenbutrazate	...	56	69	91	261	98	119	84	190	367
Phenbutrazate*	...	69	56	71	91	261	84	119	70	367
Phencyclidine	...	200	91	84	242	243	115	129	117	243
Phencyclidine	...	200	91	243	84	242	186	166	201	243
Phencyclidine	...	91	84	200	41	55	117	42	115	243
Phendimetrazine	...	85	57	42	44	56	191	77	0	191
Phendimetrazine	...	57	42	85	56	54	58	70	44	191
Phendimetrazine*	...	57	85	42	56	76	191	70	77	191
Phenelzine	...	105	91	104	77	133	92	51	65	136
Phenelzine	...	45	91	105	77	65	52	136	0	136
Phenelzine	...	45	91	64	105	147	30	77	104	136
Phenelzine	...	91	147	45	64	105	65	77	48	136
Phenelzine*	...	31	45	46	29	27	59	74	43	136
Phenelzine Probe Product 1*	...	147	45	91	64	48	105	65	77	999
Phenelzine Probe Product 2*	...	45	64	147	105	91	77	79	65	999
Phenethylamine(beta-)	...	30	91	121	92	51	0	0	0	121
Phenformin*	...	91	30	92	42	29	146	44	104	205
Phenformin Probe Product*	...	91	114	43	85	139	104	72	243	999
Phenindamine	...	260	261	42	202	57	217	203	218	261
Phenindamine	...	260	261	202	42	203	182	115	215	261
Phenindamine*	...	260	261	42	57	184	215	217	262	261
Phenindione*	...	222	165	76	223	166	105	104	90	222
Pheniramine	...	169	58	149	168	170	57	167	43	240
Pheniramine	...	58	169	168	42	167	170	72	51	240
Pheniramine*	...	169	58	168	170	72	167	44	42	240
Phenkapton	...	45	121	186	97	65	93	214	153	376
Phenmetrazine	...	71	42	43	56	177	77	70	72	177
Phenmetrazine	...	71	56	177	77	70	105	51	72	177
Phenmetrazine	...	71	42	56	43	44	177	30	117	177
Phenmetrazine*	...	71	42	56	43	177	77	178	105	177
Phenobarbitone	...	204	232	117	161	146	77	118	115	232
Phenobarbitone	...	204	63	146	232	117	143	174	89	232
Phenobarbitone	...	204	117	77	115	51	91	103	118	232
Phenobarbitone	...	204	31	117	51	77	161	146	118	232

Compound Name				PEAKS					MWT
Phenobarbitone*	204	117	146	161	77	103	115	118	232
Phenol	94	66	39	65	40	95	38	55	94
Phenol	94	65	66	95	55	51	63	47	94
Phenolphthalein	274	273	225	318	181	257	121	197	318
Phenolphthalein*	274	225	318	273	275	226	257	121	318
Phenomorphan	256	58	257	182	199	157	105	91	347
Phenoperidine	246	367	247	42	91	103	77	57	367
Phenoperidine*	246	42	367	247	57	56	91	77	367
Phenothiazine	199	167	198	166	99	154	69	77	199
Phenoxyacetic Acid	107	92	152	121	65	0	0	0	152
Phenoxybenzamine	91	65	77	196	39	56	51	90	303
Phenprobamate	91	117	118	42	92	65	77	136	179
Phenprobamate*	118	117	91	92	119	65	77	103	179
Phenprocoumon*	251	280	118	91	121	119	252	189	280
Phensuximide(i)*	104	189	103	78	51	77	105	52	189
Phensuximide(ii)*	189	103	190	104	188	102	77	174	189
Phentermine	58	91	59	44	42	134	65	57	149
Phentermine	58	42	91	41	30	39	65	59	149
Phentermine*	58	91	42	41	134	65	59	40	149
Phentermine TFA	154	59	45	91	69	132	114	117	245
Phenthoate	274	93	125	121	91	107	135	246	320
Phentolamine	120	122	281	91	160	65	280	162	281
Phenylacetaldehyde	91	92	120	65	63	51	39	89	120
Phenylacetic Acid	91	136	92	65	39	63	45	51	136
Phenylacetone	43	91	134	92	65	39	63	135	134
Phenylacetonitrile	117	90	116	89	51	63	39	118	117
Phenylacetonitrile	117	116	90	89	51	77	118	63	117
Phenylacetylsalicylate	197	256	240	198	152	106	257	241	256
1-Phenylazonapth-2-ol	143	248	115	77	69	171	81	249	248
Phenylbutazone	77	183	308	184	105	55	51	41	308
Phenylbutazone	183	77	308	184	105	93	309	91	308
Phenylbutazone	77	183	105	184	55	93	91	51	308
Phenylbutazone*	77	183	308	184	105	55	51	41	308
Phenylbutyrolactone (Primidone, Phenobarb, Glutethimide Metab)	117	118	162	91	77	78	103	105	162
1-Phenyl-1-carbethoxy-2-dimethylaminocyclohex-3-ene	82	97	103	77	72	42	91	115	273
1-Phenyl-1-carbethoxy-2-methylaminocyclohex-3-ene	83	68	84	103	77	91	157	155	259
1-Phenylcyclohexene	129	158	130	143	115	128	91	77	158
Phenylcyclohexene	153	115	129	91	143	0	0	0	158
1-(1-Phenylcyclohexyl)-4-hydroxypiperidine	91	216	77	259	0	0	0	0	259
1-(1-Phenylcyclohexyl)morpholine	91	202	115	129	158	57	77	143	245
1-(1-Phenylcyclohexyl)pyrrolidine	91	70	186	152	117	77	172	229	229
1-Phenyl-1,2-dihydroxyethane	107	79	77	91	51	105	108	78	138
Phenylenediamine(ortho-)	108	80	81	107	28	0	0	0	108
Phenylephrine	44	77	95	121	141	167	168	133	167
Phenylephrine	44	77	95	65	39	45	42	43	167
1-Phenylethyl Acetate	104	43	91	39	51	103	78	27	164
Phenylethyl Acetate	104	43	91	105	65	39	51	77	164
2-Phenylethylamine	30	91	39	65	92	51	63	121	121
2-Phenyl-2-Ethylmalonamide(PEMA-Phenobarb./Primidone metab.)	163	148	91	103	117	120	44	77	206
Phenylglutarimide(alpha)	104	189	146	105	0	0	0	0	189
5-Phenylhydantoin(5-Methyl-)	104	175	77	119	51	190	42	161	190
5-Phenylhydantoin(5-methyl-) Dimethyl Derivative	118	203	56	77	218	141	132	204	218
5-Phenylhydantoin(5-(p-methylphenyl)-)	180	266	237	208	118	223	194	77	266
5-Phenylhydantoin(5-(p-methylphenyl))- Dimethyl Derivative	217	294	132	194	118	208	222	77	294

Compound Name									MWT
Phenylindanione	222	165	223	221	194	166	164	76	222
2-Phenyllactic Acid	91	43	44	41	45	148	149	103	166
2-Phenyllactic Acid	43	121	77	105	51	103	78	48	166
Phenylmercuric Nitrate	77	51	78	50	356	354	279	353	629
Phenylmethylbarbituric Acid*	104	132	218	51	103	77	78	52	218
1-Phenyl-2-nitropropene	115	91	105	116	51	39	77	63	163
5-Phenyloxazilidindione-(2,4)(Pemoline Metabolite 1)	90	177	105	77	106	51	89	50	177
3-Phenylpiperidin-2,6-dione(Glutethimide Metabolite)	104	189	103	117	78	91	51	146	189
3-Phenyl-3-piperidinocyclohexanol	124	98	105	84	216	96	259	200	259
4-Phenyl-4-piperidinocyclohexanol	200	91	84	86	186	259	0	0	259
3-Phenyl-1-propanol Acetate	118	117	43	91	119	105	65	77	178
Phenylpropanolamine	44	36	38	77	43	51	79	132	151
Phenylpropanolamine	44	77	79	105	91	51	92	132	151
Phenylpropanolamine*	44	77	79	51	45	42	107	105	151
Phenylpropionic Acid	91	104	150	105	78	77	65	51	150
Phenylpropylmethylamine	77	44	79	51	57	105	107	91	149
2-Phenylpyridine	155	154	156	77	127	128	51	78	155
Phenylsalicylate	121	65	39	93	63	92	64	122	214
Phenyltoloxamine	58	255	40	72	42	59	71	91	255
Phenyltoloxamine*	58	255	42	71	59	44	181	165	255
Phenyltoxolamine	58	42	91	56	57	43	59	165	255
Phenyramidol*	107	108	78	79	80	77	51	52	214
Phenytoin	180	209	223	252	104	73	0	0	252
Phenytoin*	180	104	223	77	209	252	51	165	252
Phenytoin Dimethyl Derivative	118	280	203	194	77	165	223	222	280
Phloroglucinol	126	69	85	52	43	97	80	51	126
Phloroglucinol	126	69	85	52	43	63	42	41	126
Pholcodeine*	114	100	56	42	115	101	70	398	398
Pholcodeine	100	114	42	56	115	70	101	55	398
Phorate	75	121	97	260	47	65	93	69	260
Phorate	75	231	45	97	153	47	46	125	260
Phosalone	182	121	184	154	367	97	241	58	367
Phosalone	182	121	97	184	111	154	65	138	367
Phosdrin	127	192	109	67	43	164	193	224	224
Phosdrin	127	15	43	67	109	79	192	39	224
Phosmet	160	61	76	77	133	104	161	50	317
Phosphamidon	127	72	264	138	109	67	193	42	299
Phosphoramide Mustard	92	63	56	94	65	57	171	219	220
Phosphorylated Amphetamine	180	124	152	91	44	40	42	105	271
Photodieldrin	81	79	281	244	246	279	271	273	378
Phoxim	77	97	129	157	125	31	103	51	298
Phthalazine	130	76	103	50	102	75	129	52	130
Phthalodinitrile(o-)	128	101	50	75	129	76	51	64	128
Phthalylsulphacetamide*	109	76	104	92	239	65	108	43	362
Phthalylsulphathiazole*	92	156	191	65	108	76	104	50	403
Physostigmine	218	275	174	160	161	0	0	0	275
Physostigmine	174	160	161	218	275	175	162	149	275
Physostigmine*	218	174	160	161	275	219	175	162	275
Picloram	196	198	161	200	163	86	242	245	240
Picloram Methyl Ester	196	198	197	195	200	225	223	256	254
2-Picoline(2-Methylpyridine)	93	66	92	65	78	39	51	67	93
3-Picoline(3-Methylpyridine)	93	66	92	65	39	67	40	94	93
Picolinic Acid	79	52	51	78	50	39	45	53	123
Picric Acid	229	30	62	91	53	63	50	230	229

Compound Name	Peaks								MWT
Pilocarpine	95	96	124	42	82	83	94	41	208
Pilocarpine	95	96	109	208	42	41	54	83	208
Piminodine	246	45	42	366	58	57	43	106	366
Piminodine	42	106	77	57	246	43	56	91	366
Piminodine (Alvodine, Pimadin, Cimadon)	246	366	106	247	133	260	234	367	366
Piminodine Ethane Sulphonate	246	366	42	106	133	247	57	260	366
Pimozide	230	187	134	217	96	83	461	109	461
Pinacolone	57	41	29	43	100	39	56	0	100
Pindolol	72	43	133	116	248	204	104	56	248
Pindolol*	72	133	30	116	248	134	56	41	248
Pinene(alpha-)	93	92	91	77	41	79	121	39	136
Pinene(beta-)	93	41	69	79	91	77	94	80	136
Pipamazine	141	123	42	41	151	55	169	96	401
Pipamazine	123	141	43	151	41	42	57	231	401
Pipamazine*	141	169	401	42	96	41	142	70	401
Pipazethate*	98	111	99	199	41	288	200	55	399
Pipenzolate Bromide	105	111	77	96	97	183	42	51	433
Piperacetazine	142	42	44	170	43	41	140	96	410
Piperazine Adipate*	44	100	41	60	29	27	30	86	232
Piperazine Phosphate	44	29	30	85	86	56	57	185	184
Piperidine	84	85	56	57	44	42	43	41	85
Piperidolate	111	96	165	42	167	43	41	56	323
Piperidolate	105	77	235	96	111	97	183	250	323
Piperidolate*	111	96	167	112	165	71	43	42	323
Piperidone	30	99	28	42	43	41	27	39	99
3-Piperidylbenzilate	83	105	77	183	165	152	139	128	311
4-Piperidylbenzilate	183	105	77	84	82	165	152	129	311
Piperine	201	115	285	173	143	84	202	174	285
Piperitone	82	95	137	54	109	152	41	110	152
Piperocaine	105	112	77	41	55	44	42	56	261
Piperocaine	112	105	246	77	55	41	98	97	261
Piperocaine*	112	246	105	77	55	41	44	247	261
Piperonal	149	150	63	121	65	62	91	39	150
Piperoxan*	98	41	99	42	55	98	233	69	233
Pipradrol	84	77	105	56	85	83	51	182	267
Pipradrol*	84	56	85	77	105	55	30	42	267
Pipradrol	84	105	77	56	85	248	249	36	267
Piridocaine	84	92	120	65	56	111	119	82	248
Pirimicarb	166	72	238	167	138	123	152	109	238
Pirimicarb	72	166	238	167	42	44	138	109	238
Pirimiphos Ethyl	333	318	57	304	168	180	71	166	333
Pirimiphos Methyl	109	137	246	290	110	305	276	81	305
Piritramide*	386	138	387	84	110	42	301	263	430
Pivalic Acid	57	41	29	39	56	27	45	87	102
Polythiazide	310	42	312	129	45	64	30	131	439
Polythiazide*	310	206	42	64	312	299	45	48	439
Practolol	72	30	151	56	109	43	108	251	266
Practolol, tri-TFA Derivative	308	266	440	43	512	152	194	126	554
Pramoxine	100	120	70	41	56	42	101	293	293
Pramoxine*	100	128	70	41	42	293	56	101	293
Prazepam	91	55	269	295	294	324	29	241	324
Prazepam	269	324	91	296	295	55	323	325	324
Prazepam	269	91	55	295	296	241	324	271	324
Prazepam*	91	269	324	55	296	295	323	297	324

Prazepam Benzophenone	285	270	77	91	105	257	166	287	285
Prazepam-N-Oxide	55	91	340	44	339	57	105	269	340
Prazosin	233	383	259	43	245	31	95	205	383
Prednisolone	121	122	91	147	225	43	135	120	360
Prenylamine	238	58	91	239	167	117	165	77	329
Prenylamine Metabolite	58	254	91	56	183	115	107	343	345
Prenylamine Metabolite	58	284	91	213	147	152	137	115	375
Prenylamine Metabolite	58	238	345	91	107	167	165	252	345
Prenylamine Metabolite	239	167	165	59	194	181	152	193	239
Prenylamine Metabolite	73	253	167	165	194	193	152	115	253
Prenylamine Metabolite	210	209	30	183	165	115	181	133	227
Prenylamine Metabolite	105	77	30	45	227	51	165	183	227
Prenylamine Metabolite	30	194	167	165	193	116	77	179	211
Pridoxamine	151	94	168	106	150	122	123	0	168
Prilocaine*	86	43	77	91	40	106	84	65	220
Prilocaine*	86	44	87	107	43	106	56	41	220
Primaquine*	201	81	98	175	259	176	202	242	259
Primidone	117	146	43	91	32	115	39	116	218
Primidone	146	117	190	118	115	103	91	161	218
Primidone*	146	190	117	118	161	189	103	91	218
Primidone Dimethyl Derivative	146	218	117	118	44	217	42	103	246
Primidone 1,3-di-TMS Derivative	146	73	232	334	247	117	246	147	362
Primidone (5-Ethylhexahydro-5-phenylpyrimidine-4,6-dione)	146	190	117	118	189	161	103	115	218
Primidone Metabolite	163	148	91	103	164	0	0	0	206
Primidone Metabolite	146	117	103	115	91	0	0	0	216
Probarbital*	141	156	41	43	39	98	155	169	198
Probarbitone	156	141	41	43	155	157	98	169	198
Probenecid*	256	121	185	43	257	65	42	214	285
Procainamide	86	30	120	99	58	65	92	42	235
Procainamide*	86	99	120	30	92	87	58	65	235
Procaine	86	120	65	92	42	43	41	56	236
Procaine	86	99	120	58	65	42	92	56	236
Procaine*	86	99	120	58	87	30	92	71	236
Procetofene Metabolite Methyl Ester	121	232	139	273	234	332	141	275	332
Prochlorperazine	70	42	113	43	141	56	71	44	373
Prochlorperazine*	113	70	373	141	43	72	42	127	373
Proclonol	264	139	266	111	141	251	152	292	292
Procyclidine	84	80	55	41	42	105	77	200	287
Procyclidine	84	55	42	70	205	43	204	41	287
Procyclidine*	84	204	205	85	42	55	105	77	287
Progesterone	124	43	314	79	91	229	272	105	314
Proheptazine	58	202	57	201	42	84	44	186	275
Prolintane	126	91	127	174	55	41	42	97	217
Prolintane*	126	127	174	91	70	69	55	42	217
Promazine	58	85	284	238	239	198	226	224	284
Promazine	58	86	42	85	199	198	238	284	284
Promazine*	58	284	86	238	198	199	85	42	284
Promethazine	72	73	198	180	213	284	42	44	284
Promethazine	72	199	167	198	71	42	56	200	284
Promethazine	72	180	198	73	42	56	70	71	284
Promethazine*	72	284	73	43	42	198	180	166	284
Promethazine*	72	73	284	198	213	199	180	56	284
Promethazine Sulphoxide	72	58	198	213	214	180	73	42	316
Prometryne	241	58	43	184	69	68	41	199	241

Pronethalol (Nethalide) TMS Ester	72	73	229	75	230	43	45	153	301
Propachlor (tech)	120	77	176	93	43	57	169	211	211
Propanidid	114	72	337	100	43	237	115	85	337
2-Propanol	45	43	27	41	29	39	31	59	60
1-Propanol	31	59	42	60	27	29	45	41	60
Propantheline	86	181	43	44	41	114	42	152	353
Proparcaine	86	99	58	87	56	100	84	70	294
Propargite	135	81	173	39	41	57	150	201	350
Propazine	214	58	229	172	43	187	216	41	229
1-Propen-2-ol	57	29	31	27	39	58	30	26	58
Properdin	71	70	218	261	57	174	219	172	302
Properidine	71	70	218	57	42	44	36	43	261
Propiomazine	72	149	269	255	197	73	254	340	340
Propiomazine	72	198	197	196	73	56	70	255	340
Propiomazine*	72	73	340	269	197	71	70	56	340
Propionaldehyde	29	27	58	26	57	30	39	43	58
Propionitrile	54	26	27	52	55	51	53	29	55
Propranolol	72	115	144	259	116	215	127	254	259
Propranolol	30	72	115	57	41	43	56	42	259
Propranolol*	72	56	30	43	98	115	144	41	259
Propyl Acetate(n-)	43	61	73	42	41	59	27	0	102
Propylaminopropane(n-)	30	72	44	43	27	28	41	86	101
Propyl Cannabinol	267	268	45	238	282	43	55	41	282
Propyl Ether(n-)	43	73	41	102	27	31	57	42	102
Propylethyldithiocarbamate (S-n-propyl-)	88	60	163	43	41	148	121	107	163
Propylhexedrine	58	140	67	72	83	81	155	156	155
Propylhexedrine	58	40	38	57	55	140	41	56	155
Propyl Hydroxybenzoate Methyl Ester	135	152	77	92	64	63	136	107	194
Propyl Hydroxybenzoate (Propylparaben) TMS Ether	73	193	43	195	210	237	75	252	252
Propyl Isothiocyanate	101	43	72	41	27	42	39	45	101
Propylmethyldithiocarbamate (S-n-propyl-)	149	107	30	74	73	60	0	0	149
Propyl Nitrate	46	29	76	30	0	0	0	0	105
Propylnorfenfluramine	86	44	159	43	41	87	42	56	245
Propylparaben	121	138	65	93	39	122	41	63	180
1-Propyl-2-phenylethylamine	72	31	91	120	65	73	39	42	163
Propyl Thiocyanate	43	41	27	101	39	42	45	28	101
5-Propyl-5-(2-thiocyanatoethyl)-barbituric Acid	124	141	154	153	170	98	125	155	255
2-(2'Propynyloxy)phenyl N-Methylcarbamate	109	148	110	81	88	58	57	39	205
3-(2'-Propynyloxy)phenyl N-Methylcarbamate	147	148	88	91	58	110	81	145	205
Propyphenazone	215	230	56	77	96	122	81	201	230
Propyphenazone*	215	230	56	77	216	96	41	39	230
Proquamazine	198	70	58	115	71	56	72	269	327
Prothipendyl*	58	285	214	200	86	227	85	212	285
Prothoate	115	97	73	43	65	121	125	93	285
Protriptyline	70	191	44	189	43	165	69	192	263
Protriptyline	191	70	44	58	189	84	165	192	263
Proxymetacaine	86	99	58	71	136	56	41	80	294
Prynachlor	186	130	77	53	120	93	221	51	221
Pseudoephedrine	58	77	30	51	49	56	59	42	165
Pseudoephedrine	58	30	77	42	56	106	105	51	165
Pseudoephedrine*	58	77	59	56	51	42	105	91	165
Psilocin	58	204	59	42	146	160	130	0	204
Psilocin	58	42	204	30	146	117	130	59	204
Psilocin (4-Hydroxy-N-dimethyltryptamine)	58	204	59	146	159	205	160	60	204

COMPOUND NAME									MWT
Psilocin-TMS Derivative	58	276	218	261	0	0	0	0	276
Psilocybin	58	204	59	146	159	205	160	57	284
Psilocybin	58	42	204	30	44	146	117	118	284
Psilocybin Metabolite	58	204	59	0	0	0	0	0	204
Pyramat	72	42	151	153	43	29	40	222	222
Pyrantel	205	42	135	173	206	123	145	45	206
Pyrathiazine	84	296	42	85	180	55	212	41	296
Pyrazole	68	41	40	28	67	69	0	0	68
Pyrazon	77	221	105	220	88	223	222	51	221
Pyridine	79	52	51	50	78	39	53	80	79
Pyridostigmine Bromide	72	39	42	38	94	56	51	81	260
Pyridoxine	151	94	122	106	51	53	149	150	169
5-(4'-Pyridylmethylene)barbituric Acid	217	128	118	103	146	119	145	174	217
5-(2'-Pyridylmethylene)barbituric Acid	216	217	128	173	118	103	130	146	217
5-(3'-Pyridylmethylene)barbituric Acid	128	217	216	173	103	146	119	130	217
Pyrilamine	121	58	72	79	71	78	42	122	285
Pyrimethamine	247	248	249	250	219	212	106	211	248
Pyrimethamine	247	248	249	42	89	219	114	123	248
Pyrogallol	126	52	80	51	108	39	79	53	126
Pyrogallol	126	80	52	108	51	63	79	39	126
Pyrovalerone	126	127	42	91	55	41	65	119	245
Pyrrobutamine	84	205	42	240	81	55	91	125	311
Pyrrobutamine	205	42	125	115	186	41	91	127	311
Pyrrobutamine*	205	240	91	84	125	242	206	186	311
Pyrrole	67	41	39	40	38	42	37	66	67
Quinacrine	86	126	259	58	112	99	400	74	399
Quinalbarbitone	168	167	41	43	124	97	169	195	238
Quinalbarbitone*	167	168	41	43	97	124	39	55	238
Quinalphos	146	298	157	156	129	118	90	97	298
Quinethazone	260	262	180	145	261	179	182	181	289
Quinethazone*	260	262	180	287	261	145	286	124	289
Quinidine	136	324	189	81	173	138	55	0	324
Quinidine	136	41	42	55	158	173	81	79	324
Quinine	136	158	159	186	81	0	0	0	324
Quinine	136	137	83	117	189	158	85	160	324
Quinine*	136	137	81	42	41	189	55	117	324
Quinoline	129	102	128	51	130	76	50	103	129
Quinolinic Acid	123	51	105	78	50	77	44	52	167
Racemethorphan	59	42	150	271	44	0	0	0	271
Racemethorphan*	59	271	150	31	270	214	42	171	271
Racemorphan	59	257	256	150	80	42	82	200	257
Racemorphan	44	43	257	59	150	200	0	0	257
Racemorphan*	257	59	150	256	31	157	42	200	257
Racumin (4-Hydroxy-3-(1,2,3,4-tetrahydro-1-naphthyl)coumarin)	292	188	121	130	115	293	129	128	292
Rauwolfine	326	43	144	182	58	311	183	145	326
RDX	28	30	46	42	29	128	120	75	222
RDX	46	30	42	75	120	44	128	56	222
RDX	46	28	42	75	120	56	29	128	222
Rectidon	43	167	237	41	39	124	168	55	316
Rescinnamine	221	199	200	186	395	251	77	214	634
Reserpine	195	199	200	186	214	152	174	251	608
Reserpine*	608	606	195	609	395	397	212	396	608
Resorcinol	110	82	39	81	53	69	55	111	110
Rifampicin	43	45	60	58	95	99	398	206	822

Compound Name									MWT
Rubazonic Acid	91	77	92	359	132	144	266	0	359
S 421	130	132	79	83	85	134	109	81	374
Sabinene	93	77	91	41	79	39	136	94	136
Sabinene	93	41	77	91	79	27	39	69	136
Sabinol	92	91	42	81	41	43	109	55	152
Saccharin	76	183	50	119	120	0	0	0	183
Saccharin*	183	76	90	50	91	120	106	92	183
Salbutamol*	30	86	57	41	77	135	29	206	239
Salbutamol TMS Derivative	68	369	41	86	370	73	40	140	455
Salicyclic Acid TMS Ester TMS Ether	267	73	268	269	45	135	193	75	282
Salicylamide*	120	92	137	65	121	39	64	53	137
Salicylic Acid	92	120	138	64	39	63	65	53	138
Salicylic acid*	120	92	138	64	39	63	121	65	138
Salicylic Acid Methyl Ester TMS Ether	209	179	210	135	59	89	161	193	224
Salicyluric Acid	121	120	69	92	195	39	93	45	195
Salicyluric Acid Methyl Ester Methyl Ether	135	77	90	92	51	64	136	63	223
Salicyluric Acid Methyl Ester Methyl Ether	135	90	77	92	136	134	51	64	223
Salicyluric Acid TMS Ester TMS Ether	324	193	206	73	325	75	194	45	339
Santonin (alpha)	246	173	41	91	135	0	0	0	246
Scopoline TMS Ether	96	73	94	42	57	227	142	212	227
Secbutobarbitone	141	156	40	69	98	38	70	67	212
Secbutobarbitone*	141	156	41	57	39	98	157	47	212
Secbutyl Vinyl Ether	41	57	29	56	28	44	27	39	100
Secobarbitone	41	168	167	43	39	55	97	124	238
Secobarbitone	168	167	41	43	97	195	124	169	238
Secobarbitone Dimethyl Derivative	196	195	181	41	138	111	53	110	266
Senecioic Acid	100	83	39	55	82	85	29	27	100
Senecionine	136	120	220	119	121	138	335	246	335
Seneciphylline	119	120	136	94	93	138	121	246	333
Sigmodal*	167	43	41	39	78	55	247	122	316
Sigmodal Methylation Artifact*	194	193	136	43	41	109	39	137	264
Simazine	201	44	186	173	68	71	55	158	201
Simazine	44	201	43	186	68	173	71	96	201
Simetryne	213	43	68	44	71	155	170	96	213
Sorbide Nitrate*	43	31	29	61	60	85	73	44	236
Sotalol, tri TFA Derivative	168	126	405	43	368	326	446	0	560
Spironolactone	55	91	269	136	107	340	41	67	416
Spironolactone*	341	43	340	374	267	107	55	342	416
Stearic Acid	44	43	73	57	60	41	55	69	284
Stearic Acid	57	73	60	43	55	71	41	69	284
STP	44	166	151	57	43	91	135	209	209
STP-NCS Derivative	165	251	135	0	0	0	0	0	251
Strychnine	334	44	120	77	41	107	55	144	334
Strychnine	334	130	120	333	143	107	162	144	334
Strychnine*	334	335	162	120	107	144	143	130	334
Styramate	107	79	120	77	75	44	43	51	181
Styramate*	107	120	79	75	77	44	91	45	181
Styrene	104	103	78	51	77	50	105	52	104
Succinic Acid	100	55	45	74	119	73	56	41	118
Succinylsulphathiazole*	92	156	65	191	108	55	45	174	355
Sucrose	31	43	57	73	61	29	60	44	342
Sulfanilamide	39	52	65	80	92	108	156	172	172
Sulfinpyrazone	77	278	51	78	105	279	91	39	404
Sulfinpyrazone	77	51	40	39	91	105	130	64	404

Sulphacetamide*	172	92	156	65	108	39	44	41	214
Sulphadiazine*	186	185	92	65	108	39	93	187	250
Sulphadimethoxine*	246	92	65	245	108	247	39	260	310
Sulphadimidine*	214	213	92	65	215	108	39	42	278
Sulphaethidole*	92	284	156	108	65	106	93	220	284
Sulphafurazole*	156	92	108	65	140	43	157	42	267
Sulphaguanidine*	92	65	108	214	156	39	109	43	214
Sulphamerazine*	199	200	92	65	108	39	66	201	264
Sulphamethizole*	92	270	65	156	108	106	93	59	270
Sulphamethoxazole*	156	92	108	65	140	253	157	43	253
Sulphamethoxydiazine*	216	215	92	65	108	54	125	39	280
Sulphamethoxypyrazine*	92	216	215	108	65	156	54	125	280
Sulphamethoxypyridazine*	215	92	216	65	108	53	69	39	280
Sulphanilamide*	172	92	156	65	108	173	39	174	172
Sulphanilic Acid	173	108	65	92	80	39	63	156	173
Sulphaphenazole*	156	158	92	77	157	314	108	65	314
Sulphapyridine	184	185	92	65	108	186	66	183	249
Sulphapyridine*	184	92	185	65	39	108	66	186	249
Sulphasalazine*	169	92	289	65	290	39	333	184	398
Sulphasomidine*	214	92	65	213	108	42	215	39	278
Sulphasomizole*	156	92	108	65	269	39	45	270	269
Sulphathiazole	92	65	108	156	191	71	63	93	255
Sulphathiazole*	92	156	65	108	191	45	39	55	255
Sulphinpyrazone*	77	278	105	78	51	279	252	130	404
Sulphonal*	77	59	29	43	78	41	27	135	228
Sulpyrid	98	70	214	111	134	341	199	326	341
Sulpyrid Trimethyl Derivative	98	134	70	111	198	242	383	382	383
Sulthiame*	290	184	185	104	77	168	291	198	290
Sulthiame Dimethyl Derivative	318	274	77	226	104	44	210	105	318
Synephrine	44	77	95	65	123	39	121	42	167
Syrosingopine	181	395	198	251	397	396	199	666	666
Talbutal	167	41	168	39	97	57	124	53	224
Talbutal*	167	168	41	97	124	39	57	53	224
Talbutal (5-Allyl-5-s-butylbarbituric Acid)	41	167	168	39	125	97	44	53	224
Tartaric Acid	76	29	58	31	44	59	70	105	150
Taurolin*	30	43	57	44	42	135	56	55	284
TDE (p,p'-)	235	237	165	236	199	239	82	238	318
Temazepam	77	104	75	51	76	111	106	138	300
Temazepam*	271	273	300	272	256	77	255	257	300
Terbacil	161	216	160	218	163	162	117	118	216
Terbutaline	30	86	41	57	65	39	42	111	225
Terbutryn	226	185	241	170	43	68	41	71	241
Terpinene(alpha-)	121	93	136	119	106	91	79	77	136
Terpinene(gamma-)	93	136	121	92	91	43	77	79	136
Terpinene(gamma-)	93	91	77	136	121	39	43	27	136
Terpineol	71	93	111	43	86	69	55	68	154
Terpineol	59	121	93	136	43	81	41	55	154
2-Terpineol	59	93	121	43	136	28	81	41	154
Terpineol(alpha-)	58	120	72	70	136	18	42	0	154
Terpineol(alpha-)	59	93	121	81	43	136	68	92	154
1-Terpin-4-ol	92	70	43	41	109	90	39	76	154
Terpinolene	93	121	136	39	41	79	91	27	136
Tertrazepam	253	288	287	289	225	259	254	41	288
1,1,2,2-Tetrabromoethane	265	267	26	186	263	269	105	107	342

Compound Name									MWT
Tetracaine	58	71	42	105	176	150	92	56	264
2,3,4,5-Tetrachloroaniline	231	229	233	158	167	169	235	131	229
1,2,3,4-Tetrachlorobenzene	216	214	218	179	181	143	109	74	214
1,2,3,5-Tetrachlorobenzene	216	214	218	181	179	109	108	143	214
1,2,4,5-Tetrachlorobenzene	216	214	218	179	181	74	108	109	214
Tetrachlorobiphenyl	220	292	290	222	150	74	294	255	290
2,2',4,4'-Tetrachlorobiphenyl	292	290	220	294	222	150	293	184	290
Tetrachloroethylene	166	164	129	131	168	94	96	47	164
2,4,5,6-Tetrachloroisophthalonitrile (Daconil)	266	264	268	124	229	231	194	159	264
Tetrachloromethane	117	119	47	121	82	84	49	35	152
Tetrachloronitrobenzene	261	215	259	213	203	201	263	217	259
Tetrachloro-p-benzoquinone (Chloranil)	246	244	209	211	87	248	218	190	244
Tetrachlorophenol	232	230	131	234	170	84	133	166	230
Tetraethylphosphate	161	163	179	235	29	162	99	207	274
1,1,1,2-Tetrafluoroethane	33	69	83	32	31	51	63	82	102
Tetrafluoroethylene	81	31	100	50	69	28	82	29	100
Tetrahydrocannabidiol	233	318	193	234	262	319	273	136	318
Tetrahydrocannabinol	314	299	231	271	243	258	41	43	314
Tetrahydrocannabinol (delta 8)	221	314	248	261	193	236	222	315	314
Tetrahydrocannabinol(Delta-9)	299	231	314	43	41	295	55	271	314
Tetrahydrocannabinol (delta 9)	314	299	231	271	243	258	315	300	314
Tetrahydrocannabinol (delta-8)	231	314	271	258	43	193	41	246	314
Tetrahydrocannabinol-9-oic Acid	299	344	329	283	276	288	273	281	344
Tetrahydrofuran	42	41	72	71	43	27	39	29	72
Tetrahydrofurfuryl Alcohol	71	43	41	27	31	42	29	0	102
Tetrahydropyran-3-one	42	100	71	41	70	55	39	45	100
Tetrahydropyridine (Methyl Phenidate Decomposition Prod)	55	83	82	41	54	39	28	42	83
Tetrahydrozoline	185	115	200	171	128	44	91	199	200
1,3,7,8-Tetramethylxanthine (or 1,3,8,9-Tetramethyl-)	208	82	123	67	209	42	55	207	208
Tetrazolophthalazine	115	114	88	62	116	63	89	51	171
Tetronic Acid	42	100	43	29	72	41	30	69	100
Tetryl	241	242	181	75	224	194	91	77	287
Tetryl	241	181	91	66	30	75	224	90	287
Thalidomide	76	173	104	111	148	50	169	130	258
Thebacon	341	298	342	242	284	299	340	326	341
Thebaine	311	296	42	255	44	310	312	253	311
Thebaine	311	44	255	296	310	312	42	253	311
Thebaine	42	296	211	239	152	165	311	139	311
Thebaine*	311	255	42	44	296	310	312	174	311
Thenyldiamine	58	97	72	71	79	42	78	40	261
Thenyldiamine	58	97	71	72	91	79	203	78	261
Thenyldiamine*	58	97	72	71	203	191	190	42	261
Theobromine	55	82	67	109	180	0	0	0	180
Theobromine	180	55	109	67	82	137	181	42	180
Theobromine	180	67	82	109	55	181	66	81	180
Theophylline	180	95	68	41	43	53	57	55	180
Theophylline	68	95	180	53	41	40	67	42	180
Theophylline*	180	95	68	41	53	181	96	40	180
Thiabendazole	174	63	64	201	90	52	39	65	201
Thiabendazole	201	174	63	64	202	65	129	90	201
Thialbarbitone*	81	223	79	41	80	157	185	77	264
Thialbarbitone oxygenated analogue	81	207	80	79	141	169	41	77	248
Thiamylal	43	41	184	168	167	97	55	53	254
Thiethylperazine	70	113	42	141	43	56	399	71	399

Compound Name									MWT
Thiethylperazine Maleate (Torecane)	399	113	70	141	43	72	400	71	399
Thiobarbituric Acid	144	43	42	116	69	59	145	41	144
Thionazin	107	96	106	97	143	248	140	79	248
Thiopentone	172	157	42	173	242	69	71	97	242
Thiopentone	157	172	41	173	43	69	55	39	242
Thiopentone*	172	157	173	43	41	55	69	71	242
Thiopentone Dimethyl Derivative	169	184	41	112	69	43	185	55	270
Thiophanate	206	133	150	370	160	134	178	151	370
Thiophanate Methyl	192	160	342	59	150	177	105	209	342
Thiophencyclidine (Thio-PCP)	97	165	164	206	84	249	135	136	249
Thiophene	84	58	45	39	57	69	83	50	84
1-Thiophenylcyclohexene	164	135	136	149	97	84	163	91	164
Thiopropazate	246	185	445	70	125	154	213	87	445
Thiopropazate	43	70	42	185	246	445	125	98	445
Thiopropazate	53	35	246	185	70	125	153	385	445
Thiopropazate*	42	43	70	246	185	55	56	445	445
Thioproperazine	70	113	43	42	127	71	44	56	446
Thioproperazine	79	96	346	81	345	70	78	113	446
Thioproperazine	70	113	127	446	198	71	445	43	446
Thioridazine	98	70	42	185	126	99	96	55	370
Thioridazine	98	126	370	99	70	258	125	42	370
Thioridazine*	98	370	126	99	40	70	371	258	370
Thioridazine (Melleril)	98	70	370	126	99	185	244	125	370
Thiothixene	113	70	42	221	56	43	222	114	443
Thiourea	76	43	60	77	59	42	44	55	76
Thonzylamine	121	58	72	71	78	215	122	77	286
Thonzylamine	58	121	72	78	42	77	122	91	286
Thonzylamine	58	121	216	72	215	71	122	59	286
Thonzylamine*	58	121	72	71	216	215	122	78	286
Thozalinone	70	98	69	77	204	89	83	105	204
Thujyl Alcohol	95	93	43	121	81	55	110	70	154
Thymine	55	126	54	127	83	52	82	56	126
Thymol	135	150	91	107	117	0	0	0	150
Thymoxamine	58	72	279	234	151	192	166	165	279
Thymoxamine Hydrolysis Product	58	72	59	57	121	164	237	77	237
Tiglic Acid(2-Methyl-but-2-enoic Acid)	55	100	27	29	39	54	85	53	100
Tilidine	103	77	82	29	97	42	51	104	273
Tolazamide*	91	155	114	65	197	42	41	85	311
Tolazoline	91	159	65	160	131	39	51	81	160
Tolbutamide	91	155	171	65	107	39	108	63	270
Tolbutamide*	91	30	155	108	65	197	39	107	270
Tolmetin*	212	213	122	198	44	53	91	65	257
Toluene	91	92	65	39	63	51	93	45	92
Toluidine(m-)	107	106	77	79	108	39	65	78	107
Toluidine(p-)	106	107	77	108	79	78	53	39	107
Tolylfluanide	137	238	181	40	37	240	44	138	346
Tranylcypromine	130	77	51	132	115	103	133	78	133
Tranylcypromine*	133	132	56	115	30	117	91	77	133
Trazodone*	205	70	231	78	135	136	42	166	371
Trazodone Metabolite	239	205	221	211	139	166	195	387	405
3,6,17-Triacetylnormorphine	87	43	209	72	86	210	44	181	397
Triamifos	160	135	44	294	104	92	161	77	294
Triamterene (2,4,7-Triamino-6-phenylpyimido(4,5-b)pyrazine)	253	252	43	254	104	235	57	251	253
Triarimol	107	173	77	175	79	253	251	105	330

Compound Name				Peaks					MWT
Triazolam	342	313	238	344	315	102	75	137	342
Triazolophthalazine	170	115	114	88	171	62	116	89	170
Tribromethyl Alcohol*	31	203	93	121	123	95	44	201	280
2,4,6-Tribromophenol	330	332	62	63	141	334	328	143	328
Tri-2-butoxyethyl Phosphate	57	45	56	85	101	125	41	199	398
Tributoxyethyl Phosphate	85	100	125	101	199	227	99	153	398
Tributyl Phosphate	99	41	56	57	39	55	29	155	266
Trichlorfon	79	109	110	139	145	80	112	95	256
Trichlormethiazide	296	298	279	205	36	64	117	62	379
Trichloroacetic Acid	44	83	85	36	28	0	0	0	162
2,4,5-Trichloroaniline	195	197	199	123	133	135	160	167	195
1,2,4-Trichlorobenzene	180	182	184	145	109	147	75	181	180
2,3,6-Trichlorobenzoate	207	209	211	238	240	179	181	109	238
2,3,5-Trichlorobiphenyl	258	256	186	260	150	151	188	259	256
Trichlorocresol(m-)	210	175	212	177	111	75	214	73	210
Trichloroethane	99	83	101	85	61	70	31	0	132
1,1,1-Trichloroethane	97	99	61	63	117	119	101	62	132
1,1,2-Trichloroethane	97	83	99	85	61	26	96	63	132
1,1,1-Trichloroethane	43	97	61	99	45	27	119	117	132
2,2,2-Trichloroethanol	31	49	77	113	82	115	51	117	148
Trichloroethylene	95	130	132	60	97	35	134	47	130
Trichloroethylene	130	95	132	97	134	60	99	62	130
2,2,2-Trichloroethyl Heptafluorobutyrate	169	69	131	133	197	227	95	97	344
Trichlorofluoromethane(Freon 11)	101	103	66	105	47	35	31	82	136
2,4,5-Trichlorophenol	198	196	97	200	132	134	99	133	196
2,4,6-Trichlorophenol	196	198	97	132	200	134	99	62	196
2,4,5-Trichlorophenoxyacetic Acid	196	198	254	256	200	197	209	211	254
2,4,5-Trichlorophenoxyacetic Acid (2,4,5-T)	196	254	209	195	0	0	0	0	254
1,1,2-Trichlorotrifluoroethane	101	151	103	153	85	87	31	105	147
1,3,5-Trichloro-2,4,6-trinitrobenzene (Bulbosan)	315	316	142	144	177	179	107	285	315
Triclocarban*	127	161	163	153	129	187	90	189	314
Triclofos*	31	49	77	29	113	51	48	115	228
Tricresyl Phosphate(m-)	368	367	91	369	165	108	90	65	368
Tricresyl Phosphate(o-)	368	165	91	179	181	90	180	107	368
Tricresyl Phosphate(p-)	368	367	107	108	369	91	198	165	368
Tridemorph	128	129	115	41	202	43	70	55	297
Tridihexethyl Chloride	86	58	88	105	206	87	55	77	353
1-(1(2-Trienyl)cyclohexl)piperidine	135	136	45	97	149	164	91	77	249
Triethylamine	86	30	27	58	29	28	42	101	101
1,3,7-Triethyl-8-methylxanthine	250	164	222	194	235	251	207	150	250
Triethyl Phosphate	99	155	109	127	45	82	81	27	182
Trifluoperazine	70	42	113	43	407	56	141	248	407
Trifluoperazine*	113	70	407	43	141	42	127	71	407
2,2,2-Trifluoroethanol	31	29	33	61	51	69	50	32	100
1,1,1-Trifluoro-2,2,2-trifluoroethane(Freon 113)	151	153	117	119	70	101	121	29	186
Triflupromazine	58	86	42	85	44	59	30	70	352
Triflupromazine	58	86	352	85	306	59	266	248	352
Trihexyphenidyl	98	218	97	55	41	219	105	77	301
Trihexyphenidyl	98	55	218	99	41	96	219	42	301
2',4,4'-Trihydroxychalcone	137	256	44	45	43	41	120	55	256
Tri-(1-isopropylphenyl)amine	91	41	44	43	190	119	65	55	369
Tri-(1-isopropylphenyl)amine	91	190	119	70	71	55	77	72	369
Trilostane*	41	147	79	93	55	105	67	91	329
Trimeperidine	186	201	42	202	187	56	57	71	275

Trimeprazine	58	45	43	298	42	41	46	57	298
Trimeprazine	58	198	42	180	30	154	57	199	298
Trimeprazine*	58	298	212	198	100	299	252	199	298
Trimethadione	43	143	58	42	41	39	128	56	143
Trimethaphan	91	65	187	92	277	259	90	273	596
Trimethobenzamide	58	195	59	72	388	89	315	42	388
Trimethobenzamide	58	195	42	388	59	57	43	56	388
Trimethoprim	290	259	275	291	243	123	200	43	290
3,4,5-Trimethoxyamphetamine	43	182	167	225	181	183	151	142	225
2,4,5-Trimethoxyamphetamine	44	182	167	151	181	139	183	136	225
3,4,5-Trimethoxybenzaldehyde (Mescaline Precursor)	196	181	125	39	110	95	93	51	196
3,4,5-Trimethoxybenzoic Acid	212	197	141	160	154	111	93	0	212
3,4,5-Trimethoxybenzoyl Cyanide	221	124	64	206	150	151	163	178	221
3,4,5-Trimethoxybenzyl Alcohol (Mescaline Precursor)	198	127	183	95	123	181	196	199	199
3,4,5-Trimethoxyphenylacetonitrile	207	192	164	74	149	134	124	0	207
Trimethylacetamide	57	41	29	101	44	36	58	27	101
Trimethylamine	58	42	59	30	43	41	0	0	59
2,2,3-Trimethylbutane	57	43	56	41	85	29	27	39	100
2,3,5-Trimethylindole	158	159	144	115	157	160	156	143	159
Trimethyl-(N, alpha, alpha-)-diphenethylamine	91	58	176	119	42	41	56	65	267
1,3,3-Trimethyl-2-norboranone	81	69	41	152	80	82	109	39	152
2,2,4-Trimethylpentane	57	56	41	43	29	58	99	27	114
Trimethyl Phosphate	110	109	79	95	80	140	47	29	140
2,3,5-Trimethylpyrazine	42	122	39	27	81	54	53	52	122
2,4,6-Trimethylpyridine	121	120	79	106	122	77	107	39	121
Trimethylquinalbarbitone Carboxylic Acid Metabolite	195	196	41	138	181	55	237	111	310
Trimethylurea	44	43	42	45	72	30	102	58	102
1,3,7-Trimethyluric Acid	210	82	67	153	125	42	55	0	210
Trimipramine	58	249	208	99	193	232	84	0	294
Trimipramine	58	249	208	193	234	99	248	194	294
Trimipramine	58	193	194	42	234	235	39	179	294
Trimipramine Metabolite	208	249	193	44	195	194	234	209	280
1,3,5-Trinitrobenzene	75	30	213	74	120	91	63	167	213
1,3,5-Trinitrobenzene	30	75	74	213	120	63	91	92	213
2,4,6-Trinitrotoluene	210	30	149	164	180	193	134	0	227
2,4,6-Trinitrotoluene	210	89	30	63	39	51	76	134	227
Tripelennamine	58	91	72	71	65	42	79	78	255
Tripelennamine*	58	91	72	71	197	185	184	92	255
Triprolidine	209	208	278	207	193	194	84	200	278
Triprolidine	208	209	42	41	207	193	39	206	278
Triprolidine	208	209	31	207	278	84	193	42	278
Triprolidine*	208	209	278	207	193	200	194	84	278
Tris-(2-chloroethyl) Phosphate	63	249	205	251	143	65	207	223	284
Trithion	157	97	121	342	153	125	159	199	342
Trithion	157	121	342	97	153	45	159	125	342
Tropicamide	92	91	65	103	93	39	163	77	284
Tropicamide	92	254	163	91	93	65	103	104	284
Tropine TMS Ether	83	82	96	124	97	42	73	184	213
Tropocaine	124	82	83	245	94	77	105	67	245
Tryptamine	130	131	36	77	108	65	160	38	160
Tryptophol	130	161	143	131	103	77	115	102	161
Tuaminoheptane	44	100	42	41	30	55	39	45	115
Tubocurarine Chloride	298	594	58	593	299	609	595	564	680
Tubocurarine Chloride	298	163	43	41	55	162	57	175	680

Compound Name	Peaks								MWT
Tybamate	55	41	56	97	72	57	62	43	274
Tybamate*	55	72	97	41	56	158	118	57	274
Tyramine	107	36	108	77	38	39	51	79	137
Tyramine	30	108	107	77	137	43	41	55	137
Tyramine	30	108	107	77	137	51	109	78	137
Undec-10-en-1-al	43	57	41	55	82	56	44	29	170
Uracil	112	69	42	40	41	68	39	70	112
Urea	60	44	28	0	0	0	0	0	60
Urea di-TMS Derivative	147	189	73	74	148	75	190	146	204
7-Ureidodiftalone	132	147	262	133	104	77	148	190	322
Uric Acid	54	125	168	69	97	53	43	0	168
Vacor	135	92	65	108	138	80	39	106	272
Valeramide	59	44	41	29	72	27	43	28	101
Valeric Acid	60	43	41	87	45	27	74	39	102
Valerolactone(gamma-)	28	56	29	41	43	85	39	57	100
Valeronitrile(n-)	43	41	54	27	55	39	29	40	83
Valproic Acid	73	102	57	41	43	55	101	115	144
Vamidithion	87	109	145	58	79	142	112	88	287
Vanillal	137	81	138	109	166	0	0	0	166
Vanillin	152	151	81	109	51	0	0	0	152
Vapona (DDVP)	109	79	47	195	197	145	83	220	220
Verbenol(trans-(-))	93	43	41	109	94	91	39	55	152
Viloxazine*	56	100	138	110	57	237	70	41	237
Vinbarbital	195	41	39	69	67	53	152	135	224
Vinbarbitone*	195	41	141	69	39	152	135	196	224
Vinyl Propionate	29	57	27	43	26	44	28	0	100
4-Vinylpyridine	105	104	78	52	51	77	50	79	105
Warfarin	265	131	43	103	121	120	145	146	308
Warfarin	265	120	249	43	103	121	162	131	308
Warfarin	103	131	43	77	51	145	146	50	308
Warfarin*	265	208	121	43	266	187	213	251	308
Warfarin Alcohol	121	189	292	310	263	265	131	251	310
Xanthine	152	54	109	53	81	0	0	0	152
Xanthurenic Acid	159	205	131	76	187	51	103	104	205
Xipamide*	121	120	354	122	43	234	106	64	354
Xylazine	205	77	41	39	220	130	145	131	220
Xylene (meta)	91	106	105	77	51	92	39	79	106
Xylene (ortho)	91	106	105	77	51	92	39	78	106
Xylene (para)	91	106	105	77	51	92	79	103	106
2,3-Xylenol	122	107	121	77	91	79	39	78	122
2,4-Xylenol	122	107	121	77	91	123	39	79	122
2,5-Xylenol	122	107	121	77	91	79	123	39	122
2,6-Xylenol	122	107	121	77	91	79	78	103	122
3,4-Xylenol	122	107	121	77	91	123	39	108	122
3,5-Xylenol	122	107	121	77	79	91	123	39	122
Xylometazoline	229	173	44	41	115	128	91	129	244
Xylometazoline	229	244	173	243	44	230	81	245	244
Yohimbine	169	156	170	143	144	115	129	142	354
Yohimbine*	353	354	169	170	355	156	184	144	354
Zectran	165	164	150	222	58	134	77	39	222
Zinochlor	239	241	178	143	75	87	99	89	274
Zolimidine	272	193	209	78	192	273	194	181	272
Zomepirac*	246	291	248	290	247	139	111	292	293
Zomepirac GC Decomposition Product*	246	247	41	42	136	67	111	39	999

COMPOUND NAME   PEAKS   MWT

Zoxazolamine 168 170 113 78 63 169 43 76   168

APPENDIX 3

MOLECULAR WEIGHT LISTING
Eight peak mass spectra in ascending order of compound
molecular weight

An asterisk after the compound name indicates that this spectrum forms
part of the collection of full spectra available at CRE to be published
separately

MWT	COMPOUND NAME								
31	Methylamine	30	31	29	27	26	32	0	0
32	Methanol	31	32	29	30	28	33	34	27
41	Acetonitrile	41	40	39	38	32	29	0	0
42	Ketene	42	41	29	25	40	24	43	21
44	Acetaldehyde	29	44	43	42	26	41	27	25
44	Nitrous oxide	44	30	45	46	31	29	0	0
46	Ethanol	31	29	45	27	26	46	43	30
46	Ethanol	31	45	29	27	46	43	30	42
46	Formic Acid	29	46	45	44	30	47	31	48
52	Difluoromethane	33	51	31	52	50	34	20	0
53	Acrylonitrile	53	26	52	43	51	27	41	31
55	Propionitrile	54	26	27	52	55	51	53	29
56	Acrolein	27	56	26	55	29	25	37	38
58	1-Propen-2-ol	57	29	31	27	39	58	30	26
58	Acetone	43	58	42	27	40	29	26	39
58	Butane(n-)	43	29	41	27	58	42	39	44
58	Propionaldehyde	29	27	58	26	57	30	39	43
59	Methylformamide (N-Methyl)	59	58	60	28	41	42	0	0
59	Trimethylamine	58	42	59	30	43	41	0	0
60	1,2-Ethanediamine	30	42	43	27	44	29	31	41
60	1-Propanol	31	59	42	60	27	29	45	41
60	2-Propanol	45	43	27	41	29	39	31	59
60	Acetic Acid	45	29	43	60	42	44	41	31
60	Carbonyl Sulphide	60	30	62	44	34	33	46	31
60	Glycoaldehyde	31	29	60	30	42	41	26	56
60	Isopropanol	45	43	27	41	31	44	29	0
60	Isopropanol or Propan-2-ol	45	43	29	41	31	0	0	0
60	Methyl Formate	31	29	60	30	33	44	45	61
60	Urea	60	44	28	0	0	0	0	0
61	Ethanolamine	42	31	61	43	29	27	44	41
62	Boric Acid	45	62	44	61	43	63	0	0
62	Ethylene Glycol	31	33	29	43	27	42	62	44
66	1,2-Difluoroethane	51	65	27	45	62	47	26	64
67	Allyl Cyanide	41	67	39	27	40	38	66	37
67	Crotononitrile(cis-)	41	67	39	40	66	38	27	37
67	Crotononitrile(trans-)	41	67	39	40	66	38	27	37
67	Pyrrole	67	41	39	40	38	42	37	66
68	Chlorofluoromethane	68	70	33	49	47	51	48	50
68	Cyclopentene	67	68	39	53	41	65	42	27
68	Furan	39	68	38	29	40	37	42	26
68	Imidazole	44	29	27	41	58	57	43	39
68	Pyrazole	68	41	40	28	67	69	0	0
69	Butyronitrile(n-)	41	29	27	39	40	0	0	0
69	Butyronitrile(t-)	42	41	68	67	39	27	40	29
70	Crotonaldehyde	41	70	39	69	42	27	38	29
70	Methyl Vinyl Ketone	55	43	27	70	97	42	41	26
72	2-Butanone	43	29	27	72	42	57	44	39
72	2-Methylpropanal	43	27	41	29	72	39	42	26
72	Butanal(n-)	27	29	44	43	41	72	39	57
72	Isobutyraldehyde	43	41	72	27	39	29	42	26
72	Pentane(n-)	43	42	41	27	29	39	57	72
72	Tetrahydrofuran	42	41	72	71	43	27	39	29
73	Butylamine(n-)	30	73	41	27	42	39	31	56
73	Butylamine(t-)	58	41	30	42	57	39	27	59

Molecular Weight Index

MWT	COMPOUND NAME							
73	Dimethylformamide	73	44	42	30	43	72	29	58
73	Isobutylamine	43	41	57	71	70	55	73	42
73	Isobutylamine	30	73	41	27	55	39	56	31
74	1-Butanol	56	31	41	43	27	42	29	0
74	1-Butanol	31	56	41	43	27	42	29	39
74	2-Butanol	43	31	42	41	33	29	39	0
74	2-Butanol	45	59	31	41	27	43	44	29
74	Butanol (tert-)	59	31	41	57	43	29	39	0
74	Diethyl Ether	31	29	59	45	74	43	0	0
74	Diethyl Ether	31	59	29	45	74	27	41	43
74	Methyl Acetate	43	74	42	59	29	31	44	45
74	Nirosodimethylamine(N-)	42	74	30	43	44	41	40	45
75	Nitroethane	29	27	30	26	28	43	46	0
76	2-Methoxyethanol	45	31	29	47	43	76	27	46
76	Carbon Disulphide	76	78	38	44	77	39	64	46
76	Thiourea	76	43	60	77	59	42	44	55
77	Methyl Nitrate	46	30	29	28	15	31	0	0
78	Benzene	78	52	51	77	50	39	79	43
78	Benzene	78	77	52	51	50	39	79	74
78	Dimethylsulfoxide	63	78	45	61	46	62	48	47
79	Pyridine	79	52	51	50	78	39	53	80
80	2-Chloroethanol	31	27	43	29	49	51	80	26
80	Ammonium Nitrate	30	44	46	0	0	0	0	0
81	1-Methylpyrrole	81	80	39	42	53	55	54	38
82	1,5-Hexadiene	67	41	54	39	27	53	81	68
82	1-Hexyne	67	41	43	27	39	54	40	53
82	2-Methylfuran	82	53	81	39	27	51	50	54
83	Tetrahydropyridine (Methyl Phenidate Decomposition Prod)	55	83	82	41	54	39	28	42
83	Valeronitrile(n-)	43	41	54	27	55	39	29	40
84	Amitrol	26	84	31	57	43	42	58	40
84	Benzene-d6	84	56	54	42	52	82	85	40
84	Cyclohexane	56	41	84	54	55	42	69	0
84	Cyclohexane	56	84	41	55	42	69	39	27
84	Deuterobenzene	84	56	54	52	42	82	40	85
84	Dichloromethane	49	84	86	51	47	48	88	50
84	Methylene Chloride	49	84	86	51	47	88	48	50
84	Thiophene	84	58	45	39	57	69	83	50
85	Piperidine	84	85	56	57	44	42	43	41
86	2-Methylbutanal	57	29	41	58	27	39	86	43
86	2-Methylpentane	43	42	71	41	29	57	39	0
86	2-Pentanone	43	86	41	58	27	71	39	42
86	2-Pentanone	43	27	29	57	86	41	39	58
86	3-Methyl-2-butanone	43	41	27	86	39	42	71	44
86	3-Methylbutanal	44	41	43	29	27	39	58	42
86	3-Methylpentane	57	56	41	0	0	0	0	0
86	3-Pentanone	57	29	86	27	56	58	26	43
86	Chlorodifluoromethane(Freon 22)	51	31	67	35	50	69	37	47
86	Hexane(n-)	57	43	41	29	27	56	42	39
86	Methyl Propyl Ketone	43	86	41	71	58	57	0	0
88	Amyl Alcohol	55	41	42	43	70	57	29	31
88	Ethyl Acetate	43	45	61	70	29	27	42	73
88	Pentanol(n-)	42	55	41	31	29	70	43	57
90	2-Ethoxyethanol	31	59	29	45	27	72	43	0
91	Ethyl Nitrate	46	29	76	30	0	0	0	0

MWT	COMPOUND NAME							
92	1-Chlorobutane	56	41	43	27	29	55	57	42
92	Methyl Fluoroacetate	59	33	61	41	44	92	31	93
92	Toluene	91	92	65	39	63	51	93	45
93	2-Methylpyridine	93	66	92	78	65	39	51	94
93	2-Picoline(2-Methylpyridine)	93	66	92	65	78	39	51	67
93	3-Methylpyridine	93	66	92	65	39	94	67	59
93	3-Picoline(3-Methylpyridine)	93	66	92	65	39	67	40	94
93	4-Methylpyridine	93	66	92	65	39	94	54	67
93	4-Methylpyridine(4-Picoline)	93	66	92	39	65	67	51	54
94	1,Chloro-3-hydroxypropane	58	57	30	36	27	26	76	29
94	Dichloroacetylene	94	96	44	47	59	35	24	98
94	Dimethyldisulphide	94	45	79	49	46	47	48	61
94	Methylpyrazine	94	67	26	39	40	53	38	42
94	Phenol	94	66	39	65	40	95	38	55
94	Phenol	94	65	66	95	55	51	63	47
95	2,4-Dimethylpyrrole	94	95	80	39	41	27	67	53
96	1,2-Dichloroethylene	61	96	98	63	26	60	62	100
96	2,5-Dimethylfuran	43	96	95	53	81	91	27	41
96	2,5-Dimethylfuran	96	95	43	53	81	27	51	50
96	Furfural	39	96	95	29	0	0	0	0
96	Furfural	96	95	39	38	29	37	97	67
98	1,1-Dichloroethane	63	27	65	83	26	85	61	98
98	1,2-Dichloroethane	62	49	64	63	51	61	0	0
98	1,2-Dichloroethane	62	27	49	64	63	98	51	61
98	1,2-Dichloroethane	27	62	49	64	63	98	51	61
98	3-Methylcyclopentanone	42	55	69	98	41	56	27	39
98	4-Methyl-3-pentene-2-one	55	83	43	98	29	27	39	53
98	Furfuryl Alcohol	98	41	81	42	39	97	53	69
99	Allyl Isothiocyanate	99	41	39	72	45	98	71	38
99	Allyl Isothiocyanate	41	99	39	72	45	38	27	40
99	Piperidone	30	99	28	42	43	41	27	39
100	1,1-Difluoro-1-chloroethane	65	45	85	64	44	31	61	81
100	2,2,2-Trifluoroethanol	31	29	33	61	51	69	50	32
100	2,2,3-Trimethylbutane	57	43	56	41	85	29	27	39
100	2,2-Dimethylpentane	57	43	41	56	85	29	27	39
100	2,3-Dimethyl-3-hydroxy-1-butene	59	43	86	85	40	41	38	42
100	2,3-Dimethylpentane	43	56	57	41	29	27	71	42
100	2,3-Pentanedione	43	29	57	27	100	26	42	44
100	2,4-Dimethylpentane	43	57	41	56	42	27	29	39
100	2,4-Dimethyltetrahydrofuran	85	41	55	56	43	42	57	70
100	2,4-Pentanedione(Acetylacetone)	43	85	100	27	29	42	41	0
100	2,5-Dihydrothiophene(2-oxy-)	100	72	39	71	55	45	46	101
100	2,5-Dimethyltetrahydrofuran	56	41	43	57	29	45	67	27
100	2-Ethyl Butyraldehyde	43	72	27	41	29	71	57	39
100	2-Hexanone	43	58	29	27	41	57	39	100
100	2-Methyl-3-tetrahydrofuranone	43	28	72	29	27	100	44	45
100	2-Methylcyclopentanol	57	41	44	82	67	71	56	39
100	2-Methylhexane	43	42	41	85	27	57	29	56
100	2-Methylpentan-1-al	43	58	41	29	27	71	39	57
100	2-Methylpentan-3-one	57	43	29	100	71	27	41	56
100	2-Methylpiperazine	44	28	85	30	42	56	57	43
100	2-Methylvaleraldehyde	43	58	41	29	27	71	57	39
100	3,3-Dimethylbutan-2-one	57	41	43	29	100	39	28	56
100	3,3-Dimethylpentane	43	71	27	41	29	70	85	39

MWT	COMPOUND NAME								
100	3,4-Dimethyltetrahydrofuran	55	70	42	100	41	29	39	43
100	3-Ethylpentane	43	71	70	29	27	41	55	39
100	3-Hexanone	43	57	29	27	71	41	100	39
100	3-Methyl-2-pentanone	43	29	57	41	72	27	56	39
100	3-Methylhexane	43	57	71	41	29	70	27	56
100	3-Methylpentan-2-one	43	57	29	41	72	56	27	100
100	3-Methylsydnone	100	42	28	41	70	45	40	30
100	3-Methyltetrahydropyran	42	68	55	45	100	41	70	39
100	4-Methyl-2-pentanone	43	58	57	41	85	100	29	39
100	4-Methylpentan-2-one	43	58	41	57	27	39	29	100
100	5-Methyl-1,2,3-thiadiazole	71	45	72	28	31	39	29	27
100	Allyl Acetate	43	58	41	39	57	29	27	0
100	Aminoheptane	44	30	41	42	100	43	45	55
100	Butyl Methyl Ketone(t-)	57	41	43	29	39	100	27	0
100	Butyl Vinyl Ether	41	29	56	57	28	43	44	27
100	Cyclohexanol	57	82	44	67	41	71	29	56
100	Cyclopentylmethanol	41	68	69	67	39	27	31	29
100	Dimethylpropiolactone(beta,beta-)	41	44	39	56	28	27	55	40
100	Ethyl Acrylate	55	27	29	56	45	28	73	26
100	Ethyl Isopropyl Ketone	57	43	29	27	41	28	100	71
100	Ethyl-2-methylallyl Ether	57	55	85	29	100	72	41	43
100	Glutaraldehyde	44	43	29	27	41	82	0	0
100	Heptane(n-)	43	41	57	71	29	56	42	27
100	Hexamethylene Oxide	42	55	41	100	39	68	29	27
100	Hexanal	44	56	43	41	29	72	82	67
100	Isobutyl Methyl Ketone	43	58	57	100	41	85	29	39
100	Isobutyl Vinyl Ether	57	41	56	29	44	27	100	39
100	Isopropenyl Acetate	43	58	41	39	42	27	0	0
100	Methyl Crotonate	69	41	39	85	100	28	59	0
100	Methyl Methacrylate	41	40	39	69	29	38	59	100
100	Methyl Methacrylate	41	69	39	100	40	59	99	0
100	Methylpiperazine(N-)	58	43	100	42	56	44	29	28
100	Nitrosopyrrolidine	41	100	42	43	30	69	39	68
100	Pentanolactone(delta-)	42	41	56	100	55	71	0	0
100	Pinacolone	57	41	29	43	100	39	56	0
100	Secbutyl Vinyl Ether	41	57	29	56	28	44	27	39
100	Senecioic Acid	100	83	39	55	82	85	29	27
100	Tetrafluoroethylene	81	31	100	50	69	28	82	29
100	Tetrahydropyran-3-one	42	100	71	41	70	55	39	45
100	Tetronic Acid	42	100	43	29	72	41	30	69
100	Tiglic Acid(2-Methyl-but-2-enoic Acid)	55	100	27	29	39	54	85	53
100	Valerolactone(gamma-)	28	56	29	41	43	85	39	57
100	Vinyl Propionate	29	57	27	43	26	44	28	0
101	2-Aminohexane	44	58	41	86	42	45	43	0
101	3,4-Dihydroxyproline	102	87	69	41	59	0	0	0
101	3-Hydroxypiperidine	44	30	28	57	29	56	43	101
101	Butylamine(N-ethyl(n-))	58	30	29	27	28	44	101	41
101	Diacetimide	42	59	73	42	44	101	41	0
101	Dimethylaminoacetone(N,N-)	58	42	30	43	57	44	0	0
101	Dimethylbutylamine	58	101	44	42	59	29	30	0
101	Dipropylamine(n-)	72	30	43	101	41	27	58	28
101	Ethyl Butylamine(n-)	58	30	101	44	29	28	41	27
101	Hexylamine(n-)	44	30	58	41	86	42	27	0
101	Methyl Acrylamide(N-)	29	30	28	27	44	71	55	26

MWT	COMPOUND NAME								
101	Methyl Amylamine(n-)	44	101	58	30	28	41	42	43
101	Methyl Isoamylamine	44	101	28	43	30	41	43	45
101	Methylethane(N-(2'-methylethyl)-1-amino-2-)	44	86	58	28	42	41	43	27
101	Methylisobutyramide(N-)	58	43	101	73	27	41	71	86
101	Methylmorpholine (N-)	43	42	101	71	27	44	29	0
101	Propyl Isothiocyanate	101	43	72	41	27	42	39	45
101	Propyl Thiocyanate	43	41	27	101	39	42	45	28
101	Propylaminopropane(n-)	30	72	44	43	27	28	41	86
101	Triethylamine	86	30	27	58	29	28	42	101
101	Trimethylacetamide	57	41	29	101	44	36	58	27
101	Valeramide	59	44	41	29	72	27	43	28
102	1,1,1,2-Tetrafluoroethane	33	69	83	32	31	51	63	82
102	1,1-Dimethylbutanol	59	45	87	43	41	31	73	55
102	1,5-Diaminopentane	30	56	28	85	43	45	41	27
102	1-Hexanol	56	43	55	42	41	69	31	29
102	1-Methylpropyl Formate	43	56	41	29	31	27	60	39
102	2,2-Dimethylbutan-2-ol	59	87	41	69	43	31	45	39
102	2,2-Dimethylbutan-3-ol	57	45	56	41	87	69	29	43
102	2,2-Dimethylbutanol	43	71	70	55	41	73	45	29
102	2,2-Dimethylpropanoic Acid	57	41	29	39	27	45	59	56
102	2-Ethyl-1,3-dioxolane	73	45	28	27	29	43	57	0
102	2-Ethyl-1-butanol	43	70	71	55	41	29	56	84
102	2-Hexanol	45	69	41	43	44	87	27	29
102	2-Hydroxy-2-methyl-3-butanone	59	31	43	41	29	39	42	0
102	2-Hydroxy-2-methyltetrahydrofuran	71	43	41	27	42	31	44	39
102	2-Hydroxytetrahydropyran	85	55	41	29	56	28	84	39
102	2-Methoxyethyl Vinyl Ether	45	58	59	29	31	43	27	0
102	2-Methoxytetrahydrofuran	71	41	42	43	61	72	101	102
102	2-Methyl-1,3-dioxane	43	87	31	45	59	28	29	44
102	2-Methyl-1,4-dioxane	43	28	58	102	29	42	45	44
102	2-Methyl-1-pentanol	43	71	70	41	55	69	56	84
102	2-Methyl-2-pentanol	59	45	87	43	41	31	73	27
102	2-Methylbutanoic Acid	74	29	57	41	27	28	87	45
102	2-Methyltetrahydrothiophene	87	102	45	41	59	39	60	74
102	3,3-Dimethylbutan-1-ol	57	69	41	56	43	29	45	31
102	3-Dimethylaminopropylamine	58	30	42	85	28	44	56	57
102	3-Ethoxypropionaldehyde	31	29	45	27	28	58	43	74
102	3-Hexanol	59	55	73	31	43	41	27	29
102	3-Hydroxy-3-methyl-2-butanone	59	31	43	41	71	87	39	60
102	3-Methyl-1-pentanol	56	69	55	41	43	29	57	31
102	3-Methyl-3-Pentanol	73	55	43	45	87	41	69	29
102	3-Methyltetrahydrothiophene	60	102	41	45	74	87	56	39
102	4-Chloropyrazole	102	104	75	48	77	50	38	47
102	4-Hydroxy-3-methyl-2-butanone	43	61	42	31	41	57	84	29
102	4-Methoxy-2-buten-1-ol	41	71	45	29	39	27	70	0
102	4-Methyl-1,3-dioxane	43	31	55	72	101	28	45	32
102	4-Methyl-1,3-dioxane	43	31	55	72	101	28	45	32
102	4-Methyl-2-pentanol	45	43	69	41	44	87	57	56
102	Acetylurea(N-)	43	28	44	59	42	102	41	74
102	Butyl Formate(n-)	56	41	31	43	29	27	55	28
102	Cadaverine	56	55	41	43	45	85	42	84
102	Cadaverine	30	56	28	85	43	45	41	27
102	Cycloserine*	43	29	28	59	30	42	74	102
102	Di-isopropyl Ether	45	43	87	41	59	27	69	39

MWT	COMPOUND NAME								
102	Dichlorofluoromethane	67	69	35	47	31	48	83	49
102	Dichloromonofluoromethane	67	69	35	47	31	32	48	83
102	Diethyl Nitrosamine	102	44	42	29	56	57	27	30
102	Dimethylaminomethane(bis-)	58	28	42	30	102	45	59	0
102	Dioxepane	71	42	41	102	43	31	55	44
102	Dipropyl Ether	43	41	73	29	39	102	31	0
102	Ethyl Butyl Ether	59	31	29	57	41	56	27	73
102	Ethyl Isobutyl Ether	59	31	29	27	41	44	39	57
102	Ethyl Propionate	57	29	75	27	101	74	28	45
102	Fluorocyclohexane	67	41	82	54	56	59	55	39
102	Guanylurea	86	43	102	69	44	42	59	0
102	Isobutyl Formate	43	56	41	31	29	27	60	39
102	Isopropyl Acetate	43	61	87	41	59	42	27	39
102	Isopropyl Ether	45	43	87	59	41	27	39	69
102	Isovaleric Acid	60	43	41	27	45	29	39	74
102	Methyl Butyrate(n-)	74	43	71	59	41	87	27	0
102	Methyl Isobutanoate	43	71	41	39	27	87	102	0
102	Methyl Isopropyl Nitrosamine	102	43	42	57	56	41	40	39
102	Methyl Pyruvate	43	102	42	28	29	59	44	41
102	Methylene Oxalate	30	44	29	28	102	56	0	0
102	Pentanoic Acid	60	73	27	29	41	43	45	28
102	Pivalic Acid	57	41	29	39	56	27	45	87
102	Propyl Acetate(n-)	43	61	73	42	41	59	27	0
102	Propyl Ether(n-)	43	73	41	102	27	31	57	42
102	Tetrahydrofurfuryl Alcohol	71	43	41	27	31	42	29	0
102	Trimethylurea	44	43	42	45	72	30	102	58
102	Valeric Acid	60	43	41	87	45	27	74	39
103	Benzonitrile	103	76	50	104	51	75	77	52
104	1-Chloro-3-methyl-2-butene	68	67	53	41	39	40	27	69
104	Chlorotrifluoromethane	69	85	50	87	35	31	37	33
104	Styrene	104	103	78	51	77	50	105	52
105	4-Vinylpyridine	105	104	78	52	51	77	50	79
105	Propyl Nitrate	46	29	76	30	0	0	0	0
106	1,4-Dicyano-1-butene	66	39	106	79	52	53	51	40
106	Benzaldehyde	77	106	105	51	50	78	52	39
106	Ethylbenzene	91	106	51	65	77	92	39	78
106	Xylene (meta)	91	106	105	77	51	92	39	79
106	Xylene (ortho)	91	106	105	77	51	92	39	78
106	Xylene (para)	91	106	105	77	51	92	79	103
107	2,3-Dimethylpyridine	107	106	66	39	65	92	79	108
107	2,4-Dimethylpyridine(2,4-Lutidine)	107	106	79	92	39	65	80	77
107	2,5-Dimethylpyridine	107	106	79	77	92	39	65	108
107	2,6-Dimethylpyridine	107	106	66	92	65	39	108	79
107	2-Ethylpyridine	106	107	79	78	52	51	65	80
107	3,4-Dimethylpyridine	107	106	79	92	77	39	108	65
107	3,5-Dimethylpyridine	107	106	79	92	77	39	108	80
107	3-Ethylpyridine	107	92	106	65	79	39	108	77
107	4-Ethylpyridine	107	106	92	65	79	39	51	108
107	Toluidine(m-)	107	106	77	79	108	39	65	78
107	Toluidine(p-)	106	107	77	108	79	78	53	39
108	2,3-Dimethylpyrazine	108	67	42	40	41	26	39	109
108	2,5-Dimethylpyrazine	108	42	39	40	81	38	52	41
108	2,6-Dimethylpyrazine	42	108	39	40	38	27	37	41
108	Adiponitrile(1,4-Dicyanobutane)	41	68	54	55	39	27	40	29

MWT	COMPOUND NAME								
108	Anisole(Methyl Phenyl Ether)	108	78	65	39	77	93	79	51
108	Benzyl Alcohol	79	77	108	101	51	50	39	40
108	Benzyl Alcohol	79	108	107	77	51	91	50	78
108	Cresol(m-)	108	107	79	77	39	90	109	51
108	Cresol(o-)	108	107	77	79	90	39	51	109
108	Cresol(p-)	108	107	77	79	109	90	39	53
108	Cresol(p-)	107	108	77	51	79	39	53	50
108	Phenylenediamine(ortho-)	108	80	81	107	28	0	0	0
109	2-Aminophenol	109	80	53	52	39	64	63	81
109	3-Aminophenol	10⌐	80	81	110	41	39	53	74
109	4-Aminophenol	109	80	53	81	108	52	54	110
109	Nicotinyl Alcohol	109	108	80	57	43	55	53	71
109	Nicotinyl Alcohol*	109	108	80	53	51	39	91	27
110	2-Acetylfuran	95	110	43	39	96	68	0	0
110	Catechol	110	63	64	81	53	55	92	111
110	Catechol	110	64	63	81	39	92	55	111
110	Resorcinol	110	82	39	81	53	69	55	111
111	Ametazole	82	36	81	38	55	83	54	42
111	Ametazole	82	81	55	83	54	42	41	39
111	Ametazole*	82	30	81	83	55	54	42	27
111	Chlorodifluoroacetonitrile	76	85	50	31	92	87	77	57
111	Histamine (4-(omega-Aminoethyl)-1,3-diazole)	82	81	44	54	55	83	94	41
111	Histamine	82	30	81	54	28	55	83	41
112	1-Octene	43	41	55	56	42	70	29	39
112	Bromofluoromethane	112	114	33	93	95	91	81	79
112	Chlorobenzene	112	77	114	51	50	113	75	76
112	Maleinehydrazide	112	82	55	54	45	84	41	53
112	Uracil	112	69	42	40	41	68	39	70
114	2,2,4-Trimethylpentane	57	56	41	43	29	58	99	27
114	2,5-Hexanedione(Acetonyl Acetone)	43	99	71	57	114	27	42	72
114	2-Heptanone	43	58	71	41	59	27	29	85
114	3-Heptanone	57	29	85	72	41	43	27	114
114	4-Heptanone	43	71	27	41	39	42	29	58
114	5-Methyl-3-hexanone	57	29	27	85	41	43	114	39
114	Methimazole*	114	42	72	113	69	81	54	115
115	Methylhexaneamine	44	41	42	43	39	30	56	45
115	Tuaminoheptane	44	100	42	41	30	55	39	45
116	3-Hydroxy-1-nitrosopyrrolidine	42	44	116	41	30	57	56	68
116	Butyl Acetate(n-)	43	56	73	41	61	55	27	29
116	Chlorotrifluoroethylene	116	31	66	85	118	97	47	81
116	Diacetone Alcohol	43	59	58	101	29	0	0	0
116	Ethyl Butyrate	71	43	88	60	41	73	45	89
116	Indene	116	115	58	63	89	117	39	62
116	Laevulinic Acid	43	56	45	55	73	99	39	44
117	Amyl Nitrite	70	43	55	41	57	42	71	0
117	Amyl Nitrite	41	57	29	60	43	30	71	39
117	Benzyl Cyanide	117	90	116	89	63	51	50	39
117	Indole	117	90	89	118	116	63	59	39
117	Phenylacetonitrile	117	90	116	89	51	63	39	118
117	Phenylacetonitrile	117	116	90	89	51	77	118	63
118	Chlorocyclohexane	67	82	54	41	55	83	39	36
118	Chloroform	83	85	47	87	48	49	35	82
118	Methylstyrene(alpha-)	118	117	103	78	77	115	51	91
118	Methylstyrene(p-)	117	118	115	91	58	39	103	65

MWT	COMPOUND NAME								
118	Succinic Acid	100	55	45	74	119	73	56	41
119	Butyl Nitrate	43	46	76	41	29	57	30	0
120	1-Chlorohexane	91	55	43	56	41	42	29	93
120	Dichlorodifluoromethane(Freon 12)	85	87	50	101	103	31	35	66
120	Isopropylbenzene	105	120	40	29	44	77	79	51
120	Mesitylene(1,3,5-Trimethylbenzene)	105	120	119	77	91	39	106	79
120	Phenylacetaldehyde	91	92	120	65	63	51	39	89
121	2,4,6-Trimethylpyridine	121	120	79	106	122	77	107	39
121	2,4-Dimethylaniline	121	120	106	77	91	122	93	118
121	2-Ethylaniline	106	121	77	107	79	53	120	78
121	2-Phenylethylamine	30	91	39	65	92	51	63	121
121	Cysteine(L-)*	74	76	28	75	59	43	42	47
121	Ethylaniline(N-)	106	121	77	107	120	104	51	79
121	Methylbenzylamine (N-Methyl)	120	91	44	121	42	65	51	77
121	Methyltoluidine(N-Me,p-)	120	121	91	65	77	119	122	118
121	Phenethylamine(beta-)	30	91	121	92	51	0	0	0
122	2,3,5-Trimethylpyrazine	42	122	39	27	81	54	53	52
122	2,3-Xylenol	122	107	121	77	91	79	39	78
122	2,4-Dimethylphenol	107	122	121	77	91	79	51	39
122	2,4-Xylenol	122	107	121	77	91	123	39	79
122	2,5-Xylenol	122	107	121	77	91	79	123	39
122	2,6-Xylenol	122	107	121	77	91	79	78	103
122	2-Bromopropane	43	41	27	42	124	122	45	44
122	2-Methyl-5-ethylpyrazine	121	122	39	56	27	42	94	54
122	3,4-Xylenol	122	107	121	77	91	123	39	108
122	3,5-Xylenol	122	107	121	77	79	91	123	39
122	4-Hydroxybenzaldehyde	121	122	123	65	39	93	63	66
122	Benzoic Acid	105	122	77	51	50	39	74	78
122	Nicotinamide	122	104	78	106	51	0	0	0
122	Nicotinamide*	122	78	106	51	50	52	44	123
123	1-Butylpyrrole	81	80	123	39	53	41	67	40
123	4-Nitrosophenol	123	107	80	78	95	79	93	81
123	Anisidine(o-)	123	108	80	53	65	124	109	52
123	Isonicotinic Acid	123	51	78	106	50	52	105	39
123	Methylaminophenol (para)	123	122	94	108	81	29	60	43
123	Nicotinic Acid*	123	105	78	51	106	77	124	50
123	Nitrobenzene	77	51	123	50	65	93	30	78
123	Nitrobenzene	77	123	51	65	78	93	50	30
123	Picolinic Acid	79	52	51	78	50	39	45	53
124	1,4-Dichloro-2-butene	75	53	89	77	27	39	62	54
124	4-Methylcatechol	124	123	78	77	51	39	107	110
124	Orcinol	124	123	39	95	51	67	55	69
126	1,3-Dichlorobutane	55	27	63	90	41	39	29	54
126	2-Chlorotoluene	91	126	128	89	125	90	65	63
126	3-Chlorotoluene	91	126	128	125	65	127	92	89
126	4-Chlorotoluene	91	126	125	128	65	127	92	89
126	5-(Hydroxymethyl)-2-furfuraldehyde	97	41	126	39	69	53	51	125
126	Phloroglucinol	126	69	85	52	43	97	80	51
126	Phloroglucinol	126	69	85	52	43	63	42	41
126	Pyrogallol	126	52	80	51	108	39	79	53
126	Pyrogallol	126	80	52	108	51	63	79	39
126	Thymine	55	126	54	127	83	52	82	56
127	2-Chloroaniline	127	129	65	92	64	39	91	63
127	4-Chloroaniline	127	129	65	92	128	100	99	91

MWT	COMPOUND NAME								
127	Coniine	84	56	43	55	85	41	70	42
127	Coniine*	84	82	80	56	43	28	30	41
128	1,3-Dichloro-2-hydroxypropane	79	43	81	49	29	27	36	57
128	2,2-Dichloroethyl Methyl Ether	45	29	61	43	27	49	46	36
128	2-Chlorophenol	128	64	130	63	65	92	39	129
128	3-Chlorophenol	128	130	65	64	39	26	63	129
128	3-Octanone	43	29	57	27	72	71	99	41
128	4-Chlorophenol	128	130	65	39	64	63	129	99
128	4-Octanone	43	57	71	27	41	85	58	39
128	6-Methyl-3-heptanone	43	57	29	27	72	41	71	99
128	Barbituric Acid	42	128	85	43	44	41	69	70
128	Bromochloromethane	49	130	128	51	93	132	95	47
128	Dichloroacetic Acid	36	28	38	44	48	49	29	35
128	Isophthalonitrile	128	101	50	75	129	76	64	51
128	Methylenechlorobromide	49	130	128	51	93	81	79	95
128	Phthalodinitrile(o-)	128	101	50	75	129	76	51	64
129	Isoquinoline	129	102	128	51	130	76	75	103
129	Metformin*	43	44	85	86	42	129	68	30
129	Quinoline	129	102	128	51	130	76	50	103
130	2-Ethyl-1-hexanol	57	43	41	70	55	56	83	29
130	Amyl Acetate	43	70	55	42	41	73	61	87
130	Difluorobromomethane	51	31	130	132	79	81	111	113
130	Phthalazine	130	76	103	50	102	75	129	52
130	Trichloroethylene	95	130	132	60	97	134	47	
130	Trichloroethylene	130	95	132	97	134	60	99	62
132	1,1,1-Trichloroethane	97	99	61	63	117	119	101	62
132	1,1,1-Trichloroethane	43	97	61	99	45	27	119	117
132	1,1,2-Trichloroethane	97	83	99	85	61	26	96	63
132	1-Methyl-4-isopropenylbenzene	132	117	133	92	91	115	118	59
132	2-Methyl-3-phenylaziridine	41	91	132	105	42	92	43	105
132	Paraldehyde	43	44	87	31	45	42	131	71
132	Paraldehyde	45	43	60	44	89	42	87	117
132	Trichloroethane	99	83	101	85	61	70	31	0
133	2-Benzylaziridine	91	42	132	117	43	41	104	57
133	Methylphenylaziridine	91	132	42	105	92	43	41	77
133	Pentyl Nitrate	29	41	57	46	76	43	27	28
133	Tranylcypromine	130	77	51	132	115	103	133	78
133	Tranylcypromine*	133	132	56	115	30	117	91	77
134	3-Chloroheptane	56	41	69	55	43	57	70	29
134	Benzyl Methyl Ketone	91	134	92	43	65	77	89	51
134	Phenylacetone	43	91	134	92	65	39	63	135
135	Acetanilide	93	135	94	66	43	65	39	0
135	Acetanilide	93	135	65	43	39	66	63	92
135	Adenine	135	108	54	53	81	43	136	66
135	Amphetamine	44	91	65	42	45	120	40	92
135	Amphetamine	44	103	91	65	42	104	45	102
135	Amphetamine	44	91	65	51	63	77	0	0
135	Amphetamine	44	91	65	58	39	42	51	63
135	Amphetamine*	44	91	40	42	65	45	42	39
136	1-Bromobutane	57	41	29	56	27	39	55	43
136	1-Methyl-4-isopropenylcyclohexene	68	93	67	39	41	53	136	79
136	Allopurinol	136	135	52	28	137	109	29	18
136	Betahistine	79	105	104	51	78	52	50	77
136	Camphene	93	121	79	67	107	95	94	68

MWT	COMPOUND NAME								
136	Car-3-ene	93	77	79	91	80	41	92	136
136	Car-3-ene	93	91	92	79	80	77	43	121
136	Dipentene	68	93	67	136	41	121	79	39
136	Fenchene(alpha-)	93	79	80	121	94	107	81	136
136	Fenchene(alpha-)	93	136	111	78	79	40	94	80
136	Limonene	68	93	67	39	79	53	27	94
136	Limonene	68	93	67	136	94	41	121	53
136	Limonene	68	67	93	94	79	92	121	107
136	Myrcene	93	41	69	77	91	53	67	94
136	Myrcene	41	69	93	39	67	79	27	91
136	Myrcene	41	93	69	79	91	77	67	53
136	Ocimene	93	92	91	79	77	80	121	105
136	Ocimene	93	41	27	39	79	80	77	43
136	Ocimene(allo-)	121	136	105	93	79	91	122	107
136	Ocimene-X(beta-)	93	92	91	79	77	80	121	105
136	Ocimene-Y(beta-)	93	80	92	91	79	121	105	77
136	Phellandrene(alpha-)	93	91	77	119	92	136	41	39
136	Phellandrene(beta-)	93	136	77	94	79	91	80	92
136	Phenelzine	105	91	104	77	133	92	51	65
136	Phenelzine	45	91	105	77	65	52	136	0
136	Phenelzine	45	91	64	105	147	30	77	104
136	Phenelzine*	91	147	45	64	105	65	77	48
136	Phenelzine*	31	45	46	29	27	59	74	43
136	Phenylacetic Acid	91	136	92	65	39	63	45	51
136	Pinene(alpha-)	93	92	91	77	41	79	121	39
136	Pinene(beta-)	93	41	69	79	91	77	94	80
136	Sabinene	93	77	91	41	79	39	136	94
136	Sabinene	93	41	77	91	79	27	39	69
136	Terpinene(alpha-)	121	93	136	119	106	91	79	77
136	Terpinene(gamma-)	93	136	121	92	91	43	77	79
136	Terpinene(gamma-)	93	91	77	136	121	39	43	27
136	Terpinolene	93	121	136	39	41	79	91	27
136	Trichlorofluoromethane(Freon 11)	101	103	66	105	47	35	31	82
137	2-Aminobenzoic Acid	119	137	92	65	39	64	120	91
137	2-Methyl-4-nitrosophenol	137	121	120	94	77	80	107	92
137	2-Nitrotoluene	120	65	91	92	39	77	121	89
137	3-Aminobenzoic Acid	137	92	120	65	39	52	66	138
137	3-Methyl-4-nitrosophenol	137	92	80	77	123	109	122	107
137	4-Aminobenzoic Acid*	137	120	92	65	39	138	121	63
137	Iproniazid Metabolite	78	106	51	137	79	0	0	0
137	Isoniazid	78	106	51	137	50	79	107	52
137	Isoniazid	78	106	51	104	77	137	50	0
137	Isoniazid*	106	78	51	137	50	79	107	31
137	Methyl Pyridine-3-carboxylate	106	137	78	136	79	105	107	138
137	Methylnicotinamide (N-)	78	106	135	136	51	50	79	52
137	Methylnicotinate*	78	106	51	137	50	136	107	79
137	Nitrotoluene (meta)	137	91	107	79	65	92	138	77
137	Nitrotoluene (meta)	91	65	137	41	57	39	30	0
137	Nitrotoluene (ortho)	120	92	91	65	121	77	93	137
137	Nitrotoluene (ortho)	65	57	41	91	120	29	43	39
137	Nitrotoluene (para)	91	65	137	107	39	63	77	79
137	Nitrotoluene(p-)	91	137	65	39	107	63	89	77
137	Salicylamide*	120	92	137	65	121	39	64	53
137	Tyramine	107	36	108	77	38	39	51	79

MWT	COMPOUND NAME								
137	Tyramine	30	108	107	77	137	43	41	55
137	Tyramine	30	108	107	77	137	51	109	78
138	1-Phenyl-1,2-dihydroxyethane	107	79	77	91	51	105	108	78
138	2-Nitroaniline	138	65	92	77	104	39	90	64
138	3-Nitroaniline	92	128	65	39	66	122	80	93
138	4-Hydroxybenzoic Acid	121	138	93	65	39	63	127	53
138	4-Nitroaniline	138	65	92	39	108	106	66	52
138	Hydroxyphenylethanol (beta-m-Hydroxy)	107	108	77	138	51	53	79	78
138	Leptazol	55	82	41	39	54	42	56	109
138	Pentylenetetrazol	55	41	39	82	54	53	109	42
138	Salicylic Acid	92	120	138	64	39	63	65	53
138	Salicylic acid*	120	92	138	64	39	63	121	65
139	Nitrophenol(para)	139	65	39	109	93	53	81	63
140	1,5-Dichloropentane	55	42	41	68	27	29	69	39
140	Hexamine*	42	140	112	41	85	43	71	141
140	Methenamine	42	140	41	85	112	44	43	40
140	Trimethyl Phosphate	110	109	79	95	80	140	47	29
141	Chlormethiazole Metab (5-Acetyl-4-methylthiazole)	126	141	43	45	71	98	69	72
141	Chlormethiazole Metabolite	112	85	45	141	113	59	71	69
141	Cyclopentamine	58	36	41	59	126	38	44	56
141	Cyclopentamine	58	59	45	41	40	74	42	59
141	Cyclopentamine*	58	41	30	126	59	69	56	44
141	Ethosuximide	113	55	70	42	41	39	85	69
141	Ethosuximide(i)*	55	70	113	42	41	39	44	69
141	Ethosuximide(ii)*	113	70	55	42	41	39	85	69
141	Methamidophos	94	95	141	64	47	46	79	45
141	Methylaminomethylheptane	58	41	55	44	43	59	128	56
141	Methylpentynol Carbamate*	69	83	43	79	41	81	53	80
142	2-Nonanone	43	58	41	71	59	57	27	29
142	Alloxan	43	44	142	86	42	114	71	70
142	Bis(2-chloroethyl) Ether	93	63	27	95	65	31	94	36
142	Dalapon	62	97	99	64	61	36	107	45
143	2-Hydroxy-4-methyl-5-ethylthiazole	128	100	45	73	43	143	0	0
143	2-Methylquinoline	143	128	115	144	142	101	51	77
143	3-Methylisoquinoline	143	115	142	116	144	89	72	59
143	4-Methyl-5-b-hydroxyethiazole	112	143	113	85	45	0	0	0
143	4-Methylquinoline	143	115	142	144	116	89	77	39
143	6-Methylquinoline	143	142	115	144	141	89	116	63
143	7-Methylquinoline	143	142	115	144	141	116	89	39
143	8-Methylquinoline	143	142	115	144	141	89	116	39
143	Chlormethiazole Metabolite (5-(1-hydroxyethyl)-4-methylthiazole)	128	45	100	73	43	71	97	125
143	Trimethadione	43	143	58	42	41	39	128	56
144	1-Napthanol	144	115	116	145	72	89	63	58
144	Ethchlorvynol (Placidyl)	115	117	89	53	109	51	91	39
144	Ethephon (tech)	82	77	105	51	81	50	122	109
144	Thiobarbituric Acid	144	43	42	116	69	59	145	41
144	Valproic Acid	73	102	57	41	43	55	101	115
145	3-Indolealdehyde	144	145	116	89	63	146	90	58
145	Emylcamate	73	43	84	55	69	41	44	85
145	Heptaminol*	44	43	59	56	69	55	41	113
145	Hydroxyquinoline*	145	117	122	89	105	90	63	146
146	1,2-Dichlorobenzene	146	148	111	75	113	74	50	150
146	1,3-Dichlorobenzene	146	148	111	75	50	113	150	147
146	1,4-Dichlorobenzene	146	148	111	75	50	150	113	147

MWT	COMPOUND NAME				PEAKS				
146	Chloral	82	84	111	83	29	47	113	85
146	Fluorochlorobromomethane	67	69	31	111	113	79	48	47
147	1,1,2-Trichlorotrifluoroethane	101	151	103	153	85	87	31	105
148	1-Hydroxy-1,2,3,4,-tetrahydronaphthalene	130	120	91	119	129	115	147	148
148	2,2,2-Trichloroethanol	31	49	77	113	82	115	51	117
148	Bromotrifluoromethane	69	148	150	129	131	79	81	50
148	Ethyl Phenylmethyl Ketone (ortho)	100	133	77	52	148	50	76	0
149	1-Ethyl-2-phenylethylamine	58	41	91	31	120	65	29	42
149	2,4-Dimethylformanilide	120	149	106	121	77	91	132	150
149	2-Amino-1-phenylpropanone	44	40	51	77	42	105	50	45
149	Benzyl Methyl Ketoxime	91	116	131	65	149	92	150	90
149	Benzyl Methyl Ketoxime	91	75	92	65	42	39	40	116
149	Benzyl Methyl Ketoxime	91	41	92	65	39	42	40	116
149	Dimethyl-2-phenylethylamine (N,N-)	58	42	91	31	59	65	39	51
149	Dimethylaminobenzaldehyde(p-)	148	149	77	42	51	132	105	50
149	Methamphetamine	58	91	56	65	51	77	134	0
149	Methamphetamine	58	91	56	65	42	39	51	41
149	Methylamphetamine*	58	91	59	134	65	56	42	57
149	Penicillamine*	75	41	57	70	43	59	56	47
149	Phentermine	58	91	59	44	42	134	65	57
149	Phentermine	58	42	91	41	30	39	65	59
149	Phentermine*	58	91	42	41	134	65	59	40
149	Phenylpropylmethylamine	77	44	79	51	57	105	107	91
149	Propylmethyldithiocarbamate (S-n-propyl-)	149	107	30	74	73	60	0	0
150	2-Hydroxymethyl-4-hydroxybenzolactone	121	150	65	93	0	0	0	0
150	2-Hydroxymethyl-5-hydroxybenzolactone	121	150	93	65	0	0	0	0
150	Butylphenol(p-t-)	135	107	41	150	95	136	77	91
150	Carvone	82	54	108	93	58	107	106	39
150	Methyl Phenyl Acetate (Methyl Phenidate Decomposition Prod)	91	150	65	92	89	59	0	0
150	Phenylpropionic Acid	91	104	150	105	78	77	65	51
150	Piperonal	149	150	63	121	65	62	91	39
150	Tartaric Acid	76	29	58	31	44	59	70	105
150	Thymol	135	150	91	107	117	0	0	0
151	2-Nitroxylene(m-)	134	79	77	106	105	103	151	78
151	3,4-Methylenedioxybenzylamine	150	151	121	65	93	77	135	0
151	4-Nitroxylene(o-)	151	105	77	79	103	121	78	39
151	Acetaminophen	109	80	151	108	43	81	53	52
151	Amantadine	94	41	39	42	77	151	95	93
151	Amantadine*	94	151	57	95	40	41	58	108
151	Hydroxyacetanilide(p-)	109	151	43	80	81	0	0	0
151	Hydroxyamphetamine	72	44	57	43	149	41	55	69
151	Hydroxyamphetamine	44	107	77	43	78	51	42	108
151	Methylaminobenzoate (ortho)	119	151	120	92	152	91	93	65
151	Methylaminobenzoate (para)	120	151	92	121	152	65	93	122
151	Norephedrine	44	77	79	51	45	42	78	105
151	Norpseudoephedrine	44	77	79	51	42	45	43	105
151	Norpseudoephedrine	44	77	105	106	51	91	79	132
151	Norpseudoephedrine	44	57	43	41	55	69	77	79
151	Norpseudoephedrine	44	77	79	51	105	0	0	0
151	Paracetamol	109	151	43	79	80	53	108	110
151	Paracetamol*	109	151	43	80	108	81	53	52
151	Phenylpropanolamine	44	36	38	77	43	51	79	132
151	Phenylpropanolamine	44	77	79	105	91	51	92	132
151	Phenylpropanolamine*	44	77	79	51	45	42	107	105

MWT	COMPOUND NAME								
152	1,3,3-Trimethyl-2-norboranone	81	69	41	152	80	82	109	39
152	2,5-Dimethoxytoluene	137	152	109	77	78	39	138	15
152	2-Methyl-4-nitroaniline	152	106	77	79	122	59	0	0
152	4-Methyl-2-nitroaniline	152	106	77	79	91	104	78	105
152	Bornanone-2	95	81	108	152	41	83	109	69
152	Bornanone-3	69	81	152	41	109	95	80	68
152	Camphor	95	81	41	69	55	83	67	137
152	Camphor	95	81	69	41	55	67	108	109
152	Camphor	95	41	81	69	83	108	109	67
152	Camphor	95	81	108	55	109	152	67	83
152	Camphor	95	81	108	152	69	109	83	110
152	Carbon Tetrachloride	117	119	121	82	47	84	35	49
152	Carbon Tetrachloride	117	119	35	47	121	82	84	49
152	Carveol(cis-)	118	133	93	88	91	42	27	0
152	Citral	41	69	39	84	94	27	29	67
152	Hydroxyphenylacetic Acid(meta)	107	152	77	39	108	51	79	53
152	Hydroxyphenylacetic Acid(ortho)	78	134	106	51	77	39	40	107
152	Hydroxyphenylacetic Acid(para)	107	77	51	39	53	78	50	52
152	Methyl Hydroxybenzoate	121	152	40	93	65	0	0	0
152	Methyl Salicylate	120	92	152	121	65	64	93	63
152	Methyl Salicylate	92	120	39	65	152	121	63	64
152	Methylnitroaniline(N-Me,o-)	77	152	79	105	78	104	118	51
152	Methylparaben	121	152	93	65	39	0	0	0
152	Pemoline Metabolite 2	107	79	77	51	152	105	50	78
152	Phenoxyacetic Acid	107	92	152	121	65	0	0	0
152	Piperitone	82	95	137	54	109	152	41	110
152	Sabinol	92	91	42	81	41	43	109	55
152	Tetrachloromethane	117	119	47	121	82	84	49	35
152	Vanillin	152	151	81	109	51	0	0	0
152	Verbenol(trans-(-))	93	43	41	109	94	91	39	55
152	Xanthine	152	54	109	53	81	0	0	0
153	2-Methoxy-4-nitrosophenol	139	99	124	153	78	79	95	107
153	4-(2-Aminoethyl)pyrocatechol	124	30	123	153	77	51	125	78
153	Aminosalicylic Acid(p-)	79	135	153	107	52	53	80	51
153	Aminosalicylic Acid*	109	80	44	81	53	39	52	54
154	1,4-Dihydroxybenzoic Acid	110	136	44	108	154	81	69	80
154	1,6-Dichlorohexane	55	41	69	56	82	42	67	43
154	1-Terpin-4-ol	92	70	43	41	109	90	39	76
154	2,2-Diethylcyclohexanone	55	41	126	68	97	110	81	43
154	2,4-Dihydroxybenzoic Acid*	136	154	108	80	52	95	137	69
154	2,6-Diethylcyclohexanone	55	41	83	42	126	82	56	97
154	2-Terpineol	59	93	121	43	136	28	81	41
154	3,5-Dihydroxybenzoic Acid	154	137	109	69	81	53	155	51
154	Caran-trans-2-ol((-)cis-)	93	121	136	111	43	41	55	79
154	Caran-trans-2-ol(trans-)	111	41	43	93	55	121	69	95
154	Caran-trans-3-ol(cis-)	93	121	136	81	43	55	41	107
154	Chloropentafluoroethane	85	119	69	31	87	135	29	50
154	Citronellal	41	69	29	95	55	39	43	27
154	Citronellal	69	41	55	95	67	56	81	71
154	Geraniol	69	41	39	65	27	29	53	154
154	Geraniol	69	41	68	67	93	55	39	0
154	Isoborneol	95	110	41	55	96	69	139	136
154	Isoborneol	95	41	27	43	39	93	55	29
154	Isoborneol	95	41	43	110	93	55	136	67

MWT	COMPOUND NAME								
154	Isofenchol(endo-)	139	95	109	41	121	43	55	136
154	Isopulegol	69	81	67	55	71	68	56	41
154	Lavandulol	69	111	41	68	93	123	67	81
154	Linalool	71	93	47	43	55	69	80	154
154	Linalool	71	43	41	93	55	69	80	67
154	Linalool	93	71	41	69	55	43	80	121
154	Menthone(o-)	97	111	83	112	55	41	69	56
154	Nebol	69	41	39	93	68	29	27	154
154	Neoisothujyl Alcohol	93	121	136	95	43	81	110	107
154	Nerol	69	41	93	68	80	121	67	111
154	Nerol	41	69	68	93	39	29	67	43
154	Terpineol	71	93	111	43	86	69	55	68
154	Terpineol	59	121	93	136	43	81	41	55
154	Terpineol(alpha-)	58	120	72	70	136	18	42	0
154	Terpineol(alpha-)	59	93	121	81	43	136	68	92
154	Thujyl Alcohol	95	93	43	121	81	55	110	70
155	2-Phenylpyridine	155	154	156	77	127	128	51	78
155	Arecoline	155	96	140	43	42	81	94	53
155	Bemegride	55	83	82	113	70	69	97	155
155	Bemegride	55	82	39	83	41	113	70	127
155	Bemegride*	55	83	82	41	113	70	29	69
155	Methylethosuximide (N-)	127	55	70	41	42	128	69	112
155	Methylsuximide(N-)	127	55	70	41	42	0	0	0
155	Propylhexedrine	58	140	67	72	83	81	155	156
155	Propylhexedrine	58	40	38	57	55	140	41	56
156	4-Chloro-3,5-xylenol	156	121	158	91	77	65	157	122
156	4-Chlorobenzoic Acid*	139	156	111	75	141	50	158	113
156	Citronellol	41	69	55	67	81	82	68	43
156	Ectylurea	41	43	113	96	44	98	39	69
156	Mechlorethamine	106	42	108	44	63	43	49	41
156	Menthol	81	95	71	41	67	55	138	123
156	Menthol	71	81	95	41	55	82	43	69
156	Neomenthol	71	95	81	41	55	43	69	82
157	1-Chloro-4-nitrobenzene	111	75	157	113	50	127	99	159
157	2,4-Dimethylquinoline	157	156	115	158	142	116	128	77
157	2,6-Dimethylquinoline	157	156	158	115	142	128	89	154
157	4-Methyl-5-thiazoleacetic Acid	112	157	85	45	158	0	0	0
157	Buformin*	43	114	85	30	101	86	44	72
157	Chlormequat	50	58	52	36	42	44	38	49
157	Oxanamide (Quinactin)	36	57	41	40	44	38	55	39
157	Paramethadione	43	129	57	56	41	72	39	58
158	1-Phenylcyclohexene	129	158	130	143	115	128	91	77
158	Ethyl Heptanoate	88	43	60	113	101	61	73	70
158	Ethylbutyrylurea(Carbromal Metabolite)	45	130	44	71	42	61	115	55
158	Methyl Octanoate	74	87	43	57	41	55	59	127
158	Methyl dipropylacetate	87	57	116	55	115	129	88	127
158	Phenylcyclohexene	153	115	129	91	143	0	0	0
159	2,3,5-Trimethylindole	158	159	144	115	157	160	156	143
159	3-Indoleacetaldehyde	130	159	77	64	131	103	48	51
159	Pargyline	91	82	42	39	65	159	158	68
159	Pargyline	82	91	68	42	92	159	158	65
159	Pargyline*	82	68	91	159	42	158	92	65
160	Aminozide	59	60	118	100	44	45	42	101
160	Hydrallazine	50	130	76	74	75	63	51	38

MWT	COMPOUND NAME								
160	Hydrallazine*	160	103	89	131	115	76	161	104
160	Tolazoline	91	159	65	160	131	39	51	81
160	Tryptamine	130	131	36	77	108	65	160	38
161	2,6-Dichloroaniline	161	163	90	99	63	165	162	126
161	3,4-Dichloroaniline	161	163	99	90	165	126	134	73
161	Aletamine	70	120	43	91	41	65	77	98
161	Aletamine	71	120	43	91	41	0	0	0
161	Chlormethiazole	112	45	85	161	163	0	0	0
161	Chlormethiazole*	112	161	85	45	163	113	114	59
161	Cypenamine	56	161	57	43	144	117	91	129
161	Tryptophol	130	161	143	131	103	77	115	102
162	2,4-Dichlorophenol	162	63	164	98	99	49	62	73
162	2,4-Dichlorophenol	162	164	63	98	73	166	100	99
162	2,4-Dimethylphenyl-N'-methylformamidine (N-2,4-Dimethyl-)	162	132	121	120	106	118	147	77
162	2,6-Dichlorophenol	162	164	89	126	166	73	99	
162	Bromocyclohexane	83	55	41	67	54	39	82	27
162	Bromodichloromethane	83	85	129	47	127	87	48	81
162	Dazomet	42	44	43	57	89	72	41	45
162	Hexamethyldisiloxane*	147	148	66	73	149	45	59	131
162	Methyl Cinnamate	131	162	79	103	161	52	132	163
162	Nicotine	84	42	133	162	161	51	119	65
162	Nicotine	84	133	42	162	161	39	51	41
162	Nicotine	84	44	133	42	162	161	44	85
162	Nicotine	84	133	76	42	162	161	58	119
162	Nicotine*	84	133	42	162	161	105	77	119
162	Phenylbutyrolactone (Primidone, Phenobarb, Glutethimide Metab)	117	118	162	91	77	78	103	105
162	Trichloroacetic Acid	44	83	85	36	28	0	0	0
163	1-Ethyl-N-methyl-2-phenylethylamine	72	134	91	42	57	31	65	44
163	1-Isopropyl-2-phenylethylamine	72	120	55	91	65	39	77	31
163	1-Phenyl-2-nitropropene	115	91	105	116	51	39	77	63
163	1-Propyl-2-phenylethylamine	72	31	91	120	65	73	39	42
163	2-Ethyl-2-phenylmalondiamide (Primidone Metab)	148	163	91	103	120	115	117	77
163	Acetylcholine	58	43	57	149	71	42	41	55
163	Difluorochlorobromomethane	85	87	129	131	147	79	81	50
163	Dimethylamphetamine (N,N-)	72	44	42	91	56	65	58	162
163	Dimethylamphetamine	72	42	91	44	73	56	39	70
163	Ethylamphetamine	72	44	91	73	42	56	65	39
163	Formylamphetamine (N-Formyl)	72	44	118	91	65	117	42	39
163	Mephentermine	72	39	148	56	41	71	51	73
163	Mephentermine	72	91	73	56	148	41	57	42
163	Nitrostyrene	115	91	77	116	117	129	163	0
163	Nitrostyrene	115	91	105	116	40	39	51	77
163	Propylethyldithiocarbamate (S-n-propyl-)	88	60	163	43	41	148	121	107
164	1-Phenylethyl Acetate	104	43	91	39	51	103	78	27
164	1-Thiophenylcyclohexene	164	135	136	149	97	84	163	91
164	3-Bromohexane	43	85	55	41	42	56	84	29
164	Chloral Hydrate	82	84	83	111	47	29	85	113
164	Chloral Hydrate	81	83	46	110	112	0	0	0
164	Chloral Hydrate*	82	47	84	29	111	83	113	85
164	Droperidol Hofmann Reaction Product	123	95	164	69	0	0	0	0
164	Eugenol	164	149	131	137	103	77	133	165
164	Haloperidol Hofmann Reaction Product 1	123	95	164	69	0	0	0	0
164	Methyl beta-Phenylpropionate	104	164	91	105	133	103	165	78
164	Phenylethyl Acetate	104	43	91	105	65	39	51	77

MWT	COMPOUND NAME	PEAKS							
164	Tetrachloroethylene	166	164	129	131	168	94	96	47
165	2-Methoxyamphetamine	44	91	122	45	77	66	78	107
165	3-Methoxyamphetamine	44	45	122	91	77	78	121	42
165	Benzocaine	120	92	165	65	137	0	0	0
165	Benzocaine*	120	165	92	65	137	39	121	93
165	Ephedrine	56	77	91	105	106	42	0	0
165	Ephedrine	58	77	105	56	57	51	42	79
165	Ephedrine	58	105	77	42	106	56	57	51
165	Ephedrine	58	56	77	108	42	105	146	91
165	Ephedrine GC Decomposition	58	30	77	59	56	42	51	79
165	Ephedrine GC Decomposition	58	71	56	77	42	105	146	59
165	Ephedrine GC Decomposition	58	85	70	57	42	148	56	77
165	Ephedrine GC Decomposition	58	85	42	77	56	70	105	57
165	Ephedrine*	58	146	56	105	77	42	106	40
165	Ethenzamide*	120	92	105	148	150	121	133	65
165	Ethyl-4-aminobenzoate	120	65	92	39	165	137	63	41
165	Methoxyamphetamine (para)	44	122	121	42	78	77	91	107
165	Methoxyamphetamine	44	122	121	78	77	42	51	52
165	Paracetamol Methyl Ester	108	123	165	43	52	80	53	122
165	Pseudoephedrine	58	77	30	51	49	56	59	42
165	Pseudoephedrine	58	30	77	42	56	106	105	51
165	Pseudoephedrine*	58	77	59	56	51	42	105	91
166	1-Methylxanthine	166	54	109	53	81	136	137	0
166	2-Phenyllactic Acid	91	43	44	41	45	148	149	103
166	2-Phenyllactic Acid	43	121	77	105	51	103	78	48
166	3-(4-Hydroxyphenyl)propionic Acid	107	166	77	39	108	45	65	120
166	3-Methylxanthine	166	68	95	41	53	123	0	0
166	7-Methylxanthine	166	68	123	53	42	41	95	0
166	Ethionamide*	166	165	167	138	133	105	60	106
166	Methyl methoxybenzoate(meta)	166	135	107	167	136	77	108	59
166	Methyl methoxybenzoate(ortho)	135	166	133	77	105	137	134	136
166	Methyl methoxybenzoate(para)	135	166	136	167	77	107	92	137
166	Methylparaben Dimethyl Ether	135	66	77	107	92	136	63	64
166	Vanillal	137	81	138	109	166	0	0	0
167	2,4-Diethylamino-s-triazine	43	44	167	57	55	139	71	152
167	Cinchomeronic Acid	50	43	105	44	123	77	78	51
167	Ethinamate (Ethynylcyclohexyl Carbamate)	81	91	106	95	79	68	67	78
167	Ethinamate	91	43	67	81	78	79	106	44
167	Ethinamate*	91	81	106	78	39	95	68	43
167	Isocinchomeronic Acid	123	43	44	51	78	77	50	105
167	Mecamylamine	98	71	84	56	99	124	41	167
167	Mecamylamine*	98	84	71	56	41	42	99	124
167	Metaraminol	76	77	95	105	70	121	58	44
167	Metaraminol*	44	77	76	29	95	39	58	42
167	Orthocaine*	136	167	108	80	53	52	137	51
167	Phenylephrine	44	77	95	121	141	167	168	133
167	Phenylephrine	44	77	95	65	39	45	42	43
167	Quinolinic Acid	123	51	105	78	50	77	44	52
167	Synephrine	44	77	95	65	123	39	121	42
168	2-Hydroxymethyl-'-hydroxybenzoic Acid	121	150	65	93	168	0	0	0
168	2-Hydroxymethyl-5-hydroxybenzoic Acid	121	150	93	65	168	0	0	0
168	Norharman	168	140	169	64	63	167	114	141
168	Pridoxamine	151	94	168	106	150	122	123	0
168	Uric Acid	54	125	168	69	97	53	43	0

MWT	COMPOUND NAME								
168	Zoxazolamine	168	170	113	78	63	169	43	76
169	Bemegride Methyl Derivative	55	169	83	70	127	140	41	97
169	Chlorzoxazone	169	78	113	171	51	63	44	50
169	Chlorzoxazone	169	113	78	171	115	76	63	51
169	Chlorzoxazone*	169	78	171	113	115	63	51	170
169	Ethylethosuximide (N-)	141	70	55	41	42	69	112	126
169	Methylbemegride	55	58	70	83	69	73	97	127
169	Noradrenaline	31	139	93	153	151	77	137	123
169	Noradrenaline	30	65	93	139	39	111	53	51
169	Noradrenaline*	44	45	58	76	60	43	42	46
169	Pyridoxine	151	94	122	106	51	53	149	150
170	2-Hydroxybiphenyl	170	169	141	115	171	139	142	39
170	4-Benzylpyrimidine	169	170	91	115	142	65	116	143
170	4-Chlorobenzoic Acid Methyl Ester	139	111	141	75	170	50	113	172
170	4-Methyl-5-phenylpyrimidine	170	102	169	115	77	51	60	61
170	4-Methyl-5-phenylpyrimidine	170	169	102	115	116	171	51	76
170	4-Methyl-5-phenylpyrimidine*	40	43	170	44	169	102	69	91
170	Diphenyl Ether	170	77	51	141	169	171	142	115
170	Gallic Acid	170	153	39	51	79	125	53	126
170	Triazolophthalazine	170	115	114	88	171	62	116	89
170	Undec-10-en-1-al	43	57	41	55	82	56	44	29
171	Crimidine	104	71	147	42	44	70	56	102
171	Dichlobenil	171	100	173	75	74	136	99	50
171	Metronidazole*	81	124	54	53	125	171	45	42
171	Tetrazolophthalazine	115	114	88	62	116	63	89	51
172	Dibromomethane	93	174	95	59	172	176	74	43
172	Ethyl Caprylate	88	57	101	60	61	127	73	43
172	Methyl 3-Oxodipropylacetate	57	87	130	55	116	101	143	141
172	Methylene Bromide	174	93	95	172	176	91	81	79
172	Sulfanilamide	39	52	65	80	92	108	156	172
172	Sulphanilamide*	172	92	156	65	108	173	39	174
173	1-Nitronaphthalene	127	173	115	126	77	101	128	143
173	5-Nitrobarbituric Acid	44	43	42	30	70	54	96	141
173	Sulphanilic Acid	173	108	65	92	80	39	63	156
174	3-Indoleacetamide	130	174	77	131	103	102	129	51
174	6-Nitroquinoline	174	128	101	116	77	102	75	51
174	8-Nitroquinoline	174	128	116	101	77	89	102	75
174	Methyl 3-Hydroxydipropylacetate	87	116	55	113	57	145	59	143
174	Norantipyrine	174	77	91	105	132	0	0	0
175	3-(3-Indolyl)acetic Acid	130	45	131	159	77	62	43	103
175	Debrisoquine	132	104	78	130	77	103	43	48
175	Debrisoquine*	132	104	44	175	130	117	103	43
175	Indolyl-3-acetic Acid	130	175	77	131	103	102	129	176
176	2,4-Dichlorophenyl Methyl Ether	176	161	178	163	133	135	63	180
176	6-Hydroxy-4-methylcoumarin	147	176	148	91	39	65	51	63
176	7-Hydroxy-4-methylcoumarin	148	176	147	91	39	51	120	63
176	Ascorbic Acid (Vitamin C)	29	41	39	42	69	116	167	168
176	Bromochloro-1,1-difluoroethylene	178	176	97	128	31	180	126	47
176	Bromochloro-1,2-difluoroethylene	178	176	97	61	128	180	31	177
176	Cotinine	98	42	65	78	176	0	0	0
176	Cotinine	98	176	42	118	41	119	51	39
176	Dealkylated Ethotoin	104	105	176	133	0	0	0	0
176	Dybenal	141	77	176	111	113	178	75	143
176	Pemoline	176	107	90	89	77	105	79	70

MWT	COMPOUND NAME	PEAKS							
176	Pemoline	77	107	51	89	42	105	90	79
176	Pemoline*	107	176	90	77	70	105	42	79
177	2-Ethylamino-1-phenylpropanone	72	44	134	77	42	73	105	51
177	2-Hydroxy-4-methyl-5-b-chloroethiazole	128	45	73	100	177	0	0	0
177	5-Phenyloxazilidindione-(2,4)(Pemoline Metabolite 1)	90	177	105	77	106	51	89	50
177	Amphetamine-NCS Derivative	91	86	177	65	0	0	0	0
177	Bethanidine	177	106	0	0	0	0	0	0
177	Bethanidine	71	91	57	106	65	72	55	77
177	Bethanidine*	71	91	106	177	57	72	65	30
177	Dimethylpropion	72	40	42	44	73	105	77	56
177	Ethylpropion	72	44	42	70	77	128	51	105
177	Formylmethylamphetamine (N-)	86	58	91	56	65	42	39	118
177	Modaline	84	94	177	148	42	93	134	41
177	Phenmetrazine	71	42	43	56	177	77	70	72
177	Phenmetrazine	71	56	177	77	70	105	51	72
177	Phenmetrazine*	71	42	56	43	44	177	30	117
177	Phenmetrazine*	71	42	56	43	177	77	178	105
178	2-Bromoheptane	57	41	56	55	43	69	29	42
178	3-Phenyl-1-propanol Acetate	118	117	43	91	119	105	65	77
178	Nicotine-N-oxide	84	178	161	133	118	0	0	0
178	Nikethamide	106	78	176	51	148	162	56	0
178	Nikethamide	106	78	177	51	178	107	149	163
178	Nikethamide	106	177	78	178	51	105	107	149
178	Phenacemide	118	91	44	92	178	65	0	0
178	Phenacemide	91	92	65	44	117	39	90	89
178	Phenacemide*	91	92	118	44	43	135	65	178
179	2-Ethylamino-1-phenylpropanol	72	44	73	77	105	42	132	160
179	3,4-Methylenedioxyamphetamine	44	137	136	163	179	0	0	0
179	3,4-Methylenedioxyamphetamine	44	136	135	77	51	179	45	78
179	3-Methoxy-N-methylamphetamine	58	56	91	44	77	42	78	121
179	4-Methoxy-N-methylamphetamine	58	44	59	121	91	56	77	78
179	5,6-Benzoquinoline	179	178	151	180	152	177	150	90
179	Ethylnorepinephrine	58	30	41	65	36	39	93	59
179	Fluorodichlorobromomethane	101	103	147	145	105	31	149	66
179	Hippuric Acid	105	135	51	134	77	106	50	78
179	Hippuric Acid	105	77	51	135	134	50	106	117
179	Iproniazid	123	106	43	58	78	79	51	164
179	Iproniazid	106	162	78	51	79	123	43	41
179	Iproniazid*	123	31	58	106	79	43	78	51
179	Isopropyl-N-phenylcarbamate (IPC)	43	93	41	179	39	137	65	32
179	Methoxyphenamine (ortho-)	58	91	43	45	30	59	56	42
179	Methoxyphenamine	58	91	57	56	78	65	77	42
179	Methoxyphenamine	58	30	91	59	56	42	65	77
179	Methoxyphenamine*	58	91	59	56	30	42	121	78
179	Methylephedrine	72	44	42	77	56	73	51	70
179	Methylephedrine*	58	72	30	77	56	44	42	73
179	Mexiletine	58	44	41	83	69	77	85	43
179	Mexiletine	58	41	77	42	43	91	47	122
179	Phenacetin*	108	109	179	137	43	81	80	110
179	Phenprobamate	91	117	118	42	92	65	77	136
179	Phenprobamate*	118	117	91	92	119	65	77	103
180	1,2,4-Trichlorobenzene	180	182	184	145	109	147	75	181
180	1,7-Dimethylxanthine	55	82	67	180	109	0	0	0
180	4-Methyl-2,5-dimethoxybenzaldehyde	180	165	137	179	134	77	39	109

MWT	COMPOUND NAME								
180	4-Oxo-4,5,6,7-tetrahydrocoumarone-3-carboxylic Acid	152	136	96	124	180	51	39	52
180	Acetylsalicylic Acid	120	43	138	92	42	121	39	45
180	Acetylsalicylic Acid	120	138	43	92	64	65	63	42
180	Acetylsalicylic Acid	120	138	43	92	121	39	63	64
180	Acetylsalicylic Acid*	120	43	138	92	121	39	64	63
180	Fructose	73	60	86	71	57	61	103	149
180	Glucose	73	60	57	71	61	74	101	98
180	Methyl Methoxyphenyl-(meta)-acetate	180	121	181	122	59	91	148	182
180	Methyl Methoxyphenyl-(ortho)-acetate	121	180	91	181	122	148	107	93
180	Methyl Methoxyphenyl-(para)-acetate	121	180	122	181	0	0	0	0
180	Paraxanthine	180	68	123	53	42	95	150	151
180	Propylparaben	121	138	65	93	39	122	41	63
180	Theobromine	55	82	67	109	180	0	0	0
180	Theobromine	180	55	109	67	82	137	181	42
180	Theobromine	180	67	82	109	55	181	66	81
180	Theophylline	180	95	68	41	43	53	57	55
180	Theophylline	68	95	180	53	41	40	67	42
180	Theophylline*	180	95	68	41	53	181	96	40
181	Dicyclohexylamine	138	56	55	41	44	30	28	82
181	Etilefrine*	58	30	59	77	29	95	65	57
181	Hexapropymate	99	81	55	44	41	43	120	79
181	Hydroxyephedrine(p-)	58	56	77	107	160	65	38	57
181	Hydroxyephedrine*	58	30	59	56	77	0	0	0
181	Styramate	107	79	120	77	75	44	43	51
181	Styramate*	107	120	79	75	77	44	91	45
182	2,4-Dinitrotoluene	165	119	30	78	90	92	91	79
182	2,4-Dinitrotoluene	165	89	63	30	39	119	182	78
182	2,6-Dinitrotoluene	165	119	30	118	92	91	90	78
182	3,4-Dinitrotoluene	30	182	94	66	65	78	92	53
182	Bibenzyl	91	65	182	92	39	51	63	77
182	Carbachol	43	58	42	44	30	129	36	143
182	Harman	182	57	43	55	40	69	41	181
182	Hydrophenyllactic Acid(para)	107	44	77	40	108	39	51	43
182	Mannitol	73	61	103	74	56	60	57	133
182	Mephenesin	108	118	91	107	43	57	182	75
182	Mephenesin*	108	107	91	182	109	77	79	31
182	Triethyl Phosphate	99	155	109	127	45	82	81	27
183	Acephate	136	42	94	43	95	96	47	79
183	Adrenaline	44	93	65	43	166	137	111	139
183	Adrenaline*	44	124	165	163	123	93	65	
183	Chlorphentermine	58	125	42	168	60	41	89	127
183	Chlorphentermine	58	42	41	125	59	168	89	63
183	Methyprylone	155	140	83	98	55	168	183	0
183	Methyprylone	140	155	83	38	69	98	55	183
183	Methyprylone*	155	140	83	98	55	41	84	69
183	Saccharin	76	183	50	119	120	0	0	0
183	Saccharin*	183	76	90	50	91	120	106	92
184	2,4-Dinitrophenol	184	44	91	63	53	107	92	51
184	3-Methyl-s-triazolopthalazine (Hydralazine Metab)	184	115	88	62	185	114	51	50
184	3-Methyltriazolophthalazine	184	115	185	114	183	88	129	155
184	Barbitone	156	141	98	155	55	112	41	83
184	Barbitone	141	156	41	39	55	98	112	44
184	Barbitone*	156	141	55	155	98	39	82	43
184	Piperazine Phosphate	44	29	30	85	86	56	57	185

MWT	COMPOUND NAME								
185	Chlorophenoxyacetamide (p-Chloro) (Iproclozide Metab)	185	58	141	111	128	185	113	75
185	Ecgonine	82	97	42	83	96	57	94	55
185	Ecgonine	82	96	83	124	138	168	185	186
185	Methyl-p-tolylsulphonamide (N-Methyl)	91	92	65	155	185	58	77	121
185	Methyl-p-tolylsulphonamide (N-Methyl)	91	155	185	65	92	0	0	0
186	1,1,1-Trifluoro-2,2,2-trifluoroethane(Freon 113)	151	153	117	119	70	101	121	29
186	1,2-Dibromoethane	27	107	109	26	45	108	105	95
186	1-Methyl-1,2,3,4-tetrahydro-beta-carboline	171	186	156	155	185	130	116	144
186	5-(2-Hydroxyethyl)-5-methylbarbituric Acid	142	100	99	125	143	98	124	156
186	Alclofenac Metabolite	141	143	77	51	105	0	0	0
186	Carbimazole*	186	114	29	72	42	113	27	109
186	Ethylene dibromide	109	107	73	108	106	123	43	121
186	Mafenide*	106	30	77	185	105	104	89	141
186	Methyl Decanoate	74	87	43	55	41	75	143	57
187	2-Carbomethoxyquinoline	129	157	128	130	187	158	156	188
188	2-Chlorobiphenyl	188	152	190	153	76	189	151	150
188	Antipyrine	77	96	188	105	38	82	93	106
188	Chlorobenzylidenemalononitrile (o-Chloro - CS gas)	153	188	126	190	161	154	137	189
188	Dimethyltryptamine (N,N-)	58	44	188	130	42	143	59	77
188	Dimethyltryptamine (N,N-)	58	59	42	188	72	115	145	104
188	Dimethyltryptamine	58	42	130	77	30	103	115	102
188	Dimethyltryptamine(N,N-)*	58	188	130	59	42	143	129	115
188	Ethyltryptamine (N-)	58	131	130	30	77	132	103	59
188	Phenazone*	188	96	77	56	105	189	55	51
189	3-Phenylpiperidin-2,6-dione(Glutethimide Metabolite)	104	189	103	117	78	91	51	146
189	Kynurenic Acid	55	43	45	143	57	189	145	73
189	Methyl 3-Indoleacetate	130	189	131	79	190	52	103	51
189	Phensuximide(i)*	104	189	103	78	51	77	105	52
189	Phensuximide(ii)*	189	103	190	104	188	102	77	174
189	Phenylglutarimide(alpha)	104	189	146	105	0	0	0	0
190	1-Bromo-2-chlorobenzene	192	190	111	75	194	50	113	74
190	5-Phenylhydantoin(5-Methyl-)	104	175	77	119	51	190	42	161
190	Dibromofluoromethane	111	113	192	41	43	190	194	79
191	5-Hydroxyindole-3-acetic Acid	146	191	147	91	117	63	65	39
191	5-Hydroxyindole-3-acetic Acid	146	191	147	130	57	145	117	89
191	Amiphenazole*	191	121	77	104	122	43	51	192
191	Phendimetrazine	85	57	42	44	56	191	77	0
191	Phendimetrazine	57	42	85	56	54	58	70	44
191	Phendimetrazine*	57	85	42	56	76	191	70	77
192	Citric Acid	42	43	84	56	60	27	44	102
192	Citric Acid	42	84	39	56	43	40	60	102
192	Dibenzo(a,d)cycloheptene (5H-Dibenzo)	192	191	189	165	152	190	193	139
192	Ionone(beta-)	177	43	41	135	91	178	93	95
193	2-Bromo-2-ethylbutyramide(Carbromal Metabolite)	69	43	41	44	71	167	165	55
193	3-Isopropylphenyl-N-methylcarbamate (UC-10854, H 8757)	121	136	91	77	58	39	41	193
193	Butylaminobenzoate	120	137	193	92	65	121	138	41
193	Etafedrine	58	86	87	77	56	42	30	44
193	Etafedrine*	86	58	87	42	56	77	44	43
193	Ethylephedrine	86	58	42	87	77	30	56	51
193	Isobutylaminobenzoate(p-Amino-)*	120	137	193	65	92	39	41	121
193	Methyl Hippurate	105	77	134	51	106	193	161	50
194	Butylmethoxyphenol(3-t-) Methyl Ether	179	194	164	151	91	77	121	149
194	Caffeine	194	109	55	67	82	42	137	41
194	Caffeine	194	109	67	82	55	193	195	81

MWT	COMPOUND NAME								
194	Caffeine	194	109	55	67	82	193	42	40
194	Caffeine*	194	109	55	67	82	195	42	110
194	Dimethyl Isophthalate	163	194	135	76	164	103	120	77
194	Dimethyl Phthalate	163	77	164	135	194	76	92	50
194	Ethyl Bromopyruvate	29	42	43	120	122	27	93	95
194	Meconin	165	194	147	77	121	176	118	51
194	Methyl Acetylsalicylate	120	152	43	121	92	63	65	64
194	Methyl Acetylsalicylate	135	194	179	136	91	137	180	40
194	Methyl Methoxyphenyl-(para)-propionate	121	194	134	163	122	195	135	108
194	Propyl Hydroxybenzoate Methyl Ester	135	152	77	92	64	63	136	107
195	2,3-Dimethoxyamphetamine	44	65	91	152	77	51	164	137
195	2,3-Dimethoxyamphetamine	44	40	152	91	65	165	42	164
195	2,4,5-Trichloroaniline	195	197	199	123	133	135	160	167
195	2,4-Diisopropylamino-s-triazine	58	138	195	180	43	41	152	69
195	2,4-Dimethoxyamphetamine	44	152	151	121	153	77	91	78
195	2,4-Dimethoxyamphetamine	44	152	40	121	151	91	77	153
195	2,5-Dimethoxy-4-methylphenethylamine	166	151	30	135	195	165	44	167
195	2,5-Dimethoxyamphetamine	44	152	137	77	65	91	78	121
195	2,5-Dimethoxyamphetamine	44	152	137	121	153	65	77	91
195	2,5-Dimethoxyamphetamine	44	152	40	137	91	77	65	121
195	2,6-Dimethoxyamphetamine	44	152	91	77	65	153	151	195
195	2,6-Dimethoxyamphetamine	44	152	91	77	65	151	121	78
195	3,4-Dimethoxyamphetamine	44	152	151	65	51	137	78	77
195	3,4-Dimethoxyamphetamine	44	152	151	137	153	107	65	43
195	3,5-Dimethoxyamphetamine	44	152	137	195	179	0	0	0
195	3,5-Dimethoxyamphetamine	44	152	195	151	77	78	65	51
195	3,5-Dimethoxyamphetamine	44	152	40	77	42	91	151	78
195	Hydroxyhippuric Acid(ortho)	121	120	92	65	195	39	93	149
195	Iminodibenzyl(Imipramine Metabolite)	195	194	180	96	193	196	83	167
195	Impramine Degradation Product	194	195	180	193	89	90	77	63
195	Salicyluric Acid	121	120	69	92	195	39	93	45
196	2,4,5-Trichlorophenol	198	196	97	200	132	134	99	133
196	2,4,6-Trichlorophenol	196	198	97	132	200	134	99	62
196	3,4,5-Trimethoxybenzaldehyde (Mescaline Precursor)	196	181	125	39	110	95	93	51
196	Bethanechol	43	58	42	143	85	171	157	102
196	Bromotrichloromethane	117	119	163	82	161	121	47	165
196	Methyl 3,4-dimethoxybenzoate	196	165	197	181	166	59	94	121
197	Beclamide*	91	106	197	162	107	148	27	63
197	Halothane	116	197	196	119	67	129	69	98
197	Levodopa	123	124	77	74	197	152	105	179
197	Levodopa	123	77	51	124	39	74	53	78
197	Levodopa*	123	124	77	44	51	74	53	39
197	Methylsaccharin	133	104	76	105	197	0	0	0
197	Methyprylone Methyl Derivative	154	169	83	55	41	42	98	69
197	Phenacetin Metabolite	197	155	140	0	0	0	0	0
198	Cycloate	72	44	41	45	89	42	43	55
198	Ditolyl Ether(p-)	198	91	65	199	197	155	107	92
198	Glycerol guiacolate	124	198	109	77	81	110	125	167
198	Guaiacol Glyceryl Ether	124	109	81	52	77	51	65	95
198	Guaiphenesin	124	109	198	77	81	43	180	122
198	Guaiphenesin	124	109	198	31	81	77	95	0
198	Guanethidine	126	44	55	58	139	42	43	96
198	Guanethidine	126	139	44	58	43	55	42	41
198	Hydroxyharman	198	168	180	140	197	169	179	196

MWT	COMPOUND NAME								
198	Metharbitone (Gemonil)	170	155	41	169	55	112	39	83
198	Metharbitone*	155	170	112	169	55	82	41	39
198	Methylbarbitone (N-Methyl)	170	155	169	55	112	83	98	69
198	Monuron	72	198	44	200	73	42	99	199
198	Probarbital*	141	156	41	43	39	98	155	169
198	Probarbitone	156	141	41	43	155	157	98	169
199	3,4,5-Trimethoxybenzyl Alcohol (Mescaline Precursor)	198	127	183	95	123	181	196	199
199	Diethylcarbamazine	71	28	100	72	58	199	129	83
199	Diethylcarbamazine*	71	72	58	100	83	56	70	44
199	Methaprylon Metabolite	83	55	98	155	126	0	0	0
199	Methaprylon Metabolite	153	55	183	138	69	0	0	0
199	Methylecgonine	82	96	83	97	42	199	94	168
199	Methylecgonine	96	82	83	42	97	55	68	199
199	Methyprylon Metabolite (6-Hydroxy)	83	98	55	166	153	41	84	155
199	Methyprylon Metabolite	83	55	98	153	166	41	155	152
199	Phenothiazine	199	167	198	166	99	154	69	77
200	1,2-Dibromopropane	41	121	123	39	27	42	40	38
200	1,3-Dibromopropane	41	121	123	39	202	27	204	200
200	2-Methyl-4-chlorophenoxyacetic Acid (MCPA)	200	141	155	0	0	0	0	0
200	5-Ethyl-5-(2-hydroxyethyl)barbituric Acid	124	141	156	139	98	99	101	114
200	Acetonide	185	200	186	89	57	42	103	131
200	Ethyl Caprate	88	101	43	61	60	41	73	70
200	Harmalol	199	200	198	170	172	171	63	42
200	Hydrallazine Metabolite	200	171	129	117	115	145	183	199
200	Lauric acid(n-Dodecanoic Acid)	73	60	43	57	41	55	129	71
200	MCPA	141	200	77	143	155	125	142	202
200	Meconic Acid	44	36	102	69	156	38	45	71
200	Methyl Undecanoate(n-)	74	87	169	55	157	143	75	43
200	Pentachloroethane	167	165	117	119	83	169	130	132
200	Pethidine Intermediate A	57	43	70	71	42	200	199	91
200	Pethidine Intermediate A	57	42	43	70	71	200	199	44
200	Tetrahydrozoline	185	115	200	171	128	44	91	199
201	1-Naphthyl-N-methylcarbamate (Carbaryl, Sevin)	144	115	116	58	55	63	201	40
201	Simazine	201	44	186	173	68	71	55	158
201	Simazine	44	201	43	186	68	173	71	96
201	Thiabendazole	174	63	64	201	90	52	39	65
201	Thiabendazole	201	174	63	64	202	65	129	90
202	Chlorphenesin*	128	130	202	31	29	65	43	111
202	Fendroprox	97	56	32	91	68	98	65	57
202	Methyl 2-Propylglutarate	142	128	121	120	171	83	87	143
203	4-Aminoantipyrine	56	84	57	203	42	83	77	93
203	Aminoantipyrine	56	57	84	83	203	77	0	0
203	Ethyl-1-phenylcyclohexylamine (N-)	160	203	91	146	161	117	44	104
203	Ethyl-1-phenylcyclohexylamine (N-)	160	91	129	130	117	115	104	203
203	Methsuximide	118	117	203	103	77	78	119	91
204	2-Isopropylquinoxaline-1,4-dioxide	171	173	169	188	76	160	170	129
204	Bufotenine	58	204	146	59	42	160	43	159
204	Cedrene(alpha-)	119	93	41	69	105	161	55	91
204	Cedrene(alpha-)	119	41	93	204	69	105	91	55
204	Ethambutol*	102	30	72	44	116	55	173	71
204	Ethotoin (Peganone)	204	105	104	133	77	72	205	78
204	Ethotoin	104	105	204	77	78	133	51	132
204	Methoin Metabolite (Phenylethylhydantoin)	175	104	77	176	105	132	51	204
204	Psilocin (4-Hydroxy-N-dimethyltryptamine)	58	204	59	146	159	205	160	60

MWT	COMPOUND NAME	PEAKS							
204	Psilocin	58	204	59	42	146	160	130	0
204	Psilocin	58	42	204	30	146	117	130	59
204	Psilocybin Metabolite	58	204	59	0	0	0	0	0
204	Thozalinone	70	98	69	77	204	89	83	105
204	Urea di-TMS Derivative	147	189	73	74	148	75	190	146
205	2-(2'Propynyloxy)phenyl N-Methylcarbamate	109	148	110	81	88	58	57	39
205	3-(2'-Propynyloxy)phenyl N-Methylcarbamate	147	148	88	91	58	110	81	145
205	3-Indolelactic Acid	130	44	205	77	129	131	103	102
205	3-Indolelactic Acid	130	131	205	129	77	43	41	45
205	Chlorthiamid	170	205	172	171	207	173	60	136
205	Diethylpropion	100	77	44	42	72	105	101	51
205	Diethylpropion*	100	44	72	101	77	56	42	105
205	Methyl-p-chlorobenzenesulphonamide (N-Methyl)	111	75	141	205	171	112	50	113
205	Methyl-p-chlorobenzenesulphonamide (N-Methyl)	111	175	205	75	113	207	84	177
205	Norpethidinic Acid	57	56	205	103	91	158	77	115
205	Phenformin*	91	30	92	42	29	146	44	104
205	Xanthurenic Acid	159	205	131	76	187	51	103	104
206	1,4-Dichloro-2,5-dimethoxybenzene (Chloroneb)	191	206	193	208	141	195	163	143
206	2,6-Dichloro-4-nitroaniline (DCNA)	206	124	176	208	160	178	162	126
206	2-Phenyl-2-Ethylmalonamide(PEMA-Phenobarb./Primidone metab.)	163	148	91	103	117	120	44	77
206	5-Furfurylidenebarbituric Acid	206	128	107	119	92	135	178	106
206	Chlorodibromoethane	129	127	131	48	47	91	93	81
206	Dibromochloromethane	129	127	131	48	47	93	91	79
206	Ibuprofen	91	163	161	119	107	117	41	118
206	Ibuprofen*	163	161	119	91	206	117	107	164
206	Primidone Metabolite	163	148	91	103	164	0	0	0
206	Pyrantel	205	42	135	173	206	123	145	45
207	1,2-Methylenedioxy-4-(2-nitropropenyl)benzene	103	160	77	207	41	39	150	51
207	1-(3,4-Methylenedioxyphenyl)-2-nitropropene	103	28	160	77	207	41	150	0
207	1-Methyl-2-phenylindole	207	206	208	204	103	205	177	102
207	2-(Allyloxyphenyl) N-Methylcarbamate	109	150	78	41	110	81	58	39
207	3,4,5-Trimethoxyphenylacetonitrile	207	192	164	74	149	134	124	0
208	1,3,7,8-Tetramethylxanthine (or 1,3,8,9-Tetramethyl-)	208	82	123	67	209	42	55	207
208	1-Ethyltheobromine	208	180	109	67	42	137	55	179
208	4-Dimethylamino-3-methylphenyl n-methylcarbamate (Bayer 44646)	151	150	136	208	58	77	45	40
208	7-Ethyltheophylline	208	95	193	67	180	123	73	43
208	9,10-Diaminophenanthrene	208	180	205	207	206	209	104	77
208	Allobarbitone	167	41	124	80	39	32	166	53
208	Allobarbitone*	41	167	124	39	80	53	68	141
208	Dextropropoxyphene Metabolite	208	115	117	91	193	0	0	0
208	Dibenzo(a,d)cycloheptene-10,11-epoxide (5H-Dibenzo)	179	178	180	208	152	177	176	206
208	Dibenzosuberone	208	89	179	76	165	77	209	88
208	Difluorodibromomethane	129	131	191	81	79	50	31	189
208	Isoamyl Salicylate	120	43	121	138	39	65	41	29
208	Pilocarpine	95	96	124	42	82	83	94	41
208	Pilocarpine	95	96	109	208	42	41	54	83
209	2,5-Dimethoxy-4-methylamphetamine(STP)	44	166	151	91	77	79	39	42
209	2,5-Dimethoxy-b-nitrostyrene	209	162	133	148	77	51	147	105
209	2-Bromo-2-ethyl-3-hydroxybutyramide(Carbromal Metabolite)	150	152	165	41	167	43	44	130
209	3-Methoxy-4,5-methylenedioxyamphetamine	44	166	165	42	77	39	64	194
209	Bamethan*	86	30	57	108	44	29	84	41
209	Dimethirimol (tech)	166	209	96	167	71	180	93	55
209	Ethirimol	166	96	209	55	167	71	42	69
209	Hydroxyphenamate	135	57	91	77	43	119	105	180

MWT	COMPOUND NAME								
209	Isopropoxyphenyl N-methylcarbamate (o-Isopropoxy-)	110	152	111	81	58	43	41	52
209	STP	44	166	151	57	43	91	135	209
210	1,3,7-Trimethyluric Acid	210	82	67	153	125	42	55	0
210	Aprobarbitone (Allypropymal, Aprozal, Alurate)	167	41	39	124	168	29	43	169
210	Aprobarbitone	167	41	168	124	43	97	169	96
210	Aprobarbitone*	167	41	124	168	97	39	169	45
210	Crotylbarbitone	181	156	55	141	180	138	155	39
210	Dibenzyl Ketone	91	65	39	119	92	63	89	51
210	Methyl 3,4-dimethoxyphenylacetate	151	210	152	211	107	195	59	153
210	Naphazoline	209	115	141	210	152	153	139	208
210	Naphazoline*	209	210	141	115	153	208	46	181
210	Trichlorocresol(m-)	210	175	212	177	111	75	214	73
211	1,3-Diphenylguanidine	93	77	119	51	211	0	0	0
211	2-Hydroxyiminodibenzyl(Imipramine Metabolite)	211	210	196	212	105	180	167	77
211	2-Methylbenzanilide	119	91	211	65	120	212	92	90
211	Benzylphenethylemine(alpha)	120	91	103	121	77	65	42	39
211	Isolan	72	41	45	43	42	29	44	211
211	Isoprenaline*	72	44	43	124	123	30	42	41
211	Mescaline	182	30	167	181	51	107	151	211
211	Mescaline	182	181	149	44	167	211	40	41
211	Mescaline	182	181	167	211	151	148	139	136
211	Mescaline*	182	30	181	167	211	183	151	148
211	Methoxamine	168	137	44	139	43	152	124	167
211	Methoxamine	44	166	168	137	151	63	53	95
211	Methyldopa	88	123	44	42	124	41	122	77
211	Methyldopa	88	42	123	77	51	124	43	166
211	Methyldopa	88	42	123	124	89	77	51	40
211	Orciprenaline	72	43	65	41	111	93	39	56
211	Orciprenaline*	72	43	73	41	70	65	40	39
211	Prenylamine Metabolite	30	194	167	165	193	116	77	179
211	Propachlor (tech)	120	77	176	93	43	57	169	211
212	1,2-Dimethylbarbitone (or 1,4-Dimethyl-)	184	40	169	140	126	183	44	55
212	1,3-Dimethylbarbitone	169	184	183	126	112	83	41	40
212	1,4-Dibromo-2-butene	53	133	135	54	39	27	89	51
212	2,5-Diaminobenzophenone	211	212	107	77	43	57	106	139
212	3,4,5-Trimethoxybenzoic Acid	212	197	141	160	154	111	93	0
212	Butabarbitone	156	141	41	57	157	39	98	55
212	Butethal	141	156	55	41	29	142	39	98
212	Butobarbitone	156	141	184	55	155	142	170	98
212	Butobarbitone	41	141	39	156	57	55	53	44
212	Butobarbitone	141	156	41	55	98	142	184	40
212	Butobarbitone*	141	156	41	55	98	39	142	155
212	Chlorotoluron	72	44	212	45	77	42	132	214
212	Dimethylbarbitone (N,N-Dimethyl-)	169	184	59	55	112	183	126	69
212	Harmine	212	197	169	213	211	170	106	168
212	Harmine*	212	169	197	213	106	211	170	168
212	Permethylated Barbitone*	169	184	126	183	41	112	83	55
212	Permethylated Metharbitone*	169	184	126	183	112	83	55	170
212	Secbutobarbitone	141	156	40	69	98	38	70	67
212	Secbutobarbitone*	141	156	41	57	39	98	157	47
213	1,3,5-Trinitrobenzene	75	30	213	74	120	91	63	167
213	1,3,5-Trinitrobenzene	30	75	74	213	120	63	91	92
213	2-Methylsulphonylacetanilide	93	213	79	77	106	121	119	104
213	Azopyridine	213	108	81	36	44	136	54	77

MWT	COMPOUND NAME	PEAKS							
213	Baclofen*	30	138	195	140	103	197	77	196
213	Banol	156	121	158	93	58	65	141	51
213	Chlorpropham	43	127	213	171	129	41	154	153
213	Clorprenaline*	72	30	43	77	73	51	41	27
213	Isopropyl-N-(3-chlorophenyl)carbamate (CIPC)	43	41	127	129	213	154	105	171
213	Omethoate	156	110	58	109	79	80	126	47
213	Phenazopyridine	81	54	108	77	51	213	39	52
213	Phenazopyridine	108	81	213	54	77	136	51	43
213	Simetryne	213	43	68	44	71	155	170	96
213	Tropine TMS Ether	83	82	96	124	97	42	73	184
214	1,2,3,4-Tetrachlorobenzene	216	214	218	179	181	143	109	74
214	1,2,3,5-Tetrachlorobenzene	216	214	218	181	179	109	108	143
214	1,2,4,5-Tetrachlorobenzene	216	214	218	179	181	74	108	109
214	3-(4-Chlorophenyl)-1-methyl-1-methoxyurea	61	214	185	126	153	140	93	127
214	4-Chloro-2-methylphenoxypropionic Acid (CMPP)	142	214	141	169	107	77	0	0
214	Dimexan	75	76	47	60	214	107	121	77
214	Harmaline	213	214	170	198	169	115	63	143
214	Harmaline	213	214	198	170	199	215	169	172
214	MCPA Methyl Ester	214	141	155	125	216	77	45	143
214	Methyl Laurate	74	87	41	43	55	75	57	69
214	Metribuzin	200	42	58	61	43	47	75	104
214	Monolinuron	61	46	126	99	214	60	63	45
214	Niridazole*	158	214	57	145	124	45	70	96
214	Phenaglycodol	59	43	156	121	155	157	158	139
214	Phenaglycodol	43	59	31	155	121	156	77	41
214	Phenylsalicylate	121	65	39	93	63	92	64	122
214	Phenyramidol*	107	108	78	79	80	77	51	52
214	Sulphacetamide*	172	92	156	65	108	39	44	41
214	Sulphaguanidine*	92	65	108	214	156	39	109	43
215	1-Naphthyl-N,N-dimethyl carbamate	74	215	115	42	39	127	0	0
215	Atrazine	200	215	149	173	202	58	217	92
215	Atrazine	200	215	58	202	43	68	173	69
215	Fencamfamin*	98	215	58	84	91	56	71	186
215	Fencamfamine	98	91	84	41	56	58	39	115
215	Furcarbanil	123	215	43	81	53	124	216	94
215	Furfurylmethylamphetamine	81	138	53	82	91	139	56	42
216	3-Indolylglyoxyldimethylamide	144	216	116	89	72	63	0	0
216	5-Benzylidenebarbituric Acid	215	216	172	102	129	118	128	117
216	Diethyltryptamine (N,N-)	86	58	30	87	42	130	77	56
216	Diethyltryptamine	86	30	58	130	29	77	87	42
216	Dioxyparaquat	216	147	158	57	217	131	160	187
216	Primidone Metabolite	146	117	103	115	91	0	0	0
216	Terbacil	161	216	160	218	163	162	117	118
217	4-Methylaminoantipyrine	56	217	83	42	57	98	77	218
217	5-(2'-Pyridylmethylene)barbituric Acid	216	217	128	173	118	103	130	146
217	5-(3'-Pyridylmethylene)barbituric Acid	128	217	216	173	103	146	119	130
217	5-(4'-Pyridylmethylene)barbituric Acid	217	128	118	103	146	119	145	174
217	Ethoxyquin	202	108	174	137	109	80	203	145
217	Glutethimide (Doriden)	117	189	132	115	91	160	77	39
217	Glutethimide	189	117	132	160	91	115	77	103
217	Glutethimide*	189	132	117	160	91	115	103	77
217	Methyl-3-indolebutyrate	130	217	143	131	186	218	144	117
217	Methylaminophenazone	56	123	217	83	57	77	215	216
217	Noraminophenazone	56	123	215	217	83	119	77	57

MWT	COMPOUND NAME	PEAKS							
217	Norfenfluramine	44	42	159	43	45	184	41	109
217	Prolintane	126	91	127	174	55	41	42	97
217	Prolintane*	126	127	174	91	70	69	55	42
218	2-(Methyl-2'-propynylamino)phenyl N-Methylcarbamate	122	161	94	160	57	120	123	162
218	5-Phenylhydantoin(5-methyl-) Dimethyl Derivative	118	203	56	77	218	141	132	204
218	Acetylserotonin (N-)	159	146	218	160	147	219	0	0
218	Mephenytoin	189	104	77	190	51	105	103	132
218	Mephenytoin	104	189	77	51	105	103	190	56
218	Meprobamate	83	55	71	96	114	144	62	56
218	Meprobamate	83	55	71	41	44	62	56	45
218	Meprobamate	44	41	55	83	31	43	39	56
218	Meprobamate*	83	84	55	56	43	71	41	62
218	Methetoin	189	104	77	190	51	0	0	0
218	Methoin*	189	104	190	77	44	105	132	103
218	Phenylmethylbarbituric Acid*	104	132	218	51	103	77	78	52
218	Primidone (5-Ethylhexahydro-5-phenylpyrimidine-4,6-dione)	146	190	117	118	189	161	103	115
218	Primidone	117	146	43	91	32	115	39	116
218	Primidone	146	117	190	118	115	103	91	161
218	Primidone*	146	190	117	118	161	189	103	91
219	Encyprate	91	146	65	219	92	41	39	190
219	Methyl-5-methoxyindoleacetate	160	219	161	220	145	74	69	83
219	Norpethidinic Acid Methyl Ester	57	56	42	219	43	103	91	158
219	Pethidine Intermediate C	42	43	44	57	70	71	103	91
219	Pethidinic acid*	71	70	57	43	219	42	218	44
220	2,3-D	162	164	220	175	185	111	147	222
220	2,4-D	162	164	220	161	63	133	111	222
220	2,4-Dichlorophenoxyacetic Acid	162	220	161	175	0	0	0	0
220	2,6-Di-t-butyl-4-methylphenol (Antioxidant)	205	220	57	206	41	145	219	55
220	3,4-D	220	175	162	222	177	145	147	164
220	Dichlorvos	109	185	79	202	145	187	200	199
220	Dichlorvos	109	96	15	79	185	47	45	95
220	Dichlorvos	109	185	79	145	187	47	220	110
220	Hydroxylated Ethotoin	121	120	220	148	0	0	0	0
220	Ibuprofen Methyl Ester	161	117	177	91	119	118	220	121
220	Ionol	205	57	220	41	206	145	29	81
220	Phosphoramide Mustard	92	63	56	94	65	57	171	219
220	Prilocaine	86	43	77	91	40	106	84	65
220	Prilocaine*	86	44	87	107	43	106	56	41
220	Vapona (DDVP)	109	79	47	195	197	145	83	220
220	Xylazine	205	77	41	39	220	130	145	131
221	2-Amino-4,6-di-t-butylphenol	206	221	57	41	207	150	222	68
221	3,4,5-Trimethoxybenzoyl Cyanide	221	124	64	206	150	151	163	178
221	6-Hydroxy-3-(2-dimethylaminoethyl)benzo(b)thiophene	58	30	148	115	59	77	91	89
221	Carbofuran	164	149	57	122	123	131	165	121
221	Cyclofuramid	123	139	221	122	43	140	81	124
221	Formetanate	221	163	164	44	149	122	42	36
221	Hydrocotarnine	220	178	221	205	163	177	179	42
221	Ketamine Metabolite	153	118	138	102	130	0	0	0
221	Metaxalone	122	77	107	105	91	79	123	30
221	Methabenzthiazuron	164	136	135	163	96	108	165	57
221	Prynachlor	186	130	77	53	120	93	221	51
221	Pyrazon	77	221	105	220	88	223	222	51
222	1,3-Diethyl-7-methylxanthine	222	136	166	194	150	123	207	67
222	1,3-Diethyl-9-methylxanthine	123	42	222	207	179	149	150	166

MWT	COMPOUND NAME								
222	1,7-Diethyl-3-methylxanthine	222	95	194	166	207	179	123	67
222	3,7-Diethyl-1-methylxanthine	222	150	194	179	166	207	109	43
222	4,4'-Dichlorobiphenyl	222	152	224	151	223	150	75	153
222	Acetazolamide*	43	180	42	45	44	64	100	222
222	Apiol	222	207	149	177	223	221	195	191
222	Bromvaletone	44	83	180	182	137	143	139	41
222	Diethyl Phthalate	149	177	150	65	76	105	176	104
222	Diethyl Phthalate	149	178	65	177	150	76	121	105
222	Diethyl Terephthalate	177	149	166	65	194	178	104	121
222	Flavone	222	120	92	194	64	63	221	97
222	Methyl 3,4-Dimethoxycinnamate	222	191	223	207	79	147	164	190
222	Nicametate*	86	30	58	123	78	51	29	42
222	Phenindione*	222	165	76	223	166	105	104	90
222	Phenylindanione	222	165	223	221	194	166	164	76
222	Pyramat	72	42	151	153	43	29	40	222
222	RDX	28	30	46	42	29	128	120	75
222	RDX	46	30	42	75	120	44	128	56
222	RDX	46	28	42	75	120	56	29	128
222	Zectran	165	164	150	222	58	134	77	39
223	2,5-Dimethoxy-b-methyl-b-nitrostyrene	223	162	161	147	176	91	77	119
223	3-Butynyl N-(3-chlorophenyl)carbamate	53	223	153	127	171	181	225	155
223	4-Acetylaminophenol(Mono-TMS-)	181	166	223	73	43	45	208	75
223	Bendiocarb	151	126	166	31	51	43	58	223
223	Bufexamac*	107	163	223	108	29	164	41	57
223	Ethamivan	151	52	123	108	42	65	51	72
223	Ethamivan	151	223	222	72	123	152	108	224
223	Ethamivan*	151	72	223	123	222	152	52	29
223	Ketamine Metabolite	166	131	102	138	77	91	195	0
223	Mephenoxalone (Trepidone)	124	223	109	77	122	123	52	95
223	Mephenoxalone	124	109	77	52	223	95	81	122
223	Paracetamol TMS Ester	181	223	166	73	43	45	208	75
223	Salicyluric Acid Methyl Ester Methyl Ether	135	77	90	92	51	64	136	63
223	Salicyluric Acid Methyl Ester Methyl Ether	135	90	77	92	136	134	51	64
224	Butalbital	168	41	124	181	167	141	97	98
224	Butalbital	168	167	41	181	124	97	169	141
224	Butalbital	168	167	41	39	124	97	43	141
224	Butalbital*	41	167	168	39	124	97	141	181
224	Butylvinal	154	83	71	55	155	43	70	67
224	Diethylamyl Phosphate	99	155	127	81	109	0	0	0
224	Enallylpropymal*	181	41	182	39	124	53	138	97
224	Idobutal*	167	41	168	124	39	97	141	67
224	Methyl 3,4-Dimethoxyphenylpropionate	151	224	164	152	225	149	165	193
224	Mevinphos	127	192	109	67	43	193	39	79
224	Mevinphos GC peak 2	127	192	109	67	43	193	79	39
224	Oxyethyltheophylline	180	95	224	123	68	109	193	194
224	Phosdrin	127	192	109	67	43	164	193	224
224	Phosdrin	127	15	43	67	109	79	192	39
224	Salicylic Acid Methyl Ester TMS Ether	209	179	210	135	59	89	161	193
224	Talbutal (5-Allyl-5-s-butylbarbituric Acid)	41	167	168	39	125	97	44	53
224	Talbutal	167	41	168	39	97	57	124	53
224	Talbutal*	167	168	41	97	124	39	57	53
224	Vinbarbital	195	41	39	69	67	53	152	135
224	Vinbarbitone*	195	41	141	69	39	152	135	196
225	(4-(Methylthio)-3,5-dimethylphenyl N-methylcarbamate (Bayer 37344)	168	153	225	91	58	45	39	77

MWT	COMPOUND NAME	PEAKS							
225	(4-(Methylthio)-3,5-dimethylphenyl N-methylcarbamate (Methiocarb)	168	153	109	91	225	45	169	154
225	2,4,5-Trimethoxyamphetamine	44	182	167	151	181	139	183	136
225	2-Butylamino-4-ethylamino-6-methoxy-s-triazine (s-Butyl)	196	169	225	210	43	94	197	44
225	3,4,5-Trimethoxyamphetamine	43	182	167	225	181	183	151	142
225	Benzyl-N-methylphenethylamine (alpha-)	134	91	42	119	65	135	58	86
225	Dinitrorhodane Benzene	30	63	69	74	75	79	62	131
225	Furazolidone*	87	79	51	225	42	50	86	80
225	Haloperidol Hofmann Reaction Product	207	225	96	111	0	0	0	0
225	Mephenesin Carbamate	108	118	91	107	57	43	44	75
225	Mephenesin Carbamate*	118	108	182	91	225	107	57	75
225	Methiocarb	168	153	109	65	58	139	91	39
225	Terbutaline	30	86	41	57	65	39	42	111
226	2,3-Dichloro-1,4-naphthoquinone (Dichlone)	191	226	228	163	193	192	165	99
226	Alclofenac	41	141	226	143	181	0	0	0
226	Alclofenac*	41	226	77	143	181	141	39	145
226	Amobarbitone (Amytal)	156	141	157	41	43	142	197	55
226	Amobarbitone	156	158	141	143	157	159	142	0
226	Amylobarbitone	156	141	55	157	142	197	69	98
226	Amylobarbitone	141	156	157	40	69	98	142	70
226	Amylobarbitone*	156	141	157	41	55	142	98	39
226	Antimony Trichloride	193	191	121	195	123	158	156	228
226	Carbidopa*	123	57	42	103	44	85	124	51
226	Hydroxyaprobarbitone (N-Hydroxy)	41	43	183	167	140	184	124	109
226	Methyl 3,4,5-trimethoxybenzoate	226	211	59	155	66	53	195	183
226	Methylbutobarbitone (N-Methyl)	155	170	156	98	169	112	197	55
226	Metyrapone*	120	106	78	92	51	226	41	39
226	Nimorazole*	100	56	42	101	55	54	41	30
226	Pentobarbitone	156	141	157	55	69	155	98	197
226	Pentobarbitone	156	141	98	112	197	226	0	0
226	Pentobarbitone*	141	156	43	41	157	55	39	98
226	Permethylated Probarbital*	169	184	183	112	197	41	126	170
227	2,4,6-Trinitrotoluene	210	30	149	164	180	193	134	0
227	2,4,6-Trinitrotoluene	210	89	30	63	39	51	76	134
227	2-Methyl-4'-hydroxybenzanilide	119	91	227	65	120	228	92	63
227	Bromodiethylaniline(p-Br,N,N-diEt)	212	214	227	229	184	186	105	118
227	Buclosamide*	155	157	227	154	185	44	30	156
227	Dodine	73	43	72	86	100	87	59	128
227	Nitroglycerin	46	30	76	31	45	43	42	47
227	Nitroglycerin	46	30	29	76	28	43	31	44
227	Prenylamine Metabolite	210	209	30	183	165	115	181	133
227	Prenylamine Metabolite	105	77	30	45	227	51	165	183
227	Scopoline TMS Ether	96	73	94	42	57	227	142	212
228	1,2-Dibromopentane	69	41	43	42	27	149	29	151
228	1,5-Dibromopentane	69	41	27	55	43	42	39	29
228	4-Chloro-2-methylphenoxybutyric Acid (MCPB)	142	228	141	0	0	0	0	0
228	5-Butyl-5-(2-hydroxyethyl)barbituric Acid	124	99	142	141	154	98	129	125
228	Allantoin Pentamethyl Derivative	42	156	127	228	142	140	83	72
228	Clofibrate Derivative	128	130	169	41	129	228	75	69
228	Ethyl Laurate	88	101	43	41	73	61	70	55
228	Hydroxybutobarbitone	156	141	157	45	211	199	181	55
228	Indomethacin Derivative (4-Chlorobenzoic Acid TMS Ester)	213	139	75	169	111	215	141	77
228	MCPB	142	107	144	87	43	77	45	108
228	Mecoprop Methyl Ester	142	169	228	107	141	144	171	77
228	Metoxuron	72	228	183	230	44	229	168	45

MWT	COMPOUND NAME	PEAKS							
228	Sulphonal*	77	59	29	43	78	41	27	135
228	Triclofos*	31	49	77	29	113	51	48	115
229	(3,4-Dichlorophenyl)methylacrylamide (N-)	69	229	231	161	230	133	162	233
229	1-(1-Phenylcyclohexyl)pyrrolidine	91	70	186	152	117	77	172	229
229	1-Methoxyphenothiazine	229	214	186	114	215	154	230	199
229	2,3,4,5-Tetrachloroaniline	231	229	233	158	167	169	235	131
229	3-Methoxyphenothiazine	229	214	186	114	215	230	154	199
229	Amiloride*	43	187	42	171	229	86	144	170
229	Clonidine	229	231	30	172	194	174	200	230
229	Clonidine	229	231	172	194	174	230	36	228
229	Clonidine	30	229	172	42	231	194	174	43
229	Clonidine*	229	30	231	172	194	174	200	230
229	Desalkylclomipramine(N-)	229	228	231	194	230	193	214	232
229	Dimethoate (Cygon)	87	93	125	58	47	79	63	229
229	Ergothioneine	57	71	43	85	40	55	41	149
229	Picric Acid	229	30	62	91	53	63	50	230
229	Propazine	214	58	229	172	43	187	216	41
230	2,5-Diamino-2'-Fluorobenzophenone	230	229	107	86	43	211	95	123
230	Bromocamphor	123	83	55	41	151	81	69	39
230	Demeton Methyl	110	109	79	156	80	47	125	126
230	Diazoxide*	125	189	230	127	191	63	232	90
230	Metasystox	169	109	125	79	59	0	0	0
230	Naproxen	185	115	141	230	170	45	153	142
230	Naproxen*	185	230	141	186	184	115	170	153
230	Propyphenazone	215	230	56	77	96	122	81	201
230	Propyphenazone*	215	230	56	77	216	96	41	39
230	Tetrachlorophenol	232	230	131	234	170	84	133	166
231	2-Amino-5-chlorobenzophenone	77	230	231	105	154	232	126	51
231	2-Amino-5-chlorobenzophenone*	230	231	77	232	105	154	233	126
231	2-Chlorobenzanilide	139	231	141	111	233	75	140	113
231	Amidopyrine	56	97	231	42	111	77	71	112
231	Amidopyrine	56	231	97	204	77	111	112	0
231	Amidopyrine*	56	231	97	111	112	42	77	71
231	Aminopyrine	56	97	77	231	42	91	51	55
231	Amphetamine TFA Derivative	140	118	91	29	46	69	65	39
231	Amphetamine Trifluoroacetate	140	118	91	69	45	0	0	0
231	Chlordiazepoxide Benzophenone	230	231	77	232	105	154	233	126
231	Fenfluramine*	72	44	159	73	58	42	109	56
231	Glutethimide Methyl Derivative	203	132	117	174	115	91	231	103
231	Isocarboxazid	91	106	127	110	57	92	65	104
231	Isocarboxazid	91	43	65	127	110	39	51	106
231	Isocarboxazid*	91	127	106	110	43	92	65	120
231	Metazocine	231	216	84	124	59	42	72	174
231	Methylnorfenfluramine	58	30	42	159	56	59	31	202
231	Oxazepam Benzophenone	230	231	77	232	105	154	51	233
232	Aminoglutethimide*	203	232	132	175	204	233	160	118
232	Chlorprothixene Metabolite	98	231	232	197	0	0	0	0
232	Diuron	72	232	234	161	73	45	163	124
232	Mebutamate	97	72	55	71	62	110	69	158
232	Mebutamate*	97	55	69	72	71	98	43	62
232	Melatonin	160	173	232	161	174	145	158	0
232	Nalidixic Acid*	188	189	159	132	53	173	131	145
232	Phenobarbitone	204	232	117	161	146	77	118	115
232	Phenobarbitone	204	63	146	232	117	143	174	89

MWT	COMPOUND NAME								
232	Phenobarbitone	204	117	77	115	51	91	103	118
232	Phenobarbitone	204	31	117	51	77	161	146	118
232	Phenobarbitone*	204	117	146	161	77	103	115	118
232	Piperazine Adipate*	44	100	41	60	29	27	30	86
233	2-Chlorophenothiazine	233	198	201	199	116	166	171	197
233	2-Ethyl-2-(4-hydroxyphenyl)glutarimide (Glutethimide Metab)	204	133	176	233	148	131	188	119
233	2-Methyl-5-methoxyindole-3-acetic Acid Methyl Ester	174	233	175	159	131	130	234	158
233	4-Hydroxyglutethimide	146	233	103	133	91	117	115	77
233	Ethyl-N-demethyl-4-phenylpiperidine-4-carboxylate	57	233	42	56	43	158	131	160
233	Glutethimide Metabolite	233	146	188	104	103	0	0	0
233	Glutethimide Metabolite	91	161	117	103	115	189	143	55
233	Levophacetoperane	84	43	126	77	56	91	105	55
233	Methylphenidate	84	85	56	91	36	55	30	118
233	Methylphenidate	84	91	56	55	121	118	117	85
233	Methylphenidate	84	91	55	150	146	56	85	83
233	Methylphenidate	84	91	56	90	65	89	77	115
233	Methylphenidate	84	56	91	85	55	30	41	42
233	Methylphenidate	84	91	85	56	55	150	41	118
233	Norpethidine	57	42	56	43	233	158	91	103
233	Norpethidine	57	233	56	158	103	91	160	77
233	Pethidine Metabolite	57	233	42	56	158	0	0	0
233	Pethidinic Acid Methyl Ester	71	70	42	57	233	232	43	44
233	Piperoxan*	98	41	99	42	55	98	233	69
234	2,4-Dichlorophenoxymethyl Acetate	175	234	45	199	177	236	73	161
234	2,4-Dichlorophenoxypropionic Acid (2,4-DP)	162	234	189	161	0	0	0	0
234	Brallobarbitone Methylation Artifact*	39	138	195	136	119	53	91	120
234	Cyclopal	193	169	67	66	57	233	65	205
234	Cyclopentobarbitone*	67	193	66	41	169	39	65	77
234	Dibromochloropropane	57	157	75	155	39	49	77	41
234	Dicamba Methyl Ester	203	205	234	236	188	201	204	190
234	Hexachloroethane	201	117	119	203	199	166	94	47
234	Hexachloroethane	117	119	201	166	164	203	199	168
234	Lenacil	153	154	77	94	110	53	40	51
234	Lignocaine	86	85	58	87	120	91	77	56
234	Lignocaine	86	58	87	42	56	72	234	120
234	Lignocaine*	86	87	58	44	72	42	120	85
234	Methisazone*	234	146	179	206	131	91	117	118
235	Azapetine	165	41	234	193	179	42	39	178
235	Azapetine*	165	234	194	235	193	196	166	179
235	Butyl Hippurate (n-)	105	134	77	106	51	135	235	160
235	Meprylcaine	100	58	105	77	56	101	41	70
235	Procainamide	86	30	120	99	58	65	92	42
235	Procainamide*	86	99	120	30	92	87	58	65
236	(3-Butynyl)-N'-(4-chlorophenyl) N-Methylcarbamate (N-)	53	236	110	127	153	68	238	58
236	1,2-Dibromobenzene	236	234	238	155	157	75	50	76
236	1,2-Dimethylallobarbital (or 1,4-Dimethyl-)	195	41	194	53	138	70	137	79
236	1,3-Dimethylallobarbitone	195	138	41	53	194	80	110	58
236	Butethamine	120	86	58	99	56	44	57	36
236	Butethamine	120	65	86	92	56	99	41	44
236	Butethamine	120	85	58	99	56	44	57	149
236	Butethamine*	120	99	30	86	56	44	57	193
236	Carbamazepine	193	192	165	191	194	167	190	63
236	Carbamazepine*	193	192	236	191	194	165	190	237
236	Carbromal	44	69	41	55	43	39	71	53

MWT	COMPOUND NAME								
236	Carbromal	44	69	41	55	43	29	208	71
236	Carbromal*	44	69	41	208	210	55	71	43
236	Cyclobarbitone	207	67	79	81	141	77	55	91
236	Cyclobarbitone	207	141	40	81	79	44	77	41
236	Cyclobarbitone*	207	141	81	79	67	80	41	77
236	Hexobarbitone	81	221	79	80	39	77	157	41
236	Hexobarbitone*	221	81	157	80	79	155	41	77
236	Ibomal Methylation Artifact*	194	136	193	81	137	109	39	197
236	Methazolamide*	43	221	83	236	56	223	222	55
236	Permethylated Allobarbitone*	195	138	41	80	53	194	39	110
236	Procaine	86	120	65	92	42	43	41	56
236	Procaine	86	99	120	58	65	42	92	56
236	Procaine*	86	99	120	58	87	30	92	71
236	Sorbide Nitrate*	43	31	29	61	60	85	73	44
237	3-Hydroxycarbofuran	137	180	147	151	162	58	134	65
237	Cartap	104	71	147	42	44	70	56	102
237	Dicrotophos	127	67	72	193	237	44	109	111
237	Ephedrine TMS Ether	58	73	59	88	45	75	43	56
237	Ketamine	180	182	209	152	138	211	102	154
237	Ketamine	180	36	102	115	152	138	146	125
237	Ketamine Impurity	152	180	138	102	154	182	209	0
237	Ketamine*	180	209	182	152	181	30	211	138
237	Thymoxamine Hydrolysis Product	58	72	59	57	121	164	237	77
237	Viloxazine*	56	100	138	110	57	237	70	41
238	1,3-Dimethylaprobarbitone	195	196	41	138	53	111	181	58
238	1-Chloro-2-iodobenzene	238	111	240	75	113	50	74	127
238	2,3,6-Trichlorobenzoate	207	209	211	238	240	179	181	109
238	3,5-Diacetoxybenzoic Acid	154	43	196	137	155	69	238	42
238	Nealbarbitone	57	141	167	41	181	182	83	223
238	Nealbarbitone*	57	41	141	167	39	83	55	182
238	Nomifensine*	194	195	238	193	72	178	45	196
238	Permethylated Aprobarbitone*	195	196	138	41	111	181	110	53
238	Pirimicarb	166	72	238	167	138	123	152	109
238	Pirimicarb	72	166	238	167	42	44	138	109
238	Quinalbarbitone	168	167	41	43	124	97	169	195
238	Quinalbarbitone*	167	168	41	43	97	124	39	55
238	Secobarbitone	41	168	167	43	39	55	97	124
238	Secobarbitone	168	167	41	43	97	195	124	169
239	3,5-Dimethylpyrrole-2,4-dicarboxylic Acid Ethyl Ester	239	148	194	164	210	193	182	147
239	Benzphetamine*	91	148	149	65	92	42	56	39
239	Formyl-1,3-diphenylisopropylamine (N-)	120	91	148	103	194	77	65	0
239	Isoetharine	100	58	41	65	43	30	192	56
239	Isoetharine Mesylate	56	100	98	58	43	123	41	192
239	Isoetharine*	100	58	41	101	43	56	65	30
239	Mescaline Precursor	239	177	192	149	63	77	92	134
239	Methyldiphenethylamine (N-)	148	105	91	44	77	79	65	56
239	Monalide	85	43	127	41	197	57	239	129
239	Prenylamine Metabolite	239	167	165	59	194	181	152	193
239	Salbutamol*	30	86	57	41	77	135	29	206
240	1,2-Dibromocyclohexane(trans-)	81	79	80	41	82	163	161	77
240	1,3-Dimethylbutabarbitone	169	184	41	112	185	69	183	55
240	1,3-Dimethylbutethal	184	112	42	55	170	183	212	58
240	Benzathine	91	120	121	92	65	106	30	135
240	Budralazine	225	149	185	240	103	131	89	129

MWT	COMPOUND NAME	PEAKS							
240	Cyanazine	225	68	44	173	240	198	172	43
240	Danthron	240	184	241	212	92	138	63	223
240	Danthron*	240	212	241	184	138	92	128	63
240	Dibromodichloromethane	163	161	165	79	81	82	47	36
240	Dimethylbutobarbitone (N,N-Dimethyl)	169	184	183	170	112	212	55	185
240	Dimetilan	72	240	42	39	40	41	44	29
240	Dimetilan	72	240	169	44	42	73	170	56
240	Dinoseb	211	163	147	117	240	77	89	205
240	Dinoterb	225	177	131	41	77	240	38	103
240	GS 29696	183	58	94	42	182	40	196	73
240	Hexethal*	156	141	55	41	157	43	98	39
240	Methyl-6-n-pentyl-4-hydroxy-2-oxocyclohex-3-ene-1-carboxylate	169	137	125	95	55	41	84	43
240	Methylamobarbitone (N-methyl)	155	157	170	172	156	171	158	173
240	Methylamylobarbitone (N-Methyl)	170	155	55	156	171	69	212	112
240	Permethylated Butobarbitone*	169	184	55	112	41	170	58	183
240	Permethylated Secbutobarbitone*	169	184	41	58	112	57	128	183
240	Pheniramine	169	58	149	168	170	57	167	43
240	Pheniramine	58	169	168	42	167	170	72	51
240	Pheniramine*	169	58	168	170	72	167	44	42
240	Picloram	196	198	161	200	163	86	242	245
241	Mefenamic Acid	180	194	77	222	241	179	242	224
241	Mefenamic Acid*	223	241	208	222	194	180	77	224
241	Methocarbamol	118	124	109	44	57	77	81	45
241	Methocarbamol	124	109	118	43	198	122	123	125
241	Methocarbamol	118	109	124	77	81	62	95	75
241	Methocarbamol*	124	118	198	109	43	57	125	77
241	Prometryne	241	58	43	184	69	68	41	199
241	Terbutryn	226	185	241	170	43	68	41	71
242	2-Amino-5-nitrobenzophenone	241	77	242	105	44	43	195	57
242	3'-Hydroxyamobarbitone	141	143	156	158	0	0	0	0
242	3'-Hydroxyamylobarbitone	59	157	156	141	43	41	71	69
242	3-Hydroxyamylobarbitone	59	156	157	69	56	43	41	55
242	5-(2-Hydroxyethyl)-5-isopentylbarbituric Acid	124	142	154	141	99	172	155	98
242	5-Cinnamylidenebarbituric Acid	242	128	127	155	171	143	154	156
242	Cambendazole GC Decomposition Product	242	215	243	214	111	121	216	0
242	Clofibrate*	128	130	169	87	41	129	242	171
242	Cruformate Metabolite	183	185	155	143	77	89	242	148
242	Fenoprofen*	197	241	198	77	242	104	91	103
242	Hydroxyamylobarbitone (N-Hydroxy)	200	143	185	169	126	201	170	155
242	Hydroxypentobarbitone	156	141	157	69	45	197	195	98
242	MCPB Methyl Ester	101	59	69	41	142	107	77	102
242	Methyl Myristate	74	87	43	41	55	57	75	69
242	Methyl Tetradecanoate	74	87	43	75	55	57	41	143
242	Nitrazepam Benzophenone	241	242	77	105	195	211	212	51
242	Thiopentone	172	157	42	173	242	69	71	97
242	Thiopentone	157	172	41	173	43	69	55	39
242	Thiopentone*	172	157	173	43	41	55	69	71
243	4'-Methoxy-2-(methylsulphonyl)acetanilide	243	108	123	122	164	136	95	79
243	Benazolin	170	134	199	243	172	198	201	200
243	Norlevorphanol	45	243	136	200	159	157	198	242
243	Phencyclidine	200	91	84	242	243	115	129	117
243	Phencyclidine	200	91	243	84	242	166	201	0
243	Phencyclidine	91	84	200	41	55	117	42	115
244	2-Methylamino-5-amino-2'-Fluorobenzophenone	244	227	55	57	95	243	69	43

MWT	COMPOUND NAME							
244	Alphenal	215	41	104	77	132	39	128	244
244	Amidephrine*	44	42	147	120	65	43	45	39
244	Diisopropyltryptamine (N,N-)	144	72	30	115	43	130	144	56
244	Flurbiprofen*	199	244	200	178	179	184	183	245
244	Indol-3-yl-N,N-diethylglyoxamide	144	72	145	116	89	29	100	244
244	Naproxen Methyl Ester	185	244	141	115	170	186	153	154
244	Tetrachloro-p-benzoquinone (Chloranil)	246	244	209	211	87	248	218	190
244	Xylometazoline	229	173	44	41	115	128	91	129
244	Xylometazoline	229	244	173	243	44	230	81	245
245	1-(1-Phenylcyclohexyl)morpholine	91	202	115	129	158	57	77	143
245	2-Methylamino-5-chlorobenzophenone	77	245	244	105	228	246	247	51
245	2-Methylamino-5-chlorobenzophenone	245	244	77	105	193	246	228	44
245	2-Methylamino-5-chlorobenzophenone	77	245	244	105	193	228	168	246
245	2-Methylamino-5-chlorobenzophenone*	245	77	246	247	228	105	51	193
245	4-Acetylaminoantipyrine	56	84	245	57	203	83	43	42
245	Bisnortilidine	69	68	70	77	103	56	54	51
245	Chlorphenesin Carbamate	128	118	130	75	57	43	44	61
245	Chlorphenesin Carbamate	118	128	75	130	44	57	62	61
245	Chlorphenesin Carbamate*	128	130	202	43	129	111	75	204
245	Diazepam Benzophenone	245	77	244	105	228	246	51	247
245	Flurazepam Metabolite	245	183	89	247	0	0	0	0
245	Flurazepam N1-Acetic Acid Decomposition Product	245	247	183	210	89	105	122	246
245	Medazepam Benzophenone	245	77	244	228	105	246	193	247
245	Methylamphetamine TFA	154	118	110	69	91	45	0	0
245	Mydocalm (Tolperisone)	98	119	160	84	85	91	245	230
245	Phentermine TFA	154	59	45	91	69	132	114	117
245	Propylnorfenfluramine	86	44	159	43	41	87	42	56
245	Pyrovalerone	126	127	42	91	55	41	65	119
245	Tropocaine	124	82	83	245	94	77	105	67
246	2,5-Diamino-2'-Chlorobenzophenone	246	211	107	43	245	57	248	80
246	4-(Methyl-2'-propynylamino)-3,5-dimethylphenyl N-Methylcarbamate	150	189	246	57	151	174	188	190
246	5-(2-Bromoallyl)-barbituric Acid	168	167	43	97	169	41	124	153
246	5-Butyl-5-(2-chloroethyl)-barbituric Acid	124	141	167	154	125	112	98	190
246	5-Methoxybenzylidene-(para)-barbituric Acid	246	245	202	117	159	132	145	215
246	Cruformate Metabolite	108	246	151	229	95	152	153	92
246	Diallyl Phthalate	189	41	190	104	149	103	39	115
246	Mephobarbitone (Mebaral, Prominal)	218	117	146	118	219	161	39	51
246	Mephobarbitone	218	118	146	117	103	77	219	91
246	Mephobarbitone	218	217	117	118	146	103	115	91
246	Mepivacaine	98	70	99	36	42	96	99	38
246	Mepivacaine*	98	99	70	42	96	55	41	40
246	Methylphenobarbitone*	218	117	118	146	103	77	91	115
246	Permethylated Phenylmethylbarbituric Acid*	132	104	246	103	78	77	51	133
246	Primidone Dimethyl Derivative	146	218	117	118	44	217	42	103
246	Santonin (alpha)	246	173	41	91	135	0	0	0
247	1,2-Dimethyl-5-methoxyindole-3-acetic Acid Me Ester (Indomethacin Der)	247	188	160	63	215	43	106	216
247	2-Amino-3-Hydroxy-5-Chlorobenzophenone	77	246	247	105	78	45	51	248
247	2-Chloro-4'-hydroxybenzanilide	139	247	141	111	75	249	140	248
247	3-Desmethylprodine	96	44	36	173	172	56	70	129
247	Benzamine*	110	126	44	77	41	58	42	105
247	Betacaine	109	44	77	125	58	105	51	0
247	Ketobemidone	70	44	71	57	247	190	119	191
247	Meperidine	71	42	70	57	44	43	103	91
247	Pethidine (Demerol)	71	70	247	57	42	246	91	103

MWT	COMPOUND NAME								
247	Pethidine (N-Methyl-4-phenyl-4-carboxypiperidine Ethyl Ester)	71	70	42	247	57	44	96	43
247	Pethidine	71	70	42	57	43	172	247	91
247	Pethidine	71	247	172	218	174	103	96	91
247	Pethidine	71	70	247	246	57	172	42	218
247	Pethidine*	71	247	246	70	218	57	174	248
247	Pethidine*	71	70	44	57	42	247	43	246
248	1-Phenylazonapth-2-ol	143	248	115	77	69	171	81	249
248	2,4-Dichloroprop Methyl Ester	162	164	189	59	191	55	248	87
248	4-Hydroxyphenobarbitone	219	248	148	220	120	218	133	65
248	Dapsone	108	248	140	65	92	141	109	80
248	Dimethylcyclophosphoramide Mustard	108	199	163	201	92	149	132	170
248	Ethyl 2,4-dichlorophenoxyacetate	175	248	177	250	185	145	69	57
248	Linuron	61	46	160	60	248	133	45	124
248	Methylcyclopal	221	67	196	41	164	181	222	0
248	Pentachlorobenzene	250	252	248	108	215	73	213	85
248	Pentachlorobenzene	250	248	252	254	215	108	213	178
248	Pindolol	72	43	133	116	248	204	104	56
248	Pindolol*	72	133	30	116	248	134	56	41
248	Piridocaine	84	92	120	65	56	111	119	82
248	Pyrimethamine	247	248	249	250	219	212	106	211
248	Pyrimethamine	247	248	249	42	89	219	114	123
248	Thialbarbitone oxygenated analogue	81	207	80	79	141	169	41	77
248	Thionazin	107	96	106	97	143	248	140	79
249	1-(1(2-Trienyl)cyclohexl)piperidine	135	136	45	97	149	164	91	77
249	2-Amino-5-Chloro-2'-Fluorobenzophenone	249	248	123	154	250	95	251	230
249	2-Ethyl-2-(3,4-dihydroxyphenyl)glutarimide (Glutethimide Metab)	249	220	192	149	164	131	135	147
249	Alprenolol*	72	30	56	73	249	98	234	102
249	Benzoctamine*	218	44	191	221	219	178	42	180
249	Flurazepam Benzophenone	248	249	123	250	251	154	95	230
249	Sulphapyridine	184	185	92	65	108	186	66	183
249	Sulphapyridine*	184	92	185	65	39	108	66	186
249	Thiophencyclidine (Thio-PCP)	97	165	164	206	84	249	135	136
250	1,3,7-Triethyl-8-methylxanthine	250	164	222	194	235	251	207	150
250	2-Methylhexobarbitone (or 4-Methyl-)	235	150	165	236	79	137	250	164
250	3'-Ketohexobarbitone	250	95	39	235	66	207	41	193
250	3-Methylhexobarbitone	235	81	169	171	79	236	170	91
250	Amphetaminil	132	100	91	77	65	89	79	0
250	Diflunisal*	232	250	175	204	176	233	102	78
250	Diisopropyl Phthalate	149	104	76	50	150	41	43	42
250	Heptabarbitone	221	141	81	79	222	41	67	93
250	Heptabarbitone	221	141	79	81	77	40	38	67
250	Heptabarbitone*	221	43	78	93	80	41	141	39
250	Methaqualone	235	91	65	132	76	233	90	250
250	Methaqualone*	235	250	91	233	236	65	76	132
250	Permethylated Hexobarbitone*	235	81	79	169	171	170	41	80
250	Sulphadiazine*	186	185	92	65	108	39	93	187
251	3-Cyano-3,3-Diphenylpropionic Acid(Difenoxin metab.)	192	165	251	193	190	77	51	166
251	7-Aminonitrazepam	251	222	223	250	252	195	110	97
251	7-Aminonitrazepam	251	44	223	222	43	84	98	111
251	C 10015	122	121	56	58	194	55	45	73
251	Dextropropoxyphene Metabolite	44	91	129	0	0	0	0	0
251	Dextropropoxyphene Metabolite	44	91	129	115	178	205	220	251
251	Dextropropoxyphene Metabolite 1	44	220	100	59	205	57	91	129
251	STP-NCS Derivative	165	251	135	0	0	0	0	0

MWT	COMPOUND NAME	PEAKS							
252	1,2-Dimethylbutalbital (or 1,4-Dimethyl-)	196	41	195	209	138	181	67	43
252	1,2-Diphenyl-3,5-pyrazolidinedione	183	77	252	184	105	91	51	64
252	1,3-Dimethylbutalbital	196	195	41	138	181	111	209	169
252	10,11-Epoxycarbamazepine	180	223	252	179	152	0	0	0
252	2,4-Dimethylphenylformamidine (N,N'-bis-2,4-Dimethyl-)	121	252	105	77	126	79	237	251
252	2,4-Dimethyltalbutal (or 4,6-Dimethyl-)	95	115	196	41	67	43	96	209
252	3-Ketoquinalbarbitone	43	69	168	41	85	167	86	97
252	Acetylsalicylic Acid TMS Ester	195	120	43	210	135	75	73	92
252	Carbamazepine Metabolite	180	179	223	252	0	0	0	0
252	Chlorodinitronaphthalene	252	160	206	254	125	162	217	148
252	Cimetidine	53	94	57	30	82	99	67	111
252	Cimetidine*	30	57	82	116	99	53	55	42
252	Diphenylhydantoin	180	77	104	209	223	252	51	181
252	Diphenylhydantoin	180	209	104	223	252	77	181	165
252	Diphenylhydantoin	180	104	77	51	209	223	165	181
252	Methyl 3,4,5-trimethoxycinnamate	252	237	253	221	238	177	209	149
252	Methylquinalbarbitone (N-Methyl)	182	181	97	209	167	53	55	124
252	Permethylated Butalbital*	196	195	41	138	169	111	112	181
252	Permethylated Enallylpropymal*	195	196	41	138	111	58	181	53
252	Permethylated Idobutal*	195	196	41	138	181	169	111	39
252	Permethylated Talbutal*	195	196	41	138	181	111	58	110
252	Permethylated Vinbarbitone*	223	41	166	224	169	67	138	39
252	Phenytoin	180	209	223	252	104	73	0	0
252	Phenytoin*	180	104	223	77	209	252	51	165
252	Propyl Hydroxybenzoate (Propylparaben) TMS Ether	73	193	43	195	210	237	75	252
253	Di-(1-isopropylphenyl)amine	91	44	162	119	41	65	43	42
253	Di-(1-isopropylphenyl)amine*	162	91	44	119	163	41	70	65
253	Emepromium Bromide GC Breakdown Product	72	167	73	165	152	253	166	168
253	Mescaline-NCS Derivative	181	253	72	0	0	0	0	0
253	Nefopam	179	180	178	165	225	89	195	210
253	Nefopam*	58	179	180	225	178	165	42	210
253	Prenylamine Metabolite	73	253	167	165	194	193	152	115
253	Sulphamethoxazole*	156	92	108	65	140	253	157	43
253	Triamterene (2,4,7-Triamino-6-phenylpyimido(4,5-b)pyrazine)	253	252	43	254	104	235	57	251
254	1,2-Dimethylamobarbital (or 1,4-Dimethyl-)	169	184	140	185	126	226	55	41
254	1,3-Dimethylamobarbitone	169	184	185	170	55	112	183	69
254	1,3-Dimethylpentobarbitone	169	184	41	43	183	69	112	55
254	2,4,5-Trichlorophenoxyacetic Acid	196	198	254	256	200	197	209	211
254	2,4,5-Trichlorophenoxyacetic Acid (2,4,5-T)	196	254	209	195	0	0	0	0
254	Butanilacaine*	86	30	72	44	141	42	29	57
254	Dimethylamobarbitone	169	184	120	112	121	126	225	226
254	Dimethylamylobarbitone (N,N-Dimethyl)	169	184	170	207	57	55	69	225
254	Dimethylpentobarbitone (N,N-Dimethyl)	169	184	112	183	69	55	185	58
254	Diphylline	180	194	223	109	95	42	254	193
254	Diprophylline*	180	223	194	254	109	95	193	166
254	Dyphylline	180	194	223	109	95	193	166	42
254	Ketoprofen	105	254	177	209	77	210	181	255
254	Ketoprofen*	105	177	77	209	254	210	103	181
254	Lysergol*	254	192	193	154	223	255	180	221
254	Paracetamol Cysteine Conjugate	141	43	183	44	140	108	80	52
254	Permethylated Amylobarbitone*	169	184	170	41	185	112	55	183
254	Permethylated Pentobarbitone*	169	184	41	43	112	183	58	55
254	Picloram Methyl Ester	196	198	197	195	200	225	223	256
254	Thiamylal	43	41	184	168	167	97	55	53

MWT	COMPOUND NAME								
255	3-Chloro-6-cyano-2-norbornanon-o-(methylcarbamoyl) Oxime	184	148	58	118	186	154	149	104
255	5-Butyl-5-(2-dimethylaminoethyl)-barbituric Acid	58	124	71	199	125	167	154	141
255	5-Propyl-5-(2-thiocyanatoethyl)-barbituric Acid	124	141	154	153	170	98	125	155
255	Diphenhydramine	58	73	165	45	105	167	44	77
255	Diphenhydramine	58	42	73	165	45	77	57	152
255	Diphenhydramine	58	73	45	44	59	165	167	42
255	Diphenhydramine*	58	73	45	167	165	166	44	152
255	Mefenamic Acid Methyl Ester	223	255	208	77	222	180	194	96
255	Phenyltoloxamine	58	255	40	72	42	59	71	91
255	Phenyltoloxamine*	58	255	42	71	59	44	181	165
255	Phenyltoxolamine	58	42	91	56	57	43	59	165
255	Sulphathiazole	92	65	108	156	191	71	63	93
255	Sulphathiazole*	92	156	65	108	191	45	39	55
255	Tripelennamine	58	91	72	71	65	42	79	78
255	Tripelennamine*	58	91	72	71	197	185	184	92
256	2',4,4'-Trihydroxychalcone	137	256	44	45	43	41	120	55
256	2,3,5-Trichlorobiphenyl	258	256	186	260	150	151	188	259
256	3,5-Diacetoxybenzoyl Chloride	43	135	170	212	219	69	172	15
256	5-Indanmethylenebarbituric Acid	242	256	228	255	115	129	141	116
256	Bromoform	173	171	175	91	93	92	94	79
256	Clorindione	256	165	258	193	257	76	89	104
256	Desmethylmedazepam (N-)	193	228	89	255	77	165	51	110
256	Desmethylmedazepam	193	255	228	256	257	165	230	258
256	Ethyl Myristate	88	101	43	41	73	55	57	70
256	Palmitic Acid	73	71	83	129	256	98	85	97
256	Paraquat Dichloride	156	50	155	52	51	128	157	76
256	Phenylacetylsalicylate	197	256	240	198	152	106	257	241
256	Trichlorfon	79	109	110	139	145	80	112	95
257	3-Hydroxy-N-methylmorphinan	257	59	150	256	200	157	76	189
257	Barban	222	51	87	143	153	104	224	69
257	Benserazide*	60	42	88	30	31	43	29	70
257	Levorphanol	31	32	29	257	59	256	150	200
257	Levorphanol	59	42	150	157	44	257	256	41
257	Levorphanol*	59	257	150	256	44	31	200	157
257	Meclofenoxate	58	42	71	36	59	111	30	75
257	Meclofenoxate*	58	111	71	42	75	59	141	113
257	Racemorphan	59	257	256	150	80	42	82	200
257	Racemorphan	44	43	257	59	150	200	0	0
257	Racemorphan*	257	59	150	256	31	157	42	200
257	Tolmetin*	212	213	122	198	44	53	91	65
258	Demeton	88	60	89	126	61	115	114	170
258	Demeton O (technical)	88	89	60	61	171	97	115	59
258	Demeton PO	88	101	73	60	81	93	109	111
258	Demeton PS	88	89	60	61	171	97	115	59
258	Hexachloro-1,3-butadiene	225	227	223	190	260	141	118	188
258	Metobromuron	61	46	258	91	260	172	170	63
258	Thalidomide	76	173	104	111	148	50	169	130
259	(3,4-Dichlorophenyl)-2'-methylvaleramide (N-)	161	163	259	261	217	165	219	162
259	1-(1-Phenylcyclohexyl)-4-hydroxypiperidine	91	216	77	259	0	0	0	0
259	1-Phenyl-1-carbethoxy-2-methylaminocyclohex-3-ene	83	68	84	103	77	91	157	155
259	2,4-Dimethyl-3,5-diphenylpyrimidine	259	260	244	115	108	85	91	215
259	2,6-Dimethyl-3,5-diphenylpyrimidine	259	260	115	244	116	95	108	215
259	2-Methyl-3-phenyl-6-(benzyl)pyrimidine	258	259	244	260	91	115	65	72
259	3-Phenyl-3-piperidinocyclohexanol	124	98	105	84	216	96	259	200

MWT	COMPOUND NAME								
259	4-Bromo-2,5-dimethoxyphenethylamine	30	230	232	44	215	217	77	156
259	4-Methyl-5-phenyl-2-(benzyl)pyridine	258	259	243	260	115	91	116	108
259	4-Phenyl-4-piperidinocyclohexanol	200	91	84	86	186	259	0	0
259	5-Dimethylaminobenzylidene-(para)-barbituric Acid	259	258	260	257	216	173	217	172
259	Nortilidine	83	68	82	72	84	77	103	115
259	Primaquine*	201	81	98	175	259	176	202	242
259	Propranolol	72	115	144	259	116	215	127	254
259	Propranolol	30	72	115	57	41	43	56	42
259	Propranolol*	72	56	30	43	98	115	144	41
259	Tetrachloronitrobenzene	261	215	259	213	203	201	263	217
260	1,2-Dimethylphenobarbital (or 1,4-Dimethyl-)	203	188	232	117	204	40	115	70
260	1,3-Dimethylphenobarbitone	232	146	117	175	118	233	103	188
260	2-Amino-5-Nitro-2'-Fluorobenzophenone	260	43	123	259	95	77	57	165
260	5-(2-Chloroethyl)-5-isopentyl-barbituric Acid	190	141	154	124	191	155	142	181
260	Alclofenac Metabolite	141	186	77	143	105	0	0	0
260	Bromacil	205	207	161	163	110	112	260	262
260	Bromacil	205	207	40	39	70	206	190	162
260	Bromacil	205	207	42	162	70	164	231	188
260	Carisoprodol	58	55	97	43	104	158	62	56
260	Carisoprodol	55	58	43	104	56	41	97	62
260	Carisoprodol	104	55	43	56	44	58	41	62
260	Carisoprodol*	55	57	43	97	41	56	158	44
260	Dimethylphenobarbitone (N,N-)	232	118	117	146	175	233	188	260
260	Dimethylphenobarbitone (N,N-Dimethyl)	232	118	146	117	103	77	175	115
260	Dimethylphenobarbitone	232	146	118	175	117	120	121	188
260	Methazole	159	124	260	161	262	126	163	88
260	Oxymetazoline*	245	260	44	217	218	246	261	259
260	Permethylated Methylphenobarbitone*	232	118	117	146	175	103	233	77
260	Permethylated Phenobarbitone*	232	118	117	146	175	103	77	91
260	Phorate	75	121	97	260	47	65	93	69
260	Phorate	75	231	45	97	153	47	46	125
260	Pyridostigmine Bromide	72	39	42	38	94	56	51	81
261	Alphaprodine (alpha-Prodine)	172	187	42	84	129	144	44	91
261	Alphaprodine	172	187	84	57	42	188	44	43
261	Alphaprodine	172	187	144	84	42	57	188	44
261	Alphaprodine	172	187	84	42	57	171	144	186
261	Aminoparathion	261	125	109	108	97	80	233	205
261	Betaprodine	172	187	144	84	42	57	188	44
261	Betaprodine*	172	187	84	42	44	57	29	43
261	Desmethylcyclobenzaprine	44	218	217	58	57	261	216	219
261	Ethoheptazine	107	149	57	78	108	79	77	72
261	Ethoheptazine	57	42	186	58	44	70	83	84
261	Ethoheptazine	57	58	44	42	39	70	40	68
261	Ethoheptazine*	57	58	70	42	44	188	84	43
261	Ethylnorpethidine (N-)	246	261	260	103	186	91	77	0
261	Hexylcaine	112	77	105	139	55	41	96	56
261	Methapyrilene	58	97	72	84	71	191	192	261
261	Methapyrilene	97	58	72	191	71	78	79	190
261	Methapyrilene	58	97	72	71	121	224	149	191
261	Methapyrilene*	58	97	72	71	42	191	79	78
261	Nortriptyline Metabolite	44	215	202	216	45	213	91	189
261	Phenindamine	260	261	42	202	57	217	203	218
261	Phenindamine	260	261	202	42	203	182	115	215
261	Phenindamine*	260	261	42	57	184	215	217	262

MWT	COMPOUND NAME								
261	Piperocaine	105	112	77	41	55	44	42	56
261	Piperocaine	112	105	246	77	55	41	98	97
261	Piperocaine*	112	246	105	77	55	41	44	247
261	Properidine	71	70	218	57	42	44	36	43
261	Thenyldiamine	58	97	72	71	79	42	78	40
261	Thenyldiamine	58	97	71	72	91	79	203	78
261	Thenyldiamine*	58	97	72	71	203	191	190	42
262	5-(4'-Hydroxy-2'-methoxybenzylidene)-barbituric Acid	260	173	259	244	201	216	128	145
262	Allylcyclopentenylbarbitone Dimethyl Derivative	221	67	196	181	41	164	111	107
262	Demeton S-Methyl Sulphone	169	109	125	168	79	110	142	170
262	Methohexital	79	53	221	81	41	77	178	93
262	Methohexitone*	41	81	53	221	79	39	178	233
262	Permethylated Cyclopentobarbitone*	67	221	196	41	164	111	39	181
263	(3,4-Dichlorophenyl)-2'-methyl-2',3'-dihydroxypropionamide (N-)	75	161	163	263	265	127	126	162
263	2-Ethyl-2-(4-hydroxy-3-methoxyphenyl)glutarimide (Glutethimide Metab)	234	263	206	163	178	131	149	161
263	5-(2-Bromoethyl)-5-ethylbarbituric Acid	156	141	124	139	98	157	155	112
263	Butethamate*	86	99	91	191	87	119	58	248
263	Cinchophen Methyl Ester	205	263	204	75	102	101	51	88
263	Dimethylthiambutene*	248	97	219	218	111	249	217	263
263	Hydroxypethidine	71	140	70	263	262	189	57	42
263	Hydroxypethidine(p-)	71	70	263	57	42	44	262	43
263	Methyl Parathion	109	125	263	79	93	63	47	264
263	Nortriptyline	44	57	43	71	41	55	45	56
263	Nortriptyline	44	203	202	45	220	219	215	204
263	Nortriptyline	44	202	259	203	218	215	217	42
263	Nortriptyline	44	45	202	203	91	215	189	115
263	Nortriptyline	44	202	45	215	203	42	220	204
263	Nortriptyline	44	45	202	203	91	99	189	165
263	Nortriptyline*	44	202	45	220	218	215	91	0
263	Parathion-methyl	263	125	109	79	93	47	63	264
263	Parathion-methyl	109	263	125	79	63	93	264	64
263	Protriptyline	70	191	44	189	43	165	69	192
263	Protriptyline	191	70	44	58	189	84	165	192
264	2,4,5,6-Tetrachloroisophthalonitrile (Daconil)	266	264	268	124	229	231	194	159
264	3-Ethylhexobarbitone	249	81	183	250	79	185	184	264
264	Amethocaine	58	71	36	176	150	72	193	59
264	Amethocaine*	58	71	150	176	72	193	105	59
264	Carbocaine	98	70	41	77	91	96	120	
264	Cyclobarbitone Dimethyl Derivative	235	169	79	236	77	91	81	112
264	Diftalone	132	264	90	133	89	105	118	265
264	Mianserin (Bolvidon)	193	264	194	43	71	72	220	109
264	Mianserin*	193	264	43	72	71	220	192	194
264	Pentachlorophenol (PCP)	266	264	268	165	167	130	202	200
264	Pentachlorophenol	266	268	264	165	167	270	202	130
264	Pentifylline*	180	264	193	109	194	181	67	42
264	Permethylated Cyclobarbitone*	235	170	40	236	180	81	73	121
264	Permethylated Nealbarbitone*	169	195	57	209	41	112	250	138
264	Sigmodal Methylation Artifact*	194	193	136	43	41	109	39	137
264	Sulphamerazine*	199	200	92	65	108	39	66	201
264	Tetracaine	58	71	42	105	176	150	92	56
264	Thialbarbitone*	81	223	79	41	80	157	185	77
265	2-Amino-5,2'-Dichlorobenzophenone	139	156	111	44	75	141	230	50
265	2-Ethyl-5-Methyl-3,3-Diphenyl-1-Pyrroline	208	193	130	115	91	165	179	207
265	3,3-Diphenyl-1,5-dimethyl-2-pyrrolidone(Methadone metab.)	265	193	115	130	42	208	56	264

MWT	COMPOUND NAME								
265	Antazoline	84	91	55	77	85	182	83	65
265	Antazoline	84	91	77	65	104	85	83	182
265	Antazoline	84	91	36	51	55	0	0	0
265	Antazoline*	84	91	77	55	182	85	65	104
265	Dextropropoxyphene Metabolite	58	91	59	42	129	0	0	0
265	Ethomoxane*	86	44	30	265	41	180	29	87
265	Hydrolysed Dextropropoxyphene	58	91	191	129	128	205	178	174
265	Lorazepam Benzophenone	230	265	139	267	111	232	264	266
265	Oxprenolol	72	30	221	41	56	73	150	43
265	Oxprenolol*	72	41	56	43	221	73	57	45
265	Paraethoxycaine	86	121	99	30	58	65	42	56
266	((3-Methylaminopropyl)-iminodibenzyl)(N-)(Imipramine Metabolite)	235	195	234	208	193	266	194	45
266	2-Methyl-3-(4-hydroxy-2'-methylphenyl)-4-(3H)-quinazolinone	251	266	249	77	143	76	252	39
266	3-Methyldilantin	180	104	266	77	237	57	209	71
266	4-Methyl-1,2-diphenyl-3,5-pyrazolinedione	77	183	266	105	51	91	78	65
266	5-Methylphenyl-5-phenylhydantoin(p-5-phenylhydantoin)(Metabolite)	180	77	118	91	104	65	89	51
266	5-Phenylhydantoin(5-(p-methylphenyl)-)	180	266	237	208	118	223	194	77
266	Atenolol	72	30	107	56	43	73	222	57
266	Atenolol*	72	30	56	98	43	107	41	73
266	Chlorfenethol	139	43	251	253	178	111	141	248
266	Chlorprocaine	86	99	154	58	56	87	42	84
266	Cyclizine	56	99	165	167	42	0	0	0
266	Cyclizine	99	56	167	207	194	164	195	208
266	Cyclizine	99	56	42	43	167	70	44	207
266	Cyclizine	99	165	56	194	167	207	152	208
266	Cyclizine*	99	56	167	207	194	266	195	165
266	Desipramine	195	235	234	193	208	266	194	84
266	Desipramine	71	85	193	195	42	130	70	194
266	Desipramine	44	193	195	194	208	234	71	235
266	Desipramine*	235	195	208	44	234	193	194	71
266	Dimethylnealbarbitone (N,N-Dimethyl)	169	195	57	209	251	112	210	196
266	Dimethylquinalbarbitone (N,N-Dimethyl)	196	195	181	111	138	55	197	266
266	Diphenazoline*	84	160	167	159	165	152	55	77
266	Hydroxymethaqualone	160	266	235	251	77	146	58	247
266	Methaqualone Metabolite	251	266	249	77	143	252	0	0
266	Methaqualone Metabolite	235	160	266	77	235	251	146	0
266	Methyldilantin	180	77	104	266	237	209	165	0
266	Nealbarbitone Dimethyl Derivative	169	195	209	57	41	112	138	251
266	Permethylated Quinalbarbitone*	196	195	41	138	181	111	40	43
266	Practolol	72	30	151	56	109	43	108	251
266	Secobarbitone Dimethyl Derivative	196	195	181	41	138	111	53	110
266	Tributyl Phosphate	99	41	56	57	39	55	29	155
267	Anisotropine	124	82	83	57	41	94	42	43
267	Apomorphine	266	267	42	220	152	224	165	178
267	Apomorphine*	266	267	224	220	268	44	250	248
267	Azacyclonol	84	105	85	77	55	56	183	42
267	Azacyclonol	85	84	183	107	77	55	56	184
267	Azacyclonol*	85	84	183	105	56	77	55	30
267	Di-(1-isopropylphenyl)methylamine	91	58	176	90	41	42	119	65
267	Emepromium Bromide GC Breakdown Product	86	167	58	87	165	152	263	42
267	Lysergamide*	267	221	207	180	223	154	196	268
267	Metoprolol	72	30	56	107	223	98	73	43
267	Metoprolol*	72	30	107	56	45	41	44	43
267	Pipradrol	84	77	105	56	85	83	51	182

MWT	COMPOUND NAME	PEAKS							
267	Pipradrol	84	105	77	56	85	248	249	36
267	Pipradrol*	84	56	85	77	105	55	30	42
267	Sulphafurazole*	156	92	108	65	140	43	157	42
267	Trimethyl-(N, alpha, alpha-)-diphenethylamine	91	58	176	119	42	41	56	65
268	5-(4-Hydroxyphenyl)-5-Phenylhydantoin(Para HPPH)	239	196	268	120	197	77	225	104
268	Dichlorophen*	128	141	268	130	270	77	143	233
268	Diethylstilbestrol	107	145	268	238	121	133	159	224
268	Dimethyl-3'-hydroxyquinalbarbitone	196	195	237	181	138	45	41	69
268	Fenson	77	51	141	99	63	73	50	268
268	Ketoprofen Methyl Ester	209	105	77	268	191	103	210	51
268	Lysergic Acid (d)	268	154	224	180	207	192	221	223
268	Lysergic Acid*	268	224	154	180	207	223	192	179
268	Michler's Ketone	148	268	267	224	269	251	120	118
268	Permethylated Hexethal*	169	184	55	112	41	185	170	183
269	5-Butyl-5-(2-thiocyanatoethyl)-barbituric Acid	124	154	141	181	198	142	153	155
269	7-Aminodesmethylflunitrazepam	269	240	241	268	270	107	121	213
269	Alachlor	45	160	188	237	162	224	146	161
269	Bufylline*	180	95	68	41	58	53	123	96
269	Norpropoxyphene Carbinol	44	105	91	77	178	208	65	130
269	Orphenadrine	58	73	165	45	166	42	181	59
269	Orphenadrine	58	73	45	42	165	59	46	44
269	Orphenadrine*	58	73	44	45	165	42	40	181
269	Sulphasomizole*	156	92	108	65	269	39	45	270
270	2,3-Dihydroxyquinalbarbitone	171	43	143	41	128	55	141	159
270	2-Nitrotartranilic Acid	138	44	92	43	195	134	90	106
270	Chloroprocaine*	86	99	154	30	87	58	29	156
270	Desmethyldiazepam	242	270	269	241	243	271	244	103
270	Desmethyldiazepam	241	242	269	77	243	103	270	51
270	Desmethyldiazepam*	242	269	270	241	243	271	244	272
270	Desmethylpromethazine	58	213	198	180	214	57	212	270
270	Dimethyl-3-hydroxyamylobarbitone (N,N'-)	184	169	0	0	0	0	0	0
270	Doxylamine	71	58	72	167	182	42	59	0
270	Doxylamine	71	58	167	180	72	42	182	78
270	Doxylamine*	58	71	72	167	182	42	180	59
270	Medazepam	207	242	244	165	270	243	208	269
270	Medazepam*	242	207	244	270	243	271	269	165
270	Methyl Hexadecanoate	74	87	43	75	57	55	143	69
270	Methyl Palmitate	74	87	270	75	143	43	55	41
270	Methyl Palmitate	74	87	43	41	55	75	57	69
270	Permethylated Thiopentone*	200	120	185	121	40	41	69	44
270	Sulphamethizole*	92	270	65	156	108	106	93	59
270	Thiopentone Dimethyl Derivative	169	184	41	112	69	43	185	55
270	Tolbutamide	91	155	171	65	107	39	108	63
270	Tolbutamide*	91	30	155	108	65	197	39	107
271	Acetaminosalol (Phenetsal, Salophen)	121	109	151	65	93	43	39	271
271	Amylocaine	58	105	77	42	59	113	51	30
271	Amylocaine*	58	105	77	98	122	42	113	51
271	Cyclazocine*	230	271	55	256	164	270	71	124
271	Desomorphine	271	214	44	270	32	228	272	42
271	Dextromethorphan (Racemethorphan)	59	42	150	271	44	171	115	128
271	Dextromethorphan	59	150	271	270	214	171	128	212
271	Dextromethorphan*	59	271	150	270	31	214	42	171
271	Methorphan	59	271	150	214	270	171	112	213
271	Normorphine	271	81	150	148	45	110	42	82

MWT	COMPOUND NAME							
271	Phosphorylated Amphetamine	180	124	152	91	44	40	42 105
271	Racemethorphan	59	42	150	271	44	0	0 0
271	Racemethorphan*	59	271	150	31	270	214	42 171
272	1,3-Dimethylalphenal	118	243	104	77	129	231	130 128
272	Alphenal Dimethyl Derivative	118	243	272	104	77	129	128 89
272	Dimethisoquin	58	71	159	56	72	41	42 115
272	Dimethisoquin*	71	58	72	43	159	56	42 201
272	Indol-3-yl-N,N-disopropylglyoxamide	144	43	100	86	116	145	128 89
272	Vacor	135	92	65	108	138	80	39 106
272	Zolimidine	272	193	209	78	192	273	194 181
273	1-Phenyl-1-carbethoxy-2-dimethylaminocyclohex-3-ene	82	97	103	77	72	42	91 115
273	2,4-Dimethyl-3-phenyl-6-(benzyl)pyrimidine	272	273	258	259	115	243	91 260
273	Bromo STP	44	230	232	273	275	77	42 45
273	Chlormezanone (Trancopal)	152	42	98	154	153	174	69 56
273	Chlormezanone	152	153	154	42	155	125	111 56
273	Chlormezanone*	98	152	154	42	69	174	208 153
273	Dichlozoline	186	41	201	188	43	152	273 203
273	Tilidine	103	77	82	29	97	42	51 104
274	19-Nortesterone	274	110	91	79	105	147	215 256
274	2,4-Dichloro-6-(o-chloroaniline)-s-triazine (Dyrene)	239	241	178	143	274	276	240 242
274	2-Methylamino-5-nitro-2'-fluorobenzophenone	274	273	211	257	123	199	275 95
274	3-Ethylmephobarbitone	246	146	117	118	247	175	103 77
274	4-Ethylmephobarbitone	217	218	132	146	117	103	246 118
274	5-Methoxydiisopropyltryptamine	114	72	30	115	160	43	174 145
274	Chlorpheniramine	203	58	205	204	72	167	202 168
274	Chlorpheniramine	58	203	167	72	205	202	168 204
274	Chlorpheniramine(dex)	203	58	43	57	205	71	72 32
274	Chlorpheniramine*	203	58	44	205	54	204	72 202
274	Desmethyltetrazepam	239	274	273	275	240	276	245 211
274	Disulfoton	88	89	60	61	97	274	142 186
274	Disulfoton	88	89	29	97	61	60	27 65
274	Phenampromide*	98	125	42	41	29	99	120 190
274	Phenampromide	98	125	99	120	41	42	57 77
274	Tetraethylphosphate	161	163	179	235	29	162	99 207
274	Tybamate	55	41	56	97	72	57	62 43
274	Tybamate*	55	72	97	41	56	158	118 57
274	Zinochlor	239	241	178	143	75	87	99 89
275	(3,4-Dichlorophenyl)-2-methyl-3'-hydroxyvaleramide (N-)	161	163	275	277	217	165	219 165
275	Alphameprodine	97	172	91	105	57	42	77 201
275	Alphameprodine	172	42	98	57	44	201	91 36
275	Alphameprodine	172	98	201	91	202	275	96 200
275	Amitriptyline Metabolite	58	215	59	229	228	227	226 218
275	Betameprodine	97	172	91	105	57	42	77 201
275	Bromoxynil	88	62	61	53	37	277	63 89
275	Cyclobenzaprine	58	59	42	215	202	57	189 43
275	Diethyl-p-nitrophenyl Phosphate	109	81	149	81	275	139	99 65
275	Dihydroxyphencyclidine	91	115	128	117	129	219	157 0
275	Dimethyldemerol	186	70	83	42	85	71	57 56
275	Homatropine	124	82	83	275	94	96	67 80
275	Homatropine	124	82	94	77	42	44	105 106
275	Homatropine	124	94	82	77	79	123	96 67
275	Homatropine	124	42	82	83	94	67	96 77
275	Homatropine*	124	107	82	83	42	77	79 94
275	Paraoxone	109	81	275	149	99	139	127 247

MWT	COMPOUND NAME								
275	Physostigmine	218	275	174	160	161	0	0	0
275	Physostigmine	174	160	161	218	275	175	162	149
275	Physostigmine*	218	174	160	161	275	219	175	162
275	Proheptazine	58	202	57	201	42	84	44	186
275	Trimeperidine	186	201	42	202	187	56	57	71
276	2-(2-Amino-5-bromobenzoyl)pyridine(Bromazepam Benzophenone analogue)	45	62	43	61	44	63	249	247
276	2-Amino-2'-chloro-5-nitrobenzophenone*	241	276	139	165	195	111	278	242
276	2-Amino-5-Nitro-2'-Chlorobenzophenone	241	139	276	195	111	44	165	119
276	4-Hydroxycyclophosphamide	92	56	94	63	84	86	184	93
276	Bromazepam Benzophenone	249	247	248	250	276	278	198	200
276	Chlorpropamide*	111	175	75	85	30	276	127	113
276	Clonazepam Benzophenone	241	276	139	165	195	111	242	277
276	Cyclandelate	107	69	125	83	79	55	41	77
276	Mebhydroline	60	73	233	91	232	276	275	0
276	Mebhydroline*	91	233	30	232	31	276	275	65
276	Methohexital Methyl Derivative	79	235	53	178	81	41	195	261
276	Molindone	100	56	42	176	98	120	70	189
276	Permethylated Methohexitone*	81	41	53	235	178	79	39	195
276	Permethylated Thialbarbitone Oxygen Analogue*	235	81	169	41	79	54	197	196
276	Psilocin-TMS Derivative	58	276	218	261	0	0	0	0
277	2-Benzyl-2-methyl-5-phenyl-2,3-dihydropyrimid-4-one	186	91	150	187	143	65	115	116
277	Acetylprocainamide (N-Acetyl)	86	58	99	56	162	132	149	205
277	Amitriptyline	58	59	202	91	203	215	218	217
277	Amitriptyline	58	42	59	57	202	203	43	91
277	Amitriptyline	58	59	42	30	202	91	203	115
277	Amitriptyline	58	59	42	202	215	203	91	189
277	Amitriptyline*	58	59	202	42	203	214	217	0
277	Azar	220	57	41	221	206	58	29	55
277	Azathioprine	42	231	119	152	247	0	0	0
277	Azathioprine*	231	42	119	232	92	65	67	74
277	Di-2-Ethyl-1,5-Dimethyl-3,3-Diphenyl-1-Pyrrolinium(Methadone metab.)	277	276	262	42	115	105	91	56
277	Ethylmethylthiambutene	262	135	111	219	97	263	56	0
277	Ethylmethylthiambutene	262	219	111	97	263	56	218	42
277	Ethylmethylthiambutene*	262	111	219	97	263	86	42	264
277	Fenitrothion	127	109	125	277	260	79	192	93
277	Hydroxydesmethylcyclobenzaprine	44	234	233	58	231	202	215	218
277	Maprotiline (Ludiomil)	44	59	70	277	71	191	203	178
277	Maprotiline*	44	70	59	277	71	191	278	203
277	Methadone Metabolite	40	277	276	44	262	57	43	41
277	Methadone Metabolite 1	277	276	262	105	278	220	200	91
277	Norpethidinic Acid TMS Ester	57	42	73	277	103	56	187	43
277	Perhexiline*	84	194	55	85	56	41	30	99
278	2,3,4,5,6-Pentachlorobenzyl Alcohol (Blastin)	245	243	247	278	262	179	244	181
278	2-Ethylhexyl Phthalate	149	43	167	57	104	112	279	113
278	Absisic Acid	190	125	134	162	91	41	39	135
278	Acetylcarbromal	41	43	69	44	39	55	53	70
278	Acetylcarbromal	43	41	129	69	55	39	149	42
278	Acetylcarbromal*	43	129	69	41	86	97	55	44
278	Acetylprocaine Acetate	86	58	162	120	99	43	71	278
278	Acetylprocaine Hydrochloride	86	99	30	58	87	120	162	206
278	Amydricaine	58	112	105	96	42	111	77	51
278	Dibutyl Phthalate	149	41	29	57	56	104	32	65
278	Dibutyl Phthalate(n-)	149	150	41	57	223	205	56	104
278	Dibutyl Terephthalate(n-)	56	205	223	167	149	41	57	65

MWT	COMPOUND NAME	PEAKS							
278	Diisobutyl Phthalate	149	57	41	150	56	223	104	76
278	Dimethocaine*	86	120	30	87	58	92	84	56
278	Fenthion	278	125	109	153	168	169	93	79
278	Methadone Intermediate	58	72	192	42	165	73	71	59
278	Permethylated Heptabarbitone*	249	169	250	41	133	67	79	93
278	Sulphadimidine*	214	213	92	65	215	108	39	42
278	Sulphasomidine*	214	92	65	213	108	42	215	39
278	Triprolidine	209	208	278	207	193	194	84	200
278	Triprolidine	208	209	42	41	207	193	39	206
278	Triprolidine	208	209	31	207	278	84	193	42
278	Triprolidine*	208	209	278	207	193	200	194	84
279	10,11-Epoxyprotriptyline	44	70	178	250	179	279	0	0
279	2-Methylamino-2',5-Dichlorobenzophenone	62	244	279	45	229	57	61	111
279	Diethylaminoethyltheophylline	86	87	99	58	71	207	84	56
279	Doxepin	58	220	59	219	277	179	191	193
279	Doxepin	58	42	59	57	165	43	115	178
279	Doxepin	58	59	42	178	165	277	219	202
279	Doxepin*	58	220	219	59	191	189	42	205
279	Etamphylline*	86	109	30	151	87	81	99	58
279	Hydroxyprotriptyline	70	44	207	178	279	249	0	0
279	Karbutilate	72	180	84	135	44	57	45	41
279	Thymoxamine	58	72	279	234	151	192	166	165
280	1,3-Dimethyldilantin	251	280	72	134	77	265	208	175
280	3-Ethyl-5,5-diphenylhydantoin	180	209	280	77	181	104	251	165
280	7-Hydroxydiftalone	133	147	280	105	77	132	134	90
280	7-Hydroxydiftalone	132	133	280	262	235	104	77	89
280	Dimethyl-3'-ketoquinalbarbitone	196	195	43	138	181	237	197	69
280	Hydroxystilbamidine*	96	44	43	31	45	27	78	264
280	Imipramine	235	58	234	85	280	195	193	35
280	Imipramine	58	85	173	193	194	195	234	235
280	Imipramine	58	234	235	85	193	194	195	192
280	Imipramine	58	235	85	234	195	280	193	194
280	Imipramine	58	85	235	234	42	280	208	193
280	Imipramine*	58	235	85	234	236	195	193	208
280	Naphthylethyldiethyl Phosphate (beta-)	280	144	115	224	154	252	29	281
280	Oxo-2-phthalazinylmethyl Benzoic Acid (2-(1-(2H))) (Diftalone Metab)	132	280	133	104	235	262	77	89
280	Phenprocoumon*	251	280	118	91	121	119	252	189
280	Phenytoin Dimethyl Derivative	118	280	203	194	77	165	223	222
280	Sulphamethoxydiazine*	216	215	92	65	108	54	125	39
280	Sulphamethoxypyrazine*	92	216	215	108	65	156	54	125
280	Sulphamethoxypyridazine*	215	92	216	65	108	53	69	39
280	Tribromethyl Alcohol*	31	203	93	121	123	95	44	201
280	Trimipramine Metabolite	208	249	193	44	195	194	234	209
281	Alverine	72	176	91	58	177	65	281	42
281	Alverine*	72	176	91	58	177	41	42	30
281	Chloral Betaine	81	83	46	110	82	112	84	117
281	Diminazine*	30	43	72	102	73	42	118	99
281	Diphenylpyraline	99	42	98	70	114	96	44	43
281	Diphenylpyraline*	99	114	98	167	70	165	57	43
281	Dodemorph	154	55	41	141	42	155	281	70
281	Flufenamic Acid	263	281	166	92	145	167	235	139
281	Formyldi-(1-isopropylphenyl)amine* (N-)	91	190	119	72	191	41	44	162
281	Menazon	156	93	281	125	43	157	55	63
281	Nitrazepam*	280	253	281	206	234	252	254	264

MWT	COMPOUND NAME								
281	Phentolamine	120	122	281	91	160	65	280	162
282	1-Bromo-2-iodobenzene	282	284	155	157	76	75	50	74
282	Dimethylglycol Phthalate	59	58	31	45	149	104	43	76
282	Hexachlorobenzene (HCB)	284	286	288	282	249	247	251	214
282	Niflumic Acid	282	236	237	281	263	145	44	93
282	Oleic Acid	41	55	43	69	83	57	67	54
282	Propyl Cannabinol	267	268	45	238	282	43	55	41
282	Salicyclic Acid TMS Ester TMS Ether	267	73	268	269	45	135	193	75
283	5-Pentyl-5-(2-thiocyanatoethyl)-barbituric Acid	124	154	181	125	141	153	142	112
283	7-Aminoflunitrazepam	283	44	255	282	254	284	264	256
283	Dextropropoxyphene Carbinol	58	45	91	77	105	192	59	44
283	Dextropropoxyphene Carbinol	58	44	91	77	105	59	192	43
283	Levallorphan	41	85	84	157	56	43	57	70
283	Levallorphan	283	44	282	256	43	176	85	57
283	Levallorphan*	283	282	256	176	157	43	41	57
283	Methyldesorphine	283	282	160	42	284	268	110	44
283	Nitrofen	283	285	202	139	50	75	63	76
284	Diazepam	256	77	283	221	255	257	89	165
284	Diazepam*	256	283	284	285	257	255	258	286
284	Ethyl Palmitate	88	101	43	57	55	41	89	73
284	Lynestrenol	91	79	67	201	77	105	93	120
284	Mazindol	266	88	102	75	115	128	131	176
284	Mazindol*	266	268	267	255	231	102	88	176
284	Methyl Enol Ether of Tolbutamide	91	155	129	41	284	229	163	184
284	Methyl Heptadecanoate	74	87	43	41	55	75	57	69
284	Methyltolbutamide (N-Methyl)	91	155	185	65	41	92	121	56
284	Promazine	58	85	284	238	239	198	226	224
284	Promazine	58	86	42	85	199	198	238	284
284	Promazine*	58	284	86	238	198	199	85	42
284	Promethazine	72	73	198	180	213	284	42	44
284	Promethazine	72	199	167	198	71	42	56	200
284	Promethazine	72	180	198	73	42	56	70	71
284	Promethazine	72	284	73	43	42	198	180	166
284	Promethazine*	72	73	284	198	213	199	180	56
284	Psilocybin	58	204	59	146	159	205	160	57
284	Psilocybin	58	42	204	30	44	146	117	118
284	Stearic Acid	44	43	73	57	60	41	55	69
284	Stearic Acid	57	73	60	43	55	71	41	69
284	Sulphaethidole*	92	284	156	108	65	106	93	220
284	Taurolin*	30	43	57	44	42	135	56	55
284	Tris-(2-chloroethyl) Phosphate	63	249	205	251	143	65	207	223
284	Tropicamide	92	91	65	103	93	39	163	77
284	Tropicamide	92	254	163	91	93	65	103	104
285	2-Cyclopropylmethylamino-5-Chlorobenzophenone	55	285	77	270	91	105	166	56
285	2-Desmethylchlordiazepoxide	285	268	284	77	42	286	233	287
285	4'-Methoxy-N-isopropyl-2-(methylsulphonyl)acetanilide	150	134	164	123	136	79	122	285
285	4-Bromo-2,5-dimethoxyamphetamine Impurity	56	254	256	229	231	199	201	0
285	7-Aminoclonazepam	285	256	43	110	257	84	287	111
285	Dihydromorphinone	285	36	229	42	96	228	44	286
285	Hydromorphone	285	96	229	228	70	214	115	200
285	Hydromorphone	42	285	115	44	96	58	228	229
285	Isothipendyl	72	73	181	86	214	42	56	200
285	Isothipendyl*	72	73	214	200	44	285	86	56
285	Mepyramine	121	58	72	71	214	122	215	78

MWT	COMPOUND NAME								
285	Morphine	285	162	42	44	31	215	70	200
285	Morphine	285	215	162	286	124	284	268	174
285	Morphine	285	42	70	162	44	284	124	59
285	Morphine	115	285	131	128	162	127	77	152
285	Morphine	285	42	215	44	162	124	70	115
285	Morphine*	285	162	42	215	286	124	44	284
285	Norcodeine	285	215	81	148	110	164	286	132
285	Norcodeine*	285	81	215	148	286	164	110	115
285	Pentazocine	217	45	41	70	69	110	202	72
285	Pentazocine	217	70	69	110	202	270	285	284
285	Pentazocine	41	69	70	45	110	217	159	145
285	Pentazocine	45	41	217	69	70	110	285	202
285	Pentazocine*	45	217	70	41	69	110	285	202
285	Piperine	201	115	285	173	143	84	202	174
285	Prazepam Benzophenone	285	270	77	91	105	257	166	287
285	Probenecid*	256	121	185	43	257	65	42	214
285	Prothipendyl*	58	285	214	200	86	227	85	212
285	Prothoate	115	97	73	43	65	121	125	93
285	Pyrilamine	121	58	72	79	71	78	42	122
286	5-Hydroxymecloqualone	251	286	252	111	288	271	152	273
286	6-Hydroxymecloqualone	251	286	252	288	111	154	271	152
286	7-Hydroxymecloqualone	251	286	252	78	288	111	152	271
286	8-Hydroxymecloqualone	251	286	152	252	288	111	154	160
286	Alloxantin	44	45	101	58	42	86	142	128
286	Boldenone	122	121	91	123	147	108	77	286
286	Brallobarbitone*	207	41	39	124	91	165	122	44
286	Demoxepam*	285	286	269	287	241	242	77	270
286	Desalkyl-3-hydroxyflurazepam Dehydration Product (N1-Desalkyl)	258	223	286	257	259	75	122	251
286	Desmethylclobazam	286	244	77	218	51	217	288	215
286	Oxazepam	257	259	77	228	286	0	0	0
286	Oxazepam	268	257	267	239	233	77	205	269
286	Oxazepam	77	75	205	239	104	233	76	177
286	Oxazepam	77	205	104	51	177	75	229	151
286	Oxazepam	77	257	268	205	239	267	233	241
286	Oxazepam*	257	77	268	239	205	267	233	259
286	Thonzylamine	121	58	72	71	78	215	122	77
286	Thonzylamine	58	121	72	78	42	77	122	91
286	Thonzylamine	58	121	216	72	215	71	122	59
286	Thonzylamine*	58	121	72	71	216	215	122	78
287	Allylprodine*	172	214	42	110	57	173	91	44
287	Chlorproguanil*	127	43	229	44	161	231	85	186
287	Cycrimine	98	99	218	85	131	219	84	69
287	Cycrimine	98	58	79	129	91	77	52	172
287	Cycrimine	98	41	42	99	55	69	77	105
287	Cyproheptadine	287	96	215	286	229	213	228	243
287	Cyproheptadine	215	287	42	96	229	189	286	216
287	Cyproheptadine*	287	96	286	215	70	44	58	42
287	Dihydromorphine	287	70	44	42	163	59	288	286
287	Dihydromorphine	287	70	164	44	42	286	285	288
287	Dihydromorphine*	287	70	44	164	42	288	59	230
287	Procyclidine	84	80	55	41	42	105	77	200
287	Procyclidine	84	55	42	70	205	43	204	41
287	Procyclidine*	84	204	205	85	42	55	105	77
287	Tetryl	241	242	181	75	224	194	91	77

MWT	COMPOUND NAME								
287	Tetryl	241	181	91	66	30	75	224	90
287	Vamidithion	87	109	145	58	79	142	112	88
288	1,3-Diethylphenobarbitone	146	260	118	117	103	261	91	232
288	Azathioprine Metabolite	43	44	87	111	129	167	209	159
288	BHC(A-)	181	183	219	217	109	111	221	51
288	Bupivacaine	140	141	84	98	56	138	41	96
288	Bupivacaine*	140	141	84	41	29	96	56	55
288	Desalkylflurazepam (N1-Desalkyl)	259	260	288	287	261	289	262	290
288	Desalkylflurazepam(N1-)	259	260	288	287	261	289	262	290
288	Desalkylflurazepam(N1-)	260	288	287	259	261	289	102	262
288	Flurazepam Metabolite	42	44	259	260	261	0	0	0
288	Ibomal*	167	209	43	124	39	41	53	140
288	Lindane	183	181	109	111	219	217	51	221
288	Lindane(G-BHC)	181	183	109	219	111	217	51	221
288	Lindane(beta)	109	219	181	183	111	193	288	0
288	Tertrazepam	253	288	287	289	225	259	254	41
289	Atropine (dl-Hyoscyamine)	124	82	42	83	94	67	96	39
289	Atropine	124	83	82	94	289	42	96	125
289	Atropine	124	42	82	94	289	67	83	96
289	Atropine*	124	82	94	83	42	96	103	67
289	Benzoylecgonine	82	105	122	77	51	42	94	81
289	Benzoylecgonine	82	124	105	77	168	122	42	83
289	Benzoylecgonine	105	77	122	51	50	44	78	52
289	Benzoylecgonine*	124	82	168	77	105	42	94	83
289	Caramiphen	86	99	91	144	58	56	41	87
289	Caramiphen	86	91	99	145	115	58	87	56
289	Chlophedianol	58	139	254	77	178	105	179	111
289	Chlophedianol*	58	254	45	44	77	42	59	72
289	Quinethazone	260	262	180	145	261	179	182	181
289	Quinethazone*	260	262	180	287	261	145	286	124
290	1,3-Dimethylhydroxyphenobarbitone Methyl Ether	261	290	233	148	262	133	176	260
290	2,2',4,4'-Tetrachlorobiphenyl	292	290	220	294	222	150	293	184
290	Acetyldapsone	290	248	108	140	43	65	93	92
290	Androsterone	290	67	108	107	79	55	41	93
290	Carbinoxamine	167	71	166	139	41	78	140	168
290	Carbinoxamine	58	71	167	72	42	59	202	45
290	Carbinoxamine	58	71	42	44	167	72	43	57
290	Carbinoxamine(1-)	58	71	167	72	42	202	203	45
290	Carbinoxamine*	58	71	26	54	167	72	42	44
290	Chloroxuron	72	245	44	290	45	40	75	247
290	Methyl Enol Ether of Chlorpropamide	115	175	111	58	177	290	113	75
290	Methylchlorpropamide (N-Methyl)	111	175	205	141	113	75	112	56
290	Sulthiame*	290	184	185	104	77	168	291	198
290	Tetrachlorobiphenyl	220	292	290	222	150	74	294	255
290	Trimethoprim	290	259	275	291	243	123	200	43
291	10,11-Epoxycyclobenzaprine	58	203	202	232	84	85	101	215
291	5-(2-Bromoethyl)-5-butylbarbituric Acid	124	141	154	167	155	184	142	98
291	Cruformate	256	182	276	108	169	184	291	278
291	Cruformate Metabolite	42	94	77	250	249	291	198	51
291	Cyclobenzaprine N-oxide	229	230	215	228	101	202	226	227
291	Cyclopentolate	58	55	89	90	118	42	71	41
291	Cyclopentolate	58	42	91	71	65	30	57	56
291	Cyclopentolate*	58	71	72	207	42	91	59	118
291	Diethylthiambutene	276	111	219	277	42	97	100	135

MWT	COMPOUND NAME								
291	Diethylthiambutene	276	219	111	277	278	42	100	97
291	Eucatropine*	124	276	58	140	56	72	125	41
291	Hydroxycyclobenzaprine	58	202	231	215	289	0	0	0
291	Indomethacin Derivative TMS Ester	174	291	73	175	159	131	75	158
291	Parathion	109	97	291	139	125	137	155	123
291	Parathion	97	291	109	137	139	155	125	65
291	Parathion	291	109	97	137	29	139	78	292
291	Parathion	97	109	291	139	125	137	155	123
291	Parathion-ethyl	291	109	97	137	139	155	125	123
291	Pethidinic Acid TMS Ester	71	70	42	103	73	291	57	44
292	5-Benzylidene-1-phenylbarbituric Acid	292	291	119	248	193	249	220	221
292	Ambucetamide*	248	44	249	121	136	29	164	192
292	Chlorbromuron	61	46	294	206	292	60	45	63
292	Dimethindene	58	42	59	57	128	115	218	78
292	Dimethindene*	58	59	72	45	292	218	42	0
292	Leucinocaine*	142	120	143	86	142	92	72	56
292	Permethylated Thialbarbitone*	81	47	251	41	79	42	178	53
292	Proclonol	264	139	266	111	141	251	152	253
292	Racumin (4-Hydroxy-3-(1,2,3,4-tetrahydro-1-naphthyl)coumarin)	292	188	121	130	115	293	129	128
293	2-Hydroxyethylamino-5-Chloro-2'-Fluorobenzophenone	262	109	166	264	293	123	168	95
293	7-Acetoamidonitrazepam	293	265	264	292	43	222	223	294
293	7-Acetoamidonitrazepam	293	264	43	292	263	213	212	294
293	Amitraz	162	121	132	147	293	344	120	106
293	Butriptyline	58	45	59	42	36	115	178	91
293	Butriptyline*	58	293	45	59	193	100	178	294
293	Flurazepam Benzophenone 2	262	109	166	264	293	95	123	75
293	Flurazepam Hydroxyethyl Metabolite Benzophenone	262	264	293	263	295	265	0	0
293	Penfluridol Hofmann Reaction Product 1	275	293	96	179	0	0	0	0
293	Pentachloronitrobenzene (PCNB)	295	249	237	293	297	214	212	142
293	Pramoxine	100	120	70	41	56	42	101	293
293	Pramoxine*	100	128	70	41	42	293	56	101
293	Zomepirac*	246	291	248	290	247	139	111	292
294	1,3-Diethylsecobarbitone	224	223	41	225	125	209	43	109
294	1-(2-Ethylhexyl)-3-hydroxyphthalate	164	120	92	83	70	112	221	165
294	1-(2-Ethylhexyl)-4-hydroxyphthalate	57	165	70	120	43	41	92	83
294	2-(2-Ethylhexyl)-4-hydroxyphthalate	120	57	164	165	70	69	84	98
294	5-Phenylhydantoin(5-(p-methylphenyl))- Dimethyl Derivative	217	294	132	194	118	208	222	77
294	Cinchonidine	136	137	81	79	42	41	55	159
294	Cinchonidine	136	42	41	81	55	79	130	77
294	Cinchonidine*	136	81	137	42	41	55	130	128
294	Cinchonine	136	42	58	55	81	294	41	159
294	Cinchonine	136	42	41	55	81	79	130	77
294	Cinchonine*	136	294	81	159	55	42	41	143
294	Cyanthoate	111	138	81	109	82	93	97	68
294	Dimethacrine	86	58	278	42	87	294	193	44
294	MK-251 Metabolite	167	127	168	294	295	285	0	0
294	Methyl Octadecadienoate	67	81	54	55	95	68	82	69
294	Noxiptilin	207	206	91	179	180	178	89	208
294	Noxiptyline*	58	71	208	72	59	42	89	57
294	Proparcaine	86	99	58	87	56	100	84	70
294	Proxymetacaine	86	99	58	71	136	56	41	80
294	Triamifos	160	135	44	294	104	92	161	77
294	Trimipramine	58	249	208	99	193	232	84	0
294	Trimipramine	58	249	208	193	234	99	248	194

MWT	COMPOUND NAME								
294	Trimipramine	58	193	194	42	234	235	39	179
295	Acetylsulphamethoxazole	43	198	134	65	108	295	92	135
295	Apomorphine Dimethyl Ether	294	295	221	252	264	280	165	237
295	Chloropyrilene (Chlorothen, Tagathen)	58	131	72	71	79	42	30	78
295	Chlorothen	58	131	71	72	42	79	133	78
295	Chlorthiazide	295	268	297	97	57	270	62	64
295	Dibenzepin*	58	224	209	71	225	72	210	180
295	Dothiepin (Prothiaden)	58	59	221	204	202	203	293	234
295	Dothiepin	58	59	236	202	203	221	42	57
295	Dothiepin	58	59	295	57	221	202	204	203
295	Dothiepin*	58	236	40	202	235	203	42	44
295	Flufenamic Acid Methyl Ester	263	295	264	166	92	235	145	243
295	Lysergic Acid Monoethylamide	295	221	181	207	196	180	223	167
295	Mebendazole*	186	218	77	295	263	105	51	158
295	Methylnitrazepam	267	77	51	91	63	117	294	248
295	Normethadone	58	72	71	224	59	42	152	57
295	Normethadone	58	72	71	42	59	163	91	179
296	((3-Dimethylaminopropyl)-2-hydroxyiminodibenzyl)(N-)(Imipramine Metab)	58	251	250	296	211	85	224	209
296	2-Methyl-3-(2'-methyl-4'-hydroxy-5'-methoxyphenyl)-quinazoline	296	281	143	279	76	297	39	77
296	3-Methyl-?-hydroxydilantin Methyl Ether	296	180	267	210	104	77	134	297
296	3-Methyl-p-hydroxydilantin Methyl Ether	296	267	210	219	134	180	77	297
296	7,14-Dihydroxydiftalone	133	146	89	278	105	77	132	118
296	Disulfiram	116	88	43	148	60	56	72	117
296	Disulfiram*	116	88	29	44	60	148	56	27
296	Dithianon	296	76	104	240	50	295	268	297
296	Ethynylestradiol	213	160	296	133	159	145	146	214
296	MK-251 Metabolite	127	296	169	281	183	154	141	128
296	Medinoterbacetate	254	239	44	41	77	91	191	296
296	Methdilazine	97	98	55	82	199	198	180	296
296	Niflumic Acid Methyl Ester	236	295	296	263	145	235	0	0
296	Pyrathiazine	84	296	42	85	180	55	212	41
297	Chlorthion	109	125	79	297	128	47	63	93
297	Dihydroxyprotriptyline	44	70	179	178	207	280	250	236
297	Hydrochlorothiazide	269	64	205	297	271	43	44	31
297	Hydrochlorothiazide*	269	205	221	297	271	62	285	124
297	Tridemorph	128	129	115	41	202	43	70	55
298	4'-Hydroxyniflumic Acid	298	232	204	252	251	279	233	280
298	5-Hydroxyniflumic Acid	298	44	252	253	297	279	280	145
298	Benzestrol	121	107	135	163	177	298	77	136
298	Diethazine	86	298	83	58	87	30	85	180
298	Diethazine	86	180	58	87	198	298	41	212
298	Diethazine*	86	298	87	30	58	299	212	180
298	Methyl Stearate	74	87	43	55	298	143	75	57
298	Nialamide	91	106	44	78	51	177	79	107
298	Nialamide	91	106	78	51	58	44	79	149
298	Nialamide*	91	177	44	106	45	78	123	51
298	Norethindrone	91	110	79	231	272	298	41	77
298	Norethynodrel	91	215	79	105	77	55	41	298
298	Phoxim	77	97	129	157	125	31	103	51
298	Quinalphos	146	298	157	156	129	118	90	97
298	Trimeprazine	58	45	43	298	42	41	46	57
298	Trimeprazine	58	198	42	180	30	154	57	199
298	Trimeprazine*	58	298	212	198	100	299	252	199
299	Buphenine	91	176	148	133	44	174	65	107

MWT	COMPOUND NAME								
299	Buphenine*	91	133	176	174	44	107	92	177
299	Captan (Orthocide)	79	80	151	149	77	114	117	107
299	Chlordiazepoxide	282	283	284	299	241	247	77	253
299	Chlordiazepoxide	44	282	283	77	284	56	91	57
299	Chlordiazepoxide	282	283	77	284	89	91	285	163
299	Chlordiazepoxide	282	299	218	284	283	219	241	214
299	Chlordiazepoxide	282	299	284	283	241	77	55	91
299	Chlordiazepoxide*	282	299	284	283	241	56	301	253
299	Codeine	299	162	229	124	300	214	298	42
299	Codeine	299	81	42	229	162	124	300	44
299	Codeine	299	162	229	298	124	214	297	282
299	Codeine	299	42	162	229	44	124	59	300
299	Codeine	299	42	124	162	300	297	44	59
299	Codeine	299	162	298	229	115	42	214	124
299	Codeine*	299	42	162	124	229	59	300	69
299	Desmethylflunitrazepam	298	271	299	224	272	270	252	280
299	Dihydrocodeinone	299	242	243	59	57	214	185	300
299	Dihydrocodienone	299	185	242	115	42	214	128	243
299	Dowco 199	130	148	243	299	209	102	194	76
299	Hydrocodone	299	242	256	96	243	58	60	214
299	Hydrocodone	299	44	242	59	96	42	76	243
299	Hydrocodone*	299	242	59	243	42	96	70	214
299	Metoclopramide*	86	99	184	58	30	87	201	186
299	Metopon	299	96	242	243	185	228	214	300
299	Neopine	299	162	229	123	59	42	44	300
299	Nylidrin	91	160	65	174	175	279	146	124
299	Phosphamidon	127	72	264	138	109	67	193	42
300	3-Hydroxydiazepam	271	77	273	300	255	0	0	0
300	3-Hydroxydiazepam	271	56	43	41	55	42	273	77
300	3-Hydroxydiazepam(N-Methyloxazepam)	271	273	77	272	256	300	255	257
300	Azapropazone*	160	300	189	145	188	301	161	42
300	Chlorcyclizine	99	56	72	165	300	228	229	242
300	Chlorcyclizine	99	56	165	43	228	229	241	242
300	Clobazam	300	77	258	51	255	259	256	283
300	Clobazam*	300	258	77	259	283	302	231	256
300	Desmethylclomipramine	269	229	268	232	71	44	242	211
300	Diazepam Metabolite	271	77	273	300	0	0	0	0
300	Diazepam-N-Oxide	299	300	301	283	77	43	256	302
300	Ethinylestradiol	213	160	159	296	133	145	212	157
300	Ethisterone	122	121	91	147	161	43	107	120
300	Temazepam	77	104	75	51	76	111	106	138
300	Temazepam*	271	273	300	272	256	77	255	257
301	Benzhexol	98	99	218	55	41	77	96	219
301	Benzhexol	98	218	99	219	55	41	42	85
301	Benzhexol*	98	105	55	41	99	77	218	84
301	Broxyquinoline*	303	301	305	115	194	196	114	87
301	Dihydrocodeine	301	300	302	164	244	242	284	286
301	Dihydrocodeine*	301	44	42	59	164	70	302	242
301	Dimethyl-s-(4-oxo-1,2,3-benzotriazin-3(4h)-ylmethyl) Phosphorothiolate	160	132	77	104	105	76	159	109
301	Isoxsuprine (Duvadilan)	178	135	176	107	70	77	84	0
301	Isoxsuprine	77	174	107	135	176	65	281	146
301	Isoxsuprine*	178	44	135	179	77	84	107	41
301	Methyldihydromorphine	57	43	149	58	55	41	69	71
301	Morphine N-oxide	285	162	284	215	286	124	268	226

MWT	COMPOUND NAME								
301	Morphine-N-oxide	58	285	72	42	71	59	186	44
301	Oxymorphone	44	42	301	115	70	216	91	43
301	Oxymorphone*	301	216	44	42	70	302	203	57
301	Pentaquine*	187	241	242	174	30	126	175	301
301	Pronethalol (Nethalide) TMS Ester	72	73	229	75	230	43	45	153
301	Trihexyphenidyl	98	218	97	55	41	219	105	77
301	Trihexyphenidyl	98	55	218	99	41	96	219	42
302	Abietic Acid	302	43	41	135	287	91	105	121
302	Cambendazole	260	216	302	215	242	0	0	0
302	Cambendazole	260	302	216	215	243	189	242	188
302	Cannabielsoin (C3)	177	43	176	219	147	302	69	111
302	Chlorfenson	111	175	75	99	113	177	73	63
302	Ethacrynic Acid*	247	189	249	191	243	55	29	245
302	Methidathion (Technical)	145	85	93	125	146	69	58	302
302	Methyltestosterone	302	124	43	91	79	121	105	122
302	Neostigmine	72	42	208	108	65	73	66	39
302	Norethandolone	302	57	85	231	91	110	79	215
302	Properdin	71	70	218	261	57	174	219	172
303	2,3,4,5,6-Pentachloromandelonitrile (Oryzon)	277	279	275	278	280	276	281	249
303	3-Hydroxycyproheptadine	303	96	302	231	202	245	259	304
303	Chlorambucil*	254	256	118	255	303	305	63	45
303	Chloromorphide (beta)	268	303	269	305	304	0	0	0
303	Chlorphenoxamine*	58	59	179	42	178	72	77	30
303	Cocaine(beta)	182	82	83	105	94	77	96	303
303	Cocaine*	82	182	83	105	303	77	94	96
303	Cyproheptadine Metabolite	303	96	302	231	202	245	259	288
303	Hydromorphinol	303	70	58	44	57	42	216	286
303	Hyoscine	94	138	42	108	154	303	136	137
303	Hyoscine	138	94	108	136	301	154	77	42
303	Hyoscine*	94	138	42	108	136	41	96	97
303	Phenoxybenzamine	91	65	77	196	39	56	51	90
304	Cannabichromanon (C3)	289	179	43	150	231	219	55	
304	Chlorphenethiazine	58	304	214	42	59	306	232	233
304	Chlorpromazine Metabolite	72	71	233	232	0	0	0	0
304	Desalkyl-3-hydroxyflurazepam (N1-Desalkyl)	275	277	258	257	259	75	223	122
304	Diazinon	179	137	152	304	93	153	97	135
304	Dichlorphenamide*	304	306	64	74	109	177	176	48
304	Methandriol	253	271	91	105	213	145	107	147
305	5-(2-Bromoethyl)-5-isopentylbarbituric Acid	124	154	155	141	142	181	125	198
305	5-(2-Bromoethyl)-5-pentylbarbituric Acid	124	155	154	141	142	181	153	98
305	Cliquinol*	305	150	307	115	152	114	306	123
305	Cruformate Metabolite	182	276	108	169	184	278	171	305
305	Desmethyloxetorone	44	262	215	202	218	189	231	203
305	Normeperidine Ethylcarbamate	232	202	129	56	42	158	217	305
305	Norpethidinic Acid Acid Ethyl Ester N-TMS Derivative	73	129	305	128	304	276	103	232
305	Pirimiphos Methyl	109	137	246	290	110	305	276	81
306	Butacaine	120	41	100	142	178	42	92	44
306	Butacaine*	120	263	142	178	100	264	41	29
306	Dihydroxyzolimidine (Metab 1)	277	306	288	278	209	198	222	237
306	Dihydroxyzolimidine (Metab 2)	249	306	250	170	307	251	171	277
306	Dipentyl Phthalate	149	43	150	55	41	42	237	71
306	Methylated Coumatetralyl*	306	175	291	115	91	202	187	129
306	Perthane	223	224	117	167	179	178	165	193
307	Amprotropine*	86	87	58	30	103	121	72	56

307	Benztropine	140	83	82	124	96	97	167	125
307	Benztropine	83	82	42	140	124	96	77	67
307	Benztropine*	83	140	82	124	96	97	42	125
307	Dehydrated Norpropoxyphene Amide	44	220	100	57	205	129	91	307
307	Desmethyldoxepin N-Acetyl	234	86	221	219	233	235	178	217
307	Dextropropoxyphene Metabolite	44	220	100	205	57	0	0	0
307	Hydroxycyclobenzaprine N-oxide	246	245	231	215	202	116	217	203
308	1,3-Diethyl-5,5-diphenylhydantoin	180	308	279	208	77	251	165	104
308	2,3-Diethyl-5,5-diphenylhydantoin	279	208	77	308	280	104	149	180
308	7-Ethoxydiftalone	263	133	132	308	264	105	77	279
308	Benoxinate	86	99	71	58	87	140	84	100
308	Benoxinate	86	99	30	58	71	41	136	56
308	Chloralose (alpha)	71	73	85	61	113	36	43	60
308	Oxybuprocaine*	86	99	29	30	100	71	87	192
308	Phenylbutazone	77	183	308	184	105	55	51	41
308	Phenylbutazone	183	77	308	184	105	93	309	91
308	Phenylbutazone	77	183	105	184	55	93	91	51
308	Phenylbutazone*	77	183	308	184	105	55	51	41
308	Warfarin	265	131	43	103	121	120	145	146
308	Warfarin	265	120	249	43	103	121	162	131
308	Warfarin	103	131	43	77	51	145	146	50
308	Warfarin*	265	208	121	43	266	187	213	251
309	Amolanone*	86	30	87	294	58	42	57	309
309	Benzydamine*	85	58	86	91	84	70	42	225
309	Diclofenac Methyl Ester	214	242	216	309	215	311	179	151
309	Dicyclomine	86	99	55	100	87	41	44	165
309	Dicyclomine	86	55	99	41	44	100	83	56
309	Dicyclomine*	86	71	99	58	55	56	100	87
309	Diphenidol	98	105	77	41	42	55	99	84
309	Diphenidol*	98	99	105	77	55	41	127	111
309	Glymidine*	244	59	77	29	43	31	45	55
309	Isomethadone	58	128	127	59	193	179	178	115
309	Isomethadone	58	43	128	127	59	0	0	0
309	Meclofenamic Acid Methyl Ester	242	244	309	311	243	214	178	277
309	Methadone (6-Dimethylamino-4,4-diphenyl-3-heptanone)	91	223	294	57	42	56	165	44
309	Methadone	72	165	42	180	178	179	91	73
309	Methadone	72	73	58	57	223	165	70	0
309	Methadone	72	73	91	223	165	71	294	57
309	Methadone	72	71	73	57	91	165	70	56
309	Methadone	72	57	43	69	71	55	81	41
309	Methadone*	72	73	91	293	223	165	85	71
309	Methixene	197	58	99	165	309	112	152	198
309	Methixene*	99	197	44	58	112	309	41	42
309	Nor LSD (6-desmethyl-LSD)	207	309	208	182	181	209	180	167
310	Cannabinol	295	296	310	238	31	59	311	297
310	Cannabinol	295	296	43	310	238	299	58	41
310	Cannabinol	295	296	310	238	251	223	165	119
310	Flunixin Methyl Ester	295	263	310	251	294	249	277	181
310	Ibogaine	136	135	310	149	122	225	155	186
310	Ibogaine*	136	310	135	225	149	122	155	311
310	Mepazine	58	42	41	198	96	112	44	180
310	Mestranol	227	310	174	284	147	160	173	199
310	Pecazine (Mepazine)	310	111	112	58	199	212	96	41
310	Sulphadimethoxine*	246	92	65	245	108	247	39	260

MWT	COMPOUND NAME								
310	Trimethylquinalbarbitone Carboxylic Acid Metabolite	195	196	41	138	181	55	237	111
310	Warfarin Alcohol	121	189	292	310	263	265	131	251
311	3-Piperidylbenzilate	83	105	77	183	165	152	139	128
311	4-Piperidylbenzilate	183	105	77	84	82	165	152	129
311	Adiphenine	42	44	58	86	56	167	165	152
311	Adiphenine	86	165	167	99	152	166	58	56
311	Adiphenine*	86	30	167	99	87	58	165	29
311	Alphamethadol	72	91	73	193	165	58	115	178
311	Alphamethadol	72	56	91	129	105	130	73	57
311	Betamethadol	72	73	91	58	44	115	105	165
311	Betamethadol	72	73	91	58	253	193	42	115
311	Biperiden	98	218	99	55	130	41	42	85
311	Biperiden*	98	218	99	55	41	42	77	84
311	Flurazepam Chloroethyl Artifact Benzophenone	262	109	166	264	275	311	313	123
311	Fomocaine	100	131	311	218	0	0	0	0
311	MK-251 (Dimethyl-4-(a,a,b,b-tetrafluorophenethyl)benzylamine)	296	169	297	127	310	311	0	0
311	Nalorphine (N-Allylnormorphine)	271	44	270	214	272	42	43	70
311	Nalorphine	311	43	60	41	312	45	241	188
311	Nalorphine*	311	312	41	188	80	82	81	241
311	Nalorphone	41	311	39	188	81	115	42	77
311	Pyrrobutamine	84	205	42	240	81	55	91	125
311	Pyrrobutamine	205	42	125	115	186	41	91	127
311	Pyrrobutamine*	205	240	91	84	125	242	206	186
311	Thebaine	311	296	42	255	44	310	312	253
311	Thebaine	311	44	255	296	310	312	42	253
311	Thebaine	42	296	211	239	152	165	311	139
311	Thebaine*	311	255	42	44	296	310	312	174
311	Tolazamide*	91	155	114	65	197	42	41	85
312	6-Deoxy-6-azidodihydroisomorphine	123	312	59	58	70	115	171	270
312	Dydrogesterone	43	91	227	268	312	79	55	77
312	Ethopropazine	100	101	72	197	312	84	179	212
312	Ethopropazine	100	44	101	72	42	56	180	198
312	Ethopropazine*	100	101	44	72	198	180	42	29
312	Ethyl Stearate	88	101	43	57	89	55	41	73
312	MK-251 Metabolite	297	127	185	170	298	312	311	293
313	Benorylate*	121	163	151	109	43	122	108	65
313	Didesethylflurazepam Dehydration Product	313	315	314	312	137	250	183	273
313	Dimethylmorphine	313	229	138	162	42	0	0	0
313	Ethylmorphine	313	42	162	36	81	124	59	44
313	Ethylmorphine	313	243	162	124	59	42	112	0
313	Ethylmorphine	42	162	313	44	124	59	115	81
313	Ethylmorphine*	313	162	314	124	284	59	42	243
313	Flunitrazepam	285	63	75	109	312	183	238	266
313	Flunitrazepam	313	285	312	286	266	238	294	239
313	Flunitrazepam*	285	312	313	286	266	238	294	284
313	Flurazepam Metabolite	313	137	315	164	312	0	0	0
313	Flurazepam Metabolite 1	313	137	315	314	0	0	0	0
313	Mefenamic Acid TMS Derivative	223	313	208	224	222	180	194	298
313	Morphine Dimethyl Derivative	313	138	42	282	229	176	115	146
314	3-Pentylcannabidiol	231	41	67	43	232	68	91	55
314	Cannabichromene	231	41	43	232	69	55	174	57
314	Cannabidiol	231	246	314	232	193	121	74	174
314	Chlorimipramine	58	268	269	85	193	229	228	192
314	Clomipramine	58	269	85	268	270	271	314	242

MWT	COMPOUND NAME	PEAKS							
314	Clomipramine	58	85	269	268	270	271	229	227
314	Clomipramine*	58	85	269	268	270	271	314	242
314	Crotoxyphos	127	105	193	104	121	43	179	166
314	Desmethyltrimeprazine	314	229	242	44	269	86	283	71
314	Dichlofenthion	279	97	223	88	251	162	281	164
314	Diethyl-O-(2,5-dichlorophenyl) Phosphorothioate (O,O-Diethyl)	223	279	97	162	251	225	164	281
314	Methotrimeprazine Metabolite	314	229	242	269	44	57	71	70
314	Methoxypromazine*	58	86	314	229	228	185	42	44
314	Methyl Trithion	157	125	159	314	93	45	316	171
314	Permethylated Brallobarbitone*	235	193	236	136	194	39	138	121
314	Progesterone	124	43	314	79	91	229	272	105
314	Sulphaphenazole*	156	158	92	77	157	314	108	65
314	Tetrahydrocannabinol (delta 8)	221	314	248	261	193	236	222	315
314	Tetrahydrocannabinol (delta 9)	314	299	231	271	243	258	315	300
314	Tetrahydrocannabinol (delta-8)	231	314	271	258	43	193	41	246
314	Tetrahydrocannabinol	314	299	231	271	243	258	41	43
314	Tetrahydrocannabinol(Delta-9)	299	231	314	43	41	295	55	271
314	Triclocarban*	127	161	163	153	129	187	90	189
315	1,3,5-Trichloro-2,4,6-trinitrobenzene (Bulbosan)	315	316	142	144	177	179	107	285
315	Bromazepam	236	315	317	44	77	91	287	316
315	Bromazepam	236	315	317	286	288	316	208	179
315	Bromazepam*	236	317	315	288	316	286	208	78
315	Chlorprothixene	58	42	221	59	57	43	222	189
315	Chlorprothixene*	58	59	221	30	42	222	255	43
315	Clonazepam	280	314	315	234	289	240	75	76
315	Clonazepam*	280	314	315	286	234	288	316	240
315	Codeine N-oxide	299	229	162	297	240	241	242	298
315	Diethyl-O-(3,6-dichloro-2-pyridyl) Phosphorothioate	163	165	280	97	224	128	252	174
315	Oxycodone	315	230	314	316	258	201	70	229
315	Oxycodone*	315	230	316	70	44	42	258	140
316	Bupirimate	273	208	166	193	150	316	108	96
316	Butalamine*	142	143	100	155	44	112	57	29
316	Cannabigerol	193	231	247	123	194	233	136	316
316	DDE (p,p'-)	318	246	316	248	320	176	105	247
316	DDE (p,p-)	246	318	316	248	320	176	210	0
316	DDE	246	75	318	248	316	73	176	55
316	Dioxopromethazine	72	73	56	71	70	231	152	180
316	Ethacrynic Acid Methyl Ester	261	45	263	243	55	73	245	316
316	Pentaerythritoltetranitrate	46	76	57	55	56	60	47	97
316	Pentaerythritoltetranitrate	46	76	240	194	316	58	169	84
316	Pentaerythritoltetranitrate	76	46	57	55	56	60	47	43
316	Permethylated Ibomal*	195	237	138	238	196	43	39	110
316	Promethazine Sulphoxide	72	58	198	213	214	180	73	42
316	Rectidon	43	167	237	41	39	124	168	55
316	Sigmodal*	167	43	41	39	78	55	247	122
317	Azinphos Methyl	160	132	77	93	125	104	147	105
317	Guthion	160	132	77	93	105	76	104	51
317	Impramine Degradation Product	77	89	248	69	63	102	51	65
317	Phosmet	160	61	76	77	133	104	161	50
318	5-(3-Bromopropyl)-5-isopropylbarbituric Acid	138	124	141	169	168	247	249	98
318	Brompheniramine (dex) (d-Parabromdylamine)	58	249	247	72	167	248	168	42
318	Brompheniramine	249	247	58	72	167	248	168	250
318	Brompheniramine	58	42	167	249	247	168	72	248
318	Brompheniramine*	247	249	58	72	248	167	250	168

MWT	COMPOUND NAME								
318	Cambendazole Metabolite	205	219	231	318	0	0	0	0
318	Chlorpromazine	58	318	86	272	85	320	232	42
318	Chlorpromazine	58	319	86	321	85	273	36	274
318	Chlorpromazine	58	86	42	85	44	57	59	43
318	Chlorpromazine*	58	86	318	85	320	272	319	273
318	Diphenyl Phthalate	225	77	226	104	76	149	153	43
318	Phenolphthalein	274	273	225	318	181	257	121	197
318	Phenolphthalein*	274	225	318	273	275	226	257	121
318	Sulthiame Dimethyl Derivative	318	274	77	226	104	44	210	105
318	TDE (p,p'-)	235	237	165	236	199	239	82	238
318	Tetrahydrocannabidiol	233	318	193	234	262	319	273	136
319	Chloroquine	86	58	73	87	319	41	99	245
319	Chloroquine	86	81	98	58	41	57	43	319
319	Chloroquine	86	84	169	170	41	205	99	155
319	Chloroquine*	86	58	319	87	73	247	245	112
319	Cruformate Metabolite	263	42	265	164	278	319	304	264
319	Oxetorone	58	98	202	231	189	215	218	205
320	Dicyclohexyl Phthalate	104	76	50	67	54	82	41	148
320	Fenchlorvos (Ronnel)	285	287	125	109	289	79	47	93
320	Fenchlorvos	285	125	287	109	93	79	47	289
320	Fenchlorvos	287	125	285	79	109	93	289	47
320	Lorazepam	239	274	302	276	304	275	241	75
320	Lorazepam	77	137	101	177	176	204	100	203
320	Lorazepam*	291	239	274	293	75	302	276	138
320	Phenthoate	274	93	125	121	91	107	135	246
321	Clamoxyquin*	86	72	191	58	30	128	193	163
321	Dihexyverine	98	111	55	41	112	99	96	97
321	Etoxeridine	246	247	42	36	45	91	56	219
321	Phenazocine	230	58	231	44	173	105	42	159
321	Phenazocine*	230	231	58	44	105	173	159	
322	7-Ureidodiftalone	132	147	262	133	104	77	148	190
322	Chloramphenicol	70	150	151	153	77	51	118	60
322	Chloramphenicol	70	153	36	118	117	60	155	150
322	Chloramphenicol	153	170	155	152	172	136	118	60
322	Fenpipramide*	98	112	99	55	42	41	211	84
322	Methylwarfarin	279	91	280	322	247	121	201	189
323	Dimefline*	279	163	323	58	277	308	280	322
323	Ethyl-3-piperidyldiphenyl Acetate (N-)	111	96	167	165	152	105	128	139
323	Iso-lysergide*	221	72	207	181	43	44	323	222
323	LSD (d) Tartrate	221	323	181	222	207	72	223	180
323	LSD (iso)	323	221	207	181	222	196	223	72
323	LSD	323	221	44	199	222	223	207	76
323	LSD	221	72	207	181	42	44	222	58
323	LSD*	323	221	181	222	207	72	223	324
323	Piperidolate	111	96	165	42	167	43	41	56
323	Piperidolate	105	77	235	96	111	97	183	250
323	Piperidolate*	111	96	167	112	165	71	43	42
324	4'-Hydroxywarfarin	281	187	324	121	43	0	0	0
324	Acetohexamide*	210	56	43	184	211	75	99	76
324	Clofibrate Derivative	128	69	143	73	169	75	159	41
324	Diampromide	162	105	44	163	57	190	106	58
324	Diampromide*	162	105	190	163	106	29	134	77
324	Furaltadone*	100	56	101	42	185	184	41	128
324	Oxyphenbutazone	93	109	199	77	119	162	55	324

MWT	COMPOUND NAME							
324	Oxyphenbutazone	199	92	135	76	324	119	93	134
324	Oxyphenbutazone*	199	324	93	77	65	55	121	135
324	Pentachlorobiphenyl	326	324	328	330	254	256	184	258
324	Prazepam	91	55	269	295	294	324	29	241
324	Prazepam	269	324	91	296	295	55	323	325
324	Prazepam	269	91	55	295	296	241	324	271
324	Prazepam*	91	269	324	55	296	295	323	297
324	Quinidine	136	324	189	81	173	138	55	0
324	Quinidine	136	41	42	55	158	173	81	79
324	Quinine	136	158	159	186	81	0	0	0
324	Quinine	136	137	83	117	189	158	85	160
324	Quinine*	136	137	81	42	41	189	55	117
325	2,3-Dihydro-LSD	325	223	225	224	326	226	182	167
325	7-Acetamidoflunitrazepam	325	297	324	43	256	306	296	326
325	Clemizole	255	131	254	125	257	256	57	58
325	Clemizole	255	131	60	73	255	43	125	258
325	Dextropropoxyphene Metabolite 1	44	220	100	57	205	91	129	221
325	Dextropropoxyphene Metabolite	44	57	105	91	88	0	0	0
325	Ergometrine	325	221	207	196	181	44	223	42
325	Ergometrine Maleate	221	196	222	307	112	181	205	154
325	Ergometrine*	221	72	325	54	196	55	207	181
325	Ergonovine	221	181	207	196	223	325	222	180
325	Ethiazide*	296	298	205	221	64	63	41	125
325	Methadone-N-Oxide	207	129	72	91	208	105	174	128
325	Methyl-3-piperidylbenzilate (N-)	97	105	77	183	84	36	42	51
325	Methyl-3-piperidylbenzilate (N-Methyl)	97	105	77	183	165	114	167	152
325	Methyl-3-piperidylbenzilate(N-)	97	105	77	96	84	98	183	70
325	Methyl-4-piperidylbenzilate (N-Methyl)	98	105	183	77	96	55	325	114
325	Norpropoxyphene	44	208	58	117	57	193	130	115
325	Norpropoxyphene Amide	44	43	41	69	57	55	81	75
325	Norpropoxyphene Amide	44	100	234	88	57	91	105	129
326	6-Deoxy-6-azidodihydroisocodeine	123	326	59	298	70	198	185	115
326	Acepromazine	58	197	86	42	43	44	85	196
326	Acepromazine	58	86	85	326	197	43	280	241
326	Butropipazone	175	188	132	123	326	95	165	311
326	Clozapine*	243	256	70	245	192	227	258	326
326	Hydroquinidine*	138	326	55	110	189	82	160	139
326	MK-251 Metabolite	311	127	169	312	295	153	199	184
326	Rauwolfine	326	43	144	182	58	311	183	145
327	3-Acetylmorphine	162	285	327	215	124	81	115	267
327	6-Acetylmorphine	327	268	215	59	204	70	146	81
327	6-Acetylmorphine	327	43	268	42	215	81	146	59
327	6-Acetylmorphine*	327	268	328	42	43	215	44	59
327	7-Acetoamidoclonazepam	43	327	299	298	292	329	328	256
327	Acetylmorphine	327	268	215	328	269	267	146	124
327	Acetylmorphine	327	268	43	42	215	44	328	146
327	Aminopromazine	198	70	115	199	71	269	72	41
327	Azaperone	107	165	123	95	121	0	0	0
327	Benactyzine	86	183	182	105	77	165	116	312
327	Benactyzine*	86	105	77	87	182	99	183	58
327	Diloxanide Furoate*	95	327	39	329	96	122	244	67
327	Dimenoxadole	58	105	57	43	71	55	41	167
327	Fomocaine Metabolite	219	131	100	218	0	0	0	0
327	Loxapine	70	83	42	257	193	56	228	164

MWT	COMPOUND NAME								
327	Naloxone	327	242	328	286	229	96	41	201
327	Naloxone	41	55	42	115	70	39	96	68
327	Naloxone*	327	328	41	242	286	96	229	70
327	Proquamazine	198	70	58	115	71	56	72	269
328	2,4,6-Tribromophenol	330	332	62	63	141	334	328	143
328	6-Deoxy-6-azido-14-hydroxydihydroisomorphine	328	70	216	58	187	115	286	285
328	Clorexolone*	247	285	328	249	41	287	330	55
328	Methotrimeprazine	58	328	100	42	135	228	229	242
328	Methotrimeprazine	58	328	100	269	229	283	242	243
328	Methotrimeprazine	58	42	185	30	100	228	57	328
328	Methotrimeprazine Metabolite	58	328	100	43	71	269	229	207
328	Methotrimeprazine Metabolite	58	328	100	229	269	207	242	283
328	Methotrimeprazine*	58	328	100	228	185	329	242	229
329	Cinnamoylcocaine	82	182	83	96	103	147	148	131
329	Cinnamoylcocaine	96	83	82	182	93	36	42	131
329	Cinnamoylcocaine	152	83	42	82	85	181	57	122
329	Mecarbam	131	97	58	159	125	160	329	65
329	Methyclonazepam	75	63	89	51	248	125	302	328
329	Prenylamine	238	58	91	239	167	117	165	77
329	Trilostane*	41	147	79	93	55	105	67	91
330	1,2-Diiodobenzene	203	76	50	330	74	75	127	165
330	11-HydroxyTHC (delta 9)	299	330	59	43	300	0	0	0
330	Chlorthal Methyl	301	299	303	332	45	330	221	334
330	Dipiproverine	174	247	98	175	173	248	112	245
330	Frusemide (Fursemide)	81	53	330	82	96	332	222	250
330	Frusemide*	81	53	330	96	82	332	64	63
330	Furosemide	81	53	80	64	52	330	51	63
330	Furosemide	81	53	64	96	251	82	39	48
330	Malathion	173	125	127	93	158	99	143	0
330	Malathion	125	93	127	173	158	99	55	79
330	Triarimol	107	173	77	175	79	253	251	105
331	3-Hydroxybromazepam	79	78	52	105	304	314	316	51
331	Bromazepam-N(Py)-Oxide	79	80	225	182	107	63	78	91
331	Didesethylflurazepam	246	30	302	36	274	211	273	248
331	Didesethylflurazepam	314	255	302	246	316	211	315	273
331	Ethyl-3-piperidylcyclopentyl Glycolate (N-)	111	96	175	128	157	316	262	331
331	Flurazepam(N1-Ethanol-)	288	273	331	287	304	290	289	275
331	Hydroflumethiazide*	303	331	239	255	30	158	64	159
331	Methyl-3-piperidylphenylcyclohexyl Glycolate (N-Methyl)	97	107	105	189	128	249	248	331
332	1,3-Diethylhydroxyphenobarbitone Ethyl Ether	332	303	275	304	190	276	134	333
332	Chlorazepate	242	43	270	269	241	103	243	76
332	Dichlofluanid	123	77	167	92	44	91	224	42
332	Hydroxyethylflurazepam	288	273	331	287	304	289	290	253
332	Procetofene Metabolite Methyl Ester	121	232	139	273	234	332	141	275
333	Bromodiphenhydramine	58	73	45	57	43	44	167	165
333	Bromodiphenhydramine	58	165	41	73	57	166	44	59
333	Bromodiphenhydramine	58	73	45	165	59	42	166	44
333	Bromodiphenhydramine*	58	73	45	165	59	42	166	149
333	Carbetapentane	86	91	145	84	144	85	58	87
333	Carbetapentane*	86	91	87	145	58	144	30	44
333	Deptropine*	83	140	82	124	43	96	97	42
333	Dipyrone	56	83	42	217	123	64	57	119
333	Dipyrone	56	83	230	217	123	97	77	64
333	Pirimiphos Ethyl	333	318	57	304	168	180	71	166

MWT	COMPOUND NAME								
333	Seneciphylline	119	120	136	94	93	138	121	246
334	3-Hydroxychlorpromazine	58	86	42	334	220	59	87	44
334	8-Hydroxychlorpromazine	58	86	334	42	85	336	44	243
334	Chlorpromazine Metabolite	58	246	233	318	86	272	248	232
334	Clotiazepam	289	318	291	320	275	290	319	317
334	Lormetazepam	304	44	75	306	51	57	50	305
334	Strychnine	334	44	120	77	41	107	55	144
334	Strychnine	334	130	120	333	143	107	162	144
334	Strychnine*	334	335	162	120	107	144	143	130
335	6-Cyano-6-desmethyl-LSD	335	193	192	207	234	233	167	180
335	Benefin	294	266	42	43	278	58	295	207
335	Hydroxychloroquine	64	48	247	102	58	245	304	231
335	Hydroxychloroquine*	102	245	247	304	305	306	58	126
335	Norpipanone*	98	111	99	42	55	41	29	112
335	Oxeladin*	86	91	105	87	144	58	100	56
335	Oxetorone N-oxide	274	60	273	259	181	202	218	215
335	Senecionine	136	120	220	119	121	138	335	246
336	Acebutolol*	72	43	30	56	151	221	41	98
336	Butylcarbobutoxymethyl Phthalate	149	150	41	57	56	76	104	205
336	Cruformate Metabolite	336	109	227	249	115	338	0	0
336	Dicoumarol	121	92	120	65	162	63	93	64
336	Dicoumarol*	336	121	120	215	162	92	337	187
336	Fentanyl	121	152	93	65	39	122	63	153
336	Fentanyl	146	57	245	189	42	91	77	132
336	Fluoxymesterone	43	71	55	79	91	109	123	336
337	2-Dimethylamino-4,4-diphenyl-5-nonane	72	73	322	251	165	167	337	0
337	Aromatised 2-oxo-LSD	235	237	337	236	221	338	209	238
337	Lobeline	105	77	120	43	51	96	106	97
337	Lobeline*	96	105	77	97	216	42	218	51
337	Propanidid	114	72	337	100	43	237	115	85
338	3-Methylhydroxyhexobarbitone TMS Ether	169	73	170	75	249	233	79	171
338	Bioresmethrin	123	171	143	81	128	91	172	41
338	Chlorthalidone	239	76	240	104	285	241	177	102
338	Chlorthalidone	148	76	130	75	102	104	239	50
338	Chlorthalidone*	76	239	104	240	285	241	50	75
338	Methaqualone Metabolite TMS Derivative	73	338	323	247	179	139	235	154
338	Methaqualone Metabolite TMS Derivative	323	338	73	321	154	143	163	249
338	Methaqualone Metabolite TMS Derivative	323	338	73	321	143	154	0	0
338	Methaqualone Metabolite TMS Derivative	323	338	73	75	249	143	0	0
338	Methaqualone Metabolite TMS Derivative	323	73	338	307	75	249	154	143
338	Methaqualone Metabolite TMS Derivative	323	91	154	338	251	266	309	307
338	Methaqualone Metabolite TMS Derivative	323	338	91	73	321	154	132	149
338	Methaqualone Metabolite TMS Derivative	323	338	91	321	75	154	73	149
338	Methaqualone Metabolite TMS Derivative	323	91	132	338	154	0	0	0
338	Methaqualone Metabolite TMS Derivative	323	235	338	73	176	91	75	247
339	12-Hydroxy-LSD	339	237	197	238	239	223	196	212
339	13-Hydroxy-LSD	339	237	197	223	238	212	229	340
339	2-Oxo-LSD	237	239	339	238	209	240	196	167
339	Bromothen	58	72	71	177	175	96	42	78
339	Dextropropoxyphene	58	57	59	91	42	77	105	115
339	Dextropropoxyphene	58	57	59	285	286	91	115	208
339	Dextropropoxyphene	58	205	42	191	178	128	91	129
339	Dextropropoxyphene	58	91	57	115	208	117	59	42
339	Dextropropoxyphene*	58	117	208	115	193	91	179	130

MWT	COMPOUND NAME								
339	Disopyramide*	195	212	114	30	194	72	44	43
339	Ethyl-3-piperidyl Benzilate (N-)	111	36	77	105	96	112	183	42
339	Ethyl-3-piperidyl Benzilate (N-)	111	105	96	77	183	165	128	324
339	Hexetidine	142	57	42	197	185	339	240	226
339	Methylergometrine Maleate	221	222	196	321	126	205	207	181
339	Methylergometrine Maleate	339	221	223	207	196	181	222	340
339	Methylergometrine*	339	221	196	181	207	223	222	72
339	Methylergonovine Hydrogen Maleate	221	54	339	72	55	196	207	181
339	Moramide Intermediate	100	36	56	101	42	38	165	91
339	Moramide Intermediate	100	56	101	42	165	91	115	70
339	Noracymethadol	58	36	134	43	99	59	91	44
339	Noracymethadol*	58	43	134	222	91	44	59	56
339	Papaverine	324	338	339	308	293	325	220	340
339	Papaverine	154	324	89	338	77	51	339	102
339	Papaverine*	339	324	338	325	340	308	154	292
339	Perazine*	113	339	44	70	141	43	60	340
339	Salicyluric Acid TMS Ester TMS Ether	324	193	206	73	325	75	194	45
340	3-Ethoxy-6-deoxy-6-azidodihydroisomorphine	123	340	59	283	115	322	199	70
340	3-Hydroxyprazepam	257	55	311	77	259	313	44	312
340	Dimethisterone	67	137	91	138	55	79	41	95
340	Diphenadione*	173	340	168	167	165	341	174	322
340	Norethisterone Acetate	43	340	298	325	91	41	231	280
340	Prazepam-N-Oxide	55	91	340	44	339	57	105	269
340	Propiomazine	72	149	269	255	197	73	254	340
340	Propiomazine	72	198	197	196	73	56	70	255
340	Propiomazine*	72	73	340	269	197	71	70	56
341	6-Acetylcodeine	341	282	229	204	81	70	149	124
341	6-Acetylcodeine*	341	282	229	42	43	59	342	204
341	Acetylcodeine	58	43	73	341	282	229	342	299
341	Acetylcodeine	341	282	43	229	42	342	204	162
341	Acetylcodeine	341	282	42	43	229	59	204	342
341	Fenethylline	250	207	70	91	251	119	148	65
341	Fenethylline*	250	70	207	91	251	119	148	56
341	Lumi-LSD (10-hydroxy-9,10-dihydro-LSD)	170	171	167	142	172	129	341	323
341	MK-251 Metabolite	295	127	168	153	167	296	276	311
341	Naltrexone	341	55	36	300	342	110	243	256
341	Sulpyrid	98	70	214	111	134	341	199	326
341	Thebacon	341	298	342	242	284	299	340	326
342	1,1,2,2-Tetrabromoethane	265	267	26	186	263	269	105	107
342	1,3-Dimethylhydroxyamobarbitone TMS Ether	131	327	73	143	75	132	328	169
342	1,3-Dimethylhydroxypentobarbitone TMS Ether	117	73	327	75	143	256	118	69
342	3-Acetoxydiazepam	271	61	43	45	70	300	256	273
342	6-Deoxy-6-azido-14-hydroxydihydroisocodeine	342	70	230	201	314	115	58	300
342	Desmethyldiazepam TMS Derivative	73	341	342	343	45	344	327	91
342	Diquat	156	82	80	128	81	155	78	79
342	Lactose	73	85	60	103	71	57	61	58
342	Megesterol	43	91	342	281	79	55	256	241
342	Sucrose	31	43	57	73	61	29	60	44
342	Thiophanate Methyl	192	160	342	59	150	177	105	209
342	Triazolam	342	313	238	344	315	102	75	137
342	Trithion	157	97	121	342	153	125	159	199
342	Trithion	157	121	342	97	153	45	159	125
343	Acetyldihydrocodeine	343	300	184	43	70	59	344	326
343	Acetyldihydrocodeine	343	184	300	70	43	59	334	226

MWT	COMPOUND NAME	PEAKS							
343	Acetyldihydrocodeine*	343	43	70	284	59	300	344	42
343	Cinchocaine	86	87	58	228	326	116	57	113
343	Cinchocaine	86	116	41	58	73	56	57	228
343	Clemastine	84	128	42	43	77	139	82	179
343	Clemastine*	84	128	179	42	85	178	214	98
343	Clothiapine	83	70	42	244	71	43	273	56
343	Clothiapine*	83	70	273	244	209	42	71	43
343	Hydroxynaltrexone (alpha)	343	55	110	36	98	302	84	344
344	2,2,2-Trichloroethyl Heptafluorobutyrate	169	69	131	133	197	227	95	97
344	Methotrimeprazine Sulphoxide	58	328	229	242	100	269	283	0
344	Methoxychlor	227	228	274	238	152	36	308	153
344	Oxyphencyclimine	129	112	55	41	42	44	262	105
344	Oxyphencyclimine	189	105	107	77	55	91	79	190
344	Oxyphencyclimine*	105	129	112	77	42	313	41	55
344	Permethylated Sigmodal*	195	265	43	196	41	44	39	110
344	Tetrahydrocannabinol-9-oic Acid	299	344	329	283	276	288	273	281
345	Azinphos Ethyl	132	77	160	104	105	76	97	129
345	Clopamide*	111	127	55	83	59	41	112	42
345	Prenylamine Metabolite	58	254	91	56	183	115	107	343
345	Prenylamine Metabolite	58	238	345	91	107	167	165	252
346	5-(2-Ethoxythiocarbonylthioethyl)-5-isopentylbarbituric Acid	124	55	139	99	142	141	125	154
346	Flurazepam N1-Acetic Acid	346	318	259	345	347	273	348	320
346	Flurazepam N1-Acetic Acid	245	44	247	209	183	246	271	259
346	Glenbar	301	299	303	221	142	223	317	315
346	Morpheridine	246	42	36	247	100	218	56	232
346	Morpheridine	246	36	100	42	38	82	91	56
346	Morpheridine	246	247	36	100	42	84	232	56
346	Morpheridine	246	100	42	82	91	56	232	41
346	Morpheridine*	246	247	42	100	232	56	218	172
346	Nifedipine*	329	284	224	268	330	285	225	270
346	Tolylfluanide	137	238	181	40	37	240	44	138
347	Embramine*	58	59	42	165	103	72	30	0
347	Phenomorphan	256	58	257	182	199	157	105	91
348	1,3-Dimethylhydroxyphenobarbitone TMS Ether	73	319	291	348	206	320	45	333
348	2-Diethylaminoethylamino-5-Chloro-2'-Fluorobenzophenone	86	87	58	78	56	109	95	123
348	Flurazepam Benzophenone	86	87	58	30	109	43	95	123
348	Flurazepam Benzophenone	86	87	109	95	123	348	166	262
348	Flurazepam Benzophenone 4	86	30	58	87	109	123	166	262
349	Ampicillin	44	100	34	55	115	147	41	0
349	Chlorpyrifos	97	197	199	314	316	258	286	125
349	Chlorpyriphos	197	199	97	314	248	316	286	250
349	Dipipanone	112	113	56	57	110	91	55	42
349	Dipipanone	112	113	41	56	69	334	55	91
349	Dipipanone	112	41	113	334	56	44	55	69
349	Dipipanone*	112	264	113	91	179	110	178	115
349	Norpethidinic Acid TMS Ester N-TMS Derivative	73	129	128	114	334	349	115	130
350	Ethyl Phenylmalondiamide (Primidone Derivative) di TMS Derivative	235	220	73	75	204	145	236	130
350	Propargite	135	81	173	39	41	57	150	201
351	Allyl-3-piperidylbenzilate (N-Allyl)	123	105	77	96	183	110	165	351
351	Nortriptyline TMS Derivative	44	218	203	73	217	202	215	219
351	Phenadoxone	162	105	57	163	114	190	44	77
351	Phenadoxone	114	115	70	57	56	336	91	223
352	5-(2-Iodoethyl)-5-isopentylbarbituric Acid	124	225	154	155	128	127	125	141
352	Ajmalicine	156	352	351	184	169	209	353	129

352	Anileridine	246	247	42	120	218	172	106	91
352	Anileridine	246	42	247	120	172	80	218	106
352	DDT (o,p'-)	235	237	165	75	199	246	352	0
352	DDT (p,p'-)	235	237	165	75	50	51	352	0
352	DDT	235	237	165	212	246	75	176	36
352	DDT(p,p'-)	235	237	165	212	176	282	236	284
352	Dicophane*	235	237	165	236	199	239	238	75
352	Fluopromazine*	58	352	86	353	85	306	42	266
352	Griseofulvin	138	352	215	310	214	69	321	354
352	Triflupromazine	58	86	42	85	44	59	30	70
352	Triflupromazine	58	86	352	85	306	59	266	248
353	12-Methoxy-LSD	353	251	211	237	253	252	338	354
353	13-Methoxy-LSD	353	251	211	252	237	253	226	196
353	14-Methoxy-LSD	353	251	237	211	249	338	252	351
353	Acenocoumarin	310	121	92	43	120	311	63	65
353	Alphacetylmethadol(d)	72	43	73	36	46	91	71	42
353	Alphacetylmethadol(l)	72	73	43	91	224	56	71	129
353	Alphacetylmethadol(l)	72	43	73	36	91	56	42	225
353	Berberine	321	278	320	292	306	191	322	304
353	Betacetylmethadol	72	43	44	91	278	0	0	0
353	Chelidonine	332	333	304	335	176	303	334	162
353	Dioxaphetyl Butyrate	100	72	56	165	42	101	91	115
353	Dioxaphetyl butyrate	100	42	56	101	114	353	165	91
353	Dioxaphetyl butyrate	100	114	353	91	165	70	161	178
353	Hydrochlorothiazide Tetramethyl Derivative	42	353	310	138	44	75	288	218
353	Methysergide	235	210	195	221	0	0	0	0
353	Methysergide	55	43	195	41	235	353	221	72
353	Methysergide Hydrogen Maleate	353	235	210	72	54	45	221	195
353	Methysergide*	353	210	235	336	72	54	236	195
353	Nicoumalone*	310	121	353	311	43	120	92	296
353	Propantheline	86	181	43	44	41	114	42	152
353	Tridihexethyl Chloride	86	58	88	105	206	87	55	77
354	5-(3-Hydroxyphenyl)-3-methyl-5-phenylhydantoin TMS Ether	354	73	104	268	325	282	355	77
354	5-(4-Hydroxyphenyl)-3-methyl-5-phenylhydantoin TMS Ether	354	277	325	104	268	73	269	355
354	Xipamide*	121	120	354	122	43	234	106	64
354	Yohimbine	169	156	170	143	144	115	129	142
354	Yohimbine*	353	354	169	170	355	156	184	144
355	Amodiaquine*	58	282	30	284	355	73	283	44
355	Boldine Dimethyl Ether	354	355	340	324	281	162	297	312
355	Moperone	204	217	123	186	205	0	0	0
355	Succinylsulphathiazole*	92	156	65	191	108	55	45	174
356	Conessine*	84	71	85	82	80	341	70	356
356	Diphenyl Mercury	77	51	50	356	78	354	353	279
356	Fluanisone	205	218	123	356	219	162	95	190
357	Indomethacin	44	28	29	31	43	60	73	41
357	Indomethacin	139	111	141	138	140	75	158	113
357	Indomethacin*	139	141	357	111	359	140	113	75
358	2,4,5,2',3',4'-Hexachlorobiphenyl	360	362	290	358	288	364	145	292
358	Carbethyl Salicylate	193	165	121	120	194	93	65	92
358	Chlorfenvinphos (alpha)	267	323	269	325	81	109	295	170
358	Chlorfenvinphos (beta)	267	269	323	81	325	109	295	170
358	Dimethoxanate	58	202	116	198	59	44	72	42
358	Dimethoxanate*	58	116	198	199	72	59	42	44
358	Hexachlorobiphenyl(bis-2,4,5-)	360	362	358	290	288	364	292	145

MWT	COMPOUND NAME	PEAKS							
358	Methylated THC-9-oic Acid	299	343	358	231	283	290	315	326
359	Captodiamine	58	165	255	359	166	73	199	45
359	Cyclomethycaine	112	121	344	55	110	41	179	96
359	Cyclomethycaine*	112	344	121	41	67	55	54	345
359	Methychlothiazide (Endurone)	310	64	36	312	42	43	62	63
359	Monodesethylflurazepam	341	245	246	274	343	313	58	299
359	Monodesethylflurazepam	58	246	302	71	289	273	274	341
359	Monodesethylflurazepam	246	302	58	71	289	341	56	274
359	Nortriptyline N-Trifluoroacetate	232	219	217	204	233	203	91	202
359	Rubazonic Acid	91	77	92	359	132	144	266	0
360	Prednisolone	121	122	91	147	225	43	135	120
361	Atropine TMS Ether	124	83	82	73	94	96	125	42
361	Bisacodyl	361	276	277	199	319	43	318	362
361	Bisacodyl*	361	277	319	276	199	318	362	43
361	Chromonar*	86	87	58	30	29	84	56	42
361	Furethidine	246	247	42	71	43	56	232	0
361	Furethidine*	246	42	43	247	71	41	56	91
361	Levophenacylmorphan	43	58	256	45	51	55	44	56
362	Aldrin	66	263	79	91	265	101	261	65
362	Coumaphos	362	109	97	226	210	125	364	29
362	Phthalylsulphacetamide*	109	76	104	92	239	65	108	43
362	Primidone 1,3-di-TMS Derivative	146	73	232	334	247	117	246	147
363	Opipramol*	363	206	143	42	70	207	218	113
364	Bromophos	331	329	125	333	93	109	332	62
364	Bromophos	331	329	125	333	79	109	47	93
364	Bumetanide*	321	364	304	240	168	91	322	365
365	1-Acetyllysergic Acid Diethylamide	365	263	221	264	223	0	0	0
365	Amitriptyline TMS Derivative	58	59	73	75	215	202	203	217
365	Indapamide*	147	131	130	132	119	148	365	218
365	Metolazone*	350	91	259	352	348	65	351	107
365	Pericyazine	114	44	42	41	142	43	223	56
365	Pericyazine*	114	44	142	365	42	223	115	205
366	Dibutylglycol Phthalate(n-)	57	45	56	29	101	149	85	41
366	Glibornuride*	91	155	197	65	84	39	95	41
366	Piminodine (Alvodine, Pimadin, Cimadon)	246	366	106	247	133	260	234	367
366	Piminodine	246	45	42	366	58	57	43	106
366	Piminodine	42	106	77	57	246	43	56	91
366	Piminodine Ethane Sulphonate	246	366	42	106	133	247	57	260
367	Acetylnorpropoxyphene (N-)	44	220	86	57	74	91	43	115
367	Acetylnorpropoxyphene (N-)	44	220	86	74	58	91	43	115
367	Benzethidine	246	91	42	247	233	57	218	56
367	Benzethidine	246	91	42	233	162	57	56	149
367	Dimoxyline	352	338	367	366	353	336	322	339
367	Phenbutrazate	56	69	91	261	98	119	84	190
367	Phenbutrazate*	69	56	71	91	261	84	119	70
367	Phenoperidine	246	367	247	42	91	103	77	57
367	Phenoperidine*	246	42	367	247	57	56	91	77
367	Phosalone	182	121	184	154	367	97	241	58
367	Phosalone	182	121	97	184	111	154	65	138
368	3,5-Cholestadiene	43	57	41	55	44	0	0	0
368	Cinnarizine*	201	117	167	202	251	165	118	115
368	Flunixin TMS Ester	263	353	251	368	73	249	75	277
368	Ketazolam	256	43	44	283	84	69	284	257
368	Ketazolam*	256	284	283	285	84	257	258	255

MWT	COMPOUND NAME	PEAKS							
368	Tricresyl Phosphate(m-)	368	367	91	369	165	108	90	65
368	Tricresyl Phosphate(o-)	368	165	91	179	181	90	180	107
368	Tricresyl Phosphate(p-)	368	367	107	108	369	91	198	165
369	Diamorphine	42	81	204	215	146	115	162	94
369	Diamorphine	327	369	43	268	310	42	215	128
369	Diamorphine*	327	43	369	268	310	42	215	204
369	Methoxypropoxyphene	58	121	134	127	148	238	270	284
369	Methoxypropoxyphene	58	121	134	59	57	127	105	77
369	Tri-(1-isopropylphenyl)amine	91	41	44	43	190	119	65	55
369	Tri-(1-isopropylphenyl)amine	91	190	119	70	71	55	77	72
370	Dioctyl Adipate*	129	57	70	71	55	112	43	41
370	Heptachlor	100	272	274	270	237	102	65	276
370	Thiophanate	206	133	150	370	160	134	178	151
370	Thioridazine (Melleril)	98	70	370	126	99	185	244	125
370	Thioridazine	98	70	42	185	126	99	96	55
370	Thioridazine	98	126	370	99	70	258	125	42
370	Thioridazine*	98	370	126	99	40	70	371	258
371	Camazepam	58	72	43	78	271	57	44	77
371	Demecoline*	207	371	312	342	208	372	42	328
371	Flurazepam Metabolite	84	70	83	68	288	260	273	289
371	Indomethacin Methyl Ester	139	141	111	371	140	75	113	158
371	Ioxynil	57	127	41	43	55	88	243	42
371	Trazodone*	205	70	231	78	135	136	42	166
372	3-Hydroxydiazepam TMS Derivative	73	343	257	256	345	372	283	45
372	3-Hydroxydiazepam TMS Derivative	343	372	345	357	257	0	0	0
372	4'-Hydroxydiazepam TMS Derivative	344	73	372	371	346	345	373	374
372	Deoxycortone Acetate	43	55	299	91	147	79	253	271
372	Furosemide Trimethyl Derivative	81	53	96	372	82	374	339	357
372	Methyloxazepam TMS Derivative	343	257	256	372	357	283	371	0
373	Methylrubazonic Acid	56	77	91	84	42	40	373	203
373	Prochlorperazine	70	42	113	43	141	56	71	44
373	Prochlorperazine*	113	70	373	141	43	72	42	127
374	Bromhexine	70	293	264	112	305	291	262	295
374	Bromhexine*	70	112	293	264	44	42	305	41
374	Hydroxyzine	201	165	45	166	56	203	42	58
374	Hydroxyzine	201	170	203	188	165	166	202	142
374	Hydroxyzine*	201	203	165	45	299	166	202	56
374	S 421	130	132	79	83	85	134	109	81
375	Benzyl Morphine	284	91	375	81	42	36	285	175
375	Haloperidol	224	238	226	123	340	0	0	0
375	Haloperidol	224	42	237	123	95	57	226	206
375	Haloperidol	123	95	224	42	237	206	56	84
375	Haloperidol*	224	42	237	226	123	206	239	56
375	Hyoscine TMS Derivative	138	94	73	108	42	154	137	136
375	Prenylamine Metabolite	58	284	91	213	147	152	137	115
376	1,3-Diethylhydroxyphenobarbitone TMS Ether	376	347	319	348	377	361	73	192
376	3-Acetoxychlorpromazine	58	86	376	43	378	42	87	59
376	8-Acetoxychlorpromazine	58	86	289	43	291	85	248	288
376	Phenkapton	45	121	186	97	65	93	214	153
377	Morazone	56	201	70	176	77	202	71	55
377	Morazone*	201	176	202	56	258	70	71	42
378	Dieldrin	79	108	263	277	279	345	378	0
378	Dieldrin	79	82	81	263	77	108	277	80
378	Doxapram	100	113	91	101	87	165	115	56

MWT	COMPOUND NAME	PEAKS							
378	Doxapram	100	378	113	56	101	87	379	194
378	Doxapram*	100	113	56	101	87	378	194	91
378	Endrin	67	81	263	36	79	82	261	265
378	Naled	109	145	79	185	147	35	187	47
378	Naled	109	145	15	79	96	185	73	47
378	Photodieldrin	81	79	281	244	246	279	271	273
379	Cinchocaine	86	58	87	29	30	116	41	56
379	Cinchocaine*	86	87	58	149	111	99	57	41
379	Cyclopenthiazide*	296	41	44	110	298	285	55	268
379	Droperidol	165	123	246	199	95	42	108	214
379	Droperidol*	246	165	42	123	199	247	214	108
379	Morphine Methobromide	45	58	73	285	72	80	82	42
379	Oxypertine*	175	70	176	132	379	204	56	217
379	Trichlormethiazide	296	298	279	205	36	64	117	62
380	Chlorotrianisene	380	223	382	238	152	345	215	113
380	Chlorotrianisene	380	382	345	381	190	223	238	113
380	Chlorthalidone Trimethyl Derivative	287	363	176	255	365	289	351	220
381	Benperidol*	230	109	82	187	243	363	42	123
382	3,5-Cholestadiene-7-one	174	382	161	187	159	175	383	41
382	3-Acetoxyprazepam	257	43	55	311	340	259	77	313
382	Daunomycin Metabolite	43	41	339	42	45	44	382	364
382	Dimethylaminoethyl-3-piperidylbenzilate (N-Dimethyl)	58	324	154	114	105	96	72	77
382	Mefruside*	85	43	42	86	44	41	75	110
383	BTS 29101	121	252	120	132	106	122	105	383
383	Prazosin	233	383	259	43	245	31	95	205
383	Sulpyrid Trimethyl Derivative	98	134	70	111	198	242	383	382
384	Daunomycin Metabolite	339	384	45	43	323	321	322	366
384	Ethion	231	97	153	121	125	29	65	93
384	Megesterol Acetate	281	43	282	187	107	91	55	105
385	Bamifylline*	102	84	58	56	42	30	103	91
386	Cholesterol	43	57	41	55	44	81	93	105
386	Cholesterol	81	107	105	91	95	79	93	67
386	Cholesterol	368	386	275	149	353	147	145	247
386	Cholesterol	315	81	43	55	368	41	107	95
386	Cholesterol*	386	368	43	55	275	81	57	95
386	Clonitazene	86	57	43	71	55	69	58	41
386	Heptachlor Epoxide	81	353	355	351	357	237	386	0
386	Mesoridazine	98	77	94	99	126	0	0	0
386	Mesoridazine	98	70	42	94	77	51	99	126
386	Mesoridazine*	98	70	99	42	386	126	55	41
387	Flurazepam	86	99	58	87	56	85	183	387
387	Flurazepam	86	58	99	87	56	42	71	84
387	Flurazepam*	86	87	99	58	84	387	315	56
387	Hydroflumethiazide Tetramethyl Derivative	42	44	387	172	145	252	236	188
388	Trimethobenzamide	58	195	59	72	388	89	315	42
388	Trimethobenzamide	58	195	42	388	59	57	43	56
389	Cyclothiazide*	66	120	39	65	269	205	118	77
390	Alphadolone Acetate*	317	43	271	147	289	390	95	81
390	Bromodan	359	357	361	237	355	272	251	239
390	Di-2-ethylhexyl Phthalate	149	167	57	71	70	43	150	41
390	Di-iso-octyl Phthalate	57	149	71	43	41	69	55	113
390	Dioctyl Phthalate(n-)	149	57	167	71	70	43	113	41
390	Isooctyl Phthalate	149	57	167	43	71	70	41	55
390	Meclizine	105	189	165	42	166	106	79	77

MWT	COMPOUND NAME								
390	Meclozine	105	189	36	38	201	165	285	166
390	Meclozine*	189	105	201	285	165	166	190	134
390	Permethrin	183	163	184	165	91	77	127	89
391	Dimethothiazine	72	73	320	71	70	56	210	198
391	Flavoxate	98	147	111	42	55	115	70	96
391	Flavoxate*	98	111	99	147	55	41	42	96
392	Betamethasone	122	121	91	312	342	0	0	0
392	Betamethasone	121	43	122	223	147	91	41	135
392	Bromophos Ethyl	97	359	242	303	357	331	240	301
392	Bromophos Ethyl	303	359	97	357	331	242	125	109
392	Deoxycholic Acid	91	282	79	105	77	81	67	338
392	Dexamethason	121	122	315	43	147	223	135	41
392	Dextromoramide	100	128	265	55	56	41	42	40
392	Dextromoramide*	100	265	128	266	44	98	56	101
392	Levomoramide	100	265	128	266	56	129	55	101
392	Moramide	100	128	265	266	129	246	236	306
393	Dialifor	208	210	94	40	77	44	76	209
394	Brucine	394	379	395	197	107	120	203	55
394	Brucine	394	395	120	107	146	91	134	79
394	Brucine*	394	395	379	392	120	197	203	393
394	Chlorthalidone Tetramethyl Derivative	363	176	286	255	365	288	192	220
395	Lsd-TMS Derivative	395	253	396	294	268	74	221	128
396	Aceperone	245	123	258	165	186	95	246	0
396	Etonitazene	86	36	162	58	87	107	105	0
397	3,6,17-Triacetylnormorphine	87	43	209	72	86	210	44	181
397	Diiodohydroxyquinoline*	397	115	242	398	88	143	271	62
397	Diperodon*	98	119	91	99	124	64	41	55
397	Methscopolamine Bromide	103	148	94	96	108	41	77	91
397	Octaverine*	397	368	382	398	354	29	340	312
398	Adriamycin Metabolite	338	339	77	75	296	310	78	76
398	Clefamide*	228	182	363	88	229	76	276	257
398	Pholcodeine	100	114	42	56	115	70	101	55
398	Pholcodeine*	114	100	56	42	115	101	70	398
398	Sulphasalazine*	169	92	289	65	290	39	333	184
398	Tri-2-butoxyethyl Phosphate	57	45	56	85	101	125	41	199
398	Tributoxyethyl Phosphate	85	100	125	101	199	227	99	153
399	Acetylmorphine TMS Ether	399	340	73	287	43	400	342	341
399	Colchicine	312	371	43	297	399	281	298	313
399	Colchicine	43	312	297	281	254	298	152	139
399	Colchicine*	312	43	399	297	356	281	371	311
399	Pipazethate*	98	111	99	199	41	288	200	55
399	Quinacrine	86	126	259	58	112	99	400	74
399	Thiethylperazine	70	113	42	141	43	56	399	71
399	Thiethylperazine Maleate (Torecane)	399	113	70	141	43	72	400	71
400	Clopenthixol*	143	70	100	144	42	56	98	221
401	Oxacillin	44	41	45	58	57	144	241	213
401	Oxyphenisatin Acetate*	317	288	359	401	43	318	289	196
401	Pipamazine	141	123	42	41	151	55	169	96
401	Pipamazine	123	141	43	151	41	42	57	231
401	Pipamazine*	141	169	401	42	96	41	142	70
402	Acetyl Tri-n-butyl Citrate	185	43	129	41	57	259	157	112
402	Acetyl Tri-n-butylcitrate	185	57	43	129	29	41	259	157
403	Amotriphene*	58	359	358	403	360	343	388	227
403	Perphenazine	42	246	232	70	56	143	214	43

MWT	COMPOUND NAME								
403	Perphenazine	246	42	70	143	403	56	248	112
403	Perphenazine	42	56	246	70	232	233	198	98
403	Perphenazine*	246	143	403	70	404	42	248	113
403	Phthalylsulphathiazole*	92	156	191	65	108	76	104	50
404	Benzquinamide*	205	244	191	345	206	72	272	246
404	Endosulphan	195	36	237	41	241	75	239	170
404	Nicocodeine	282	106	229	267	78	42	124	81
404	Nicocodeine	404	282	106	78	229	267	42	124
404	Sulfinpyrazone	77	278	51	78	105	279	91	39
404	Sulfinpyrazone	77	51	40	39	91	105	130	64
404	Sulphinpyrazone*	77	278	105	78	51	279	252	130
405	Clomiphene	86	87	100	30	58	44	56	42
405	Dihydroxytrazodone (Trazodone Metabolite)	239	70	209	210	166	139	138	211
405	Trazodone Metabolite	239	205	221	211	139	166	195	387
406	Bis(2-ethylhexyl)-3-hydroxyphthalate	165	112	71	185	113	164	182	149
406	Bis-(2-ethylhexyl)-4-hydroxyphthalate	165	183	70	59	71	43	295	112
406	Nicodicodeine	106	78	70	59	42	284	300	44
407	Gentian Violet	253	239	359	373	252	238	237	36
407	Trifluoperazine	70	42	113	43	407	56	141	248
407	Trifluoperazine*	113	70	407	43	141	42	127	71
408	Clobetasone*	331	43	71	332	121	147	131	41
408	Ethyl Biscoumacetate*	121	318	317	173	120	362	44	31
409	Butaperazine	70	113	43	42	71	141	56	72
410	Piperacetazine	142	42	44	170	43	41	140	96
411	Acetophenazine	254	143	411	70	255	42	380	157
411	Acetophenazine	72	45	55	54	44	254	43	53
411	Acetophenazine	254	42	70	43	56	143	222	44
411	Etorphine*	44	215	411	324	45	164	42	216
413	Narcotine	220	221	205	206	218	222	118	148
413	Narcotine	220	42	77	205	51	147	44	53
413	Noscapine	220	205	221	28	147	77	178	42
413	Noscapine	205	220	42	77	147	221	119	118
413	Noscapine*	220	221	205	147	42	193	77	118
414	Acetoxytetrahydrocannabinol (8-alpha-Acetoxy-delta 9-)	149	312	270	354	297	269	167	256
414	Diosgenin*	139	282	69	55	41	271	91	105
414	Miconazole*	159	161	81	335	333	163	337	205
416	Spironolactone	55	91	269	136	107	340	41	67
416	Spironolactone*	341	43	340	374	267	107	55	342
417	Cinepazide	221	82	80	98	81	97	319	417
418	Dinonyl Phthalate	57	149	71	70	293	69	111	43
418	Octachlorokepone	238	240	236	203	201	242	205	182
419	Mepenzolate Bromide	97	105	77	96	42	183	98	82
419	Methantheline Bromide (Banthine)	181	86	72	182	85	152	108	99
419	Methantheline Bromide	72	181	85	152	42	44	58	43
420	Aminophylline*	180	95	181	68	41	123	53	96
421	Bendrofluazide*	330	118	91	319	219	64	92	421
423	Cyprenorphine*	423	55	364	255	84	59	424	121
424	Difenoxin	218	42	219	91	155	165	115	56
425	Carphenazine (Proketazine)	268	143	425	55	70	41	40	269
425	Carphenazine	268	425	143	269	197	394	157	171
425	Carphenazine*	268	143	425	70	269	42	394	157
427	Cinnamyl-3-piperidylbenzilate (N-Cinnamyl)	117	199	105	77	183	427	91	336
427	Dixyrazine	212	42	187	45	70	180	56	98
427	Isoprenaline TMS-Derivative	73	68	140	98	41	125	74	99

MWT	COMPOUND NAME								
427	Morphine Methoiodide	285	226	72	58	45	42	44	162
427	Orciprenaline TMS Derivative	68	73	140	98	125	350	405	412
429	Indomethacin Derivative TMS Derivative	139	141	429	111	73	140	431	113
429	Indomethacin TMS Ester	139	141	73	111	312	429	140	75
429	Mebeverine*	308	165	309	121	55	154	98	56
429	Morphine bis TMS Ether	73	429	236	196	146	414	430	287
430	3-Hydroxydesmethyldiazepam	73	429	430	45	431	147	432	75
430	Amiperone	279	123	412	165	206	281	292	95
430	Piritramide*	386	138	387	84	110	42	301	263
431	Benzthiazide*	91	121	65	122	309	64	230	123
431	Clidinium Bromide	105	77	96	183	51	42	182	94
432	Buclizine	147	165	201	167	166	105	203	117
432	Buclizine	36	231	147	38	285	132	201	165
432	Buclizine	147	231	165	285	132	166	201	117
432	Buclizine*	231	147	285	232	201	132	165	166
433	Pipenzolate Bromide	105	111	77	96	97	183	42	51
434	Flupenthixol*	143	70	100	144	42	98	58	56
436	Bialamicol*	58	30	363	44	72	29	27	364
437	Fluphenazine	42	280	70	143	56	113	72	100
437	Fluphenazine	280	70	143	42	113	406	281	437
437	Fluphenazine*	280	42	143	70	56	113	281	265
437	Fluphenazine*	280	143	42	70	437	406	113	56
438	Diphenoxylic Acid	232	438	378	380	246	91	407	423
439	Polythiazide	310	42	312	129	45	64	30	131
439	Polythiazide*	310	206	42	64	312	299	45	48
441	Dopamine TFA Derivative	328	69	329	126	441	315	0	0
442	Acepromazine Maleate*	100	72	240	340	44	197	254	43
443	Thiothixene	113	70	42	221	56	43	222	114
444	1,3-Dimethyldihydroxysecobarbitone di-TMS Ether	73	341	43	271	75	147	41	342
445	Glipizide*	150	121	56	93	39	151	66	94
445	Metopimazine*	141	445	169	123	155	96	42	317
445	Narceine	58	234	44	59	41	42	36	427
445	Narceine	58	42	59	234	77	427	178	133
445	Narceine*	58	427	234	59	50	42	428	91
445	Thiopropazate	246	185	445	70	125	154	213	87
445	Thiopropazate	43	70	42	185	246	445	125	98
445	Thiopropazate	53	35	246	185	70	125	153	385
445	Thiopropazate*	42	43	70	246	185	55	56	445
446	Thioproperazine	70	113	43	42	127	71	44	56
446	Thioproperazine	79	96	346	81	345	70	78	113
446	Thioproperazine	70	113	127	446	198	71	445	43
452	Diphenoxylate	246	42	247	377	91	172	47	165
452	Diphenoxylate	246	91	165	377	452	0	0	0
452	Diphenoxylate	246	375	42	247	91	376	156	184
452	Diphenoxylate*	246	42	247	91	103	165	115	56
452	Nonachlorokepone	272	274	270	237	276	251	203	182
455	Salbutamol TMS Derivative	68	369	41	86	370	73	40	140
456	Dioxathion	97	125	65	153	93	45	121	73
457	Amygdalin	43	31	57	73	60	29	55	44
459	Bromophenoxim	277	184	88	279	275	63	278	53
461	Pimozide	230	187	134	217	96	83	461	109
464	Ascorbic Acid(Tetra TMS)	73	147	332	117	205	45	74	133
465	Monopentafluorobenzylmorphine	284	465	81	285	42	175	181	162
466	Cephaeline*	178	192	272	466	244	288	191	273

MWT	COMPOUND NAME							
466	Chloramphenicol TMS Ether	225	242	244	364	362	208	451	453
467	Oxethazaine	145	91	74	72	56	86	277	43
467	Oxethazaine*	72	91	114	145	75	160	117	92
469	Dimenhydrinate	58	73	45	43	57	167	44	165
470	Dienochlor	237	239	235	241	332	334	404	402
472	4-Chlorodiphenoxylic Acid	232	411	472	91	413	474	0	0
472	5-Aceto-8-hydroxyquinoline Sulphate	94	77	138	95	51	66	172	45
472	Clofazimine*	455	457	472	474	459	456	458	473
475	Fluspirilene*	244	42	72	475	109	245	85	476
476	Loperamide*	238	42	239	224	240	56	72	226
478	Dehydroemetine*	192	193	287	176	191	286	270	285
480	Emetine*	192	206	272	480	288	246	205	191
480	Isopropamide Iodide	86	114	100	44	238	115	56	72
483	Etorphine TMS Ether	44	73	272	45	164	162	396	250
486	Kepone	272	274	270	237	235	239	218	216
486	Kepone	272	274	237	270	143	276	239	235
487	Indomethacin Derivative TMS Derivative	139	73	141	111	487	370	75	140
491	Lidoflazine*	343	70	344	109	42	113	491	56
492	Benzitramide*	286	300	96	42	244	230	301	82
495	Codeine Heptafluorobutyrate Derivative	43	282	45	59	58	169	115	495
495	Nicomorphine	36	172	106	42	57	78	44	214
504	Dipyridamole*	504	473	429	505	221	474	84	430
511	Norbormide*	91	58	86	106	231	45	77	230
516	5-(3,4-Dihydroxycyclohexa-1,5-dienyl)-3-me-5-phenylhydantoin TMS Deriv	73	191	75	45	74	104	147	167
516	Dichloralphenazone*	188	47	82	96	29	77	84	56
518	Benziodarone*	518	173	264	519	373	376	520	249
523	3-Acetylmorphine 6-Heptafluorobutyrate Derivative	481	43	44	53	268	310	523	69
523	6-Acetylmorphine (3-Heptafluorobutyrate Derivative)	464	43	81	523	204	69	70	411
538	Morfamquat	170	114	156	155	113	197	83	42
540	Mirex	272	274	237	239	270	276	235	241
546	Amiodarone Probe Product*	546	86	420	517	547	391	263	250
547	Ergosine*	43	70	57	154	71	55	69	85
547	Ergosinine	70	43	40	154	41	44	267	69
547	Ergosinine*	70	154	69	224	210	55	196	209
553	6-Hydroxydopamine TFA Derivative	440	69	126	441	0	0	0	0
553	Noradrenaline TFA Derivative	440	126	69	427	0	0	0	0
554	Practolol, tri-TFA Derivative	308	266	440	43	512	152	194	126
557	Paracetamol Glucuronide Methyl Ester TMS Ether	317	223	75	147	217	43	318	73
560	Sotalol, tri TFA Derivative	168	126	405	43	368	326	446	0
561	Ergocornine Hydrogen Maleate	70	43	154	71	41	267	195	221
561	Ergocornine*	43	70	71	54	44	154	267	55
561	Ergotoxine	40	43	44	70	41	149	71	55
562	Acepifylline*	44	194	109	67	86	238	56	85
563	Dihydroergocornine	43	70	41	71	269	154	195	167
563	Dihydroergocornine*	70	71	269	154	195	55	59	57
575	Ergocryptine*	43	70	71	154	41	209	69	267
575	Ergocryptinine*	43	70	71	154	267	209	221	69
577	Dihydroergocryptine	154	246	70	349	43	223	225	167
577	Dihydroergocryptine*	154	70	155	167	223	225	349	153
578	Deserpidine (Canescine, Harmonyl)	365	195	221	366	31	29	212	197
578	Deserpidine	195	169	351	184	156	221	170	365
578	Deserpidine*	578	195	577	367	351	579	366	365
581	Ergotamine Tartrate	125	70	91	153	244	40	43	314
581	Ergotamine*	125	44	70	91	41	40	244	153

MWT	COMPOUND NAME	PEAKS							
583	Dihydroergotamine	43	70	40	269	125	91	153	131
583	Dihydroergotamine*	70	125	91	153	43	41	44	244
584	Bilirubin	43	286	299	41	44	227	300	211
585	Myrophine	585	494	91	43	57	41	73	55
585	Myrophine	91	43	57	41	73	55	60	81
587	3-Acetamido-2,4,6-triiodophenoxyacetic Acid	460	43	401	359	487	545	402	461
592	Hexobendine*	296	195	58	297	253	196	212	84
596	Chloralose (alpha) TMS Ether	73	117	75	103	205	147	145	129
596	Trimethaphan	91	65	187	92	277	259	90	273
600	Nicofuranose*	123	106	78	51	105	50	77	52
601	3-Acetamido-2,4,6-triiodophenoxyacetic Acid Methyl Ester	474	43	401	359	432	487	418	475
608	Reserpine	195	199	200	186	214	152	174	251
608	Reserpine*	608	606	195	609	395	397	212	396
609	Ergocristine	57	43	71	85	99	113	243	244
609	Ergocristine*	70	125	71	91	153	267	154	221
611	Dihydroergocristine Mesilate	125	70	91	153	41	43	244	349
611	Dihydroergocristine*	125	70	91	153	41	244	43	71
614	Heptacosafluorobutylamine	69	131	100	502	119	414	464	614
614	Heptacosafluorotributylamine	69	219	131	100	264	502	119	414
620	Clofibrate Glucuronide Methyl Ester TMS Ether	73	317	169	171	41	318	217	75
629	Phenylmercuric Nitrate	77	51	78	50	356	354	279	353
634	Rescinnamine	221	199	200	186	395	251	77	214
645	Amiodarone*	86	36	87	84	58	56	44	38
653	Bromocriptine*	70	43	154	71	41	209	86	195
654	Cortolone(Tetra TMS)	73	449	359	450	147	269	75	243
660	Hexafluorenium Bromide	165	58	163	42	166	164	139	63
666	Syrosingopine	181	395	198	251	397	396	199	666
673	Iopronic Acid	43	101	546	428	331	402	275	487
676	1,3-Dimethylhydroxypentobarbitone Glucuronide Me Ester TMS Derivative	73	253	217	185	317	75	204	69
676	Hydroxyamobarbitone Glucuronide Me TMS Derivative (peak 2)	73	217	204	75	147	185	253	45
676	Hydroxyamobarbitone Glucuronide Me TMS Derivative (peak 1)	73	253	75	217	147	185	317	45
677	Morphine Diheptafluorobutyrate Derivative	69	43	169	45	464	70	58	677
680	Tubocurarine Chloride	298	594	58	593	299	609	595	564
680	Tubocurarine Chloride	298	163	43	41	55	162	57	175
687	Iopronic Acid Methyl Ester	59	560	331	428	528	402	433	373
688	1,3-Dimethylhydroxysecobarbitone Me Ester TMS Ether	73	265	217	75	204	69	41	147
688	5-(4-Hydroxyphenyl)-3-methyl-5-phenylhydantoin Glucuronide me Ester TMS	73	317	217	75	147	318	43	79
780	Digoxin*	68	39	29	43	94	44	336	95
780	Digoxin*	73	58	57	43	41	39	29	45
822	Oxazepam Glucuronide TMS Derivative	217	269	375	241	204	341	342	329
822	Rifampicin	43	45	60	58	95	99	398	206
864	Decafentin*	154	309	197	77	155	195	307	353
999	Aloxiprin*	43	45	60	120	42	92	138	44
999	Bamethan Probe Product*	30	73	107	27	108	41	57	39
999	Benactyzine Probe Product*	105	77	182	44	86	165	51	30
999	Butethamate Prode Product*	44	43	39	40	68	38	27	37
999	Dichloralphenazone Probe Artifact*	217	218	215	235	108	36	363	247
999	Diethylcarbamazine Probe Product*	44	43	39	40	71	58	68	72
999	Methyprylone Probe Product (Dimer)*	98	321	83	55	140	41	155	182
999	Oxethazaine Probe Product 1*	91	120	121	92	65	167	119	106
999	Oxethazaine Probe Product 2*	91	92	425	72	160	44	114	65
999	Phenelzine Probe Product 1*	147	45	91	64	48	105	65	77
999	Phenelzine Probe Product 2*	45	64	147	105	91	77	79	65
999	Phenformin Probe Product*	91	114	43	85	139	104	72	243

999   Zomepirac GC Decomposition Product* 246 247 41 42 136 67 111 39

APPENDIX 4

EIGHT PEAK LISTING

Eight peak mass spectra indexed under both the base peak
(indicated by an asterisk) and the second most intense
peak

An asterisk after the compound name indicates that this spectrum forms
part of the collection of full spectra available at CRE to be published
separately

							MWT	COMPOUND
15	127*	43	67	109	79	192	39	224	Phosdrin
26*	84	31	57	43	42	58	40	84	Amitrol
26	53*	52	43	51	27	41	31	53	Acrylonitrile
26	54*	27	52	55	51	53	29	55	Propionitrile
27*	29	44	43	41	72	39	57	72	Butanal(n-)
27*	56	26	55	29	25	37	38	56	Acrolein
27*	62	49	64	63	98	51	61	98	1,2-Dichloroethane
27*	107	109	26	45	108	105	95	186	1,2-Dibromoethane
27	29*	30	26	28	43	46	0	75	Nitroethane
27	29*	58	26	57	30	39	43	58	Propionaldehyde
27	31*	43	29	49	51	80	26	80	2-Chloroethanol
27	43*	41	29	72	39	42	26	72	2-Methylpropanal
27	43*	29	57	86	41	39	58	86	2-Pentanone
27	55*	63	90	41	39	29	54	126	1,3-Dichlorobutane
27	55*	29	56	45	28	73	26	100	Ethyl Acrylate
27	62*	49	64	63	98	51	61	98	1,2-Dichloroethane
27	63*	65	83	26	85	61	98	98	1,1-Dichloroethane
28*	30	46	42	29	128	120	75	222	RDX
28*	56	29	41	43	85	39	57	100	Valerolactone(gamma-)
28	36*	38	44	48	49	29	35	128	Dichloroacetic Acid
28	43*	44	59	42	102	41	74	102	Acetylurea(N-)
28	43*	58	102	29	42	45	44	102	2-Methyl-1,4-dioxane
28	43*	72	29	27	100	44	45	100	2-Methyl-3-tetrahydrofuranone
28	44*	29	31	43	60	73	41	357	Indomethacin
28	44*	85	30	42	56	57	43	100	2-Methylpiperazine
28	46*	42	75	120	56	29	128	222	RDX
28	58*	42	30	102	45	59	0	102	Dimethylaminomethane(bis-)
28	71*	100	72	58	199	129	83	199	Diethylcarbamazine
28	103*	160	77	207	41	150	0	207	1-(3,4-Methylenedioxyphenyl)-2-nitropropene
29*	27	30	26	28	43	46	0	75	Nitroethane
29*	27	58	26	57	30	39	43	58	Propionaldehyde
29*	30	28	27	44	71	55	26	101	Methyl Acrylamide(N-)
29*	41	39	42	69	116	167	168	176	Ascorbic Acid (Vitamin C)
29*	41	57	46	76	43	27	28	133	Pentyl Nitrate
29*	42	43	120	122	27	93	95	194	Ethyl Bromopyruvate
29*	44	43	42	26	41	27	25	44	Acetaldehyde
29*	46	45	44	30	47	31	48	46	Formic Acid
29*	57	27	43	26	44	28	0	100	Vinyl Propionate
29	27*	44	43	41	72	39	57	72	Butanal(n-)
29	31*	59	45	74	43	0	0	74	Diethyl Ether
29	31*	45	27	26	46	43	30	46	Ethanol
29	31*	45	27	28	58	43	74	102	3-Ethoxypropionaldehyde
29	31*	60	30	42	41	26	56	60	Glycoaldehyde
29	31*	60	30	33	44	45	61	60	Methyl Formate
29	31*	33	61	51	69	50	32	100	2,2,2-Trifluoroethanol
29	41*	56	57	28	43	44	27	100	Butyl Vinyl Ether
29	41*	27	39	40	0	0	0	69	Butyronitrile(n-)
29	43*	41	27	58	42	39	44	58	Butane(n-)
29	43*	27	72	42	57	44	39	72	2-Butanone
29	43*	28	59	30	42	74	102	102	Cycloserine*
29	43*	57	41	72	27	56	39	100	3-Methyl-2-pentanone
29	43*	57	27	72	71	99	41	128	3-Octanone
29	43*	57	27	100	26	42	44	100	2,3-Pentanedione
29	44*	27	41	58	57	43	39	68	Imidazole

							MWT	COMPOUND
29	44*	30	85	86	56	57	185	184 Piperazine Phosphate
29	45*	43	60	42	44	41	31	60 Acetic Acid
29	45*	61	43	27	49	46	36	128 2,2-Dichloroethyl Methyl Ether
29	46*	76	30	0	0	0	0	91 Ethyl Nitrate
29	46*	76	30	0	0	0	0	105 Propyl Nitrate
29	57*	75	27	101	74	28	45	102 Ethyl Propionate
29	57*	85	72	41	43	27	114	114 3-Heptanone
29	57*	41	58	27	39	86	43	86 2-Methylbutanal
29	57*	27	85	41	43	114	39	114 5-Methyl-3-hexanone
29	57*	86	27	56	58	26	43	86 3-Pentanone
29	57*	31	27	39	58	30	26	58 1-Propen-2-ol
29	74*	57	41	27	28	87	45	102 2-Methylbutanoic Acid
29	76*	58	31	44	59	70	105	150 Tartaric Acid
30*	31	29	27	26	32	0	0	31 Methylamine
30*	42	43	27	44	29	31	41	60 1,2-Ethanediamine
30*	43	72	102	73	42	118	99	281 Diminazine*
30*	43	57	44	42	135	56	55	284 Taurolin*
30*	44	46	0	0	0	0	0	80 Ammonium Nitrate
30*	44	29	28	102	56	0	0	102 Methylene Oxalate
30*	56	28	85	43	45	41	27	102 Cadaverine
30*	56	28	85	43	45	41	27	102 1,5-Diaminopentane
30*	57	82	116	99	53	55	42	252 Cimetidine*
30*	63	69	74	75	79	62	131	225 Dinitrorhodane Benzene
30*	65	93	139	39	111	53	51	169 Noradrenaline
30*	72	115	57	41	43	56	42	259 Propranolol
30*	72	44	43	27	28	41	86	101 Propylaminopropane(n-)
30*	73	107	27	108	41	57	39	999 Bamethan Probe Product*
30*	73	41	27	42	39	31	56	73 Butylamine(n-)
30*	73	41	27	55	39	56	31	73 Isobutylamine
30*	75	74	213	120	63	91	92	213 1,3,5-Trinitrobenzene
30*	86	57	41	77	135	29	206	239 Salbutamol*
30*	86	41	57	65	39	42	111	225 Terbutaline
30*	91	121	92	51	0	0	0	121 Phenethylamine(beta-)
30*	91	39	65	92	51	63	121	121 2-Phenylethylamine
30*	99	28	42	43	41	27	39	99 Piperidone
30*	108	107	77	137	43	41	55	137 Tyramine
30*	108	107	77	137	51	109	78	137 Tyramine
30*	138	195	140	103	197	77	196	213 Baclofen*
30*	182	94	66	65	78	92	53	182 3,4-Dinitrotoluene
30*	194	167	165	193	116	77	179	211 Prenylamine Metabolite
30*	229	172	42	231	194	174	43	229 Clonidine
30*	230	232	44	215	217	77	156	259 4-Bromo-2,5-dimethoxyphenethylamine
30	28*	46	42	29	128	120	75	222 RDX
30	29*	28	27	44	71	55	26	101 Methyl Acrylamide(N-)
30	44*	41	42	100	43	45	55	100 Aminoheptane
30	44*	58	41	86	42	27	0	101 Hexylamine(n-)
30	44*	28	57	29	56	43	101	101 3-Hydroxypiperidine
30	44*	45	46	31	29	0	0	44 Nitrous oxide
30	46*	29	28	15	31	0	0	77 Methyl Nitrate
30	46*	76	31	45	43	42	47	227 Nitroglycerin
30	46*	29	76	28	43	31	44	227 Nitroglycerin
30	46*	42	75	120	44	128	56	222 RDX
30	58*	363	44	72	29	27	364	436 Bialamicol*
30	58*	29	27	28	44	101	41	101 Butylamine(N-ethyl(n-))

			PEAKS					MWT	COMPOUND
30	58*	42	85	28	44	56	57	102	3-Dimethylaminopropylamine
30	58*	77	59	56	42	51	79	165	Ephedrine GC Decomposition
30	58*	101	44	29	28	41	27	101	Ethyl Butylamine(n-)
30	58*	41	65	36	39	93	59	179	Ethylnorepinephrine
30	58*	59	77	29	95	65	57	181	Etilefrine*
30	58*	148	115	59	77	91	89	221	6-Hydroxy-3-(2-dimethylaminoethyl)benzo(b)thiophene
30	58*	59	56	77	0	0	0	181	Hydroxyephedrine*
30	58*	91	59	56	42	65	77	179	Methoxyphenamine
30	58*	42	159	56	59	31	202	231	Methylnorfenfluramine
30	58*	77	42	56	106	105	51	165	Pseudoephedrine
30	60*	62	44	34	33	46	31	60	Carbonyl Sulphide
30	72*	56	73	249	98	234	102	249	Alprenolol*
30	72*	107	56	43	73	222	57	266	Atenolol
30	72*	56	98	43	107	41	73	266	Atenolol*
30	72*	43	77	73	51	41	27	213	Clorprenaline*
30	72*	43	101	41	27	58	28	101	Dipropylamine(n-)
30	72*	56	107	223	98	73	43	267	Metoprolol
30	72*	107	56	45	41	44	43	267	Metoprolol*
30	72*	221	41	56	73	150	43	265	Oxprenolol
30	72*	151	56	109	43	108	251	266	Practolol
30	75*	213	74	120	91	63	167	213	1,3,5-Trinitrobenzene
30	82*	81	83	55	54	42	27	111	Ametazole*
30	82*	81	54	28	55	83	41	111	Histamine
30	86*	167	99	87	58	165	29	311	Adiphenine*
30	86*	87	294	58	42	57	309	309	Amolanone*
30	86*	57	108	44	29	84	41	209	Bamethan*
30	86*	72	44	141	42	29	57	254	Butanilacaine*
30	86*	58	130	29	77	87	42	216	Diethyltryptamine
30	86*	58	87	109	123	166	262	348	Flurazepam Benzophenone 4
30	86*	58	123	78	51	29	42	222	Nicametate*
30	86*	120	99	58	65	92	42	235	Procainamide
30	86*	27	58	29	28	42	101	101	Triethylamine
30	91*	92	42	29	146	44	104	205	Phenformin*
30	91*	155	108	65	197	39	107	270	Tolbutamide*
30	102*	72	44	116	55	173	71	204	Ethambutol*
30	106*	77	185	105	104	89	141	186	Mafenide*
30	124*	123	153	77	51	125	78	153	4-(2-Aminoethyl)pyrocatechol
30	182*	167	181	51	107	151	211	211	Mescaline
30	182*	181	167	211	183	151	148	211	Mescaline*
30	210*	149	164	180	193	134	0	227	2,4,6-Trinitrotoluene
30	229*	231	172	194	174	200	230	229	Clonidine*
30	229*	62	91	53	63	50	230	229	Picric Acid
30	246*	302	36	274	211	273	248	331	Didesethylflurazepam
31*	27	43	29	49	51	80	26	80	2-Chloroethanol
31*	29	59	45	74	43	0	0	74	Diethyl Ether
31*	29	45	27	26	46	43	30	46	Ethanol
31*	29	45	27	28	58	43	74	102	3-Ethoxypropionaldehyde
31*	29	60	30	42	41	26	56	60	Glycoaldehyde
31*	29	60	30	33	44	45	61	60	Methyl Formate
31*	29	33	61	51	69	50	32	100	2,2,2-Trifluoroethanol
31*	32	29	257	59	256	150	200	257	Levorphanol
31*	32	29	30	28	33	34	27	32	Methanol
31*	33	29	43	27	42	62	44	62	Ethylene Glycol
31*	43	57	73	61	29	60	44	342	Sucrose

								MWT	COMPOUND
31*	45	29	27	46	43	30	42	46	Ethanol
31*	45	46	29	27	59	74	43	136	Phenelzine*
31*	49	77	113	82	115	51	117	148	2,2,2-Trichloroethanol
31*	49	77	29	113	51	48	115	228	Triclofos*
31*	56	41	43	27	42	29	39	74	1-Butanol
31*	59	29	45	74	27	41	43	74	Diethyl Ether
31*	59	29	45	27	72	43	0	90	2-Ethoxyethanol
31*	59	42	60	27	29	45	41	60	1-Propanol
31*	139	93	153	151	77	137	123	169	Noradrenaline
31*	203	93	121	123	95	44	201	280	Tribromethyl Alcohol*
31	30*	29	27	26	32	0	0	31	Methylamine
31	42*	61	43	29	27	44	41	61	Ethanolamine
31	43*	57	73	60	29	55	44	457	Amygdalin
31	43*	42	41	33	29	39	0	74	2-Butanol
31	43*	55	72	101	28	45	32	102	4-Methyl-1,3-dioxane
31	43*	55	72	101	28	45	32	102	4-Methyl-1,3-dioxane
31	43*	29	61	60	85	73	44	236	Sorbide Nitrate*
31	45*	29	47	43	76	27	46	76	2-Methoxyethanol
31	51*	67	35	50	69	37	47	86	Chlorodifluoromethane(Freon 22)
31	51*	130	132	79	81	111	113	130	Difluorobromomethane
31	56*	41	43	27	42	29	0	74	1-Butanol
31	59*	41	57	43	29	39	0	74	Butanol (tert-)
31	59*	29	57	41	56	27	73	102	Ethyl Butyl Ether
31	59*	29	27	41	44	39	57	102	Ethyl Isobutyl Ether
31	59*	43	41	71	87	39	60	102	3-Hydroxy-3-methyl-2-butanone
31	59*	43	41	29	39	42	0	102	2-Hydroxy-2-methyl-3-butanone
31	72*	91	120	65	73	39	42	163	1-Propyl-2-phenylethylamine
31	81*	100	50	69	28	82	29	100	Tetrafluoroethylene
31	116*	66	85	118	97	47	81	116	Chlorotrifluoroethylene
31	123*	58	106	79	43	78	51	179	Iproniazid*
31	204*	117	51	77	161	146	118	232	Phenobarbitone
32	31*	29	257	59	256	150	200	257	Levorphanol
32	31*	29	30	28	33	34	27	32	Methanol
33*	51	31	52	50	34	20	0	52	Difluoromethane
33*	69	83	32	31	51	63	82	102	1,1,1,2-Tetrafluoroethane
33	31*	29	43	27	42	62	44	62	Ethylene Glycol
33	59*	61	41	44	92	31	93	92	Methyl Fluoroacetate
35	53*	246	185	70	125	153	385	445	Thiopropazate
36*	28	38	44	48	49	29	35	128	Dichloroacetic Acid
36*	57	41	40	44	38	55	39	157	Oxanamide (Quinactin)
36*	172	106	42	57	78	44	214	495	Nicomorphine
36*	231	147	38	285	132	201	165	432	Buclizine
36	44*	102	69	156	38	45	71	200	Meconic Acid
36	44*	38	77	43	51	79	132	151	Phenylpropanolamine
36	58*	41	59	126	38	44	56	141	Cyclopentamine
36	58*	134	43	99	59	91	44	339	Noracymethadol
36	82*	81	38	55	83	54	42	111	Ametazole
36	86*	87	84	58	56	44	38	645	Amiodarone*
36	86*	162	58	87	107	105	0	396	Etonitazene
36	100*	56	101	42	38	165	91	339	Moramide Intermediate
36	107*	108	77	38	39	51	79	137	Tyramine
36	111*	77	105	96	112	183	42	339	Ethyl-3-piperidyl Benzilate (N-)
36	180*	102	115	152	138	146	125	237	Ketamine
36	195*	237	41	241	75	239	170	404	Endosulphan

36	246*	100	42	38	82	91	56	346	Morpheridine
36	285*	229	42	96	228	44	286	285	Dihydromorphinone
39*	52	65	80	92	108	156	172	172	Sulfanilamide
39*	68	38	29	40	37	42	26	68	Furan
39*	96	95	29	0	0	0	0	96	Furfural
39*	138	195	136	119	53	91	120	234	Brallobarbitone Methylation Artifact*
39	66*	106	79	52	53	51	40	106	1,4-Dicyano-1-butene
39	68*	29	43	94	44	336	95	780	Digoxin*
39	72*	148	56	41	71	51	73	163	Mephentermine
39	72*	42	38	94	56	51	81	260	Pyridostigmine Bromide
40*	43	44	70	41	149	71	55	561	Ergotoxine
40*	43	170	44	169	102	69	91	170	4-Methyl-5-phenylpyrimidine*
40*	277	276	44	262	57	43	41	277	Methadone Metabolite
40	41*	39	38	32	29	0	0	41	Acetonitrile
40	41*	39	69	29	38	59	100	100	Methyl Methacrylate
40	44*	51	77	42	105	50	45	149	2-Amino-1-phenylpropanone
40	44*	152	91	65	165	42	164	195	2,3-Dimethoxyamphetamine
40	58*	38	57	55	140	41	56	155	Propylhexedrine
40	72*	42	44	73	105	77	56	177	Dimethylpropion
40	184*	169	140	126	183	44	55	212	1,2-Dimethylbarbitone (or 1,4-Dimethyl-)
41*	29	56	57	28	43	44	27	100	Butyl Vinyl Ether
41*	29	27	39	40	0	0	0	69	Butyronitrile(n-)
41*	40	39	38	32	29	0	0	41	Acetonitrile
41*	40	39	69	29	38	59	100	100	Methyl Methacrylate
41*	43	69	44	39	55	53	70	278	Acetylcarbromal
41*	43	113	96	44	98	39	69	156	Ectylurea
41*	43	183	167	140	184	124	109	226	Hydroxyaprobarbitone (N-Hydroxy)
41*	44	39	56	28	27	55	40	100	Dimethylpropiolactone(beta,beta-)
41*	55	42	115	70	39	96	68	327	Naloxone
41*	55	43	69	83	57	67	54	282	Oleic Acid
41*	57	29	60	43	30	71	39	117	Amyl Nitrite
41*	57	29	56	28	44	27	39	100	Secbutyl Vinyl Ether
41*	67	39	27	40	38	66	37	67	Allyl Cyanide
41*	67	39	40	66	38	27	37	67	Crotononitrile(cis-)
41*	67	39	40	66	38	27	37	67	Crotononitrile(trans-)
41*	68	54	55	39	27	40	29	108	Adiponitrile(1,4-Dicyanobutane)
41*	68	69	67	39	27	31	29	100	Cyclopentylmethanol
41*	69	39	84	94	27	29	67	152	Citral
41*	69	29	95	55	39	43	27	154	Citronellal
41*	69	55	67	81	82	68	43	156	Citronellol
41*	69	39	100	40	59	99	0	100	Methyl Methacrylate
41*	69	93	39	67	79	27	91	136	Myrcene
41*	69	68	93	39	29	67	43	154	Nerol
41*	69	70	45	110	217	159	145	285	Pentazocine
41*	70	39	69	42	27	38	29	70	Crotonaldehyde
41*	71	45	29	39	27	70	0	102	4-Methoxy-2-buten-1-ol
41*	81	53	221	79	39	178	233	262	Methohexitone*
41*	85	84	157	56	43	57	70	283	Levallorphan
41*	91	132	105	42	92	43	105	132	2-Methyl-3-phenylaziridine
41*	93	69	79	91	77	67	53	136	Myrcene
41*	99	39	72	45	38	27	40	99	Allyl Isothiocyanate
41*	100	42	43	30	69	39	68	100	Nitrosopyrrolidine
41*	121	123	39	202	27	204	200	200	1,3-Dibromopropane
41*	121	123	39	27	42	40	38	200	1,2-Dibromopropane

```
            PEAKS                         MWT    COMPOUND


41*  141  226  143  181    0    0    0    226  Alclofenac
41*  141   39  156   57   55   53   44    212  Butobarbitone
41*  147   79   93   55  105   67   91    329  Trilostane*
41*  167  124   39   80   53   68  141    208  Allobarbitone*
41*  167  168   39  124   97  141  181    224  Butalbital*
41*  167  168   39  125   97   44   53    224  Talbutal (5-Allyl-5-s-butylbarbituric Acid)
41*  168  167   43   39   55   97  124    238  Secobarbitone
41*  226   77  143  181  141   39  145    226  Alclofenac*
41*  311   39  188   81  115   42   77    311  Nalorphone
41   29*   39   42   69  116  167  168    176  Ascorbic Acid (Vitamin C)
41   29*   57   46   76   43   27   28    133  Pentyl Nitrate
41   42*   68   67   39   27   40   29     69  Butyronitrile(t-)
41   42*   29   25   40   24   43   21     42  Ketene
41   42*   56  100   55   71    0    0    100  Pentanolactone(delta-)
41   42*   72   71   43   27   39   29     72  Tetrahydrofuran
41   43*  129   69   55   39  149   42    278  Acetylcarbromal
41   43*   27   42  124  122   45   44    122  2-Bromopropane
41   43*  339   42   45   44  382  364    382  Daunomycin Metabolite
41   43*   73   29   39  102   31    0    102  Dipropyl Ether
41   43*   57   71   29   56   42   27    100  Heptane(n-)
41   43*   57   71   70   55   73   42     73  Isobutylamine
41   43*   72   27   39   29   42   26     72  Isobutyraldehyde
41   43*  127  129  213  154  105  171    213  Isopropyl-N-(3-chlorophenyl)carbamate (CIPC)
41   43*   27   86   39   42   71   44     86  3-Methyl-2-butanone
41   43*   55   56   42   70   29   39    112  1-Octene
41   43*   27  101   39   42   45   28    101  Propyl Thiocyanate
41   43*  184  168  167   97   55   53    254  Thiamylal
41   43*   54   27   55   39   29   40     83  Valeronitrile(n-)
41   44*   55   83   31   43   39   56    218  Meprobamate
41   44*   43   29   27   39   58   42     86  3-Methylbutanal
41   44*   42   43   39   30   56   45    115  Methylhexaneamine
41   44*   45   58   57  144  241  213    401  Oxacillin
41   45*  217   69   70  110  285  202    285  Pentazocine
41   55*   42   43   70   57   29   31     88  Amyl Alcohol
41   55*   69   56   82   42   67   43    154  1,6-Dichlorohexane
41   55*  126   68   97  110   81   43    154  2,2-Diethylcyclohexanone
41   55*   83   42  126   82   56   97    154  2,6-Diethylcyclohexanone
41   55*   39   82   54   53  109   42    138  Pentylenetetrazol
41   55*   56   97   72   57   62   43    274  Tybamate
41   56*   31   43   29   27   55   28    102  Butyl Formate(n-)
41   56*   43   27   29   55   57   42     92  1-Chlorobutane
41   56*   69   55   43   57   70   29    134  3-Chloroheptane
41   56*   84   54   55   42   69    0     84  Cyclohexane
41   56*   43   57   29   45   67   27    100  2,5-Dimethyltetrahydrofuran
41   57*   29   56   27   39   55   43    136  1-Bromobutane
41   57*   56   55   43   69   29   42    178  2-Bromoheptane
41   57*   43   29  100   39   28   56    100  3,3-Dimethylbutan-2-one
41   57*   29   39   27   45   59   56    102  2,2-Dimethylpropanoic Acid
41   57*   56   29   44   27  100   39    100  Isobutyl Vinyl Ether
41   57*   43   29   39  100   27    0    100  Butyl Methyl Ketone(t-)
41   57*   44   82   67   71   56   39    100  2-Methylcyclopentanol
41   57*  141  167   39   83   55  182    238  Nealbarbitone*
41   57*   29   43  100   39   56    0    100  Pinacolone
41   57*   29   39   56   27   45   87    102  Pivalic Acid
```

								MWT	COMPOUND
41	57*	29	101	44	36	58	27	101	Trimethylacetamide
41	58*	30	42	57	39	27	59	73	Butylamine(t-)
41	58*	30	126	59	69	56	44	141	Cyclopentamine*
41	58*	91	31	120	65	29	42	149	1-Ethyl-2-phenylethylamine
41	58*	55	44	43	59	128	56	141	Methylaminomethylheptane
41	58*	77	42	43	91	47	122	179	Mexiletine
41	67*	82	54	56	59	55	39	102	Fluorocyclohexane
41	67*	54	39	27	53	81	68	82	1,5-Hexadiene
41	67*	43	27	39	54	40	53	82	1-Hexyne
41	67*	39	40	38	42	37	66	67	Pyrrole
41	68*	40	28	67	69	0	0	68	Pyrazole
41	69*	55	95	67	56	81	71	154	Citronellal
41	69*	43	42	27	149	29	151	228	1,2-Dibromopentane
41	69*	27	55	43	42	39	29	228	1,5-Dibromopentane
41	69*	39	65	27	29	53	154	154	Geraniol
41	69*	68	67	93	55	39	0	154	Geraniol
41	69*	39	85	100	28	59	0	100	Methyl Crotonate
41	69*	39	93	68	29	27	154	154	Nebol
41	69*	93	68	80	121	67	111	154	Nerol
41	71*	42	43	61	72	101	102	102	2-Methoxytetrahydrofuran
41	72*	45	43	42	29	44	211	211	Isolan
41	72*	56	43	221	73	57	45	265	Oxprenolol*
41	75*	57	70	43	59	56	47	149	Penicillamine*
41	81*	53	235	178	79	39	195	276	Permethylated Methohexitone*
41	85*	55	56	43	42	57	70	100	2,4-Dimethyltetrahydrofuran
41	91*	92	65	39	42	40	116	149	Benzyl Methyl Ketoxime
41	91*	44	43	190	119	65	55	369	Tri-(1-isopropylphenyl)amine
41	93*	69	77	91	53	67	94	136	Myrcene
41	93*	27	39	79	80	77	43	136	Ocimene
41	93*	69	79	91	77	94	80	136	Pinene(beta-)
41	93*	77	91	79	27	39	69	136	Sabinene
41	94*	39	42	77	151	95	93	151	Amantadine
41	95*	81	69	83	108	109	67	152	Camphor
41	95*	27	43	39	93	55	29	154	Isoborneol
41	95*	43	110	93	55	136	67	154	Isoborneol
41	97*	126	39	69	53	51	125	126	5-(Hydroxymethyl)-2-furfuraldehyde
41	98*	42	99	55	69	77	105	287	Cycrimine
41	98*	81	42	39	97	53	69	98	Furfuryl Alcohol
41	98*	99	42	55	98	233	69	233	Piperoxan*
41	99*	39	72	45	98	71	38	99	Allyl Isothiocyanate
41	99*	56	57	39	55	29	155	266	Tributyl Phosphate
41	111*	43	93	55	121	69	95	154	Caran-trans-2-ol(trans-)
41	112*	113	334	56	44	55	69	349	Dipipanone
41	119*	93	204	69	105	91	55	204	Cedrene(alpha-)
41	120*	100	142	178	42	92	44	306	Butacaine
41	136*	42	55	158	173	81	79	324	Quinidine
41	149*	29	57	56	104	32	65	278	Dibutyl Phthalate
41	165*	234	193	179	42	39	178	235	Azapetine
41	167*	124	80	39	32	166	53	208	Allobarbitone
41	167*	168	124	43	97	169	96	210	Aprobarbitone
41	167*	124	168	97	39	169	45	210	Aprobarbitone*
41	167*	39	124	168	29	43	169	210	Aprobarbitone (Allypropymal, Aprozal, Alurate)
41	167*	168	124	39	97	141	67	224	Idobutal*
41	167*	168	39	97	57	124	53	224	Talbutal

								MWT	COMPOUND
41	168*	124	181	167	141	97	98	224	Butalbital
41	181*	182	39	124	53	138	97	224	Enallylpropymal*
41	186*	201	188	43	152	273	203	273	Dichlozoline
41	189*	190	104	149	103	39	115	246	Diallyl Phthalate
41	195*	194	53	138	70	137	79	236	1,2-Dimethylallobarbital (or 1,4-Dimethyl-)
41	195*	39	69	67	53	152	135	224	Vinbarbital
41	195*	141	69	39	152	135	196	224	Vinbarbitone*
41	196*	195	209	138	181	67	43	252	1,2-Dimethylbutalbital (or 1,4-Dimethyl-)
41	207*	39	124	91	165	122	44	286	Brallobarbitone*
41	215*	104	77	132	39	128	244	244	Alphenal
41	223*	166	224	169	67	138	39	252	Permethylated Vinbarbitone*
41	231*	43	232	69	55	174	57	314	Cannabichromene
41	231*	67	43	232	68	91	55	314	3-Pentylcannabidiol
41	296*	44	110	298	285	55	268	379	Cyclopenthiazide*
42*	31	61	43	29	27	44	41	61	Ethanolamine
42*	41	68	67	39	27	40	29	69	Butyronitrile(t-)
42*	41	29	25	40	24	43	21	42	Ketene
42*	41	56	100	55	71	0	0	100	Pentanolactone(delta-)
42*	41	72	71	43	27	39	29	72	Tetrahydrofuran
42*	43	84	56	60	27	44	102	192	Citric Acid
42*	43	44	57	70	71	103	91	219	Pethidine Intermediate C
42*	43	70	246	185	55	56	445	445	Thiopropazate*
42*	44	58	86	56	167	165	152	311	Adiphenine
42*	44	43	57	89	72	41	45	162	Dazomet
42*	44	259	260	261	0	0	0	288	Flurazepam Metabolite
42*	44	387	172	145	252	236	188	387	Hydroflumethiazide Tetramethyl Derivative
42*	44	116	41	30	57	56	68	116	3-Hydroxy-1-nitrosopyrrolidine
42*	55	41	100	39	68	29	27	100	Hexamethylene Oxide
42*	55	69	98	41	56	27	39	98	3-Methylcyclopentanone
42*	55	41	31	29	70	43	57	88	Pentanol(n-)
42*	56	246	70	232	233	198	98	403	Perphenazine
42*	59	73	42	44	101	41	0	101	Diacetimide
42*	68	55	45	100	41	70	39	100	3-Methyltetrahydropyran
42*	74	30	43	44	41	40	45	74	Nirosodimethylamine(N-)
42*	81	204	215	146	115	162	94	369	Diamorphine
42*	84	39	56	43	40	60	102	192	Citric Acid
42*	94	77	250	249	291	198	51	291	Cruformate Metabolite
42*	100	71	41	70	55	39	45	100	Tetrahydropyran-3-one
42*	100	43	29	72	41	30	69	100	Tetronic Acid
42*	106	77	57	246	43	56	91	366	Piminodine
42*	108	39	40	38	27	37	41	108	2,6-Dimethylpyrazine
42*	122	39	27	81	54	53	52	122	2,3,5-Trimethylpyrazine
42*	128	85	43	44	41	69	70	128	Barbituric Acid
42*	140	112	41	85	43	71	141	140	Hexamine*
42*	140	41	85	112	44	43	40	140	Methenamine
42*	156	127	228	142	140	83	72	228	Allantoin Pentamethyl Derivative
42*	162	313	44	124	59	115	81	313	Ethylmorphine
42*	231	119	152	247	0	0	0	277	Azathioprine
42*	246	232	70	56	143	214	43	403	Perphenazine
42*	280	70	143	56	113	72	100	437	Fluphenazine
42*	285	115	44	96	58	228	229	285	Hydromorphone
42*	296	211	239	152	165	311	139	311	Thebaine
42*	353	310	138	44	75	288	218	353	Hydrochlorothiazide Tetramethyl Derivative
42	29*	43	120	122	27	93	95	194	Ethyl Bromopyruvate

		PEAKS					MWT	COMPOUND
42	30*	43	27	44	29	31	41	60	1,2-Ethanediamine
42	43*	41	85	27	57	29	56	100	2-Methylhexane
42	43*	101	71	27	44	29	0	101	Methylmorpholine (N-)
42	43*	71	41	29	57	39	0	86	2-Methylpentane
42	43*	41	27	29	39	57	72	72	Pentane(n-)
42	44*	124	165	163	123	93	65	183	Adrenaline*
42	44*	147	120	65	43	45	39	244	Amidephrine*
42	44*	159	43	45	184	41	109	217	Norfenfluramine
42	44*	301	115	70	216	91	43	301	Oxymorphone
42	55*	41	68	27	29	69	39	140	1,5-Dichloropentane
42	57*	186	58	44	70	83	84	261	Ethoheptazine
42	57*	56	43	233	158	91	103	233	Norpethidine
42	57*	73	277	103	56	187	43	277	Norpethidinic Acid TMS Ester
42	57*	43	70	71	200	199	44	200	Pethidine Intermediate A
42	57*	85	56	54	58	70	44	191	Phendimetrazine
42	58*	59	57	202	203	43	91	277	Amitriptyline
42	58*	167	249	247	168	72	248	318	Brompheniramine
42	58*	41	125	59	168	89	63	183	Chlorphentermine
42	58*	221	59	57	43	222	189	315	Chlorprothixene
42	58*	91	71	65	30	57	56	291	Cyclopentolate
42	58*	59	57	128	115	218	78	292	Dimethindene
42	58*	30	43	57	44	0	0	101	Dimethylaminoacetone(N,N-)
42	58*	91	31	59	65	39	51	149	Dimethyl-2-phenylethylamine (N,N-)
42	58*	130	77	30	103	115	102	188	Dimethyltryptamine
42	58*	73	165	45	77	57	152	255	Diphenhydramine
42	58*	59	57	165	43	115	178	279	Doxepin
42	58*	71	36	59	111	30	75	257	Meclofenoxate
42	58*	41	198	96	112	44	180	310	Mepazine
42	58*	185	30	100	228	57	328	328	Methotrimeprazine
42	58*	59	234	77	427	178	133	445	Narceine
42	58*	91	41	30	39	65	59	149	Phentermine
42	58*	91	56	57	43	59	165	255	Phenyltoxolamine
42	58*	204	30	146	117	130	59	204	Psilocin
42	58*	204	30	44	146	117	118	284	Psilocybin
42	58*	59	30	43	41	0	0	59	Trimethylamine
42	59*	150	271	44	171	115	128	271	Dextromethorphan (Racemethorphan)
42	59*	150	157	44	257	256	41	257	Levorphanol
42	59*	150	271	44	0	0	0	271	Racemethorphan
42	60*	88	30	31	43	29	70	257	Benserazide*
42	70*	113	43	141	56	71	44	373	Prochlorperazine
42	70*	113	43	407	56	141	248	407	Trifluoperazine
42	71*	41	102	43	31	55	44	102	Dioxepane
42	71*	70	57	44	43	103	91	247	Meperidine
42	71*	43	56	177	77	70	72	177	Phenmetrazine
42	71*	56	43	44	177	30	117	177	Phenmetrazine
42	71*	56	43	177	77	178	105	177	Phenmetrazine*
42	72*	91	44	73	56	39	70	163	Dimethylamphetamine
42	72*	208	108	65	73	66	39	302	Neostigmine
42	72*	151	153	43	29	40	222	222	Pyramat
42	84*	133	162	161	51	119	65	162	Nicotine
42	88*	123	77	51	124	43	166	211	Methyldopa
42	88*	123	124	89	77	51	40	211	Methyldopa
42	91*	132	117	43	41	104	57	133	2-Benzylaziridine
42	98*	65	78	176	0	0	0	176	Cotinine

	PEAKS							MWT	COMPOUND
42	99*	98	70	114	96	44	43	281	Diphenylpyraline
42	100*	56	101	114	353	165	91	353	Dioxaphetyl butyrate
42	100*	28	41	70	45	40	30	100	3-Methylsydnone
42	106*	108	44	63	43	49	41	156	Mechlorethamine
42	108*	39	40	81	38	52	41	108	2,5-Dimethylpyrazine
42	114*	72	113	69	81	54	115	114	Methimazole*
42	123*	222	207	179	149	150	166	222	1,3-Diethyl-9-methylxanthine
42	124*	82	94	289	67	83	96	289	Atropine
42	124*	82	83	94	67	96	77	275	Homatropine
42	136*	94	43	95	96	47	79	183	Acephate
42	136*	41	81	55	79	130	77	294	Cinchonidine
42	136*	58	55	81	294	41	159	294	Cinchonine
42	136*	41	55	81	79	130	77	294	Cinchonine
42	142*	44	170	43	41	140	96	410	Piperacetazine
42	152*	98	154	153	174	69	56	273	Chlormezanone (Trancopal)
42	172*	98	57	44	201	91	36	275	Alphameprodine
42	200*	58	61	43	47	75	104	214	Metribuzin
42	205*	135	173	206	123	145	45	206	Pyrantel
42	205*	125	115	186	41	91	127	311	Pyrrobutamine
42	212*	187	45	70	180	56	98	427	Dixyrazine
42	218*	219	91	155	165	115	56	424	Difenoxin
42	220*	77	205	51	147	44	53	413	Narcotine
42	224*	237	123	95	57	226	206	375	Haloperidol
42	224*	237	226	123	206	239	56	375	Haloperidol*
42	231*	119	232	92	65	67	74	277	Azathioprine*
42	238*	239	224	240	56	72	226	476	Loperamide*
42	244*	72	475	109	245	85	476	475	Fluspirilene*
42	246*	247	120	172	80	218	106	352	Anileridine
42	246*	247	377	91	172	47	165	452	Diphenoxylate
42	246*	247	91	103	165	115	56	452	Diphenoxylate*
42	246*	43	247	71	41	56	91	361	Furethidine*
42	246*	36	247	100	218	56	232	346	Morpheridine
42	246*	70	143	403	56	248	112	403	Perphenazine
42	246*	367	247	57	56	91	77	367	Phenoperidine*
42	254*	70	43	56	143	222	44	411	Acetophenazine
42	263*	265	164	278	319	304	264	319	Cruformate Metabolite
42	280*	143	70	56	113	281	265	437	Fluphenazine
42	285*	70	162	44	284	124	59	285	Morphine
42	285*	215	44	162	124	70	115	285	Morphine
42	299*	162	229	44	124	59	300	299	Codeine
42	299*	124	162	300	297	44	59	299	Codeine
42	299*	162	124	229	59	300	69	299	Codeine*
42	310*	312	129	45	64	30	131	439	Polythiazide
42	313*	162	36	81	124	59	44	313	Ethylmorphine
43*	27	41	29	72	39	42	26	72	2-Methylpropanal
43*	27	29	57	86	41	39	58	86	2-Pentanone
43*	28	44	59	42	102	41	74	102	Acetylurea(N-)
43*	28	58	102	29	42	45	44	102	2-Methyl-1,4-dioxane
43*	28	72	29	27	100	44	45	100	2-Methyl-3-tetrahydrofuranone
43*	29	41	27	58	42	39	44	58	Butane(n-)
43*	29	27	72	42	57	44	39	72	2-Butanone
43*	29	28	59	30	42	74	102	102	Cycloserine*
43*	29	57	41	72	27	56	39	100	3-Methyl-2-pentanone
43*	29	57	27	72	71	99	41	128	3-Octanone

Asterisk Indicates True Base Peak

```
          PEAKS                    MWT    COMPOUND


43*  29   57   27  100   26   42   44  100  2,3-Pentanedione
43*  31   57   73   60   29   55   44  457  Amygdalin
43*  31   42   41   33   29   39    0   74  2-Butanol
43*  31   55   72  101   28   45   32  102  4-Methyl-1,3-dioxane
43*  31   55   72  101   28   45   32  102  4-Methyl-1,3-dioxane
43*  31   29   61   60   85   73   44  236  Sorbide Nitrate*
43*  41  129   69   55   39  149   42  278  Acetylcarbromal
43*  41   27   42  124  122   45   44  122  2-Bromopropane
43*  41  339   42   45   44  382  364  382  Daunomycin Metabolite
43*  41   73   29   39  102   31    0  102  Dipropyl Ether
43*  41   57   71   29   56   42   27  100  Heptane(n-)
43*  41   57   71   70   55   73   42   73  Isobutylamine
43*  41   72   27   39   29   42   26   72  Isobutyraldehyde
43*  41  127  129  213  154  105  171  213  Isopropyl-N-(3-chlorophenyl)carbamate (CIPC)
43*  41   27   86   39   42   71   44   86  3-Methyl-2-butanone
43*  41   55   56   42   70   29   39  112  1-Octene
43*  41   27  101   39   42   45   28  101  Propyl Thiocyanate
43*  41  184  168  167   97   55   53  254  Thiamylal
43*  41   54   27   55   39   29   40   83  Valeronitrile(n-)
43*  42   41   85   27   57   29   56  100  2-Methylhexane
43*  42  101   71   27   44   29    0  101  Methylmorpholine (N-)
43*  42   71   41   29   57   39    0   86  2-Methylpentane
43*  42   41   27   29   39   57   72   72  Pentane(n-)
43*  44  142   86   42  114   71   70  142  Alloxan
43*  44   87  111  129  167  209  159  288  Azathioprine Metabolite
43*  44  167   57   55  139   71  152  167  2,4-Diethylamino-s-triazine
43*  44   85   86   42  129   68   30  129  Metformin*
43*  44   87   31   45   42  131   71  132  Paraldehyde
43*  45   60  120   42   92  138   44  999  Aloxiprin*
43*  45   61   70   29   27   42   73   88  Ethyl Acetate
43*  45   60   58   95   99  398  206  822  Rifampicin
43*  46   76   41   29   57   30    0  119  Butyl Nitrate
43*  55  299   91  147   79  253  271  372  Deoxycortone Acetate
43*  56   73   41   61   55   27   29  116  Butyl Acetate(n-)
43*  56   57   41   29   27   71   42  100  2,3-Dimethylpentane
43*  56   41   31   29   27   60   39  102  Isobutyl Formate
43*  56   45   55   73   99   39   44  116  Laevulinic Acid
43*  56   41   29   31   27   60   39  102  1-Methylpropyl Formate
43*  57   41   55   44    0    0    0  368  3,5-Cholestadiene
43*  57   41   55   44   81   93  105  386  Cholesterol
43*  57   41   56   42   27   29   39  100  2,4-Dimethylpentane
43*  57   29   27   71   41  100   39  100  3-Hexanone
43*  57   29   27   72   41   71   99  128  6-Methyl-3-heptanone
43*  57   71   41   29   70   27   56  100  3-Methylhexane
43*  57   29   41   72   56   27  100  100  3-Methylpentan-2-one
43*  57   71   27   41   85   58   39  128  4-Octanone
43*  57   41   55   82   56   44   29  170  Undec-10-en-1-al
43*  58   42   27   40   29   26   39   58  Acetone
43*  58   41   39   57   29   27    0  100  Allyl Acetate
43*  58   42  143   85  171  157  102  196  Bethanechol
43*  58   42   44   30  129   36  143  182  Carbachol
43*  58   71   41   59   27   29   85  114  2-Heptanone
43*  58   29   27   41   57   39  100  100  2-Hexanone
43*  58   41   39   42   27    0    0  100  Isopropenyl Acetate
```

	PEAKS							MWT	COMPOUND
43*	58	256	45	51	55	44	56	361	Levophenacylmorphan
43*	58	57	100	41	85	29	39	100	Isobutyl Methyl Ketone
43*	58	41	29	27	71	39	57	100	2-Methylpentan-1-al
43*	58	41	57	27	39	29	100	100	4-Methylpentan-2-one
43*	58	57	41	85	100	29	39	100	4-Methyl-2-pentanone
43*	58	41	29	27	71	57	39	100	2-Methylvaleraldehyde
43*	58	41	71	59	57	27	29	142	2-Nonanone
43*	59	58	101	29	0	0	0	116	Diacetone Alcohol
43*	59	31	155	121	156	77	41	214	Phenaglycodol
43*	61	42	31	41	57	84	29	102	4-Hydroxy-3-methyl-2-butanone
43*	61	87	41	59	42	27	39	102	Isopropyl Acetate
43*	61	73	42	41	59	27	0	102	Propyl Acetate(n-)
43*	69	168	41	85	167	86	97	252	3-Ketoquinalbarbitone
43*	70	55	42	41	73	61	87	130	Amyl Acetate
43*	70	41	71	269	154	195	167	563	Dihydroergocornine
43*	70	40	269	125	91	153	131	583	Dihydroergotamine
43*	70	71	54	44	154	267	55	561	Ergocornine*
43*	70	71	154	41	209	69	267	575	Ergocryptine*
43*	70	71	154	267	209	221	69	575	Ergocryptinine*
43*	70	57	154	71	55	69	85	547	Ergosine*
43*	70	71	55	41	29	56	84	102	2-Ethyl-1-butanol
43*	70	42	185	246	445	125	98	445	Thiopropazate
43*	71	70	55	41	73	45	29	102	2,2-Dimethylbutanol
43*	71	27	41	29	70	85	39	100	3,3-Dimethylpentane
43*	71	70	29	27	41	55	39	100	3-Ethylpentane
43*	71	55	79	91	109	123	336	336	Fluoxymesterone
43*	71	27	41	39	42	29	58	114	4-Heptanone
43*	71	41	39	27	87	102	0	102	Methyl Isobutanoate
43*	71	70	41	55	69	56	84	102	2-Methyl-1-pentanol
43*	72	27	41	29	71	57	39	100	2-Ethyl Butyraldehyde
43*	73	41	102	27	31	57	42	102	Propyl Ether(n-)
43*	74	42	59	29	31	44	45	74	Methyl Acetate
43*	85	55	41	42	56	84	29	164	3-Bromohexane
43*	85	100	27	29	42	41	0	100	2,4-Pentanedione(Acetylacetone)
43*	86	41	71	58	57	0	0	86	Methyl Propyl Ketone
43*	86	41	58	27	71	39	42	86	2-Pentanone
43*	87	31	45	59	28	29	44	102	2-Methyl-1,3-dioxane
43*	91	227	268	312	79	55	77	312	Dydrogesterone
43*	91	342	281	79	55	256	241	342	Megesterol
43*	91	134	92	65	39	63	135	134	Phenylacetone
43*	93	41	179	39	137	65	32	179	Isopropyl-N-phenylcarbamate (IPC)
43*	96	95	53	81	91	27	41	96	2,5-Dimethylfuran
43*	97	61	99	45	27	119	117	132	1,1,1-Trichloroethane
43*	99	71	57	114	27	42	72	114	2,5-Hexanedione(Acetonyl Acetone)
43*	101	546	428	331	402	275	487	673	Iopronic Acid
43*	102	42	28	29	59	44	41	102	Methyl Pyruvate
43*	114	85	30	101	86	44	72	157	Buformin*
43*	121	77	105	51	103	78	48	166	2-Phenyllactic Acid
43*	127	213	171	129	41	154	153	213	Chlorpropham
43*	129	69	41	86	97	55	44	278	Acetylcarbromal*
43*	129	57	56	41	72	39	58	157	Paramethadione
43*	135	170	212	219	69	172	15	256	3,5-Diacetoxybenzoyl Chloride
43*	143	58	42	41	39	128	56	143	Trimethadione
43*	167	237	41	39	124	168	55	316	Rectidon

Asterisk Indicates True Base Peak

```
              PEAKS                           MWT      COMPOUND


43*  180    42    45    44    64   100   222   222  Acetazolamide*
43*  182   167   225   181   183   151   142   225  3,4,5-Trimethoxyamphetamine
43*  187    42   171   229    86   144   170   229  Amiloride*
43*  198   134    65   108   295    92   135   295  Acetylsulphamethoxazole
43*  221    83   236    56   223   222    55   236  Methazolamide*
43*  282    45    59    58   169   115   495   495  Codeine Heptafluorobutyrate Derivative
43*  286   299    41    44   227   300   211   584  Bilirubin
43*  312   297   281   254   298   152   139   399  Colchicine
43*  327   299   298   292   329   328   256   327  7-Acetoamidoclonazepam
43*  340   298   325    91    41   231   280   340  Norethisterone Acetate
43    30*   72   102    73    42   118    99   281  Diminazine*
43    30*   57    44    42   135    56    55   284  Taurolin*
43    31*   57    73    61    29    60    44   342  Sucrose
43    40*   44    70    41   149    71    55   561  Ergotoxine
43    40*  170    44   169   102    69    91   170  4-Methyl-5-phenylpyrimidine*
43    41*   69    44    39    55    53    70   278  Acetylcarbromal
43    41*  113    96    44    98    39    69   156  Ectylurea
43    41*  183   167   140   184   124   109   226  Hydroxyaprobarbitone (N-Hydroxy)
43    42*   84    56    60    27    44   102   192  Citric Acid
43    42*   44    57    70    71   103    91   219  Pethidine Intermediate C
43    42*   70   246   185    55    56   445   445  Thiopropazate*
43    44*   39    40    68    38    27    37   999  Butethamate Prode Product*
43    44*   39    40    71    58    68    72   999  Diethylcarbamazine Probe Product*
43    44*   29    27    41    82     0     0   100  Glutaraldehyde
43    44*   59    56    69    55    41   113   145  Heptaminol*
43    44*   42    30    70    54    96   141   173  5-Nitrobarbituric Acid
43    44*   41    69    57    55    81    75   325  Norpropoxyphene Amide
43    44*  257    59   150   200     0     0   257  Racemorphan
43    44*   73    57    60    41    55    69   284  Stearic Acid
43    44*   42    45    72    30   102    58   102  Trimethylurea
43    45*   87    41    59    27    69    39   102  Di-isopropyl Ether
43    45*   27    41    31    44    29     0    60  Isopropanol
43    45*   29    41    31     0     0     0    60  Isopropanol or Propan-2-ol
43    45*   87    59    41    27    39    69   102  Isopropyl Ether
43    45*   69    41    44    87    57    56   102  4-Methyl-2-pentanol
43    45*   60    44    89    42    87   117   132  Paraldehyde
43    45*   27    41    29    39    31    59    60  2-Propanol
43    50*  105    44   123    77    78    51   167  Cinchomeronic Acid
43    55*   45   143    57   189   145    73   189  Kynurenic Acid
43    55*   27    70    97    42    41    26    70  Methyl Vinyl Ketone
43    55*  195    41   235   353   221    72   353  Methysergide
43    56*   55    42    41    69    31    29   102  1-Hexanol
43    57*   41    56    85    29    27    39   100  2,2-Dimethylpentane
43    57*   71    85    99   113   243   244   609  Ergocristine
43    57*   41    70    55    56    83    29   130  2-Ethyl-1-hexanol
43    57*   29    27    41    28   100    71   100  Ethyl Isopropyl Ketone
43    57*   41    29    27    56    42    39    86  Hexane(n-)
43    57*  149    58    55    41    69    71   301  Methyldihydromorphine
43    57*   29   100    71    27    41    56   100  2-Methylpentan-3-one
43    57*   70    71    42   200   199    91   200  Pethidine Intermediate A
43    57*   56    41    85    29    27    39   100  2,2,3-Trimethylbutane
43    58*   57   149    71    42    41    55   163  Acetylcholine
43    58*   73   341   282   229   342   299   341  Acetylcodeine
43    58*  128   127    59     0     0     0   309  Isomethadone
```

		PEAKS						MWT	COMPOUND
43	58*	101	73	27	41	71	86	101	Methylisobutyramide(N-)
43	58*	100	42	56	44	29	28	100	Methylpiperazine(N-)
43	58*	134	222	91	44	59	56	339	Noracymethadol*
43	59*	86	85	40	41	38	42	100	2,3-Dimethyl-3-hydroxy-1-butene
43	59*	156	121	155	157	158	139	214	Phenaglycodol
43	60*	41	27	45	29	39	74	102	Isovaleric Acid
43	60*	41	87	45	27	74	39	102	Valeric Acid
43	69*	41	44	71	167	165	55	193	2-Bromo-2-ethylbutyramide(Carbromal Metabolite)
43	69*	169	45	464	70	58	677	677	Morphine Diheptafluorobutyrate Derivative
43	70*	55	41	57	42	71	0	117	Amyl Nitrite
43	70*	154	71	41	209	86	195	653	Bromocriptine*
43	70*	154	71	41	267	195	221	561	Ergocornine Hydrogen Maleate
43	70*	40	154	41	44	267	69	547	Ergosinine
43	71*	88	60	41	73	45	89	116	Ethyl Butyrate
43	71*	41	27	42	31	44	39	102	2-Hydroxy-2-methyltetrahydrofuran
43	71*	41	93	55	69	80	67	154	Linalool
43	71*	41	27	31	42	29	0	102	Tetrahydrofurfuryl Alcohol
43	72*	30	56	151	221	41	98	336	Acebutolol*
43	72*	73	36	46	91	71	42	353	Alphacetylmethadol(d)
43	72*	73	36	91	56	42	225	353	Alphacetylmethadol(l)
43	72*	44	91	278	0	0	0	353	Betacetylmethadol
43	72*	65	41	111	93	39	56	211	Orciprenaline
43	72*	73	41	70	65	40	39	211	Orciprenaline*
43	72*	133	116	248	204	104	56	248	Pindolol
43	73*	72	86	100	87	59	128	227	Dodine
43	73*	84	55	69	41	44	85	145	Emylcamate
43	74*	71	59	41	87	27	0	102	Methyl Butyrate(n-)
43	76*	60	77	59	42	44	55	76	Thiourea
43	79*	81	49	29	27	36	57	128	1,3-Dichloro-2-hydroxypropane
43	84*	126	77	56	91	105	55	233	Levophacetoperane
43	85*	42	86	44	41	75	110	382	Mefruside*
43	85*	127	41	197	57	239	129	239	Monalide
43	86*	102	69	44	42	59	0	102	Guanylurea
43	86*	77	91	40	106	84	65	220	Prilocaine
43	87*	209	72	86	210	44	181	397	3,6,17-Triacetylnormorphine
43	88*	60	113	101	61	73	70	158	Ethyl Heptanoate
43	91*	67	81	78	79	106	44	167	Ethinamate
43	91*	65	127	110	39	51	106	231	Isocarboxazid
43	91*	57	41	73	55	60	81	585	Myrophine
43	91*	44	41	45	148	149	103	166	2-Phenyllactic Acid
43	93*	41	109	94	91	39	55	152	Verbenol(trans-(-))
43	101*	72	41	27	42	39	45	101	Propyl Isothiocyanate
43	102*	42	57	56	41	40	39	102	Methyl Isopropyl Nitrosamine
43	104*	91	39	51	103	78	27	164	1-Phenylethyl Acetate
43	104*	91	105	65	39	51	77	164	Phenylethyl Acetate
43	120*	138	92	42	121	39	45	180	Acetylsalicylic Acid
43	120*	138	92	121	39	64	63	180	Acetylsalicylic Acid*
43	120*	121	138	39	65	41	29	208	Isoamyl Salicylate
43	121*	122	223	147	91	41	135	392	Betamethasone
43	123*	44	51	78	77	50	105	167	Isocinchomeronic Acid
43	124*	314	79	91	229	272	105	314	Progesterone
43	127*	229	44	161	231	85	186	287	Chlorproguanil*
43	139*	251	253	178	111	141	248	266	Chlorfenethol
43	141*	183	44	140	108	80	52	254	Paracetamol Cysteine Conjugate

							MWT	COMPOUND
43	144*	100	86	116	145	128	89	272 Indol-3-yl-N,N-disopropylglyoxamide
43	144*	42	116	69	59	145	41	144 Thiobarbituric Acid
43	149*	150	55	41	42	237	71	306 Dipentyl Phthalate
43	149*	167	57	104	112	279	113	278 2-Ethylhexyl Phthalate
43	154*	196	137	155	69	238	42	238 3,5-Diacetoxybenzoic Acid
43	167*	41	39	78	55	247	122	316 Sigmodal*
43	171*	143	41	128	55	141	159	270 2,3-Dihydroxyquinalbarbitone
43	177*	176	219	147	302	69	111	302 Cannabielsoin (C3)
43	177*	41	135	91	178	93	95	192 Ionone(beta-)
43	185*	129	41	57	259	157	112	402 Acetyl Tri-n-butyl Citrate
43	213*	68	44	71	155	170	96	213 Simetryne
43	221*	78	93	80	41	141	39	250 Heptabarbitone*
43	242*	270	269	241	103	243	76	332 Chlorazepate
43	256*	44	283	84	69	284	257	368 Ketazolam
43	257*	55	311	340	259	77	313	382 3-Acetoxyprazepam
43	260*	123	259	95	77	57	165	260 2-Amino-5-Nitro-2'-Fluorobenzophenone
43	281*	282	187	107	91	55	105	384 Megesterol Acetate
43	302*	41	135	287	91	105	121	302 Abietic Acid
43	311*	60	41	312	45	241	188	311 Nalorphine
43	312*	399	297	356	281	371	311	399 Colchicine*
43	317*	271	147	289	390	95	81	390 Alphadolone Acetate*
43	326*	144	182	58	311	183	145	326 Rauwolfine
43	327*	268	42	215	81	146	59	327 6-Acetylmorphine
43	327*	369	268	310	42	215	204	369 Diamorphine*
43	331*	71	332	121	147	131	41	408 Clobetasone*
43	341*	340	374	267	107	55	342	416 Spironolactone*
43	343*	70	284	59	300	344	42	343 Acetyldihydrocodeine*
43	460*	401	359	487	545	402	461	587 3-Acetamido-2,4,6-triiodophenoxyacetic Acid
43	464*	81	523	204	69	70	411	523 6-Acetylmorphine (3-Heptafluorobutyrate Derivative)
43	474*	401	359	432	487	418	475	601 3-Acetamido-2,4,6-triiodophenoxyacetic Acid Methyl Ester
43	481*	44	53	268	310	523	69	523 3-Acetylmorphine 6-Heptafluorobutyrate Derivative
44*	28	29	31	43	60	73	41	357 Indomethacin
44*	28	85	30	42	56	57	43	100 2-Methylpiperazine
44*	29	27	41	58	57	43	39	68 Imidazole
44*	29	30	85	86	56	57	185	184 Piperazine Phosphate
44*	30	41	42	100	43	45	55	100 Aminoheptane
44*	30	58	41	86	42	27	0	101 Hexylamine(n-)
44*	30	28	57	29	56	43	101	101 3-Hydroxypiperidine
44*	30	45	46	31	29	0	0	44 Nitrous oxide
44*	36	102	69	156	38	45	71	200 Meconic Acid
44*	36	38	77	43	51	79	132	151 Phenylpropanolamine
44*	40	51	77	42	105	50	45	149 2-Amino-1-phenylpropanone
44*	40	152	91	65	165	42	164	195 2,3-Dimethoxyamphetamine
44*	41	55	83	31	43	39	56	218 Meprobamate
44*	41	43	29	27	39	58	42	86 3-Methylbutanal
44*	41	42	43	39	30	56	45	115 Methylhexaneamine
44*	41	45	58	57	144	241	213	401 Oxacillin
44*	42	124	165	163	123	93	65	183 Adrenaline*
44*	42	147	120	65	43	45	39	244 Amidephrine*
44*	42	159	43	45	184	41	109	217 Norfenfluramine
44*	42	301	115	70	216	91	43	301 Oxymorphone
44*	43	39	40	68	38	27	37	999 Butethamate Prode Product*
44*	43	39	40	71	58	68	72	999 Diethylcarbamazine Probe Product*
44*	43	29	27	41	82	0	0	100 Glutaraldehyde

Peak Index   *Asterisk Indicates True Base Peak*   Page: 159

```
                 PEAKS                              MWT    COMPOUND


44*  43   59   56   69   55   41  113   145  Heptaminol*
44*  43   42   30   70   54   96  141   173  5-Nitrobarbituric Acid
44*  43   41   69   57   55   81   75   325  Norpropoxyphene Amide
44*  43  257   59  150  200    0    0   257  Racemorphan
44*  43   73   57   60   41   55   69   284  Stearic Acid
44*  43   42   45   72   30  102   58   102  Trimethylurea
44*  45  101   58   42   86  142  128   286  Alloxantin
44*  45  122   91   77   78  121   42   165  3-Methoxyamphetamine
44*  45   58   76   60   43   42   46   169  Noradrenaline*
44*  45  202  203   91  215  189  115   263  Nortriptyline
44*  45  202  203   91   99  189  165   263  Nortriptyline
44*  56   43   41   29   72   82   67   100  Hexanal
44*  57  105   91   88    0    0    0   325  Dextropropoxyphene Metabolite
44*  57   43   41   55   69   77   79   151  Norpseudoephedrine
44*  57   43   71   41   55   45   56   263  Nortriptyline
44*  58   41   86   42   45   43    0   101  2-Aminohexane
44*  59   70  277   71  191  203  178   277  Maprotiline (Ludiomil)
44*  65   91  152   77   51  164  137   195  2,3-Dimethoxyamphetamine
44*  69   41   55   43   39   71   53   236  Carbromal
44*  69   41   55   43   29  208   71   236  Carbromal
44*  69   41  208  210   55   71   43   236  Carbromal*
44*  70  179  178  207  280  250  236   297  Dihydroxyprotriptyline
44*  70  178  250  179  279    0    0   279  10,11-Epoxyprotriptyline
44*  70   59  277   71  191  278  203   277  Maprotiline*
44*  73  272   45  164  162  396  250   483  Etorphine TMS Ether
44*  77   76   29   95   39   58   42   167  Metaraminol*
44*  77   79   51   45   42   78  105   151  Norephedrine
44*  77   79   51   42   45   43  105   151  Norpseudoephedrine
44*  77  105  106   51   91   79  132   151  Norpseudoephedrine
44*  77   79   51  105    0    0    0   151  Norpseudoephedrine
44*  77   95  121  141  167  168  133   167  Phenylephrine
44*  77   95   65   39   45   42   43   167  Phenylephrine
44*  77   79  105   91   51   92  132   151  Phenylpropanolamine
44*  77   79   51   45   42  107  105   151  Phenylpropanolamine*
44*  77   95   65  123   39  121   42   167  Synephrine
44*  83  180  182  137  143  139   41   222  Bromvaletone
44*  83   85   36   28    0    0    0   162  Trichloroacetic Acid
44*  86   58   28   42   41   43   27   101  Methylethane(N-(2'-methylethyl)-1-amino-2-)
44*  91   65   42   45  120   40   92   135  Amphetamine
44*  91   65   51   63   77    0    0   135  Amphetamine
44*  91   65   58   39   42   51   63   135  Amphetamine
44*  91   40   42   65   45   42   39   135  Amphetamine*
44*  91  129    0    0    0    0    0   251  Dextropropoxyphene Metabolite
44*  91  129  115  178  205  220  251   251  Dextropropoxyphene Metabolite
44*  91  122   45   77   66   78  107   165  2-Methoxyamphetamine
44*  93   65   43  166  137  111  139   183  Adrenaline
44* 100   34   55  115  147   41    0   349  Ampicillin
44* 100  234   88   57   91  105  129   325  Norpropoxyphene Amide
44* 100   41   60   29   27   30   86   232  Piperazine Adipate*
44* 100   42   41   30   55   39   45   115  Tuaminoheptane
44* 101   58   30   28   41   42   43   101  Methyl Amylamine(n-)
44* 101   28   43   30   41   43   45   101  Methyl Isoamylamine
44* 103   91   65   42  104   45  102   135  Amphetamine
44* 105   91   77  178  208   65  130   269  Norpropoxyphene Carbinol
```

Asterisk Indicates True Base Peak

							MWT	COMPOUND
44*	107	77	43	78	51	42	108	151	Hydroxyamphetamine
44*	122	121	78	77	42	51	52	165	Methoxyamphetamine
44*	122	121	42	78	77	91	107	165	Methoxyamphetamine (para)
44*	136	135	77	51	179	45	78	179	3,4-Methylenedioxyamphetamine
44*	137	136	163	179	0	0	0	179	3,4-Methylenedioxyamphetamine
44*	152	137	77	65	91	78	121	195	2,5-Dimethoxyamphetamine
44*	152	137	195	179	0	0	0	195	3,5-Dimethoxyamphetamine
44*	152	195	151	77	78	65	51	195	3,5-Dimethoxyamphetamine
44*	152	91	77	65	153	151	195	195	2,6-Dimethoxyamphetamine
44*	152	151	65	51	137	78	77	195	3,4-Dimethoxyamphetamine
44*	152	151	121	153	77	91	78	195	2,4-Dimethoxyamphetamine
44*	152	137	121	153	65	77	91	195	2,5-Dimethoxyamphetamine
44*	152	40	121	151	91	77	153	195	2,4-Dimethoxyamphetamine
44*	152	40	137	91	77	65	121	195	2,5-Dimethoxyamphetamine
44*	152	91	77	65	151	121	78	195	2,6-Dimethoxyamphetamine
44*	152	151	137	153	107	65	43	195	3,4-Dimethoxyamphetamine
44*	152	40	77	42	91	151	78	195	3,5-Dimethoxyamphetamine
44*	166	151	91	77	79	39	42	209	2,5-Dimethoxy-4-methylamphetamine(STP)
44*	166	168	137	151	63	53	95	211	Methoxamine
44*	166	165	42	77	39	64	194	209	3-Methoxy-4,5-methylenedioxyamphetamine
44*	166	151	57	43	91	135	209	209	STP
44*	182	167	151	181	139	183	136	225	2,4,5-Trimethoxyamphetamine
44*	193	195	194	208	234	71	235	266	Desipramine
44*	194	109	67	86	238	56	85	562	Acepifylline*
44*	201	43	186	68	173	71	96	201	Simazine
44*	202	259	203	218	215	217	42	263	Nortriptyline
44*	202	45	215	203	42	220	204	263	Nortriptyline
44*	202	45	220	218	215	9'1	0	263	Nortriptyline*
44*	203	202	45	220	219	215	204	263	Nortriptyline
44*	208	58	117	57	193	130	115	325	Norpropoxyphene
44*	215	411	324	45	164	42	216	411	Etorphine*
44*	215	202	216	45	213	91	189	261	Nortriptyline Metabolite
44*	218	217	58	57	261	216	219	261	Desmethylcyclobenzaprine
44*	218	203	73	217	202	215	219	351	Nortriptyline TMS Derivative
44*	220	86	57	74	91	43	115	367	Acetylnorpropoxyphene (N-)
44*	220	86	74	58	91	43	115	367	Acetylnorpropoxyphene (N-)
44*	220	100	57	205	129	91	307	307	Dehydrated Norpropoxyphene Amide
44*	220	100	57	205	91	129	221	325	Dextropropoxyphene Metabolite 1
44*	220	100	205	57	0	0	0	307	Dextropropoxyphene Metabolite
44*	220	100	59	205	57	91	129	251	Dextropropoxyphene Metabolite 1
44*	230	232	273	275	77	42	45	273	Bromo STP
44*	234	233	58	231	202	215	218	277	Hydroxydesmethylcyclobenzaprine
44*	262	215	202	218	189	231	203	305	Desmethyloxetorone
44*	282	283	77	284	56	91	57	299	Chlordiazepoxide
44	29*	43	42	26	41	27	25	44	Acetaldehyde
44	30*	46	0	0	0	0	0	80	Ammonium Nitrate
44	30*	29	28	102	56	0	0	102	Methylene Oxalate
44	41*	39	56	28	27	55	40	100	Dimethylpropiolactone(beta,beta-)
44	42*	58	86	56	167	165	152	311	Adiphenine
44	42*	43	57	89	72	41	45	162	Dazomet
44	42*	259	260	261	0	0	0	288	Flurazepam Metabolite
44	42*	387	172	145	252	236	188	387	Hydroflumethiazide Tetramethyl Derivative
44	42*	116	41	30	57	56	68	116	3-Hydroxy-1-nitrosopyrrolidine
44	43*	142	86	42	114	71	70	142	Alloxan

44	43*	87	111	129	167	209	159	288	Azathioprine Metabolite	
44	43*	167	57	55	139	71	152	167	2,4-Diethylamino-s-triazine	
44	43*	85	86	42	129	68	30	129	Metformin*	
44	43*	87	31	45	42	131	71	132	Paraldehyde	
44	58*	91	77	105	59	192	43	283	Dextropropoxyphene Carbinol	
44	58*	188	130	42	143	59	77	188	Dimethyltryptamine (N,N-)	
44	58*	59	121	91	56	77	78	179	4-Methoxy-N-methylamphetamine	
44	58*	41	83	69	77	85	43	179	Mexiletine	
44	59*	41	29	72	27	43	28	101	Valeramide	
44	60*	28	0	0	0	0	0	60	Urea	
44	70*	207	178	279	249	0	0	279	Hydroxyprotriptyline	
44	70*	71	57	247	190	119	191	247	Ketobemidone	
44	72*	212	45	77	42	132	214	212	Chlorotoluron	
44	72*	41	45	89	42	43	55	198	Cycloate	
44	72*	42	91	56	65	58	162	163	Dimethylamphetamine (N,N-)	
44	72*	73	77	105	42	132	160	179	2-Ethylamino-1-phenylpropanol	
44	72*	134	77	42	73	105	51	177	2-Ethylamino-1-phenylpropanone	
44	72*	91	73	42	56	65	39	163	Ethylamphetamine	
44	72*	42	70	77	128	51	105	177	Ethylpropion	
44	72*	159	73	58	42	109	56	231	Fenfluramine*	
44	72*	118	91	65	117	42	39	163	Formylamphetamine (N-Formyl)	
44	72*	57	43	149	41	55	69	151	Hydroxyamphetamine	
44	72*	43	124	123	30	42	41	211	Isoprenaline*	
44	72*	42	77	56	73	51	70	179	Methylephedrine	
44	73*	42	30	43	72	29	58	73	Dimethylformamide	
44	77*	79	51	57	105	107	91	149	Phenylpropylmethylamine	
44	84*	133	42	162	161	44	85	162	Nicotine	
44	86*	30	265	41	180	29	87	265	Ethomoxane*	
44	86*	87	107	43	106	56	41	220	Prilocaine*	
44	86*	159	43	41	87	42	56	245	Propylnorfenfluramine	
44	91*	162	119	41	65	43	42	253	Di-(1-isopropylphenyl)amine	
44	96*	36	173	172	56	70	129	247	3-Desmethylprodine	
44	96*	43	31	45	27	78	264	280	Hydroxystilbamidine*	
44	100*	72	101	77	56	42	105	205	Diethylpropion*	
44	100*	101	72	42	56	180	198	312	Ethopropazine	
44	102*	42	29	56	57	27	30	102	Diethyl Nitrosamine	
44	107*	77	40	108	39	51	43	182	Hydrophenyllactic Acid(para)	
44	109*	77	125	58	105	51	0	247	Betacaine	
44	114*	42	41	142	43	223	56	365	Pericyazine	
44	114*	142	365	42	223	115	205	365	Pericyazine*	
44	125*	70	91	41	40	244	153	581	Ergotamine*	
44	126*	55	58	139	42	43	96	198	Guanethidine	
44	130*	205	77	129	131	103	102	205	3-Indolelactic Acid	
44	138*	92	43	195	134	90	106	270	2-Nitrotartranilic Acid	
44	178*	135	179	77	84	107	41	301	Isoxsuprine*	
44	184*	91	63	53	107	92	51	184	2,4-Dinitrophenol	
44	201*	186	173	68	71	55	158	201	Simazine	
44	218*	191	221	219	178	42	180	249	Benzoctamine*	
44	245*	247	209	183	246	271	259	346	Flurazepam N1-Acetic Acid	
44	248*	249	121	136	29	164	192	292	Ambucetamide*	
44	251*	223	222	43	84	98	111	251	7-Aminonitrazepam	
44	271*	270	214	272	42	43	70	311	Nalorphine (N-Allylnormorphine)	
44	283*	255	282	254	284	264	256	283	7-Aminoflunitrazepam	
44	283*	282	256	43	176	85	57	283	Levallorphan	

		PEAKS					MWT	COMPOUND
44	298*	252	253	297	279	280	145	298	5-Hydroxyniflumic Acid
44	299*	242	59	96	42	76	243	299	Hydrocodone
44	301*	42	59	164	70	302	242	301	Dihydrocodeine*
44	304*	75	306	51	57	50	305	334	Lormetazepam
44	311*	255	296	310	312	42	253	311	Thebaine
44	334*	120	77	41	107	55	144	334	Strychnine
45*	29	43	60	42	44	41	31	60	Acetic Acid
45*	29	61	43	27	49	46	36	128	2,2-Dichloroethyl Methyl Ether
45*	31	29	47	43	76	27	46	76	2-Methoxyethanol
45*	41	217	69	70	110	285	202	285	Pentazocine
45*	43	87	41	59	27	69	39	102	Di-isopropyl Ether
45*	43	27	41	31	44	29	0	60	Isopropanol
45*	43	29	41	31	0	0	0	60	Isopropanol or Propan-2-ol
45*	43	87	59	41	27	39	69	102	Isopropyl Ether
45*	43	69	41	44	87	57	56	102	4-Methyl-2-pentanol
45*	43	60	44	89	42	87	117	132	Paraldehyde
45*	43	27	41	29	39	31	59	60	2-Propanol
45*	58	59	29	31	43	27	0	102	2-Methoxyethyl Vinyl Ether
45*	58	73	285	72	80	82	42	379	Morphine Methobromide
45*	59	31	41	27	43	44	29	74	2-Butanol
45*	62	43	61	44	63	249	247	276	2-(2-Amino-5-bromobenzoyl)pyridine(Bromazepam Benzophenone analogue)
45*	62	44	61	43	63	0	0	62	Boric Acid
45*	64	147	105	91	77	79	65	999	Phenelzine Probe Product 2*
45*	69	41	43	44	87	27	29	102	2-Hexanol
45*	91	105	77	65	52	136	0	136	Phenelzine
45*	91	64	105	147	30	77	104	136	Phenelzine
45*	121	186	97	65	93	214	153	376	Phenkapton
45*	130	44	71	42	61	115	55	158	Ethylbutyrylurea(Carbromal Metabolite)
45*	160	188	237	162	224	146	161	269	Alachlor
45*	217	70	41	69	110	285	202	285	Pentazocine*
45*	243	136	200	159	157	198	242	243	Norlevorphanol
45	31*	29	27	46	43	30	42	46	Ethanol
45	31*	46	29	27	59	74	43	136	Phenelzine*
45	43*	60	120	42	92	138	44	999	Aloxiprin*
45	43*	61	70	29	27	42	73	88	Ethyl Acetate
45	43*	60	58	95	99	398	206	822	Rifampicin
45	44*	101	58	42	86	142	128	286	Alloxantin
45	44*	122	91	77	78	121	42	165	3-Methoxyamphetamine
45	44*	58	76	60	43	42	46	169	Noradrenaline*
45	44*	202	203	91	215	189	115	263	Nortriptyline
45	44*	202	203	91	99	189	165	263	Nortriptyline
45	57*	56	29	101	149	85	41	366	Dibutylglycol Phthalate(n-)
45	57*	56	41	87	69	29	43	102	2,2-Dimethylbutan-3-ol
45	57*	56	85	101	125	41	199	398	Tri-2-butoxyethyl Phosphate
45	58*	59	42	36	115	178	91	293	Butriptyline
45	58*	91	77	105	192	59	44	283	Dextropropoxyphene Carbinol
45	58*	43	298	42	41	46	57	298	Trimeprazine
45	59*	87	43	41	31	73	55	102	1,1-Dimethylbutanol
45	59*	87	43	41	31	73	27	102	2-Methyl-2-pentanol
45	65*	85	64	44	31	61	81	100	1,1-Difluoro-1-chloroethane
45	71*	72	28	31	39	29	27	100	5-Methyl-1,2,3-thiadiazole
45	72*	55	54	44	254	43	53	411	Acetophenazine
45	73*	28	27	29	43	57	0	102	2-Ethyl-1,3-dioxolane
45	94*	79	49	46	47	48	61	94	Dimethyldisulphide

```
              PEAKS                           MWT      COMPOUND


45  112*  85  161  163    0    0    0   161  Chlormethiazole
45  128* 100   73   43   71   97  125   143  Chlormethiazole Metabolite (5-(1-hydroxyethyl)-4-methylthiazole)
45  128*  73  100  177    0    0    0   177  2-Hydroxy-4-methyl-5-b-chloroethiazole
45  130* 131  159   77   62   43  103   175  3-(3-Indolyl)acetic Acid
45  147*  91   64   48  105   65   77   999  Phenelzine Probe Product 1*
45  217*  41   70   69  110  202   72   285  Pentazocine
45  246*  42  366   58   57   43  106   366  Piminodine
45  261* 263  243   55   73  245  316   316  Ethacrynic Acid Methyl Ester
46*  28   42   75  120   56   29  128   222  RDX
46*  29   76   30    0    0    0    0    91  Ethyl Nitrate
46*  29   76   30    0    0    0    0   105  Propyl Nitrate
46*  30   29   28   15   31    0    0    77  Methyl Nitrate
46*  30   76   31   45   43   42   47   227  Nitroglycerin
46*  30   29   76   28   43   31   44   227  Nitroglycerin
46*  30   42   75  120   44  128   56   222  RDX
46*  76   57   55   56   60   47   97   316  Pentaerythritoltetranitrate
46*  76  240  194  316   58  169   84   316  Pentaerythritoltetranitrate
46   29*  45   44   30   47   31   48    46  Formic Acid
46   43*  76   41   29   57   30    0   119  Butyl Nitrate
46   61* 294  206  292   60   45   63   292  Chlorbromuron
46   61* 160   60  248  133   45  124   248  Linuron
46   61* 258   91  260  172  170   63   258  Metobromuron
46   61* 126   99  214   60   63   45   214  Monolinuron
46   76*  57   55   56   60   47   43   316  Pentaerythritoltetranitrate
47   81* 251   41   79   42  178   53   292  Permethylated Thialbarbitone*
47   82*  84   29  111   83  113   85   164  Chloral Hydrate*
47  188*  82   96   29   77   84   56   516  Dichloralphenazone*
48   64* 247  102   58  245  304  231   335  Hydroxychloroquine
49*  84   86   51   47   48   88   50    84  Dichloromethane
49*  84   86   51   47   88   48   50    84  Methylene Chloride
49* 130  128   51   93  132   95   47   128  Bromochloromethane
49* 130  128   51   93   81   79   95   128  Methylenechlorobromide
49   31*  77  113   82  115   51  117   148  2,2,2-Trichloroethanol
49   31*  77   29  113   51   48  115   228  Triclofos*
49   62*  64   63   51   61    0    0    98  1,2-Dichloroethane
50*  43  105   44  123   77   78   51   167  Cinchomeronic Acid
50*  58   52   36   42   44   38   49   157  Chlormequat
50* 130   76   74   75   63   51   38   160  Hydrallazine
50  156* 155   52   51  128  157   76   256  Paraquat Dichloride
51*  31   67   35   50   69   37   47    86  Chlorodifluoromethane(Freon 22)
51*  31  130  132   79   81  111  113   130  Difluorobromomethane
51*  65   27   45   62   47   26   64    66  1,2-Difluoroethane
51   33*  31   52   50   34   20    0    52  Difluoromethane
51   77*  50  356   78  354  353  279   356  Diphenyl Mercury
51   77* 141   99   63   73   50  268   268  Fenson
51   77* 123   50   65   93   30   78   123  Nitrobenzene
51   77*  78   50  356  354  279  353   629  Phenylmercuric Nitrate
51   77*  40   39   91  105  130   64   404  Sulfinpyrazone
51  123*  78  106   50   52  105   39   123  Isonicotinic Acid
51  123* 105   78   50   77   44   52   167  Quinolinic Acid
51  222*  87  143  153  104  224   69   257  Barban
52   39*  65   80   92  108  156  172   172  Sulfanilamide
52   78*  51   77   50   39   79   43    78  Benzene
52   79*  51   78   50   39   45   53   123  Picolinic Acid
```

		PEAKS					MWT	COMPOUND
52	79*	51	50	78	39	53	80	79 Pyridine
52	126*	80	51	108	39	79	53	126 Pyrogallol
52	151*	123	108	42	65	51	72	223 Ethamivan
53*	26	52	43	51	27	41	31	53 Acrylonitrile
53*	35	246	185	70	125	153	385	445 Thiopropazate
53*	94	57	30	82	99	67	111	252 Cimetidine
53*	133	135	54	39	27	89	51	212 1,4-Dibromo-2-butene
53*	223	153	127	171	181	225	155	223 3-Butynyl N-(3-chlorophenyl)carbamate
53*	236	110	127	153	68	238	58	236 (3-Butynyl)-N'-(4-chlorophenyl) N-Methylcarbamate (N-)
53	75*	89	77	27	39	62	54	124 1,4-Dichloro-2-butene
53	79*	221	81	41	77	178	93	262 Methohexital
53	81*	330	96	82	332	64	63	330 Frusemide*
53	81*	330	82	96	332	222	250	330 Frusemide (Fursemide)
53	81*	80	64	52	330	51	63	330 Furosemide
53	81*	64	96	251	82	39	48	330 Furosemide
53	81*	96	372	82	374	339	357	372 Furosemide Trimethyl Derivative
53	82*	81	39	27	51	50	54	82 2-Methylfuran
54*	26	27	52	55	51	53	29	55 Propionitrile
54*	125	168	69	97	53	43	0	168 Uric Acid
54	81*	108	77	51	213	39	52	213 Phenazopyridine
54	82*	108	93	58	107	106	39	150 Carvone
54	152*	109	53	81	0	0	0	152 Xanthine
54	166*	109	53	81	136	137	0	166 1-Methylxanthine
54	221*	339	72	55	196	207	181	339 Methylergonovine Hydrogen Maleate
55*	27	63	90	41	39	29	54	126 1,3-Dichlorobutane
55*	27	29	56	45	28	73	26	100 Ethyl Acrylate
55*	41	42	43	70	57	29	31	88 Amyl Alcohol
55*	41	69	56	82	42	67	43	154 1,6-Dichlorohexane
55*	41	126	68	97	110	81	43	154 2,2-Diethylcyclohexanone
55*	41	83	42	126	82	56	97	154 2,6-Diethylcyclohexanone
55*	41	39	82	54	53	109	42	138 Pentylenetetrazol
55*	41	56	97	72	57	62	43	274 Tybamate
55*	42	41	68	27	29	69	39	140 1,5-Dichloropentane
55*	43	45	143	57	189	145	73	189 Kynurenic Acid
55*	43	27	70	97	42	41	26	70 Methyl Vinyl Ketone
55*	43	195	41	235	353	221	72	353 Methysergide
55*	57	43	97	41	56	158	44	260 Carisoprodol*
55*	58	43	104	56	41	97	62	260 Carisoprodol
55*	58	70	83	69	73	97	127	169 Methylbemegride
55*	70	42	100	41	29	39	43	100 3,4-Dimethyltetrahydrofuran
55*	70	113	42	41	39	44	69	141 Ethosuximide(i)*
55*	72	97	41	56	158	118	57	274 Tybamate*
55*	82	39	83	41	113	70	127	155 Bemegride
55*	82	67	180	109	0	0	0	180 1,7-Dimethylxanthine
55*	82	41	39	54	42	56	109	138 Leptazol
55*	82	67	109	180	0	0	0	180 Theobromine
55*	83	82	113	70	69	97	155	155 Bemegride
55*	83	82	41	113	70	29	69	155 Bemegride*
55*	83	43	98	29	27	39	53	98 4-Methyl-3-pentene-2-one
55*	83	82	41	54	39	28	42	83 Tetrahydropyridine (Methyl Phenidate Decomposition Prod)
55*	91	340	44	339	57	105	269	340 Prazepam-N-Oxide
55*	91	269	136	107	340	41	67	416 Spironolactone
55*	100	27	29	39	54	85	53	100 Tiglic Acid(2-Methyl-but-2-enoic Acid)
55*	126	54	127	83	52	82	56	126 Thymine

								MWT	COMPOUND
55*	169	83	70	127	140	41	97	169	Bemegride Methyl Derivative
55*	285	77	270	91	105	166	56	285	2-Cyclopropylmethylamino-5-Chlorobenzophenone
55	41*	42	115	70	39	96	68	327	Naloxone
55	41*	43	69	83	57	67	54	282	Oleic Acid
55	42*	41	100	39	68	29	27	100	Hexamethylene Oxide
55	42*	69	98	41	56	27	39	98	3-Methylcyclopentanone
55	42*	41	31	29	70	43	57	88	Pentanol(n-)
55	43*	299	91	147	79	253	271	372	Deoxycortone Acetate
55	56*	41	43	45	85	42	84	102	Cadaverine
55	57*	85	29	100	72	41	43	100	Ethyl-2-methylallyl Ether
55	58*	97	43	104	158	62	56	260	Carisoprodol
55	58*	89	90	118	42	71	41	291	Cyclopentolate
55	59*	73	31	43	41	27	29	102	3-Hexanol
55	73*	43	45	87	41	69	29	102	3-Methyl-3-Pentanol
55	83*	41	67	54	39	82	27	162	Bromocyclohexane
55	83*	71	96	114	144	62	56	218	Meprobamate
55	83*	71	41	44	62	56	45	218	Meprobamate
55	83*	98	155	126	0	0	0	199	Methaprylon Metabolite
55	83*	98	153	166	41	155	152	199	Methyprylon Metabolite
55	84*	42	70	205	43	204	41	287	Procyclidine
55	85*	41	29	56	28	84	39	102	2-Hydroxytetrahydropyran
55	86*	99	41	44	100	83	56	309	Dicyclomine
55	91*	43	56	41	42	29	93	120	1-Chlorohexane
55	91*	269	295	294	324	29	241	324	Prazepam
55	97*	69	72	71	98	43	62	232	Mebutamate*
55	98*	218	99	41	96	219	42	301	Trihexyphenidyl
55	100*	45	74	119	73	56	41	118	Succinic Acid
55	104*	43	56	44	58	41	62	260	Carisoprodol
55	113*	70	42	41	39	85	69	141	Ethosuximide
55	124*	139	99	142	141	125	154	346	5-(2-Ethoxythiocarbonylthioethyl)-5-isopentylbarbituric Acid
55	127*	70	41	42	128	69	112	155	Methylethosuximide (N-)
55	127*	70	41	42	0	0	0	155	Methylsuximide(N-)
55	153*	183	138	69	0	0	0	199	Methaprylon Metabolite
55	154*	41	141	42	155	281	70	281	Dodemorph
55	180*	109	67	82	137	181	42	180	Theobromine
55	257*	311	77	259	313	44	312	340	3-Hydroxyprazepam
55	341*	36	300	342	110	243	256	341	Naltrexone
55	343*	110	36	98	302	84	344	343	Hydroxynaltrexone (alpha)
55	423*	364	255	84	59	424	121	423	Cyprenorphine*
56*	31	41	43	27	42	29	0	74	1-Butanol
56*	41	31	43	29	27	55	28	102	Butyl Formate(n-)
56*	41	43	27	29	55	57	42	92	1-Chlorobutane
56*	41	69	55	43	57	70	29	134	3-Chloroheptane
56*	41	84	54	55	42	69	0	84	Cyclohexane
56*	41	43	57	29	45	67	27	100	2,5-Dimethyltetrahydrofuran
56*	43	55	42	41	69	31	29	102	1-Hexanol
56*	55	41	43	45	85	42	84	102	Cadaverine
56*	57	84	83	203	77	0	0	203	Aminoantipyrine
56*	69	55	41	43	29	57	31	102	3-Methyl-1-pentanol
56*	69	91	261	98	119	84	190	367	Phenbutrazate
56*	77	91	105	106	42	0	0	165	Ephedrine
56*	77	91	84	42	40	373	203	373	Methylrubazonic Acid
56*	83	42	217	123	64	57	119	333	Dipyrone
56*	83	230	217	123	97	77	64	333	Dipyrone

								MWT	COMPOUND
56*	84	245	57	203	83	43	42	245	4-Acetylaminoantipyrine
56*	84	57	203	42	83	77	93	203	4-Aminoantipyrine
56*	84	41	55	42	69	39	27	84	Cyclohexane
56*	97	231	42	111	77	71	112	231	Amidopyrine
56*	97	77	231	42	91	51	55	231	Aminopyrine
56*	99	165	167	42	0	0	0	266	Cyclizine
56*	100	98	58	43	123	41	192	239	Isoetharine Mesylate
56*	100	138	110	57	237	70	41	237	Viloxazine*
56*	123	217	83	57	77	215	216	217	Methylaminophenazone
56*	123	215	217	83	119	77	57	217	Noraminophenazone
56*	161	57	43	144	117	91	129	161	Cypenamine
56*	201	70	176	77	202	71	55	377	Morazone
56*	205	223	167	149	41	57	65	278	Dibutyl Terephthalate(n-)
56*	217	83	42	57	98	77	218	217	4-Methylaminoantipyrine
56*	231	97	204	77	111	112	0	231	Amidopyrine
56*	231	97	111	112	42	77	71	231	Amidopyrine*
56*	254	256	229	231	199	201	0	285	4-Bromo-2,5-dimethoxyamphetamine Impurity
56	27*	26	55	29	25	37	38	56	Acrolein
56	28*	29	41	43	85	39	57	100	Valerolactone(gamma-)
56	30*	28	85	43	45	41	27	102	Cadaverine
56	30*	28	85	43	45	41	27	102	1,5-Diaminopentane
56	31*	41	43	27	42	29	39	74	1-Butanol
56	42*	246	70	232	233	198	98	403	Perphenazine
56	43*	73	41	61	55	27	29	116	Butyl Acetate(n-)
56	43*	57	41	29	27	71	42	100	2,3-Dimethylpentane
56	43*	41	31	29	27	60	39	102	Isobutyl Formate
56	43*	45	55	73	99	39	44	116	Laevulinic Acid
56	43*	41	29	31	27	60	39	102	1-Methylpropyl Formate
56	44*	43	41	29	72	82	67	100	Hexanal
56	57*	41	0	0	0	0	0	86	3-Methylpentane
56	57*	205	103	91	158	77	115	205	Norpethidinic Acid
56	57*	42	219	43	103	91	158	219	Norpethidinic Acid Methyl Ester
56	57*	41	43	29	58	99	27	114	2,2,4-Trimethylpentane
56	58*	77	108	42	105	146	91	165	Ephedrine
56	58*	77	107	160	65	38	57	181	Hydroxyephedrine(p-)
56	58*	91	44	77	42	78	121	179	3-Methoxy-N-methylamphetamine
56	69*	71	91	261	84	119	70	367	Phenbutrazate*
56	71*	177	77	70	105	51	72	177	Phenmetrazine
56	72*	91	129	105	130	73	57	311	Alphamethadol
56	72*	30	43	98	115	144	41	259	Propranolol*
56	84*	54	42	52	82	85	40	84	Benzene-d6
56	84*	43	55	85	41	70	42	127	Coniine
56	84*	54	52	42	82	40	85	84	Deuterobenzene
56	84*	91	85	55	30	41	42	233	Methylphenidate
56	84*	85	77	105	55	30	42	267	Pipradrol*
56	92*	94	63	84	86	184	93	276	4-Hydroxycyclophosphamide
56	97*	32	91	68	98	65	57	202	Fend013prox
56	99*	72	165	300	228	229	242	300	Chlorcyclizine
56	99*	165	43	228	229	241	242	300	Chlorcyclizine
56	99*	167	207	194	164	195	208	266	Cyclizine
56	99*	42	43	167	70	44	207	266	Cyclizine
56	99*	167	207	194	266	195	165	266	Cyclizine*
56	100*	101	42	185	184	41	128	324	Furaltadone*
56	100*	42	176	98	120	70	189	276	Molindone

```
                PEAKS                          MWT    COMPOUND


56  100*  101   42  165   91  115   70   339  Moramide Intermediate
56  100*   42  101   55   54   41   30   226  Nimorazole*
56  138*   55   41   44   30   28   82   181  Dicyclohexylamine
56  210*   43  184  211   75   99   76   324  Acetohexamide*
56  271*   43   41   55   42  273   77   300  3-Hydroxydiazepam
57*   29   75   27  101   74   28   45   102  Ethyl Propionate
57*   29   85   72   41   43   27  114   114  3-Heptanone
57*   29   41   58   27   39   86   43    86  2-Methylbutanal
57*   29   27   85   41   43  114   39   114  5-Methyl-3-hexanone
57*   29   86   27   56   58   26   43    86  3-Pentanone
57*   29   31   27   39   58   30   26    58  1-Propen-2-ol
57*   41   29   56   27   39   55   43   136  1-Bromobutane
57*   41   56   55   43   69   29   42   178  2-Bromoheptane
57*   41   43   29  100   39   28   56   100  3,3-Dimethylbutan-2-one
57*   41   29   39   27   45   59   56   102  2,2-Dimethylpropanoic Acid
57*   41   56   29   44   27  100   39   100  Isobutyl Vinyl Ether
57*   41   43   29   39  100   27    0   100  Butyl Methyl Ketone(t-)
57*   41   44   82   67   71   56   39   100  2-Methylcyclopentanol
57*   41  141  167   39   83   55  182   238  Nealbarbitone*
57*   41   29   43  100   39   56    0   100  Pinacolone
57*   41   29   39   56   27   45   87   102  Pivalic Acid
57*   41   29  101   44   36   58   27   101  Trimethylacetamide
57*   42  186   58   44   70   83   84   261  Ethoheptazine
57*   42   56   43  233  158   91  103   233  Norpethidine
57*   42   73  277  103   56  187   43   277  Norpethidinic Acid TMS Ester
57*   42   43   70   71  200  199   44   200  Pethidine Intermediate A
57*   42   85   56   54   58   70   44   191  Phendimetrazine
57*   43   41   56   85   29   27   39   100  2,2-Dimethylpentane
57*   43   71   85   99  113  243  244   609  Ergocristine
57*   43   41   70   55   56   83   29   130  2-Ethyl-1-hexanol
57*   43   29   27   41   28  100   71   100  Ethyl Isopropyl Ketone
57*   43   41   29   27   56   42   39    86  Hexane(n-)
57*   43  149   58   55   41   69   71   301  Methyldihydromorphine
57*   43   29  100   71   27   41   56   100  2-Methylpentan-3-one
57*   43   70   71   42  200  199   91   200  Pethidine Intermediate A
57*   43   56   41   85   29   27   39   100  2,2,3-Trimethylbutane
57*   45   56   29  101  149   85   41   366  Dibutylglycol Phthalate(n-)
57*   45   56   41   87   69   29   43   102  2,2-Dimethylbutan-3-ol
57*   45   56   85  101  125   41  199   398  Tri-2-butoxyethyl Phosphate
57*   55   85   29  100   72   41   43   100  Ethyl-2-methylallyl Ether
57*   56   41    0    0    0    0    0    86  3-Methylpentane
57*   56  205  103   91  158   77  115   205  Norpethidinic Acid
57*   56   42  219   43  103   91  158   219  Norpethidinic Acid Methyl Ester
57*   56   41   43   29   58   99   27   114  2,2,4-Trimethylpentane
57*   58   44   42   39   70   40   68   261  Ethoheptazine
57*   58   70   42   44  188   84   43   261  Ethoheptazine*
57*   69   41   56   43   29   45   31   102  3,3-Dimethylbutan-1-ol
57*   71   43   85   40   55   41  149   229  Ergothioneine
57*   73   60   43   55   71   41   69   284  Stearic Acid
57*   82   44   67   41   71   29   56   100  Cyclohexanol
57*   85   42   56   76  191   70   77   191  Phendimetrazine*
57*   87  130   55  116  101  143  141   172  Methyl 3-Oxodipropylacetate
57*  127   41   43   55   88  243   42   371  Ioxynil
57*  141  167   41  181  182   83  223   238  Nealbarbitone
```

Asterisk Indicates True Base Peak

PEAKS							MWT	COMPOUND
57*	149	71	43	41	69	55	113	390 Di-iso-octyl Phthalate
57*	149	71	70	293	69	111	43	418 Dinonyl Phthalate
57*	157	75	155	39	49	77	41	234 Dibromochloropropane
57*	165	70	120	43	41	92	83	294 1-(2-Ethylhexyl)-4-hydroxyphthalate
57*	233	42	56	43	158	131	160	233 Ethyl-N-demethyl-4-phenylpiperidine-4-carboxylate
57*	233	56	158	103	91	160	77	233 Norpethidine
57*	233	42	56	158	0	0	0	233 Pethidine Metabolite
57	29*	27	43	26	44	28	0	100 Vinyl Propionate
57	30*	82	116	99	53	55	42	252 Cimetidine*
57	36*	41	40	44	38	55	39	157 Oxanamide (Quinactin)
57	41*	29	60	43	30	71	39	117 Amyl Nitrite
57	41*	29	56	28	44	27	39	100 Secbutyl Vinyl Ether
57	43*	41	55	44	0	0	0	368 3,5-Cholestadiene
57	43*	41	55	44	81	93	105	386 Cholesterol
57	43*	41	56	42	27	29	39	100 2,4-Dimethylpentane
57	43*	29	27	71	41	100	39	100 3-Hexanone
57	43*	29	27	72	41	71	99	128 6-Methyl-3-heptanone
57	43*	71	41	29	70	27	56	100 3-Methylhexane
57	43*	29	41	72	56	27	100	100 3-Methylpentan-2-one
57	43*	71	27	41	85	58	39	128 4-Octanone
57	43*	41	55	82	56	44	29	170 Undec-10-en-1-al
57	44*	105	91	88	0	0	0	325 Dextropropoxyphene Metabolite
57	44*	43	41	55	69	77	79	151 Norpseudoephedrine
57	44*	43	71	41	55	45	56	263 Nortriptyline
57	55*	43	97	41	56	158	44	260 Carisoprodol*
57	56*	84	83	203	77	0	0	203 Aminoantipyrine
57	58*	30	36	27	26	76	29	94 1,Chloro-3-hydroxypropane
57	58*	59	91	42	77	105	115	339 Dextropropoxyphene
57	58*	59	285	286	91	115	208	339 Dextropropoxyphene
57	65*	41	91	120	29	43	39	137 Nitrotoluene (ortho)
57	72*	43	69	71	55	81	41	309 Methadone
57	85*	42	44	56	191	77	0	191 Phendimetrazine
57	86*	43	71	55	69	58	41	386 Clonitazene
57	87*	116	55	115	129	88	127	158 Methyl dipropylacetate
57	88*	101	60	61	127	73	43	172 Ethyl Caprylate
57	120*	164	165	70	69	84	98	294 2-(2-Ethylhexyl)-4-hydroxyphthalate
57	123*	42	103	44	85	124	51	226 Carbidopa*
57	129*	70	71	55	112	43	41	370 Dioctyl Adipate*
57	135*	91	77	43	119	105	180	209 Hydroxyphenamate
57	142*	42	197	185	339	240	226	339 Hexetidine
57	146*	245	189	42	91	77	132	336 Fentanyl
57	149*	41	150	56	223	104	76	278 Diisobutyl Phthalate
57	149*	167	71	70	43	113	41	390 Dioctyl Phthalate(n-)
57	149*	167	43	71	70	41	55	390 Isooctyl Phthalate
57	182*	43	55	40	69	41	181	182 Harman
57	185*	43	129	29	41	259	157	402 Acetyl Tri-n-butylcitrate
57	205*	220	41	206	145	29	81	220 Ionol
57	220*	41	221	206	58	29	55	277 Azar
57	302*	85	231	91	110	79	215	302 Norethandolone
58*	28	42	30	102	45	59	0	102 Dimethylaminomethane(bis-)
58*	30	363	44	72	29	27	364	436 Bialamicol*
58*	30	29	27	28	44	101	41	101 Butylamine(N-ethyl(n-))
58*	30	42	85	28	44	56	57	102 3-Dimethylaminopropylamine
58*	30	77	59	56	42	51	79	165 Ephedrine GC Decomposition

								MWT	COMPOUND
58*	30	101	44	29	28	41	27	101	Ethyl Butylamine(n-)
58*	30	41	65	36	39	93	59	179	Ethylnorepinephrine
58*	30	59	77	29	95	65	57	181	Etilefrine*
58*	30	148	115	59	77	91	89	221	6-Hydroxy-3-(2-dimethylaminoethyl)benzo(b)thiophene
58*	30	59	56	77	0	0	0	181	Hydroxyephedrine*
58*	30	91	59	56	42	65	77	179	Methoxyphenamine
58*	30	42	159	56	59	31	202	231	Methylnorfenfluramine
58*	30	77	42	56	106	105	51	165	Pseudoephedrine
58*	36	41	59	126	38	44	56	141	Cyclopentamine
58*	36	134	43	99	59	91	44	339	Noracymethadol
58*	40	38	57	55	140	41	56	155	Propylhexedrine
58*	41	30	42	57	39	27	59	73	Butylamine(t-)
58*	41	30	126	59	69	56	44	141	Cyclopentamine*
58*	41	91	31	120	65	29	42	149	1-Ethyl-2-phenylethylamine
58*	41	55	44	43	59	128	56	141	Methylaminomethylheptane
58*	41	77	42	43	91	47	122	179	Mexiletine
58*	42	59	57	202	203	43	91	277	Amitriptyline
58*	42	167	249	247	168	72	248	318	Brompheniramine
58*	42	41	125	59	168	89	63	183	Chlorphentermine
58*	42	221	59	57	43	222	189	315	Chlorprothixene
58*	42	91	71	65	30	57	56	291	Cyclopentolate
58*	42	59	57	128	115	218	78	292	Dimethindene
58*	42	30	43	57	44	0	0	101	Dimethylaminoacetone(N,N-)
58*	42	91	31	59	65	39	51	149	Dimethyl-2-phenylethylamine (N,N-)
58*	42	130	77	30	103	115	102	188	Dimethyltryptamine
58*	42	73	165	45	77	57	152	255	Diphenhydramine
58*	42	59	57	165	43	115	178	279	Doxepin
58*	42	71	36	59	111	30	75	257	Meclofenoxate
58*	42	41	198	96	112	44	180	310	Mepazine
58*	42	185	30	100	228	57	328	328	Methotrimeprazine
58*	42	59	234	77	427	178	133	445	Narceine
58*	42	91	41	30	39	65	59	149	Phentermine
58*	42	91	56	57	43	59	165	255	Phenyltoxolamine
58*	42	204	30	146	117	130	59	204	Psilocin
58*	42	204	30	44	146	117	118	284	Psilocybin
58*	42	59	30	43	41	0	0	59	Trimethylamine
58*	43	57	149	71	42	41	55	163	Acetylcholine
58*	43	73	341	282	229	342	299	341	Acetylcodeine
58*	43	128	127	59	0	0	0	309	Isomethadone
58*	43	101	73	27	41	71	86	101	Methylisobutyramide(N-)
58*	43	100	42	56	44	29	28	100	Methylpiperazine(N-)
58*	43	134	222	91	44	59	56	339	Noracymethadol*
58*	44	91	77	105	59	192	43	283	Dextropropoxyphene Carbinol
58*	44	188	130	42	143	59	77	188	Dimethyltryptamine (N,N-)
58*	44	59	121	91	56	77	78	179	4-Methoxy-N-methylamphetamine
58*	44	41	83	69	77	85	43	179	Mexiletine
58*	45	59	42	36	115	178	91	293	Butriptyline
58*	45	91	77	105	192	59	44	283	Dextropropoxyphene Carbinol
58*	45	43	298	42	41	46	57	298	Trimeprazine
58*	55	97	43	104	158	62	56	260	Carisoprodol
58*	55	89	90	118	42	71	41	291	Cyclopentolate
58*	56	77	108	42	105	146	91	165	Ephedrine
58*	56	77	107	160	65	38	57	181	Hydroxyephedrine(p-)
58*	56	91	44	77	42	78	121	179	3-Methoxy-N-methylamphetamine

								MWT	COMPOUND
58*	57	30	36	27	26	76	29	94	1,Chloro-3-hydroxypropane
58*	57	59	91	42	77	105	115	339	Dextropropoxyphene
58*	57	59	285	286	91	115	208	339	Dextropropoxyphene
58*	59	202	91	203	215	218	217	277	Amitriptyline
58*	59	42	30	202	91	203	115	277	Amitriptyline
58*	59	42	202	215	203	91	189	277	Amitriptyline
58*	59	202	42	203	214	217	0	277	Amitriptyline*
58*	59	73	75	215	202	203	217	365	Amitriptyline TMS Derivative
58*	59	179	42	178	72	77	30	303	Chlorphenoxamine*
58*	59	221	30	42	222	255	43	315	Chlorprothixene*
58*	59	42	215	202	57	189	43	275	Cyclobenzaprine
58*	59	45	41	40	74	42	59	141	Cyclopentamine
58*	59	72	45	292	218	42	0	292	Dimethindene*
58*	59	42	188	72	115	145	104	188	Dimethyltryptamine (N,N-)
58*	59	236	202	203	221	42	57	295	Dothiepin
58*	59	295	57	221	202	204	203	295	Dothiepin
58*	59	221	204	202	203	293	234	295	Dothiepin (Prothiaden)
58*	59	42	178	165	277	219	202	279	Doxepin
58*	59	42	165	103	72	30	0	347	Embramine*
58*	71	36	176	150	72	193	59	264	Amethocaine
58*	71	150	176	72	193	105	59	264	Amethocaine*
58*	71	167	72	42	59	202	45	290	Carbinoxamine
58*	71	42	44	167	72	43	57	290	Carbinoxamine
58*	71	26	54	167	72	42	44	290	Carbinoxamine*
58*	71	167	72	42	202	203	45	290	Carbinoxamine(1-)
58*	71	72	207	42	91	59	118	291	Cyclopentolate*
58*	71	159	56	72	41	42	115	272	Dimethisoquin
58*	71	72	167	182	42	180	59	270	Doxylamine*
58*	71	56	77	42	105	146	59	165	Ephedrine GC Decomposition
58*	71	208	72	59	42	89	57	294	Noxiptyline*
58*	71	42	105	176	150	92	56	264	Tetracaine
58*	72	71	177	175	96	42	78	339	Bromothen
58*	72	43	78	271	57	44	77	371	Camazepam
58*	72	192	42	165	73	71	59	278	Methadone Intermediate
58*	72	30	77	56	44	42	73	179	Methylephedrine*
58*	72	71	224	59	42	152	57	295	Normethadone
58*	72	71	42	59	163	91	179	295	Normethadone
58*	72	279	234	151	192	166	165	279	Thymoxamine
58*	72	59	57	121	164	237	77	237	Thymoxamine Hydrolysis Product
58*	73	45	57	43	44	167	165	333	Bromodiphenhydramine
58*	73	45	165	59	42	166	44	333	Bromodiphenhydramine
58*	73	45	165	59	42	166	149	333	Bromodiphenhydramine*
58*	73	45	43	57	167	44	165	469	Dimenhydrinate
58*	73	165	45	105	167	44	77	255	Diphenhydramine
58*	73	45	44	59	165	167	42	255	Diphenhydramine
58*	73	45	167	165	166	44	152	255	Diphenhydramine*
58*	73	59	88	45	75	43	56	237	Ephedrine TMS Ether
58*	73	165	45	166	42	181	59	269	Orphenadrine
58*	73	45	42	165	59	46	44	269	Orphenadrine
58*	73	44	45	165	42	40	181	269	Orphenadrine*
58*	77	105	56	57	51	42	79	165	Ephedrine
58*	77	30	51	49	56	59	42	165	Pseudoephedrine
58*	77	59	56	51	42	105	91	165	Pseudoephedrine*
58*	85	269	268	270	271	229	227	314	Clomipramine

PEAKS								MWT	COMPOUND

```
          PEAKS                        MWT    COMPOUND

58*  85  269  268  270  271  314  242  314  Clomipramine*
58*  85   70   57   42  148   56   77  165  Ephedrine GC Decomposition
58*  85   42   77   56   70  105   57  165  Ephedrine GC Decomposition
58*  85  173  193  194  195  234  235  280  Imipramine
58*  85  235  234   42  280  208  193  280  Imipramine
58*  85  284  238  239  198  226  224  284  Promazine
58*  86   85  326  197   43  280  241  326  Acepromazine
58*  86  289   43  291   85  248  288  376  8-Acetoxychlorpromazine
58*  86  376   43  378   42   87   59  376  3-Acetoxychlorpromazine
58*  86   42   85   44   57   59   43  318  Chlorpromazine
58*  86  318   85  320  272  319  273  318  Chlorpromazine*
58*  86   87   77   56   42   30   44  193  Etafedrine
58*  86   42  334  220   59   87   44  334  3-Hydroxychlorpromazine
58*  86  334   42   85  336   44  243  334  8-Hydroxychlorpromazine
58*  86  314  229  228  185   42   44  314  Methoxypromazine*
58*  86   42   85  199  198  238  284  284  Promazine
58*  86   42   85   44   59   30   70  352  Triflupromazine
58*  86  352   85  306   59  266  248  352  Triflupromazine
58*  91   57  115  208  117   59   42  339  Dextropropoxyphene
58*  91   59   42  129    0    0    0  265  Dextropropoxyphene Metabolite
58*  91  191  129  128  205  178  174  265  Hydrolysed Dextropropoxyphene
58*  91   56   65   51   77  134    0  149  Methamphetamine
58*  91   56   65   42   39   51   41  149  Methamphetamine
58*  91   57   56   78   65   77   42  179  Methoxyphenamine
58*  91   59   56   30   42  121   78  179  Methoxyphenamine*
58*  91   43   45   30   59   56   42  179  Methoxyphenamine (ortho-)
58*  91   59  134   65   56   42   57  149  Methylamphetamine*
58*  91   59   44   42  134   65   57  149  Phentermine
58*  91   42   41  134   65   59   40  149  Phentermine*
58*  91   72   71   65   42   79   78  255  Tripelennamine
58*  91   72   71  197  185  184   92  255  Tripelennamine*
58*  97   72   84   71  191  192  261  261  Methapyrilene
58*  97   72   71  121  224  149  191  261  Methapyrilene
58*  97   72   71   42  191   79   78  261  Methapyrilene*
58*  97   72   71   79   42   78   40  261  Thenyldiamine
58*  97   71   72   91   79  203   78  261  Thenyldiamine
58*  97   72   71  203  191  190   42  261  Thenyldiamine*
58*  98  202  231  189  215  218  205  319  Oxetorone
58* 101   44   42   59   29   30    0  101  Dimethylbutylamine
58* 105   77   42   59  113   51   30  271  Amylocaine
58* 105   77   98  122   42  113   51  271  Amylocaine*
58* 105   57   43   71   55   41  167  327  Dimenoxadole
58* 105   77   42  106   56   57   51  165  Ephedrine
58* 111   71   42   75   59  141  113  257  Meclofenoxate*
58* 112  105   96   42  111   77   51  278  Amydricaine
58* 116  198  199   72   59   42   44  358  Dimethoxanate*
58* 117  208  115  193   91  179  130  339  Dextropropoxyphene*
58* 120   72   70  136   18   42    0  154  Terpineol(alpha-)
58* 121  134  127  148  238  270  284  369  Methoxypropoxyphene
58* 121  134   59   57  127  105   77  369  Methoxypropoxyphene
58* 121   72   78   42   77  122   91  286  Thonzylamine
58* 121  216   72  215   71  122   59  286  Thonzylamine
58* 121   72   71  216  215  122   78  286  Thonzylamine*
58* 124   71  199  125  167  154  141  255  5-Butyl-5-(2-dimethylaminoethyl)-barbituric Acid
```

Asterisk Indicates True Base Peak

```
                  PEAKS                    MWT    COMPOUND


58*  125    42  168    60    41    89  127  183  Chlorphentermine
58*  128   127    59  193   179   178  115  309  Isomethadone
58*  131    72    71    79    42    30   78  295  Chloropyrilene (Chlorothen, Tagathen)
58*  131    71    72    42    79   133   78  295  Chlorothen
58*  131   130    30    77   132   103   59  188  Ethyltryptamine (N-)
58*  138   195   180    43    41   152   69  195  2,4-Diisopropylamino-s-triazine
58*  139   254    77   178   105   179  111  289  Chlophedianol
58*  140    67    72    83    81   155  156  155  Propylhexedrine
58*  146    56   105    77    42   106   40  165  Ephedrine*
58*  165    41    73    57   166    44   59  333  Bromodiphenhydramine
58*  165   255   359   166    73   199   45  359  Captodiamine
58*  169   168    42   167   170    72   51  240  Pheniramine
58*  179   180   225   178   165    42  210  253  Nefopam*
58*  188   130    59    42   143   129  115  188  Dimethyltryptamine(N,N-)*
58*  193   194    42   234   235    39  179  294  Trimipramine
58*  195    59    72   388    89   315   42  388  Trimethobenzamide
58*  195    42   388    59    57    43   56  388  Trimethobenzamide
58*  197    86    42    43    44    85  196  326  Acepromazine
58*  198    42   180    30   154    57  199  298  Trimeprazine
58*  202   116   198    59    44    72   42  358  Dimethoxanate
58*  202   231   215   289     0     0    0  291  Hydroxycyclobenzaprine
58*  202    57   201    42    84    44  186  275  Proheptazine
58*  203   167    72   205   202   168  204  274  Chlorpheniramine
58*  203   202   232    84    85   101  215  291  10,11-Epoxycyclobenzaprine
58*  204   146    59    42   160    43  159  204  Bufotenine
58*  204    59    42   146   160   130    0  204  Psilocin
58*  204    59   146   159   205   160   60  204  Psilocin (4-Hydroxy-N-dimethyltryptamine)
58*  204    59   146   159   205   160   57  284  Psilocybin
58*  204    59     0     0     0     0    0  204  Psilocybin Metabolite
58*  205    42   191   178   128    91  129  339  Dextropropoxyphene
58*  213   198   180   214    57   212  270  270  Desmethylpromethazine
58*  215    59   229   228   227   226  218  275  Amitriptyline Metabolite
58*  220    59   219   277   179   191  193  279  Doxepin
58*  220   219    59   191   189    42  205  279  Doxepin*
58*  224   209    71   225    72   210  180  295  Dibenzepin*
58*  234   235    85   193   194   195  192  280  Imipramine
58*  234    44    59    41    42    36  427  445  Narceine
58*  235    85   234   195   280   193  194  280  Imipramine
58*  235    85   234   236   195   193  208  280  Imipramine*
58*  236    40   202   235   203    42   44  295  Dothiepin*
58*  238   345    91   107   167   165  252  345  Prenylamine Metabolite
58*  246   233   318    86   272   248  232  334  Chlorpromazine Metabolite
58*  246   302    71   289   273   274  341  359  Monodesethylflurazepam
58*  249   247    72   167   248   168   42  318  Brompheniramine (dex) (d-Parabromdylamine)
58*  249   208    99   193   232    84    0  294  Trimipramine
58*  249   208   193   234    99   248  194  294  Trimipramine
58*  251   250   296   211    85   224  209  296  ((3-Dimethylaminopropyl)-2-hydroxyiminodibenzyl)(N-)(Imipramine Metab)
58*  254    45    44    77    42    59   72  289  Chlophedianol*
58*  254    91    56   183   115   107  343  345  Prenylamine Metabolite
58*  255    40    72    42    59    71   91  255  Phenyltoloxamine
58*  255    42    71    59    44   181  165  255  Phenyltoloxamine*
58*  268   269    85   193   229   228  192  314  Chlorimipramine
58*  269    85   268   270   271   314  242  314  Clomipramine
58*  276   218   261     0     0     0    0  276  Psilocin-TMS Derivative
```

								MWT	COMPOUND
58*	282	30	284	355	73	283	44	355	Amodiaquine*
58*	284	91	213	147	152	137	115	375	Prenylamine Metabolite
58*	284	86	238	198	199	85	42	284	Promazine*
58*	285	72	42	71	59	186	44	301	Morphine-N-oxide
58*	285	214	200	86	227	85	212	285	Prothipendyl*
58*	293	45	59	193	100	178	294	293	Butriptyline*
58*	298	212	198	100	299	252	199	298	Trimeprazine*
58*	304	214	42	59	306	232	233	304	Chlorphenethiazine
58*	318	86	272	85	320	232	42	318	Chlorpromazine
58*	319	86	321	85	273	36	274	318	Chlorpromazine
58*	324	154	114	105	96	72	77	382	Dimethylaminoethyl-3-piperidylbenzilate (N-Dimethyl)
58*	328	100	42	135	228	229	242	328	Methotrimeprazine
58*	328	100	269	229	283	242	243	328	Methotrimeprazine
58*	328	100	228	185	329	242	229	328	Methotrimeprazine*
58*	328	100	43	71	269	229	207	328	Methotrimeprazine Metabolite
58*	328	100	229	269	207	242	283	328	Methotrimeprazine Metabolite
58*	328	229	242	100	269	283	0	344	Methotrimeprazine Sulphoxide
58*	352	86	353	85	306	42	266	352	Fluopromazine*
58*	359	358	403	360	343	388	227	403	Amotriphene*
58*	427	234	59	50	42	428	91	445	Narceine*
58	43*	42	27	40	29	26	39	58	Acetone
58	43*	41	39	57	29	27	0	100	Allyl Acetate
58	43*	42	143	85	171	157	102	196	Bethanechol
58	43*	42	44	30	129	36	143	182	Carbachol
58	43*	71	41	59	27	29	85	114	2-Heptanone
58	43*	29	27	41	57	39	100	100	2-Hexanone
58	43*	41	39	42	27	0	0	100	Isopropenyl Acetate
58	43*	256	45	51	55	44	56	361	Levophenacylmorphan
58	43*	57	100	41	85	29	39	100	Isobutyl Methyl Ketone
58	43*	41	29	27	71	39	57	100	2-Methylpentan-1-al
58	43*	41	57	27	39	29	100	100	4-Methylpentan-2-one
58	43*	57	41	85	100	29	39	100	4-Methyl-2-pentanone
58	43*	41	29	27	71	57	39	100	2-Methylvaleraldehyde
58	43*	41	71	59	57	27	29	142	2-Nonanone
58	44*	41	86	42	45	43	0	101	2-Aminohexane
58	45*	59	29	31	43	27	0	102	2-Methoxyethyl Vinyl Ether
58	45*	73	285	72	80	82	42	379	Morphine Methobromide
58	50*	52	36	42	44	38	49	157	Chlormequat
58	55*	43	104	56	41	97	62	260	Carisoprodol
58	55*	70	83	69	73	97	127	169	Methylbemegride
58	57*	44	42	39	70	40	68	261	Ethoheptazine
58	57*	70	42	44	188	84	43	261	Ethoheptazine*
58	59*	31	45	149	104	43	76	282	Dimethylglycol Phthalate
58	59*	60	28	41	42	0	0	59	Methylformamide (N-Methyl)
58	71*	72	43	159	56	42	201	272	Dimethisoquin*
58	71*	72	167	182	42	59	0	270	Doxylamine
58	71*	167	180	72	42	182	78	270	Doxylamine
58	72*	198	213	214	180	73	42	316	Promethazine Sulphoxide
58	73*	57	43	41	39	29	45	780	Digoxin*
58	84*	45	39	57	69	83	50	84	Thiophene
58	85*	86	91	84	70	42	225	309	Benzydamine*
58	86*	99	56	162	132	149	205	277	Acetylprocainamide (N-Acetyl)
58	86*	162	120	99	43	71	278	278	Acetylprocaine Acetate
58	86*	73	87	319	41	99	245	319	Chloroquine

PEAKS								MWT	COMPOUND
58	86*	319	87	73	247	245	112	319	Chloroquine*
58	86*	87	29	30	116	41	56	379	Cinchocaine
58	86*	30	87	42	130	77	56	216	Diethyltryptamine (N,N-)
58	86*	278	42	87	294	193	44	294	Dimethacrine
58	86*	87	42	56	77	44	43	193	Etafedrine*
58	86*	42	87	77	30	56	51	193	Ethylephedrine
58	86*	99	87	56	42	71	84	387	Flurazepam
58	86*	91	56	65	42	39	118	177	Formylmethylamphetamine (N-)
58	86*	87	42	56	72	234	120	234	Lignocaine
58	86*	88	105	206	87	55	77	353	Tridihexethyl Chloride
58	91*	176	90	41	42	119	65	267	Di-(1-isopropylphenyl)methylamine
58	91*	86	106	231	45	77	230	511	Norbormide*
58	91*	176	119	42	41	56	65	267	Trimethyl-(N, alpha, alpha-)-diphenethylamine
58	97*	72	191	71	78	79	190	261	Methapyrilene
58	98*	79	129	91	77	52	172	287	Cycrimine
58	100*	41	65	43	30	192	56	239	Isoetharine
58	100*	41	101	43	56	65	30	239	Isoetharine*
58	100*	105	77	56	101	41	70	235	Meprylcaine
58	121*	72	71	214	122	215	78	285	Mepyramine
58	121*	72	79	71	78	42	122	285	Pyrilamine
58	121*	72	71	78	215	122	77	286	Thonzylamine
58	165*	163	42	166	164	139	63	660	Hexafluorenium Bromide
58	169*	149	168	170	57	167	43	240	Pheniramine
58	169*	168	170	72	167	44	42	240	Pheniramine*
58	183*	94	42	182	40	196	73	240	GS 29696
58	185*	141	111	128	185	113	75	185	Chlorophenoxyacetamide (p-Chloro) (Iproclozide Metab)
58	197*	99	165	309	112	152	198	309	Methixene
58	203*	205	204	72	167	202	168	274	Chlorpheniramine
58	203*	44	205	54	204	72	202	274	Chlorpheniramine*
58	203*	43	57	205	71	72	32	274	Chlorpheniramine(dex)
58	214*	229	172	43	187	216	41	229	Propazine
58	230*	231	44	173	105	42	159	321	Phenazocine
58	235*	234	85	280	195	193	35	280	Imipramine
58	238*	91	239	167	117	165	77	329	Prenylamine
58	241*	43	184	69	68	41	199	241	Prometryne
58	256*	257	182	199	157	105	91	347	Phenomorphan
59*	31	41	57	43	29	39	0	74	Butanol (tert-)
59*	31	29	57	41	56	27	73	102	Ethyl Butyl Ether
59*	31	29	27	41	44	39	57	102	Ethyl Isobutyl Ether
59*	31	43	41	71	87	39	60	102	3-Hydroxy-3-methyl-2-butanone
59*	31	43	41	29	39	42	0	102	2-Hydroxy-2-methyl-3-butanone
59*	33	61	41	44	92	31	93	92	Methyl Fluoroacetate
59*	42	150	271	44	171	115	128	271	Dextromethorphan (Racemethorphan)
59*	42	150	157	44	257	256	41	257	Levorphanol
59*	42	150	271	44	0	0	0	271	Racemethorphan
59*	43	86	85	40	41	38	42	100	2,3-Dimethyl-3-hydroxy-1-butene
59*	43	156	121	155	157	158	139	214	Phenaglycodol
59*	44	41	29	72	27	43	28	101	Valeramide
59*	45	87	43	41	31	73	55	102	1,1-Dimethylbutanol
59*	45	87	43	41	31	73	27	102	2-Methyl-2-pentanol
59*	55	73	31	43	41	27	29	102	3-Hexanol
59*	58	31	45	149	104	43	76	282	Dimethylglycol Phthalate
59*	58	60	28	41	42	0	0	59	Methylformamide (N-Methyl)
59*	60	118	100	44	45	42	101	160	Aminozide

PEAKS							MWT	COMPOUND
59*	87	41	69	43	31	45	39	102	2,2-Dimethylbutan-2-ol
59*	93	121	43	136	28	81	41	154	2-Terpineol
59*	93	121	81	43	136	68	92	154	Terpineol(alpha-)
59*	121	93	136	43	81	41	55	154	Terpineol
59*	150	271	270	214	171	128	212	271	Dextromethorphan
59*	156	157	69	56	43	41	55	242	3-Hydroxyamylobarbitone
59*	157	156	141	43	41	71	69	242	3'-Hydroxyamylobarbitone
59*	257	150	256	44	31	200	157	257	Levorphanol*
59*	257	256	150	80	42	82	200	257	Racemorphan
59*	271	150	270	31	214	42	171	271	Dextromethorphan*
59*	271	150	214	270	171	112	213	271	Methorphan
59*	271	150	31	270	214	42	171	271	Racemethorphan*
59*	560	331	428	528	402	433	373	687	Iopronic Acid Methyl Ester
59	31*	29	45	74	27	41	43	74	Diethyl Ether
59	31*	29	45	27	72	43	0	90	2-Ethoxyethanol
59	31*	42	60	27	29	45	41	60	1-Propanol
59	42*	73	42	44	101	41	0	101	Diacetimide
59	43*	58	101	29	0	0	0	116	Diacetone Alcohol
59	43*	31	155	121	156	77	41	214	Phenaglycodol
59	44*	70	277	71	191	203	178	277	Maprotiline (Ludiomil)
59	45*	31	41	27	43	44	29	74	2-Butanol
59	58*	202	91	203	215	218	217	277	Amitriptyline
59	58*	42	30	202	91	203	115	277	Amitriptyline
59	58*	42	202	215	203	91	189	277	Amitriptyline
59	58*	202	42	203	214	217	0	277	Amitriptyline*
59	58*	73	75	215	202	203	217	365	Amitriptyline TMS Derivative
59	58*	179	42	178	72	77	30	303	Chlorphenoxamine*
59	58*	221	30	42	222	255	43	315	Chlorprothixene*
59	58*	42	215	202	57	189	43	275	Cyclobenzaprine
59	58*	45	41	40	74	42	59	141	Cyclopentamine
59	58*	72	45	292	218	42	0	292	Dimethindene*
59	58*	42	188	72	115	145	104	188	Dimethyltryptamine (N,N-)
59	58*	236	202	203	221	42	57	295	Dothiepin
59	58*	295	57	221	202	204	203	295	Dothiepin
59	58*	221	204	202	203	293	234	295	Dothiepin (Prothiaden)
59	58*	42	178	165	277	219	202	279	Doxepin
59	58*	42	165	103	72	30	0	347	Embramine*
59	77*	29	43	78	41	27	135	228	Sulphonal*
59	101*	69	41	142	107	77	102	242	MCPB Methyl Ester
59	154*	45	91	69	132	114	117	245	Phentermine TFA
59	244*	77	29	43	31	45	55	309	Glymidine*
59	257*	150	256	200	157	76	189	257	3-Hydroxy-N-methylmorphinan
59	257*	150	256	31	157	42	200	257	Racemorphan*
60*	30	62	44	34	33	46	31	60	Carbonyl Sulphide
60*	42	88	30	31	43	29	70	257	Benserazide*
60*	43	41	27	45	29	39	74	102	Isovaleric Acid
60*	43	41	87	45	27	74	39	102	Valeric Acid
60*	44	28	0	0	0	0	0	60	Urea
60*	73	233	91	232	276	275	0	276	Mebhydroline
60*	73	27	29	41	43	45	28	102	Pentanoic Acid
60*	102	41	45	74	87	56	39	102	3-Methyltetrahydrothiophene
60	59*	118	100	44	45	42	101	160	Aminozide
60	73*	86	71	57	61	103	149	180	Fructose
60	73*	57	71	61	74	101	98	180	Glucose

Asterisk Indicates True Base Peak

```
                    PEAKS                    MWT    COMPOUND

60    73*    43    57    41    55   129    71    200  Lauric acid(n-Dodecanoic Acid)
60    88*    89   126    61   115   114   170    258  Demeton
60    88*   163    43    41   148   121   107    163  Propylethyldithiocarbamate (S-n-propyl-)
60   274*   273   259   181   202   218   215    335  Oxetorone N-oxide
61*    46   294   206   292    60    45    63    292  Chlorbromuron
61*    46   160    60   248   133    45   124    248  Linuron
61*    46   258    91   260   172   170    63    258  Metobromuron
61*    46   126    99   214    60    63    45    214  Monolinuron
61*    96    98    63    26    60    62   100     96  1,2-Dichloroethylene
61*   214   185   126   153   140    93   127    214  3-(4-Chlorophenyl)-1-methyl-1-methoxyurea
61    43*    42    31    41    57    84    29    102  4-Hydroxy-3-methyl-2-butanone
61    43*    87    41    59    42    27    39    102  Isopropyl Acetate
61    43*    73    42    41    59    27     0    102  Propyl Acetate(n-)
61    73*   103    74    56    60    57   133    182  Mannitol
61   160*    76    77   133   104   161    50    317  Phosmet
61   271*    43    45    70   300   256   273    342  3-Acetoxydiazepam
62*    27    49    64    63    98    51    61     98  1,2-Dichloroethane
62*    49    64    63    51    61     0     0     98  1,2-Dichloroethane
62*    97    99    64    61    36   107    45    142  Dalapon
62*   244   279    45   229    57    61   111    279  2-Methylamino-2',5-Dichlorobenzophenone
62    27*    49    64    63    98    51    61     98  1,2-Dichloroethane
62    45*    43    61    44    63   249   247    276  2-(2-Amino-5-bromobenzoyl)pyridine(Bromazepam Benzophenone analogue)
62    45*    44    61    43    63     0     0     62  Boric Acid
62    88*    61    53    37   277    63    89    275  Bromoxynil
63*    27    65    83    26    85    61    98     98  1,1-Dichloroethane
63*    78    45    61    46    62    48    47     78  Dimethylsulfoxide
63*   249   205   251   143    65   207   223    284  Tris-(2-chloroethyl) Phosphate
63    30*    69    74    75    79    62   131    225  Dinitrorhodane Benzene
63    75*    89    51   248   125   302   328    329  Methyclonazepam
63    92*    56    94    65    57   171   219    220  Phosphoramide Mustard
63    93*    27    95    65    31    94    36    142  Bis(2-chloroethyl) Ether
63   110*    64    81    53    55    92   111    110  Catechol
63   162*   164    98    99    49    62    73    162  2,4-Dichlorophenol
63   174*    64   201    90    52    39    65    201  Thiabendazole
63   204*   146   232   117   143   174    89    232  Phenobarbitone
63   285*    75   109   312   183   238   266    313  Flunitrazepam
64*    48   247   102    58   245   304   231    335  Hydroxychloroquine
64    45*   147   105    91    77    79    65    999  Phenelzine Probe Product 2*
64   110*    63    81    39    92    55   111    110  Catechol
64   128*   130    63    65    92    39   129    128  2-Chlorophenol
64   269*   205   297   271    43    44    31    297  Hydrochlorothiazide
64   310*    36   312    42    43    62    63    359  Methychlothiazide (Endurone)
65*    45    85    64    44    31    61    81    100  1,1-Difluoro-1-chloroethane
65*    57    41    91   120    29    43    39    137  Nitrotoluene (ortho)
65    30*    93   139    39   111    53    51    169  Noradrenaline
65    44*    91   152    77    51   164   137    195  2,3-Dimethoxyamphetamine
65    51*    27    45    62    47    26    64     66  1,2-Difluoroethane
65    91*   182    92    39    51    63    77    182  Bibenzyl
65    91*    39   119    92    63    89    51    210  Dibenzyl Ketone
65    91*   137    41    57    39    30     0    137  Nitrotoluene (meta)
65    91*   137   107    39    63    77    79    137  Nitrotoluene (para)
65    91*    77   196    39    56    51    90    303  Phenoxybenzamine
65    91*   187    92   277   259    90   273    596  Trimethaphan
65    92*   108   214   156    39   109    43    214  Sulphaguanidine*
```

								MWT	COMPOUND
65	92*	108	156	191	71	63	93	255	Sulphathiazole
65	94*	66	95	55	51	63	47	94	Phenol
65	120*	86	92	56	99	41	44	236	Butethamine
65	120*	92	39	165	137	63	41	165	Ethyl-4-aminobenzoate
65	120*	91	92	39	77	121	89	137	2-Nitrotoluene
65	121*	39	93	63	92	64	122	214	Phenylsalicylate
65	138*	92	77	104	39	90	64	138	2-Nitroaniline
65	138*	92	39	108	106	66	52	138	4-Nitroaniline
65	139*	39	109	93	53	81	63	139	Nitrophenol(para)
66*	39	106	79	52	53	51	40	106	1,4-Dicyano-1-butene
66*	120	39	65	269	205	118	77	389	Cyclothiazide*
66*	263	79	91	265	101	261	65	362	Aldrin
66	93*	92	78	65	39	51	94	93	2-Methylpyridine
66	93*	92	65	39	94	67	59	93	3-Methylpyridine
66	93*	92	65	39	94	54	67	93	4-Methylpyridine
66	93*	92	39	65	67	51	54	93	4-Methylpyridine(4-Picoline)
66	93*	92	65	78	39	51	67	93	2-Picoline(2-Methylpyridine)
66	93*	92	65	39	67	40	94	93	3-Picoline(3-Methylpyridine)
66	94*	39	65	40	95	38	55	94	Phenol
66	135*	77	107	92	136	63	64	166	Methylparaben Dimethyl Ether
67*	41	82	54	56	59	55	39	102	Fluorocyclohexane
67*	41	54	39	27	53	81	68	82	1,5-Hexadiene
67*	41	43	27	39	54	40	53	82	1-Hexyne
67*	41	39	40	38	42	37	66	67	Pyrrole
67*	68	39	53	41	65	42	27	68	Cyclopentene
67*	69	35	47	31	48	83	49	102	Dichlorofluoromethane
67*	69	35	47	31	32	48	83	102	Dichloromonofluoromethane
67*	69	31	111	113	79	48	47	146	Fluorochlorobromomethane
67*	81	263	36	79	82	261	265	378	Endrin
67*	81	54	55	95	68	82	69	294	Methyl Octadecadienoate
67*	82	54	41	55	83	39	36	118	Chlorocyclohexane
67*	137	91	138	55	79	41	95	340	Dimethisterone
67*	193	66	41	169	39	65	77	234	Cyclopentobarbitone*
67*	221	196	41	164	111	39	181	262	Permethylated Cyclopentobarbitone*
67	41*	39	27	40	38	66	37	67	Allyl Cyanide
67	41*	39	40	66	38	27	37	67	Crotononitrile(cis-)
67	41*	39	40	66	38	27	37	67	Crotononitrile(trans-)
67	68*	53	41	39	40	27	69	104	1-Chloro-3-methyl-2-butene
67	68*	93	94	79	92	121	107	136	Limonene
67	94*	26	39	40	53	38	42	94	Methylpyrazine
67	108*	42	40	41	26	39	109	108	2,3-Dimethylpyrazine
67	127*	72	193	237	44	109	111	237	Dicrotophos
67	180*	82	109	55	181	66	81	180	Theobromine
67	207*	79	81	141	77	55	91	236	Cyclobarbitone
67	221*	196	181	41	164	111	107	262	Allylcyclopentenylbarbitone Dimethyl Derivative
67	221*	196	41	164	181	222	0	248	Methylcyclopal
67	290*	108	107	79	55	41	93	290	Androsterone
68*	39	29	43	94	44	336	95	780	Digoxin*
68*	41	40	28	67	69	0	0	68	Pyrazole
68*	67	53	41	39	40	27	69	104	1-Chloro-3-methyl-2-butene
68*	67	93	94	79	92	121	107	136	Limonene
68*	70	33	49	47	51	48	50	68	Chlorofluoromethane
68*	73	140	98	125	350	405	412	427	Orciprenaline TMS Derivative
68*	93	67	136	41	121	79	39	136	Dipentene

	PEAKS						MWT	COMPOUND
68*	93	67	39	79	53	27	94	136 Limonene
68*	93	67	136	94	41	121	53	136 Limonene
68*	93	67	39	41	53	136	79	136 1-Methyl-4-isopropenylcyclohexene
68*	95	180	53	41	40	67	42	180 Theophylline
68*	369	41	86	370	73	40	140	455 Salbutamol TMS Derivative
68	39*	38	29	40	37	42	26	68 Furan
68	41*	54	55	39	27	40	29	108 Adiponitrile(1,4-Dicyanobutane)
68	41*	69	67	39	27	31	29	100 Cyclopentylmethanol
68	42*	55	45	100	41	70	39	100 3-Methyltetrahydropyran
68	67*	39	53	41	65	42	27	68 Cyclopentene
68	69*	70	77	103	56	54	51	245 Bisnortilidine
68	73*	140	98	41	125	74	99	427 Isoprenaline TMS-Derivative
68	82*	91	159	42	158	92	65	159 Pargyline*
68	83*	82	72	84	77	103	115	259 Nortilidine
68	83*	84	103	77	91	157	155	259 1-Phenyl-1-carbethoxy-2-methylaminocyclohex-3-ene
68	166*	95	41	53	123	0	0	166 3-Methylxanthine
68	166*	123	53	42	41	95	0	166 7-Methylxanthine
68	180*	123	53	42	95	150	151	180 Paraxanthine
68	225*	44	173	240	198	172	43	240 Cyanazine
69*	41	55	95	67	56	81	71	154 Citronellal
69*	41	43	42	27	149	29	151	228 1,2-Dibromopentane
69*	41	27	55	43	42	39	29	228 1,5-Dibromopentane
69*	41	39	65	27	29	53	154	154 Geraniol
69*	41	68	67	93	55	39	0	154 Geraniol
69*	41	39	85	100	28	59	0	100 Methyl Crotonate
69*	41	39	93	68	29	27	154	154 Nebol
69*	41	93	68	80	121	67	111	154 Nerol
69*	43	41	44	71	167	165	55	193 2-Bromo-2-ethylbutyramide(Carbromal Metabolite)
69*	43	169	45	464	70	58	677	677 Morphine Diheptafluorobutyrate Derivative
69*	56	71	91	261	84	119	70	367 Phenbutrazate*
69*	68	70	77	103	56	54	51	245 Bisnortilidine
69*	81	152	41	109	95	80	68	152 Bornanone-3
69*	81	67	55	71	68	56	41	154 Isopulegol
69*	83	43	79	41	81	53	80	141 Methylpentynol Carbamate*
69*	85	50	87	35	31	37	33	104 Chlorotrifluoromethane
69*	111	41	68	93	123	67	81	154 Lavandulol
69*	131	100	502	119	414	464	614	614 Heptacosafluorobutylamine
69*	148	150	129	131	79	81	50	148 Bromotrifluoromethane
69*	219	131	100	264	502	119	414	614 Heptacosafluorotributylamine
69*	229	231	161	230	133	162	233	229 (3,4-Dichlorophenyl)methylacrylamide (N-)
69	33*	83	32	31	51	63	82	102 1,1,1,2-Tetrafluoroethane
69	41*	39	84	94	27	29	67	152 Citral
69	41*	29	95	55	39	43	27	154 Citronellal
69	41*	55	67	81	82	68	43	156 Citronellol
69	41*	39	100	40	59	99	0	100 Methyl Methacrylate
69	41*	93	39	67	79	27	91	136 Myrcene
69	41*	68	93	39	29	67	43	154 Nerol
69	41*	70	45	110	217	159	145	285 Pentazocine
69	43*	168	41	85	167	86	97	252 3-Ketoquinalbarbitone
69	44*	41	55	43	39	71	53	236 Carbromal
69	44*	41	55	43	29	208	71	236 Carbromal
69	44*	41	208	210	55	71	43	236 Carbromal*
69	45*	41	43	44	87	27	29	102 2-Hexanol
69	56*	55	41	43	29	57	31	102 3-Methyl-1-pentanol

Peak Index   *Asterisk Indicates True Base Peak*   Page: 179

	PEAKS						MWT	COMPOUND
69	56*	91	261	98	119	84	190	367 Phenbutrazate
69	57*	41	56	43	29	45	31	102 3,3-Dimethylbutan-1-ol
69	67*	35	47	31	48	83	49	102 Dichlorofluoromethane
69	67*	35	47	31	32	48	83	102 Dichloromonofluoromethane
69	67*	31	111	113	79	48	47	146 Fluorochlorobromomethane
69	81*	41	152	80	82	109	39	152 1,3,3-Trimethyl-2-norboranone
69	107*	125	83	79	55	41	77	276 Cyclandelate
69	112*	42	40	41	68	39	70	112 Uracil
69	126*	85	52	43	97	80	51	126 Phloroglucinol
69	126*	85	52	43	63	42	41	126 Phloroglucinol
69	128*	143	73	169	75	159	41	324 Clofibrate Derivative
69	169*	131	133	197	227	95	97	344 2,2,2-Trichloroethyl Heptafluorobutyrate
69	328*	329	126	441	315	0	0	441 Dopamine TFA Derivative
69	440*	126	441	0	0	0	0	553 6-Hydroxydopamine TFA Derivative
70*	42	113	43	141	56	71	44	373 Prochlorperazine
70*	42	113	43	407	56	141	248	407 Trifluoperazine
70*	43	55	41	57	42	71	0	117 Amyl Nitrite
70*	43	154	71	41	209	86	195	653 Bromocriptine*
70*	43	154	71	41	267	195	221	561 Ergocornine Hydrogen Maleate
70*	43	40	154	41	44	267	69	547 Ergosinine
70*	44	207	178	279	249	0	0	279 Hydroxyprotriptyline
70*	44	71	57	247	190	119	191	247 Ketobemidone
70*	71	269	154	195	55	59	57	563 Dihydroergocornine*
70*	83	42	257	193	56	228	164	327 Loxapine
70*	98	69	77	204	89	83	105	204 Thozalinone
70*	112	293	264	44	42	305	41	374 Bromhexine*
70*	113	43	42	71	141	56	72	409 Butaperazine
70*	113	42	141	43	56	399	71	399 Thiethylperazine
70*	113	43	42	127	71	44	56	446 Thioproperazine
70*	113	127	446	198	71	445	43	446 Thioproperazine
70*	120	43	91	41	65	77	98	161 Aletamine
70*	125	91	153	43	41	44	244	583 Dihydroergotamine*
70*	125	71	91	153	267	154	221	609 Ergocristine*
70*	150	151	153	77	51	118	60	322 Chloramphenicol
70*	153	36	118	117	60	155	150	322 Chloramphenicol
70*	154	69	224	210	55	196	209	547 Ergosinine*
70*	191	44	189	43	165	69	192	263 Protriptyline
70*	293	264	112	305	291	262	295	374 Bromhexine
70	41*	39	69	42	27	38	29	70 Crotonaldehyde
70	43*	55	42	41	73	61	87	130 Amyl Acetate
70	43*	41	71	269	154	195	167	563 Dihydroergocornine
70	43*	40	269	125	91	153	131	583 Dihydroergotamine
70	43*	71	54	44	154	267	55	561 Ergocornine*
70	43*	71	154	41	209	69	267	575 Ergocryptine*
70	43*	71	154	267	209	221	69	575 Ergocryptinine*
70	43*	57	154	71	55	69	85	547 Ergosine*
70	43*	71	55	41	29	56	84	102 2-Ethyl-1-butanol
70	43*	42	185	246	445	125	98	445 Thiopropazate
70	44*	179	178	207	280	250	236	297 Dihydroxyprotriptyline
70	44*	178	250	179	279	0	0	279 10,11-Epoxyprotriptyline
70	44*	59	277	71	191	278	203	277 Maprotiline*
70	55*	42	100	41	29	39	43	100 3,4-Dimethyltetrahydrofuran
70	55*	113	42	41	39	44	69	141 Ethosuximide(i)*
70	68*	33	49	47	51	48	50	68 Chlorofluoromethane

Asterisk Indicates True Base Peak   Peak Index

								MWT	COMPOUND
70	71*	263	57	42	44	262	43	263	Hydroxypethidine(p-)
70	71*	42	57	43	172	247	91	247	Pethidine
70	71*	247	246	57	172	42	218	247	Pethidine
70	71*	44	57	42	247	43	246	247	Pethidine*
70	71*	247	57	42	246	91	103	247	Pethidine (Demerol)
70	71*	42	247	57	44	96	43	247	Pethidine (N-Methyl-4-phenyl-4-carboxypiperidine Ethyl Ester)
70	71*	57	43	219	42	218	44	219	Pethidinic acid*
70	71*	42	57	233	232	43	44	233	Pethidinic Acid Methyl Ester
70	71*	42	103	73	291	57	44	291	Pethidinic Acid TMS Ester
70	71*	218	261	57	174	219	172	302	Properdin
70	71*	218	57	42	44	36	43	261	Properidine
70	83*	42	244	71	43	273	56	343	Clothiapine
70	83*	273	244	209	42	71	43	343	Clothiapine*
70	84*	83	68	288	260	273	289	371	Flurazepam Metabolite
70	91*	186	152	117	77	172	229	229	1-(1-Phenylcyclohexyl)pyrrolidine
70	92*	43	41	109	90	39	76	154	1-Terpin-4-ol
70	98*	41	77	91	99	96	120	264	Carbocaine
70	98*	99	36	42	96	99	38	246	Mepivacaine
70	98*	42	94	77	51	99	126	386	Mesoridazine
70	98*	99	42	386	126	55	41	386	Mesoridazine*
70	98*	214	111	134	341	199	326	341	Sulpyrid
70	98*	42	185	126	99	96	55	370	Thioridazine
70	98*	370	126	99	185	244	125	370	Thioridazine (Melleril)
70	113*	55	42	41	39	85	69	141	Ethosuximide(ii)*
70	113*	373	141	43	72	42	127	373	Prochlorperazine*
70	113*	42	221	56	43	222	114	443	Thiothixene
70	113*	407	43	141	42	127	71	407	Trifluoperazine*
70	125*	91	153	41	244	43	71	611	Dihydroergocristine*
70	125*	91	153	41	43	244	349	611	Dihydroergocristine Mesilate
70	125*	91	153	244	40	43	314	581	Ergotamine Tartrate
70	141*	55	41	42	69	112	126	169	Ethylethosuximide (N-)
70	143*	100	144	42	56	98	221	400	Clopenthixol*
70	143*	100	144	42	98	58	56	434	Flupenthixol*
70	154*	155	167	223	225	349	153	577	Dihydroergocryptine*
70	175*	176	132	379	204	56	217	379	Oxypertine*
70	186*	83	42	85	71	57	56	275	Dimethyldemerol
70	191*	44	58	189	84	165	192	263	Protriptyline
70	198*	115	199	71	269	72	41	327	Aminopromazine
70	198*	58	115	71	56	72	269	327	Proquamazine
70	205*	231	78	135	136	42	166	371	Trazodone*
70	217*	69	110	202	270	285	284	285	Pentazocine
70	239*	209	210	166	139	138	211	405	Dihydroxytrazodone (Trazodone Metabolite)
70	250*	207	91	251	119	148	56	341	Fenethylline*
70	280*	143	42	113	406	281	437	437	Fluphenazine
70	287*	44	42	163	59	288	286	287	Dihydromorphine
70	287*	164	44	42	286	285	288	287	Dihydromorphine
70	287*	44	164	42	288	59	230	287	Dihydromorphine*
70	303*	58	44	57	42	216	286	303	Hydromorphinol
70	328*	216	58	187	115	286	285	328	6-Deoxy-6-azido-14-hydroxydihydroisomorphine
70	342*	230	201	314	115	58	300	342	6-Deoxy-6-azido-14-hydroxydihydroisocodeine
70	343*	344	109	42	113	491	56	491	Lidoflazine*
71*	28	100	72	58	199	129	83	199	Diethylcarbamazine
71*	41	42	43	61	72	101	102	102	2-Methoxytetrahydrofuran
71*	42	41	102	43	31	55	44	102	Dioxepane

								MWT	COMPOUND
71*	42	70	57	44	43	103	91	247	Meperidine
71*	42	43	56	177	77	70	72	177	Phenmetrazine
71*	42	56	43	44	177	30	117	177	Phenmetrazine
71*	42	56	43	177	77	178	105	177	Phenmetrazine*
71*	43	88	60	41	73	45	89	116	Ethyl Butyrate
71*	43	41	27	42	31	44	39	102	2-Hydroxy-2-methyltetrahydrofuran
71*	43	41	93	55	69	80	67	154	Linalool
71*	43	41	27	31	42	29	0	102	Tetrahydrofurfuryl Alcohol
71*	45	72	28	31	39	29	27	100	5-Methyl-1,2,3-thiadiazole
71*	56	177	77	70	105	51	72	177	Phenmetrazine
71*	58	72	43	159	56	42	201	272	Dimethisoquin*
71*	58	72	167	182	42	59	0	270	Doxylamine
71*	58	167	180	72	42	182	78	270	Doxylamine
71*	70	263	57	42	44	262	43	263	Hydroxypethidine(p-)
71*	70	42	57	43	172	247	91	247	Pethidine
71*	70	247	246	57	172	42	218	247	Pethidine
71*	70	44	57	42	247	43	246	247	Pethidine*
71*	70	247	57	42	246	91	103	247	Pethidine (Demerol)
71*	70	42	247	57	44	96	43	247	Pethidine (N-Methyl-4-phenyl-4-carboxypiperidine Ethyl Ester)
71*	70	57	43	219	42	218	44	219	Pethidinic acid*
71*	70	42	57	233	232	43	44	233	Pethidinic Acid Methyl Ester
71*	70	42	103	73	291	57	44	291	Pethidinic Acid TMS Ester
71*	70	218	261	57	174	219	172	302	Properdin
71*	70	218	57	42	44	36	43	261	Properidine
71*	72	58	100	83	56	70	44	199	Diethylcarbamazine*
71*	73	85	61	113	36	43	60	308	Chloralose (alpha)
71*	81	95	41	55	82	43	69	156	Menthol
71*	85	193	195	42	130	70	194	266	Desipramine
71*	91	57	106	65	72	55	77	177	Bethanidine
71*	91	106	177	57	72	65	30	177	Bethanidine*
71*	93	47	43	55	69	80	154	154	Linalool
71*	93	111	43	86	69	55	68	154	Terpineol
71*	95	81	41	55	43	69	82	156	Neomenthol
71*	120	43	91	41	0	0	0	161	Aletamine
71*	140	70	263	262	189	57	42	263	Hydroxypethidine
71*	247	172	218	174	103	96	91	247	Pethidine
71*	247	246	70	218	57	174	248	247	Pethidine
71	41*	45	29	39	27	70	0	102	4-Methoxy-2-buten-1-ol
71	43*	70	55	41	73	45	29	102	2,2-Dimethylbutanol
71	43*	27	41	29	70	85	39	100	3,3-Dimethylpentane
71	43*	70	29	27	41	55	39	100	3-Ethylpentane
71	43*	55	79	91	109	123	336	336	Fluoxymesterone
71	43*	27	41	39	42	29	58	114	4-Heptanone
71	43*	41	39	27	87	102	0	102	Methyl Isobutanoate
71	43*	70	41	55	69	56	84	102	2-Methyl-1-pentanol
71	57*	43	85	40	55	41	149	229	Ergothioneine
71	58*	36	176	150	72	193	59	264	Amethocaine
71	58*	150	176	72	193	105	59	264	Amethocaine*
71	58*	167	72	42	59	202	45	290	Carbinoxamine
71	58*	42	44	167	72	43	57	290	Carbinoxamine
71	58*	26	54	167	72	42	44	290	Carbinoxamine*
71	58*	167	72	42	202	203	45	290	Carbinoxamine(1-)
71	58*	72	207	42	91	59	118	291	Cyclopentolate*
71	58*	159	56	72	41	42	115	272	Dimethisoquin

								MWT	COMPOUND
71	58*	72	167	182	42	180	59	270	Doxylamine*
71	58*	56	77	42	105	146	59	165	Ephedrine GC Decomposition
71	58*	208	72	59	42	89	57	294	Noxiptyline*
71	58*	42	105	176	150	92	56	264	Tetracaine
71	70*	269	154	195	55	59	57	563	Dihydroergocornine*
71	72*	233	232	0	0	0	0	304	Chlorpromazine Metabolite
71	72*	73	57	91	165	70	56	309	Methadone
71	73*	83	129	256	98	85	97	256	Palmitic Acid
71	84*	85	82	80	341	70	356	356	Conessine*
71	86*	99	58	55	56	100	87	309	Dicyclomine*
71	93*	41	69	55	43	80	121	154	Linalool
71	98*	84	56	99	124	41	167	167	Mecamylamine
71	104*	147	42	44	70	56	102	237	Cartap
71	104*	147	42	44	70	56	102	171	Crimidine
71	167*	166	139	41	78	140	168	290	Carbinoxamine
72*	30	56	73	249	98	234	102	249	Alprenolol*
72*	30	107	56	43	73	222	57	266	Atenolol
72*	30	56	98	43	107	41	73	266	Atenolol*
72*	30	43	77	73	51	41	27	213	Clorprenaline*
72*	30	43	101	41	27	58	28	101	Dipropylamine(n-)
72*	30	56	107	223	98	73	43	267	Metoprolol
72*	30	107	56	45	41	44	43	267	Metoprolol*
72*	30	221	41	56	73	150	43	265	Oxprenolol
72*	30	151	56	109	43	108	251	266	Practolol
72*	31	91	120	65	73	39	42	163	1-Propyl-2-phenylethylamine
72*	39	148	56	41	71	51	73	163	Mephentermine
72*	39	42	38	94	56	51	81	260	Pyridostigmine Bromide
72*	40	42	44	73	105	77	56	177	Dimethylpropion
72*	41	45	43	42	29	44	211	211	Isolan
72*	41	56	43	221	73	57	45	265	Oxprenolol*
72*	42	91	44	73	56	39	70	163	Dimethylamphetamine
72*	42	208	108	65	73	66	39	302	Neostigmine
72*	42	151	153	43	29	40	222	222	Pyramat
72*	43	30	56	151	221	41	98	336	Acebutolol*
72*	43	73	36	46	91	71	42	353	Alphacetylmethadol(d)
72*	43	73	36	91	56	42	225	353	Alphacetylmethadol(1)
72*	43	44	91	278	0	0	0	353	Betacetylmethadol
72*	43	65	41	111	93	39	56	211	Orciprenaline
72*	43	73	41	70	65	40	39	211	Orciprenaline*
72*	43	133	116	248	204	104	56	248	Pindolol
72*	44	212	45	77	42	132	214	212	Chlorotoluron
72*	44	41	45	89	42	43	55	198	Cycloate
72*	44	42	91	56	65	58	162	163	Dimethylamphetamine (N,N-)
72*	44	73	77	105	42	132	160	179	2-Ethylamino-1-phenylpropanol
72*	44	134	77	42	73	105	51	177	2-Ethylamino-1-phenylpropanone
72*	44	91	73	42	56	65	39	163	Ethylamphetamine
72*	44	42	70	77	128	51	105	177	Ethylpropion
72*	44	159	73	58	42	109	56	231	Fenfluramine*
72*	44	118	91	65	117	42	39	163	Formylamphetamine (N-Formyl)
72*	44	57	43	149	41	55	69	151	Hydroxyamphetamine
72*	44	43	124	123	30	42	41	211	Isoprenaline*
72*	44	42	77	56	73	51	70	179	Methylephedrine
72*	45	55	54	44	254	43	53	411	Acetophenazine
72*	56	91	129	105	130	73	57	311	Alphamethadol

PEAKS								MWT	COMPOUND
72*	56	30	43	98	115	144	41	259	Propranolol*
72*	57	43	69	71	55	81	41	309	Methadone
72*	58	198	213	214	180	73	42	316	Promethazine Sulphoxide
72*	71	233	232	0	0	0	0	304	Chlorpromazine Metabolite
72*	71	73	57	91	165	70	56	309	Methadone
72*	73	43	91	224	56	71	129	353	Alphacetylmethadol(1)
72*	73	91	58	44	115	105	165	311	Betamethadol
72*	73	91	58	253	193	42	115	311	Betamethadol
72*	73	320	71	70	56	210	198	391	Dimethothiazine
72*	73	322	251	165	167	337	0	337	2-Dimethylamino-4,4-diphenyl-5-nonane
72*	73	56	71	70	231	152	180	316	Dioxopromethazine
72*	73	181	86	214	42	56	200	285	Isothipendyl
72*	73	214	200	44	285	86	56	285	Isothipendyl*
72*	73	58	57	223	165	70	0	309	Methadone
72*	73	91	223	165	71	294	57	309	Methadone
72*	73	91	293	223	165	85	71	309	Methadone*
72*	73	198	180	213	284	42	44	284	Promethazine
72*	73	284	198	213	199	180	56	284	Promethazine*
72*	73	229	75	230	43	45	153	301	Pronethalol (Nethalide) TMS Ester
72*	73	340	269	197	71	70	56	340	Propiomazine*
72*	91	73	193	165	58	115	178	311	Alphamethadol
72*	91	73	56	148	41	57	42	163	Mephentermine
72*	91	114	145	75	160	117	92	467	Oxethazaine*
72*	115	144	259	116	215	127	254	259	Propranolol
72*	120	55	91	65	39	77	31	163	1-Isopropyl-2-phenylethylamine
72*	133	30	116	248	134	56	41	248	Pindolol*
72*	134	91	42	57	31	65	44	163	1-Ethyl-N-methyl-2-phenylethylamine
72*	149	269	255	197	73	254	340	340	Propiomazine
72*	165	42	180	178	179	91	73	309	Methadone
72*	166	238	167	42	44	138	109	238	Pirimicarb
72*	167	73	165	152	253	166	168	253	Emepromium Bromide GC Breakdown Product
72*	176	91	58	177	65	281	42	281	Alverine
72*	176	91	58	177	41	42	30	281	Alverine*
72*	180	84	135	44	57	45	41	279	Karbutilate
72*	180	198	73	42	56	70	71	284	Promethazine
72*	181	85	152	42	44	58	43	419	Methantheline Bromide
72*	198	44	200	73	42	99	199	198	Monuron
72*	198	197	196	73	56	70	255	340	Propiomazine
72*	199	167	198	71	42	56	200	284	Promethazine
72*	228	183	230	44	229	168	45	228	Metoxuron
72*	232	234	161	73	45	163	124	232	Diuron
72*	240	42	39	40	41	44	29	240	Dimetilan
72*	240	169	44	42	73	170	56	240	Dimetilan
72*	245	44	290	45	40	75	247	290	Chloroxuron
72*	284	73	43	42	198	180	166	284	Promethazine
72	30*	115	57	41	43	56	42	259	Propranolol
72	30*	44	43	27	28	41	86	101	Propylaminopropane(n-)
72	43*	27	41	29	71	57	39	100	2-Ethyl Butyraldehyde
72	55*	97	41	56	158	118	57	274	Tybamate*
72	58*	71	177	175	96	42	78	339	Bromothen
72	58*	43	78	271	57	44	77	371	Camazépam
72	58*	192	42	165	73	71	59	278	Methadone Intermediate
72	58*	30	77	56	44	42	73	179	Methylephedrine*
72	58*	71	224	59	42	152	57	295	Normethadone

```
72    58*    71    42    59   163    91   179   295  Normethadone
72    58*   279   234   151   192   166   165   279  Thymoxamine
72    58*    59    57   121   164   237    77   237  Thymoxamine Hydrolysis Product
72    71*    58   100    83    56    70    44   199  Diethylcarbamazine*
72    86*   191    58    30   128   193   163   321  Clamoxyquin*
72    97*    55    71    62   110    69   158   232  Mebutamate
72   100*   240   340    44   197   254    43   442  Acepromazine Maleate*
72   100*    39    71    55    45    46   101   100  2,5-Dihydrothiophene(2-oxy-)
72   100*    56   165    42   101    91   115   353  Dioxaphetyl Butyrate
72   114*    30   115   160    43   174   145   274  5-Methoxydiisopropyltryptamine
72   114*   337   100    43   237   115    85   337  Propanidid
72   127*   264   138   109    67   193    42   299  Phosphamidon
72   144*    30   115    43   130   144    56   244  Diisopropyltryptamine (N,N-)
72   144*   145   116    89    29   100   244   244  Indol-3-yl-N,N-diethylglyoxamide
72   151*   223   123   222   152    52    29   223  Ethamivan*
72   166*   238   167   138   123   152   109   238  Pirimicarb
72   221*   325    54   196    55   207   181   325  Ergometrine*
72   221*   207   181    43    44   323   222   323  Iso-lysergide*
72   221*   207   181    42    44   222    58   323  LSD
73*   43    72    86   100    87    59   128   227  Dodine
73*   43    84    55    69    41    44    85   145  Emylcamate
73*   44    42    30    43    72    29    58    73  Dimethylformamide
73*   45    28    27    29    43    57     0   102  2-Ethyl-1,3-dioxolane
73*   55    43    45    87    41    69    29   102  3-Methyl-3-Pentanol
73*   58    57    43    41    39    29    45   780  Digoxin*
73*   60    86    71    57    61   103   149   180  Fructose
73*   60    57    71    61    74   101    98   180  Glucose
73*   60    43    57    41    55   129    71   200  Lauric acid(n-Dodecanoic Acid)
73*   61   103    74    56    60    57   133   182  Mannitol
73*   68   140    98    41   125    74    99   427  Isoprenaline TMS-Derivative
73*   71    83   129   256    98    85    97   256  Palmitic Acid
73*   85    60   103    71    57    61    58   342  Lactose
73*  102    57    41    43    55   101   115   144  Valproic Acid
73*  117    75   103   205   147   145   129   596  Chloralose (alpha) TMS Ether
73*  129   305   128   304   276   103   232   305  Norpethidinic Acid Acid Ethyl Ester N-TMS Derivative
73*  129   128   114   334   349   115   130   349  Norpethidinic Acid TMS Ester N-TMS Derivative
73*  147   332   117   205    45    74   133   464  Ascorbic Acid(Tetra TMS)
73*  191    75    45    74   104   147   167   516  5-(3,4-Dihydroxycyclohexa-1,5-dienyl)-3-me-5-phenylhydantoin TMS Deriv
73*  193    43   195   210   237    75   252   252  Propyl Hydroxybenzoate (Propylparaben) TMS Ether
73*  217   204    75   147   185   253    45   676  Hydroxyamobarbitone Glucuronide Me TMS Derivative (peak 2)
73*  253   217   185   317    75   204    69   676  1,3-Dimethylhydroxypentobarbitone Glucuronide Me Ester TMS Derivative
73*  253    75   217   147   185   317    45   676  Hydroxyamobarbitone Glucuronide Me TMS Derivative (peak 1)
73*  253   167   165   194   193   152   115   253  Prenylamine Metabolite
73*  265   217    75   204    69    41   147   688  1,3-Dimethylhydroxysecobarbitone Me Ester TMS Ether
73*  317   169   171    41   318   217    75   620  Clofibrate Glucuronide Methyl Ester TMS Ether
73*  317   217    75   147   318    43    79   688  5-(4-Hydroxyphenyl)-3-methyl-5-phenylhydantoin Glucuronide me Ester TMS
73*  319   291   348   206   320    45   333   348  1,3-Dimethylhydroxyphenobarbitone TMS Ether
73*  338   323   247   179   139   235   154   338  Methaqualone Metabolite TMS Derivative
73*  341   342   343    45   344   327    91   342  Desmethyldiazepam TMS Derivative
73*  341    43   271    75   147    41   342   444  1,3-Dimethyldihydroxysecobarbitone di-TMS Ether
73*  343   257   256   345   372   283    45   372  3-Hydroxydiazepam TMS Derivative
73*  429   430    45   431   147   432    75   430  3-Hydroxydesmethyldiazepam
73*  429   236   196   146   414   430   287   429  Morphine bis TMS Ether
73*  449   359   450   147   269    75   243   654  Cortolone(Tetra TMS)
```

PEAKS							MWT	COMPOUND
73	30*	107	27	108	41	57	39	999	Bamethan Probe Product*
73	30*	41	27	42	39	31	56	73	Butylamine(n-)
73	30*	41	27	55	39	56	31	73	Isobutylamine
73	43*	41	102	27	31	57	42	102	Propyl Ether(n-)
73	44*	272	45	164	162	396	250	483	Etorphine TMS Ether
73	57*	60	43	55	71	41	69	284	Stearic Acid
73	58*	45	57	43	44	167	165	333	Bromodiphenhydramine
73	58*	45	165	59	42	166	44	333	Bromodiphenhydramine
73	58*	45	165	59	42	166	149	333	Bromodiphenhydramine*
73	58*	45	43	57	167	44	165	469	Dimenhydrinate
73	58*	165	45	105	167	44	77	255	Diphenhydramine
73	58*	45	44	59	165	167	42	255	Diphenhydramine
73	58*	45	167	165	166	44	152	255	Diphenhydramine*
73	58*	59	88	45	75	43	56	237	Ephedrine TMS Ether
73	58*	165	45	166	42	181	59	269	Orphenadrine
73	58*	45	42	165	59	46	44	269	Orphenadrine
73	58*	44	45	165	42	40	181	269	Orphenadrine*
73	60*	233	91	232	276	275	0	276	Mebhydroline
73	60*	27	29	41	43	45	28	102	Pentanoic Acid
73	68*	140	98	125	350	405	412	427	Orciprenaline TMS Derivative
73	71*	85	61	113	36	43	60	308	Chloralose (alpha)
73	72*	43	91	224	56	71	129	353	Alphacetylmethadol(1)
73	72*	91	58	44	115	105	165	311	Betamethadol
73	72*	91	58	253	193	42	115	311	Betamethadol
73	72*	320	71	70	56	210	198	391	Dimethothiazine
73	72*	322	251	165	167	337	0	337	2-Dimethylamino-4,4-diphenyl-5-nonane
73	72*	56	71	70	231	152	180	316	Dioxopromethazine
73	72*	181	86	214	42	56	200	285	Isothipendyl
73	72*	214	200	44	285	86	56	285	Isothipendyl*
73	72*	58	57	223	165	70	0	309	Methadone
73	72*	91	223	165	71	294	57	309	Methadone
73	72*	91	293	223	165	85	71	309	Methadone*
73	72*	198	180	213	284	42	44	284	Promethazine
73	72*	284	198	213	199	180	56	284	Promethazine*
73	72*	229	75	230	43	45	153	301	Pronethalol (Nethalide) TMS Ester
73	72*	340	269	197	71	70	56	340	Propiomazine*
73	96*	94	42	57	227	142	212	227	Scopoline TMS Ether
73	117*	327	75	143	256	118	69	342	1,3-Dimethylhydroxypentobarbitone TMS Ether
73	139*	141	111	487	370	75	140	487	Indomethacin Derivative TMS Derivative
73	146*	232	334	247	117	246	147	362	Primidone 1,3-di-TMS Derivative
73	169*	170	75	249	233	79	171	338	3-Methylhydroxyhexobarbitone TMS Ether
73	267*	268	269	45	135	193	75	282	Salicyclic Acid TMS Ester TMS Ether
73	323*	338	307	75	249	154	143	338	Methaqualone Metabolite TMS Derivative
73	344*	372	371	346	345	373	374	372	4'-Hydroxydiazepam TMS Derivative
73	354*	104	268	325	282	355	77	354	5-(3-Hydroxyphenyl)-3-methyl-5-phenylhydantoin TMS Ether
74*	29	57	41	27	28	87	45	102	2-Methylbutanoic Acid
74*	43	71	59	41	87	27	0	102	Methyl Butyrate(n-)
74*	76	28	75	59	43	42	47	121	Cysteine(L-)*
74*	87	43	55	41	75	143	57	186	Methyl Decanoate
74*	87	43	41	55	75	57	69	284	Methyl Heptadecanoate
74*	87	43	75	57	55	143	69	270	Methyl Hexadecanoate
74*	87	41	43	55	75	57	69	214	Methyl Laurate
74*	87	43	41	55	57	75	69	242	Methyl Myristate
74*	87	43	57	41	55	59	127	158	Methyl Octanoate

Asterisk Indicates True Base Peak

		PEAKS					MWT	COMPOUND

```
          PEAKS                            MWT   COMPOUND

74*  87  270   75  143   43   55   41     270  Methyl Palmitate
74*  87   43   41   55   75   57   69     270  Methyl Palmitate
74*  87   43   55  298  143   75   57     298  Methyl Stearate
74*  87   43   75   55   57   41  143     242  Methyl Tetradecanoate
74*  87  169   55  157  143   75   43     200  Methyl Undecanoate(n-)
74* 215  115   42   39  127    0    0     215  1-Naphthyl-N,N-dimethyl carbamate
74   42*  30   43   44   41   40   45      74  Nirosodimethylamine(N-)
74   43*  42   59   29   31   44   45      74  Methyl Acetate
75*  30  213   74  120   91   63  167     213  1,3,5-Trinitrobenzene
75*  41   57   70   43   59   56   47     149  Penicillamine*
75*  53   89   77   27   39   62   54     124  1,4-Dichloro-2-butene
75*  63   89   51  248  125  302  328     329  Methyclonazepam
75*  76   47   60  214  107  121   77     214  Dimexan
75* 121   97  260   47   65   93   69     260  Phorate
75* 161  163  263  265  127  126  162     263  (3,4-Dichlorophenyl)-2'-methyl-2',3'-dihydroxypropionamide (N-)
75* 231   45   97  153   47   46  125     260  Phorate
75   30*  74  213  120   63   91   92     213  1,3,5-Trinitrobenzene
75   77* 205  239  104  233   76  177     286  Oxazepam
75   91*  92   65   42   39   40  116     149  Benzyl Methyl Ketoxime
75  111* 157  113   50  127   99  159     157  1-Chloro-4-nitrobenzene
75  111* 141  205  171  112   50  113     205  Methyl-p-chlorobenzenesulphonamide (N-Methyl)
75  246* 318  248  316   73  176   55     316  DDE
76*  29   58   31   44   59   70  105     150  Tartaric Acid
76*  43   60   77   59   42   44   55      76  Thiourea
76*  46   57   55   56   60   47   43     316  Pentaerythritoltetranitrate
76*  77   95  105   70  121   58   44     167  Metaraminol
76*  78   38   44   77   39   64   46      76  Carbon Disulphide
76*  85   50   31   92   87   77   57     111  Chlorodifluoroacetonitrile
76* 173  104  111  148   50  169  130     258  Thalidomide
76* 183   50  119  120    0    0    0     183  Saccharin
76* 239  104  240  285  241   50   75     338  Chlorthalidone*
76   46*  57   55   56   60   47   97     316  Pentaerythritoltetranitrate
76   46* 240  194  316   58  169   84     316  Pentaerythritoltetranitrate
76   74*  28   75   59   43   42   47     121  Cysteine(L-)*
76   75*  47   60  214  107  121   77     214  Dimexan
76  103*  50  104   51   75   77   52     103  Benzonitrile
76  104*  50   67   54   82   41  148     320  Dicyclohexyl Phthalate
76  109* 104   92  239   65  108   43     362  Phthalylsulphacetamide*
76  130* 103   50  102   75  129   52     130  Phthalazine
76  148* 130   75  102  104  239   50     338  Chlorthalidone
76  183*  90   50   91  120  106   92     183  Saccharin*
76  203*  50  330   74   75  127  165     330  1,2-Diiodobenzene
76  239* 240  104  285  241  177  102     338  Chlorthalidone
76  296* 104  240   50  295  268  297     296  Dithianon
77*  44   79   51   57  105  107   91     149  Phenylpropylmethylamine
77*  51   50  356   78  354  353  279     356  Diphenyl Mercury
77*  51  141   99   63   73   50  268     268  Fenson
77*  51  123   50   65   93   30   78     123  Nitrobenzene
77*  51   78   50  356  354  279  353     629  Phenylmercuric Nitrate
77*  51   40   39   91  105  130   64     404  Sulfinpyrazone
77*  59   29   43   78   41   27  135     228  Sulphonal*
77*  75  205  239  104  233   76  177     286  Oxazepam
77*  89  248   69   63  102   51   65     317  Impramine Degradation Product
77*  96  188  105   38   82   93  106     188  Antipyrine
```

Peak Index   *Asterisk Indicates True Base Peak*

							MWT	COMPOUND
77*	97	129	157	125	31	103	51	298 Phoxim
77*	104	75	51	76	111	106	138	300 Temazepam
77*	106	105	51	50	78	52	39	106 Benzaldehyde
77*	107	51	89	42	105	90	79	176 Pemoline
77*	123	51	65	78	93	50	30	123 Nitrobenzene
77*	137	101	177	176	204	100	203	320 Lorazepam
77*	152	79	105	78	104	118	51	152 Methylnitroaniline(N-Me,o-)
77*	174	107	135	176	65	281	146	301 Isoxsuprine
77*	183	266	105	51	91	78	65	266 4-Methyl-1,2-diphenyl-3,5-pyrazolinedione
77*	183	308	184	105	55	51	41	308 Phenylbutazone
77*	183	105	184	55	93	91	51	308 Phenylbutazone
77*	183	308	184	105	55	51	41	308 Phenylbutazone*
77*	205	104	51	177	75	229	151	286 Oxazepam
77*	221	105	220	88	223	222	51	221 Pyrazon
77*	230	231	105	154	232	126	51	231 2-Amino-5-chlorobenzophenone
77*	245	244	105	228	246	247	51	245 2-Methylamino-5-chlorobenzophenone
77*	245	244	105	193	228	168	246	245 2-Methylamino-5-chlorobenzophenone
77*	246	247	105	78	45	51	248	247 2-Amino-3-Hydroxy-5-Chlorbenzophenone
77*	257	268	205	239	267	233	241	286 Oxazepam
77*	278	51	78	105	279	91	39	404 Sulfinpyrazone
77*	278	105	78	51	279	252	130	404 Sulphinpyrazone*
77	44*	76	29	95	39	58	42	167 Metaraminol*
77	44*	79	51	45	42	78	105	151 Norephedrine
77	44*	79	51	42	45	43	105	151 Norpseudoephedrine
77	44*	105	106	51	91	79	132	151 Norpseudoephedrine
77	44*	79	51	105	0	0	0	151 Norpseudoephedrine
77	44*	95	121	141	167	168	133	167 Phenylephrine
77	44*	95	65	39	45	42	43	167 Phenylephrine
77	44*	79	105	91	51	92	132	151 Phenylpropanolamine
77	44*	79	51	45	42	107	105	151 Phenylpropanolamine*
77	44*	95	65	123	39	121	42	167 Synephrine
77	56*	91	105	106	42	0	0	165 Ephedrine
77	56*	91	84	42	40	373	203	373 Methylrubazonic Acid
77	58*	105	56	57	51	42	79	165 Ephedrine
77	58*	30	51	49	56	59	42	165 Pseudoephedrine
77	58*	59	56	51	42	105	91	165 Pseudoephedrine*
77	76*	95	105	70	121	58	44	167 Metaraminol
77	78*	52	51	50	39	79	74	78 Benzene
77	79*	108	101	51	50	39	40	108 Benzyl Alcohol
77	82*	105	51	81	50	122	109	144 Ethephon (tech)
77	84*	105	56	85	83	51	182	267 Pipradrol
77	91*	92	359	132	144	266	0	359 Rubazonic Acid
77	93*	79	91	80	41	92	136	136 Car-3-ene
77	93*	119	51	211	0	0	0	211 1,3-Diphenylguanidine
77	93*	91	41	79	39	136	94	136 Sabinene
77	94*	138	95	51	66	172	45	472 5-Aceto-8-hydroxyquinoline Sulphate
77	98*	94	99	126	0	0	0	386 Mesoridazine
77	100*	44	42	72	105	101	51	205 Diethylpropion
77	103*	82	29	97	42	51	104	273 Tilidine
77	105*	182	44	86	165	51	30	999 Benactyzine Probe Product*
77	105*	122	51	50	44	78	52	289 Benzoylecgonine
77	105*	96	183	51	42	182	94	431 Clidinium Bromide
77	105*	51	135	134	50	106	117	179 Hippuric Acid
77	105*	120	43	51	96	106	97	337 Lobeline

```
           PEAKS                        MWT    COMPOUND

77  105* 134   51  106  193  161   50   193  Methyl Hippurate
77  105* 235   96  111   97  183  250   323  Piperidolate
77  105*  30   45  227   51  165  183   227  Prenylamine Metabolite
77  107*  51   39   53   78   50   52   152  Hydroxyphenylacetic Acid(para)
77  112* 114   51   50  113   75   76   112  Chlorobenzene
77  112* 105  139   55   41   96   56   261  Hexylcaine
77  120* 176   93   43   57  169  211   211  Propachlor (tech)
77  122* 107  105   91   79  123   30   221  Metaxalone
77  123* 167   92   44   91  224   42   332  Dichlofluanid
77  123*  51  124   39   74   53   78   197  Levodopa
77  130*  51  132  115  103  133   78   133  Tranylcypromine
77  132* 160  104  105   76   97  129   345  Azinphos Ethyl
77  135*  90   92   51   64  136   63   223  Salicyluric Acid Methyl Ester Methyl Ether
77  141* 176  111  113  178   75  143   176  Dybenal
77  163* 164  135  194   76   92   50   194  Dimethyl Phthalate
77  170*  51  141  169  171  142  115   170  Diphenyl Ether
77  174*  91  105  132    0    0    0   174  Norantipyrine
77  180* 104  209  223  252   51  181   252  Diphenylhydantoin
77  180* 104  266  237  209  165    0   266  Methyldilantin
77  180* 118   91  104   65   89   51   266  5-Methylphenyl-5-phenylhydantoin(p-5-phenylhydantoin)(Metabolite)
77  183* 252  184  105   91   51   64   252  1,2-Diphenyl-3,5-pyrazolidinedione
77  183* 308  184  105   93  309   91   308  Phenylbutazone
77  205*  41   39  220  130  145  131   220  Xylazine
77  225* 226  104   76  149  153   43   318  Diphenyl Phthalate
77  241* 242  105   44   43  195   57   242  2-Amino-5-nitrobenzophenone
77  245* 244  105  228  246   51  247   245  Diazepam Benzophenone
77  245* 244  228  105  246  193  247   245  Medazepam Benzophenone
77  245* 246  247  228  105   51  193   245  2-Methylamino-5-chlorobenzophenone*
77  256* 283  221  255  257   89  165   284  Diazepam
77  257* 268  239  205  267  233  259   286  Oxazepam*
77  267*  51   91   63  117  294  248   295  Methylnitrazepam
77  271* 273  300    0    0    0    0   300  Diazepam Metabolite
77  271* 273  300  255    0    0    0   300  3-Hydroxydiazepam
77  300* 258   51  255  259  256  283   300  Clobazam
78*  52   51   77   50   39   79   43    78  Benzene
78*  77   52   51   50   39   79   74    78  Benzene
78* 106   51  137   79    0    0    0   137  Iproniazid Metabolite
78* 106   51  137   50   79  107   52   137  Isoniazid
78* 106   51  104   77  137   50    0   137  Isoniazid
78* 106  135  136   51   50   79   52   137  Methylnicotinamide (N-)
78* 106   51  137   50  136  107   79   137  Methylnicotinate*
78* 134  106   51   77   39   40  107   152  Hydroxyphenylacetic Acid(ortho)
78   63*  45   61   46   62   48   47    78  Dimethylsulfoxide
78   76*  38   44   77   39   64   46    76  Carbon Disulphide
78   79*  52  105  304  314  316   51   331  3-Hydroxybromazepam
78  106*  51  137   50   79  107   31   137  Isoniazid*
78  106*  70   59   42  284  300   44   406  Nicodicodeine
78  106* 176   51  148  162   56    0   178  Nikethamide
78  106* 177   51  178  107  149  163   178  Nikethamide
78  108*  65   39   77   93   79   51   108  Anisole(Methyl Phenyl Ether)
78  122* 106   51   50   52   44  123   122  Nicotinamide*
78  169* 113  171   51   63   44   50   169  Chlorzoxazone
78  169* 171  113  115   63   51  170   169  Chlorzoxazone*
79*  43   81   49   29   27   36   57   128  1,3-Dichloro-2-hydroxypropane
```

								MWT	COMPOUND
79*	52	51	78	50	39	45	53	123	Picolinic Acid
79*	52	51	50	78	39	53	80	79	Pyridine
79*	53	221	81	41	77	178	93	262	Methohexital
79*	77	108	101	51	50	39	40	108	Benzyl Alcohol
79*	78	52	105	304	314	316	51	331	3-Hydroxybromazepam
79*	80	225	182	107	63	78	91	331	Bromazepam-N(Py)-Oxide
79*	80	151	149	77	114	117	107	299	Captan (Orthocide)
79*	82	81	263	77	108	277	80	378	Dieldrin
79*	96	346	81	345	70	78	113	446	Thioproperazine
79*	105	104	51	78	52	50	77	136	Betahistine
79*	108	107	77	51	91	50	78	108	Benzyl Alcohol
79*	108	263	277	279	345	378	0	378	Dieldrin
79*	109	110	139	145	80	112	95	256	Trichlorfon
79*	135	153	107	52	53	80	51	153	Aminosalicylic Acid(p-)
79*	235	53	178	81	41	195	261	276	Methohexital Methyl Derivative
79	81*	80	41	82	163	161	77	240	1,2-Dibromocyclohexane(trans-)
79	81*	281	244	246	279	271	273	378	Photodieldrin
79	87*	51	225	42	50	86	80	225	Furazolidone*
79	91*	67	201	77	105	93	120	284	Lynestrenol
79	93*	80	121	94	107	81	136	136	Fenchene(alpha-)
79	107*	77	51	152	105	50	78	152	Pemoline Metabolite 2
79	107*	77	91	51	105	108	78	138	1-Phenyl-1,2-dihydroxyethane
79	107*	120	77	75	44	43	51	181	Styramate
79	109*	47	195	197	145	83	220	220	Vapona (DDVP)
79	134*	77	106	105	103	151	78	151	2-Nitroxylene(m-)
80	79*	225	182	107	63	78	91	331	Bromazepam-N(Py)-Oxide
80	79*	151	149	77	114	117	107	299	Captan (Orthocide)
80	81*	123	39	53	41	67	40	123	1-Butylpyrrole
80	81*	39	42	53	55	54	38	81	1-Methylpyrrole
80	84*	55	41	42	105	77	200	287	Procyclidine
80	93*	92	91	79	121	105	77	136	Ocimene-Y(beta-)
80	108*	81	107	28	0	0	0	108	Phenylenediamine(ortho-)
80	109*	151	108	43	81	53	52	151	Acetaminophen
80	109*	53	52	39	64	63	81	109	2-Aminophenol
80	109*	81	110	41	39	53	74	109	3-Aminophenol
80	109*	53	81	108	52	54	110	109	4-Aminophenol
80	109*	44	81	53	39	52	54	153	Aminosalicylic Acid*
80	126*	52	108	51	63	79	39	126	Pyrogallol
81*	31	100	50	69	28	82	29	100	Tetrafluoroethylene
81*	41	53	235	178	79	39	195	276	Permethylated Methohexitone*
81*	47	251	41	79	42	178	53	292	Permethylated Thialbarbitone*
81*	53	330	96	82	332	64	63	330	Frusemide*
81*	53	330	82	96	332	222	250	330	Frusemide (Fursemide)
81*	53	80	64	52	330	51	63	330	Furosemide
81*	53	64	96	251	82	39	48	330	Furosemide
81*	53	96	372	82	374	339	357	372	Furosemide Trimethyl Derivative
81*	54	108	77	51	213	39	52	213	Phenazopyridine
81*	69	41	152	80	82	109	39	152	1,3,3-Trimethyl-2-norboranone
81*	79	80	41	82	163	161	77	240	1,2-Dibromocyclohexane(trans-)
81*	79	281	244	246	279	271	273	378	Photodieldrin
81*	80	123	39	53	41	67	40	123	1-Butylpyrrole
81*	80	39	42	53	55	54	38	81	1-Methylpyrrole
81*	83	46	110	82	112	84	117	281	Chloral Betaine
81*	83	46	110	112	0	0	0	164	Chloral Hydrate

        Asterisk Indicates True Base Peak   Peak Index

							MWT	COMPOUND
81*	91	106	95	79	68	67	78	167 Ethinamate (Ethynylcyclohexyl Carbamate)
81*	95	71	41	67	55	138	123	156 Menthol
81*	107	105	91	95	79	93	67	386 Cholesterol
81*	124	54	53	125	171	45	42	171 Metronidazole*
81*	138	53	82	91	139	56	42	215 Furfurylmethylamphetamine
81*	207	80	79	141	169	41	77	248 Thialbarbitone oxygenated analogue
81*	221	79	80	39	77	157	41	236 Hexobarbitone
81*	223	79	41	80	157	185	77	264 Thialbarbitone*
81*	353	355	351	357	237	386	0	386 Heptachlor Epoxide
81	41*	53	221	79	39	178	233	262 Methohexitone*
81	42*	204	215	146	115	162	94	369 Diamorphine
81	67*	263	36	79	82	261	265	378 Endrin
81	67*	54	55	95	68	82	69	294 Methyl Octadecadienoate
81	69*	152	41	109	95	80	68	152 Bornanone-3
81	69*	67	55	71	68	56	41	154 Isopulegol
81	71*	95	41	55	82	43	69	156 Menthol
81	82*	55	83	54	42	41	39	111 Ametazole
81	82*	44	54	55	83	94	41	111 Histamine (4-(omega-Aminoethyl)-1,3-diazole)
81	86*	98	58	41	57	43	319	319 Chloroquine
81	91*	106	78	39	95	68	43	167 Ethinamate*
81	95*	108	152	41	83	109	69	152 Bornanone-2
81	95*	41	69	55	83	67	137	152 Camphor
81	95*	69	41	55	67	108	109	152 Camphor
81	95*	108	55	109	152	67	83	152 Camphor
81	95*	108	152	69	109	83	110	152 Camphor
81	99*	55	44	41	43	120	79	181 Hexapropymate
81	108*	213	54	77	136	51	43	213 Phenazopyridine
81	109*	149	81	275	139	99	65	275 Diethyl-p-nitrophenyl Phosphate
81	109*	275	149	99	139	127	247	275 Paraoxone
81	135*	173	39	41	57	150	201	350 Propargite
81	136*	137	42	41	55	130	128	294 Cinchonidine*
81	137*	138	109	166	0	0	0	166 Vanillal
81	201*	98	175	259	176	202	242	259 Primaquine*
81	221*	157	80	79	155	41	77	236 Hexobarbitone*
81	235*	169	171	79	236	170	91	250 3-Methylhexobarbitone
81	235*	79	169	171	170	41	80	250 Permethylated Hexobarbitone*
81	235*	169	41	79	54	197	196	276 Permethylated Thialbarbitone Oxygen Analogue*
81	249*	183	250	79	185	184	264	264 3-Ethylhexobarbitone
81	271*	150	148	45	110	42	82	271 Normorphine
81	285*	215	148	286	164	110	115	285 Norcodeine*
81	299*	42	229	162	124	300	44	299 Codeine
81	315*	43	55	368	41	107	95	386 Cholesterol
82*	30	81	83	55	54	42	27	111 Ametazole*
82*	30	81	54	28	55	83	41	111 Histamine
82*	36	81	38	55	83	54	42	111 Ametazole
82*	47	84	29	111	83	113	85	164 Chloral Hydrate*
82*	53	81	39	27	51	50	54	82 2-Methylfuran
82*	54	108	93	58	107	106	39	150 Carvone
82*	68	91	159	42	158	92	65	159 Pargyline*
82*	77	105	51	81	50	122	109	144 Ethephon (tech)
82*	81	55	83	54	42	41	39	111 Ametazole
82*	81	44	54	55	83	94	41	111 Histamine (4-(omega-Aminoethyl)-1,3-diazole)
82*	84	111	83	29	47	113	85	146 Chloral
82*	84	83	111	47	29	85	113	164 Chloral Hydrate

82*	91	68	42	92	159	158	65	159 Pargyline
82*	95	137	54	109	152	41	110	152 Piperitone
82*	96	83	124	138	168	185	186	185 Ecgonine
82*	96	83	97	42	199	94	168	199 Methylecgonine
82*	97	42	83	96	57	94	55	185 Ecgonine
82*	97	103	77	72	42	91	115	273 1-Phenyl-1-carbethoxy-2-dimethylaminocyclohex-3-ene
82*	105	122	77	51	42	94	81	289 Benzoylecgonine
82*	124	105	77	168	122	42	83	289 Benzoylecgonine
82*	182	83	96	103	147	148	131	329 Cinnamoylcocaine
82*	182	83	105	303	77	94	96	303 Cocaine*
82	55*	39	83	41	113	70	127	155 Bemegride
82	55*	67	180	109	0	0	0	180 1,7-Dimethylxanthine
82	55*	41	39	54	42	56	109	138 Leptazol
82	55*	67	109	180	0	0	0	180 Theobromine
82	57*	44	67	41	71	29	56	100 Cyclohexanol
82	67*	54	41	55	83	39	36	118 Chlorocyclohexane
82	79*	81	263	77	108	277	80	378 Dieldrin
82	83*	42	140	124	96	77	67	307 Benztropine
82	83*	96	124	97	42	73	184	213 Tropine TMS Ether
82	84*	80	56	43	28	30	41	127 Coniine*
82	91*	42	39	65	159	158	68	159 Pargyline
82	96*	83	42	97	55	68	199	199 Methylecgonine
82	110*	39	81	53	69	55	111	110 Resorcinol
82	112*	55	54	45	84	41	53	112 Maleinehydrazide
82	124*	83	57	41	94	42	43	267 Anisotropine
82	124*	94	83	42	96	103	67	289 Atropine*
82	124*	42	83	94	67	96	39	289 Atropine (dl-Hyoscyamine)
82	124*	168	77	105	42	94	83	289 Benzoylecgonine*
82	124*	83	275	94	96	67	80	275 Homatropine
82	124*	94	77	42	44	105	106	275 Homatropine
82	124*	83	245	94	77	105	67	245 Tropocaine
82	156*	80	128	81	155	78	79	342 Diquat
82	182*	83	105	94	77	96	303	303 Cocaine(beta)
82	208*	123	67	209	42	55	207	208 1,3,7,8-Tetramethylxanthine (or 1,3,8,9-Tetramethyl-)
82	210*	67	153	125	42	55	0	210 1,3,7-Trimethyluric Acid
82	221*	80	98	81	97	319	417	417 Cinepazide
83*	55	41	67	54	39	82	27	162 Bromocyclohexane
83*	55	71	96	114	144	62	56	218 Meprobamate
83*	55	71	41	44	62	56	45	218 Meprobamate
83*	55	98	155	126	0	0	0	199 Methaprylon Metabolite
83*	55	98	153	166	41	155	152	199 Methyprylon Metabolite
83*	68	82	72	84	77	103	115	259 Nortilidine
83*	68	84	103	77	91	157	155	259 1-Phenyl-1-carbethoxy-2-methylaminocyclohex-3-ene
83*	70	42	244	71	43	273	56	343 Clothiapine
83*	70	273	244	209	42	71	43	343 Clothiapine*
83*	82	42	140	124	96	77	67	307 Benztropine
83*	82	96	124	97	42	73	184	213 Tropine TMS Ether
83*	84	55	56	43	71	41	62	218 Meprobamate*
83*	85	129	47	127	87	48	81	162 Bromodichloromethane
83*	85	47	87	48	49	35	82	118 Chloroform
83*	98	55	166	153	41	84	155	199 Methyprylon Metabolite (6-Hydroxy)
83*	105	77	183	165	152	139	128	311 3-Piperidylbenzilate
83*	140	82	124	96	97	42	125	307 Benztropine*
83*	140	82	124	43	96	97	42	333 Deptropine*

								MWT	COMPOUND
83	44*	180	182	137	143	139	41	222	Bromvaletone
83	44*	85	36	28	0	0	0	162	Trichloroacetic Acid
83	55*	82	113	70	69	97	155	155	Bemegride
83	55*	82	41	113	70	29	69	155	Bemegride*
83	55*	43	98	29	27	39	53	98	4-Methyl-3-pentene-2-one
83	55*	82	41	54	39	28	42	83	Tetrahydropyridine (Methyl Phenidate Decomposition Prod)
83	56*	42	217	123	64	57	119	333	Dipyrone
83	56*	230	217	123	97	77	64	333	Dipyrone
83	69*	43	79	41	81	53	80	141	Methylpentynol Carbamate*
83	70*	42	257	193	56	228	164	327	Loxapine
83	81*	46	110	82	112	84	117	281	Chloral Betaine
83	81*	46	110	112	0	0	0	164	Chloral Hydrate
83	96*	82	182	93	36	42	131	329	Cinnamoylcocaine
83	97*	99	85	61	26	96	63	132	1,1,2-Trichloroethane
83	99*	101	85	61	70	31	0	132	Trichloroethane
83	100*	39	55	82	85	29	27	100	Senecioic Acid
83	123*	55	41	151	81	69	39	230	Bromocamphor
83	124*	82	94	289	42	96	125	289	Atropine
83	124*	82	73	94	96	125	42	361	Atropine TMS Ether
83	140*	82	124	96	97	167	125	307	Benztropine
83	152*	42	82	85	181	57	122	329	Cinnamoylcocaine
83	154*	71	55	155	43	70	67	224	Butylvinal
84*	42	133	162	161	51	119	65	162	Nicotine
84*	43	126	77	56	91	105	55	233	Levophacetoperane
84*	44	133	42	162	161	44	85	162	Nicotine
84*	55	42	70	205	43	204	41	287	Procyclidine
84*	56	54	42	52	82	85	40	84	Benzene-d6
84*	56	43	55	85	41	70	42	127	Coniine
84*	56	54	52	42	82	40	85	84	Deuterobenzene
84*	56	91	85	55	30	41	42	233	Methylphenidate
84*	56	85	77	105	55	30	42	267	Pipradrol*
84*	58	45	39	57	69	83	50	84	Thiophene
84*	70	83	68	288	260	273	289	371	Flurazepam Metabolite
84*	71	85	82	80	341	70	356	356	Conessine*
84*	77	105	56	85	83	51	182	267	Pipradrol
84*	80	55	41	42	105	77	200	287	Procyclidine
84*	82	80	56	43	28	30	41	127	Coniine*
84*	85	56	91	36	55	30	118	233	Methylphenidate
84*	85	56	57	44	42	43	41	85	Piperidine
84*	91	55	77	85	182	83	65	265	Antazoline
84*	91	77	65	104	85	83	182	265	Antazoline
84*	91	36	51	55	0	0	0	265	Antazoline
84*	91	77	55	182	85	65	104	265	Antazoline*
84*	91	56	55	121	118	117	85	233	Methylphenidate
84*	91	55	150	146	56	85	83	233	Methylphenidate
84*	91	56	90	65	89	77	115	233	Methylphenidate
84*	91	85	56	55	150	41	118	233	Methylphenidate
84*	92	120	65	56	111	119	82	248	Piridocaine
84*	94	177	148	42	93	134	41	177	Modaline
84*	105	85	77	55	56	183	42	267	Azacyclonol
84*	105	77	56	85	248	249	36	267	Pipradrol
84*	128	42	43	77	139	82	179	343	Clemastine
84*	128	179	42	85	178	214	98	343	Clemastine*
84*	133	42	162	161	39	51	41	162	Nicotine

PEAKS							MWT	COMPOUND
84*	133	76	42	162	161	58	119	162 Nicotine
84*	133	42	162	161	105	77	119	162 Nicotine*
84*	160	167	159	165	152	55	77	266 Diphenazoline*
84*	178	161	133	118	0	0	0	178 Nicotine-N-oxide
84*	194	55	85	56	41	30	99	277 Perhexiline*
84*	204	205	85	42	55	105	77	287 Procyclidine*
84*	205	42	240	81	55	91	125	311 Pyrrobutamine
84*	296	42	85	180	55	212	41	296 Pyrathiazine
84	26*	31	57	43	42	58	40	84 Amitrol
84	42*	39	56	43	40	60	102	192 Citric Acid
84	49*	86	51	47	48	88	50	84 Dichloromethane
84	49*	86	51	47	88	48	50	84 Methylene Chloride
84	56*	245	57	203	83	43	42	245 4-Acetylaminoantipyrine
84	56*	57	203	42	83	77	93	203 4-Aminoantipyrine
84	56*	41	55	42	69	39	27	84 Cyclohexane
84	82*	111	83	29	47	113	85	146 Chloral
84	82*	83	111	47	29	85	113	164 Chloral Hydrate
84	83*	55	56	43	71	41	62	218 Meprobamate*
84	85*	183	107	77	55	56	184	267 Azacyclonol
84	85*	183	105	56	77	55	30	267 Azacyclonol*
84	86*	169	170	41	205	99	155	319 Chloroquine
84	91*	200	41	55	117	42	115	243 Phencyclidine
84	98*	71	56	41	42	99	124	167 Mecamylamine*
84	102*	58	56	42	30	103	91	385 Bamifylline*
85*	41	55	56	43	42	57	70	100 2,4-Dimethyltetrahydrofuran
85*	43	42	86	44	41	75	110	382 Mefruside*
85*	43	127	41	197	57	239	129	239 Monalide
85*	55	41	29	56	28	84	39	102 2-Hydroxytetrahydropyran
85*	57	42	44	56	191	77	0	191 Phendimetrazine
85*	58	86	91	84	70	42	225	309 Benzydamine*
85*	84	183	107	77	55	56	184	267 Azacyclonol
85*	84	183	105	56	77	55	30	267 Azacyclonol*
85*	87	50	101	103	31	35	66	120 Dichlorodifluoromethane(Freon 12)
85*	87	129	131	147	79	81	50	163 Difluorochlorobromomethane
85*	100	125	101	199	227	99	153	398 Tributoxyethyl Phosphate
85*	119	69	31	87	135	29	50	154 Chloropentafluoroethane
85	41*	84	157	56	43	57	70	283 Levallorphan
85	43*	55	41	42	56	84	29	164 3-Bromohexane
85	43*	100	27	29	42	41	0	100 2,4-Pentanedione(Acetylacetone)
85	57*	42	56	76	191	70	77	191 Phendimetrazine*
85	58*	269	268	270	271	229	227	314 Clomipramine
85	58*	269	268	270	271	314	242	314 Clomipramine*
85	58*	70	57	42	148	56	77	165 Ephedrine GC Decomposition
85	58*	42	77	56	70	105	57	165 Ephedrine GC Decomposition
85	58*	173	193	194	195	234	235	280 Imipramine
85	58*	235	234	42	280	208	193	280 Imipramine
85	58*	284	238	239	198	226	224	284 Promazine
85	69*	50	87	35	31	37	33	104 Chlorotrifluoromethane
85	71*	193	195	42	130	70	194	266 Desipramine
85	73*	60	103	71	57	61	58	342 Lactose
85	76*	50	31	92	87	77	57	111 Chlorodifluoroacetonitrile
85	83*	129	47	127	87	48	81	162 Bromodichloromethane
85	83*	47	87	48	49	35	82	118 Chloroform
85	84*	56	91	36	55	30	118	233 Methylphenidate

85	84*	56	57	44	42	43	41	85	Piperidine
85	86*	58	87	120	91	77	56	234	Lignocaine
85	112*	45	141	113	59	71	69	141	Chlormethiazole Metabolite
85	120*	58	99	56	44	57	149	236	Butethamine
85	145*	93	125	146	69	58	302	302	Methidathion (Technical)
86*	30	167	99	87	58	165	29	311	Adiphenine*
86*	30	87	294	58	42	57	309	309	Amolanone*
86*	30	57	108	44	29	84	41	209	Bamethan*
86*	30	72	44	141	42	29	57	254	Butanilacaine*
86*	30	58	130	29	77	87	42	216	Diethyltryptamine
86*	30	58	87	109	123	166	262	348	Flurazepam Benzophenone 4
86*	30	58	123	78	51	29	42	222	Nicametate*
86*	30	120	99	58	65	92	42	235	Procainamide
86*	30	27	58	29	28	42	101	101	Triethylamine
86*	36	87	84	58	56	44	38	645	Amiodarone*
86*	36	162	58	87	107	105	0	396	Etonitazene
86*	43	102	69	44	42	59	0	102	Guanylurea
86*	43	77	91	40	106	84	65	220	Prilocaine
86*	44	30	265	41	180	29	87	265	Ethomoxane*
86*	44	87	107	43	106	56	41	220	Prilocaine*
86*	44	159	43	41	87	42	56	245	Propylnorfenfluramine
86*	55	99	41	44	100	83	56	309	Dicyclomine
86*	57	43	71	55	69	58	41	386	Clonitazene
86*	58	99	56	162	132	149	205	277	Acetylprocainamide (N-Acetyl)
86*	58	162	120	99	43	71	278	278	Acetylprocaine Acetate
86*	58	73	87	319	41	99	245	319	Chloroquine
86*	58	319	87	73	247	245	112	319	Chloroquine*
86*	58	87	29	30	116	41	56	379	Cinchocaine
86*	58	30	87	42	130	77	56	216	Diethyltryptamine (N,N-)
86*	58	278	42	87	294	193	44	294	Dimethacrine
86*	58	87	42	56	77	44	43	193	Etafedrine*
86*	58	42	87	77	30	56	51	193	Ethylephedrine
86*	58	99	87	56	42	71	84	387	Flurazepam
86*	58	91	56	65	42	39	118	177	Formylmethylamphetamine (N-)
86*	58	87	42	56	72	234	120	234	Lignocaine
86*	58	88	105	206	87	55	77	353	Tridihexethyl Chloride
86*	71	99	58	55	56	100	87	309	Dicyclomine*
86*	72	191	58	30	128	193	163	321	Clamoxyquin*
86*	81	98	58	41	57	43	319	319	Chloroquine
86*	84	169	170	41	205	99	155	319	Chloroquine
86*	85	58	87	120	91	77	56	234	Lignocaine
86*	87	58	30	103	121	72	56	307	Amprotropine*
86*	87	58	30	29	84	56	42	361	Chromonar*
86*	87	58	149	111	99	57	41	379	Cinchocaine*
86*	87	100	30	58	44	56	42	405	Clomiphene
86*	87	58	228	326	116	57	113	343	Cinchocaine
86*	87	58	78	56	109	95	123	348	2-Diethylaminoethylamino-5-Chloro-2'-Fluorobenzophenone
86*	87	99	58	71	207	84	56	279	Diethylaminoethyltheophylline
86*	87	99	58	84	387	315	56	387	Flurazepam*
86*	87	58	30	109	43	95	123	348	Flurazepam Benzophenone
86*	87	109	95	123	348	166	262	348	Flurazepam Benzophenone
86*	87	58	44	72	42	120	85	234	Lignocaine*
86*	91	99	145	115	58	87	56	289	Caramiphen
86*	91	145	84	144	85	58	87	333	Carbetapentane

		PEAKS					MWT	COMPOUND
86*	91	87	145	58	144	30	44	333 Carbetapentane*
86*	91	105	87	144	58	100	56	335 Oxeladin*
86*	99	30	58	87	120	162	206	278 Acetylprocaine Hydrochloride
86*	99	71	58	87	140	84	100	308 Benoxinate
86*	99	30	58	71	41	136	56	308 Benoxinate
86*	99	91	191	87	119	58	248	263 Butethamate*
86*	99	91	144	58	56	41	87	289 Caramiphen
86*	99	154	30	87	58	29	156	270 Chloroprocaine*
86*	99	154	58	56	87	42	84	266 Chlorprocaine
86*	99	55	100	87	41	44	165	309 Dicyclomine
86*	99	58	87	56	85	183	387	387 Flurazepam
86*	99	184	58	30	87	201	186	299 Metoclopramide*
86*	99	29	30	100	71	87	192	308 Oxybuprocaine*
86*	99	120	30	92	87	58	65	235 Procainamide*
86*	99	120	58	65	42	92	56	236 Procaine
86*	99	120	58	87	30	92	71	236 Procaine*
86*	99	58	87	56	100	84	70	294 Proparcaine
86*	99	58	71	136	56	41	80	294 Proxymetacaine
86*	105	77	87	182	99	183	58	327 Benactyzine*
86*	109	30	151	87	81	99	58	279 Etamphylline*
86*	114	100	44	238	115	56	72	480 Isopropamide Iodide
86*	116	41	58	73	56	57	228	343 Cinchocaine
86*	120	30	87	58	92	84	56	278 Dimethocaine*
86*	120	65	92	42	43	41	56	236 Procaine
86*	121	99	30	58	65	42	56	265 Paraethoxycaine
86*	126	259	58	112	99	400	74	399 Quinacrine
86*	165	167	99	152	166	58	56	311 Adiphenine
86*	167	58	87	165	152	263	42	267 Emepromium Bromide GC Breakdown Product
86*	180	58	87	198	298	41	212	298 Diethazine
86*	181	43	44	41	114	42	152	353 Propantheline
86*	183	182	105	77	165	116	312	327 Benactyzine
86*	298	83	58	87	30	85	180	298 Diethazine
86*	298	87	30	58	299	212	180	298 Diethazine*
86	30*	57	41	77	135	29	206	239 Salbutamol*
86	30*	41	57	65	39	42	111	225 Terbutaline
86	43*	41	71	58	57	0	0	86 Methyl Propyl Ketone
86	43*	41	58	27	71	39	42	86 2-Pentanone
86	44*	58	28	42	41	43	27	101 Methylethane(N-(2'-methylethyl)-1-amino-2-)
86	58*	85	326	197	43	280	241	326 Acepromazine
86	58*	289	43	291	85	248	288	376 8-Acetoxychlorpromazine
86	58*	376	43	378	42	87	59	376 3-Acetoxychlorpromazine
86	58*	42	85	44	57	59	43	318 Chlorpromazine
86	58*	318	85	320	272	319	273	318 Chlorpromazine*
86	58*	87	77	56	42	30	44	193 Etafedrine
86	58*	42	334	220	59	87	44	334 3-Hydroxychlorpromazine
86	58*	334	42	85	336	44	243	334 8-Hydroxychlorpromazine
86	58*	314	229	228	185	42	44	314 Methoxypromazine*
86	58*	42	85	199	198	238	284	284 Promazine
86	58*	42	85	44	59	30	70	352 Triflupromazine
86	58*	352	85	306	59	266	248	352 Triflupromazine
86	91*	177	65	0	0	0	0	177 Amphetamine-NCS Derivative
86	120*	58	99	56	44	57	36	236 Butethamine
86	181*	72	182	85	152	108	99	419 Methantheline Bromide (Banthine)
86	234*	221	219	233	235	178	217	307 Desmethyldoxepin N-Acetyl

Asterisk Indicates True Base Peak

Peak Index

	PEAKS						MWT	COMPOUND
86	546*	420	517	547	391	263	250	546 Amiodarone Probe Product*
87*	43	209	72	86	210	44	181	397 3,6,17-Triacetylnormorphine
87*	57	116	55	115	129	88	127	158 Methyl dipropylacetate
87*	79	51	225	42	50	86	80	225 Furazolidone*
87*	93	125	58	47	79	63	229	229 Dimethoate (Cygon)
87*	102	45	41	59	39	60	74	102 2-Methyltetrahydrothiophene
87*	109	145	58	79	142	112	88	287 Vamidithion
87*	116	55	113	57	145	59	143	174 Methyl 3-Hydroxydipropylacetate
87	43*	31	45	59	28	29	44	102 2-Methyl-1,3-dioxane
87	57*	130	55	116	101	143	141	172 Methyl 3-Oxodipropylacetate
87	59*	41	69	43	31	45	39	102 2,2-Dimethylbutan-2-ol
87	74*	43	55	41	75	143	57	186 Methyl Decanoate
87	74*	43	41	55	75	57	69	284 Methyl Heptadecanoate
87	74*	43	75	57	55	143	69	270 Methyl Hexadecanoate
87	74*	41	43	55	75	57	69	214 Methyl Laurate
87	74*	43	41	55	57	75	69	242 Methyl Myristate
87	74*	43	57	41	55	59	127	158 Methyl Octanoate
87	74*	270	75	143	43	55	41	270 Methyl Palmitate
87	74*	43	41	55	75	57	69	270 Methyl Palmitate
87	74*	43	55	298	143	75	57	298 Methyl Stearate
87	74*	43	75	55	57	41	143	242 Methyl Tetradecanoate
87	74*	169	55	157	143	75	43	200 Methyl Undecanoate(n-)
87	85*	50	101	103	31	35	66	120 Dichlorodifluoromethane(Freon 12)
87	85*	129	131	147	79	81	50	163 Difluorochlorobromomethane
87	86*	58	30	103	121	72	56	307 Amprotropine*
87	86*	58	30	29	84	56	42	361 Chromonar*
87	86*	58	149	111	99	57	41	379 Cinchocaine*
87	86*	100	30	58	44	56	42	405 Clomiphene
87	86*	58	228	326	116	57	113	343 Cinchocaine
87	86*	58	78	56	109	95	123	348 2-Diethylaminoethylamino-5-Chloro-2'-Fluorobenzophenone
87	86*	99	58	71	207	84	56	279 Diethylaminoethyltheophylline
87	86*	99	58	84	387	315	56	387 Flurazepam*
87	86*	58	30	109	43	95	123	348 Flurazepam Benzophenone
87	86*	109	95	123	348	166	262	348 Flurazepam Benzophenone
87	86*	58	44	72	42	120	85	234 Lignocaine*
87	102*	69	41	59	0	0	0	101 3,4-Dihydroxyproline
88*	42	123	77	51	124	43	166	211 Methyldopa
88*	42	123	124	89	77	51	40	211 Methyldopa
88*	43	60	113	101	61	73	70	158 Ethyl Heptanoate
88*	57	101	60	61	127	73	43	172 Ethyl Caprylate
88*	60	89	126	61	115	114	170	258 Demeton
88*	60	163	43	41	148	121	107	163 Propylethyldithiocarbamate (S-n-propyl-)
88*	62	61	53	37	277	63	89	275 Bromoxynil
88*	89	60	61	171	97	115	59	258 Demeton O (technical)
88*	89	60	61	171	97	115	59	258 Demeton PS
88*	89	60	61	97	274	142	186	274 Disulfoton
88*	89	29	97	61	60	27	65	274 Disulfoton
88*	101	73	60	81	93	109	111	258 Demeton PO
88*	101	43	61	60	41	73	70	200 Ethyl Caprate
88*	101	43	41	73	61	70	55	228 Ethyl Laurate
88*	101	43	41	73	55	57	70	256 Ethyl Myristate
88*	101	43	57	55	41	89	73	284 Ethyl Palmitate
88*	101	43	57	89	55	41	73	312 Ethyl Stearate
88*	123	44	42	124	41	122	77	211 Methyldopa

							MWT	COMPOUND
88	116*	43	148	60	56	72	117	296	Disulfiram
88	116*	29	44	60	148	56	27	296	Disulfiram*
88	266*	102	75	115	128	131	176	284	Mazindol
89	77*	248	69	63	102	51	65	317	Impramine Degradation Product
89	88*	60	61	171	97	115	59	258	Demeton O (technical)
89	88*	60	61	171	97	115	59	258	Demeton PS
89	88*	60	61	97	274	142	186	274	Disulfoton
89	88*	29	97	61	60	27	65	274	Disulfoton
89	165*	63	30	39	119	182	78	182	2,4-Dinitrotoluene
89	208*	179	76	165	77	209	88	208	Dibenzosuberone
89	210*	30	63	39	51	76	134	227	2,4,6-Trinitrotoluene
90*	177	105	77	106	51	89	50	177	5-Phenyloxazilidindione-(2,4)(Pemoline Metabolite 1)
90	117*	116	89	63	51	50	39	117	Benzyl Cyanide
90	117*	89	118	116	63	59	39	117	Indole
90	117*	116	89	51	63	39	118	117	Phenylacetonitrile
90	135*	77	92	136	134	51	64	223	Salicyluric Acid Methyl Ester Methyl Ether
91*	30	92	42	29	146	44	104	205	Phenformin*
91*	30	155	108	65	197	39	107	270	Tolbutamide*
91*	41	92	65	39	42	40	116	149	Benzyl Methyl Ketoxime
91*	41	44	43	190	119	65	55	369	Tri-(1-isopropylphenyl)amine
91*	42	132	117	43	41	104	57	133	2-Benzylaziridine
91*	43	67	81	78	79	106	44	167	Ethinamate
91*	43	65	127	110	39	51	106	231	Isocarboxazid
91*	43	57	41	73	55	60	81	585	Myrophine
91*	43	44	41	45	148	149	103	166	2-Phenyllactic Acid
91*	44	162	119	41	65	43	42	253	Di-(1-isopropylphenyl)amine
91*	55	43	56	41	42	29	93	120	1-Chlorohexane
91*	55	269	295	294	324	29	241	324	Prazepam
91*	58	176	90	41	42	119	65	267	Di-(1-isopropylphenyl)methylamine
91*	58	86	106	231	45	77	230	511	Norbormide*
91*	58	176	119	42	41	56	65	267	Trimethyl-(N, alpha, alpha-)-diphenethylamine
91*	65	182	92	39	51	63	77	182	Bibenzyl
91*	65	39	119	92	63	89	51	210	Dibenzyl Ketone
91*	65	137	41	57	39	30	0	137	Nitrotoluene (meta)
91*	65	137	107	39	63	77	79	137	Nitrotoluene (para)
91*	65	77	196	39	56	51	90	303	Phenoxybenzamine
91*	65	187	92	277	259	90	273	596	Trimethaphan
91*	70	186	152	117	77	172	229	229	1-(1-Phenylcyclohexyl)pyrrolidine
91*	75	92	65	42	39	40	116	149	Benzyl Methyl Ketoxime
91*	77	92	359	132	144	266	0	359	Rubazonic Acid
91*	79	67	201	77	105	93	120	284	Lynestrenol
91*	81	106	78	39	95	68	43	167	Ethinamate*
91*	82	42	39	65	159	158	68	159	Pargyline
91*	84	200	41	55	117	42	115	243	Phencyclidine
91*	86	177	65	0	0	0	0	177	Amphetamine-NCS Derivative
91*	92	65	155	185	58	77	121	185	Methyl-p-tolylsulphonamide (N-Methyl)
91*	92	425	72	160	44	114	65	999	Oxethazaine Probe Product 2*
91*	92	65	44	117	39	90	89	178	Phenacemide
91*	92	118	44	43	135	65	178	178	Phenacemide*
91*	92	120	65	63	51	39	89	120	Phenylacetaldehyde
91*	92	65	39	63	51	93	45	92	Toluene
91*	104	150	105	78	77	65	51	150	Phenylpropionic Acid
91*	106	197	162	107	148	27	63	197	Beclamide*
91*	106	51	65	77	92	39	78	106	Ethylbenzene

        Asterisk Indicates True Base Peak   Peak Index

```
              PEAKS                        MWT    COMPOUND


91* 106  127  110   57   92   65  104     231  Isocarboxazid
91* 106   44   78   51  177   79  107     298  Nialamide
91* 106   78   51   58   44   79  149     298  Nialamide
91* 106  105   77   51   92   39   79     106  Xylene (meta)
91* 106  105   77   51   92   39   78     106  Xylene (ortho)
91* 106  105   77   51   92   79  103     106  Xylene (para)
91* 110   79  231  272  298   41   77     298  Norethindrone
91* 114   43   85  139  104   72  243     999  Phenformin Probe Product*
91* 115  128  117  129  219  157    0     275  Dihydroxyphencyclidine
91* 116  131   65  149   92  150   90     149  Benzyl Methyl Ketoxime
91* 117  118   42   92   65   77  136     179  Phenprobamate
91* 120  121   92   65  106   30  135     240  Benzathine
91* 120  121   92   65  167  119  106     999  Oxethazaine Probe Product 1*
91* 121   65  122  309   64  230  123     431  Benzthiazide*
91* 126  128  125   65  127   92   89     126  3-Chlorotoluene
91* 126  128   89  125   90   65   63     126  2-Chlorotoluene
91* 126  125  128   65  127   92   89     126  4-Chlorotoluene
91* 127  106  110   43   92   65  120     231  Isocarboxazid*
91* 132   42  105   92   43   41   77     133  Methylphenylaziridine
91* 133  176  174   44  107   92  177     299  Buphenine*
91* 134   92   43   65   77   89   51     134  Benzyl Methyl Ketone
91* 136   92   65   39   63   45   51     136  Phenylacetic Acid
91* 137   65   39  107   63   89   77     137  Nitrotoluene(p-)
91* 146   65  219   92   41   39  190     219  Encyprate
91* 147   45   64  105   65   77   48     136  Phenelzine
91* 148  149   65   92   42   56   39     239  Benzphetamine*
91* 150   65   92   89   59    0    0     150  Methyl Phenyl Acetate (Methyl Phenidate Decomposition Prod)
91* 155  197   65   84   39   95   41     366  Glibornuride*
91* 155  129   41  284  229  163  184     284  Methyl Enol Ether of Tolbutamide
91* 155  185   65   92    0    0    0     185  Methyl-p-tolylsulphonamide (N-Methyl)
91* 155  185   65   41   92  121   56     284  Methyltolbutamide (N-Methyl)
91* 155  114   65  197   42   41   85     311  Tolazamide*
91* 155  171   65  107   39  108   63     270  Tolbutamide
91* 159   65  160  131   39   51   81     160  Tolazoline
91* 160   65  174  175  279  146  124     299  Nylidrin
91* 161  117  103  115  189  143   55     233  Glutethimide Metabolite
91* 163  161  119  107  117   41  118     206  Ibuprofen
91* 176  148  133   44  174   65  107     299  Buphenine
91* 177   44  106   45   78  123   51     298  Nialamide*
91* 190  119   72  191   41   44  162     281  Formyldi-(1-isopropylphenyl)amine* (N-)
91* 190  119   70   71   55   77   72     369  Tri-(1-isopropylphenyl)amine
91* 202  115  129  158   57   77  143     245  1-(1-Phenylcyclohexyl)morpholine
91* 215   79  105   77   55   41  298     298  Norethynodrel
91* 216   77  259    0    0    0    0     259  1-(1-Phenylcyclohexyl)-4-hydroxypiperidine
91* 223  294   57   42   56  165   44     309  Methadone (6-Dimethylamino-4,4-diphenyl-3-heptanone)
91* 233   30  232   31  276  275   65     276  Mebhydroline*
91* 269  324   55  296  295  323  297     324  Prazepam*
91* 282   79  105   77   81   67  338     392  Deoxycholic Acid
91   30* 121   92   51    0    0    0     121  Phenethylamine(beta-)
91   30*  39   65   92   51   63  121     121  2-Phenylethylamine
91   41* 132  105   42   92   43  105     132  2-Methyl-3-phenylaziridine
91   43* 227  268  312   79   55   77     312  Dydrogesterone
91   43* 342  281   79   55  256  241     342  Megesterol
91   43* 134   92   65   39   63  135     134  Phenylacetone
```

								MWT	COMPOUND
91	44*	65	42	45	120	40	92	135	Amphetamine
91	44*	65	51	63	77	0	0	135	Amphetamine
91	44*	65	58	39	42	51	63	135	Amphetamine
91	44*	40	42	65	45	42	39	135	Amphetamine*
91	44*	129	0	0	0	0	0	251	Dextropropoxyphene Metabolite
91	44*	129	115	178	205	220	251	251	Dextropropoxyphene Metabolite
91	44*	122	45	77	66	78	107	165	2-Methoxyamphetamine
91	45*	105	77	65	52	136	0	136	Phenelzine
91	45*	64	105	147	30	77	104	136	Phenelzine
91	55*	340	44	339	57	105	269	340	Prazepam-N-Oxide
91	55*	269	136	107	340	41	67	416	Spironolactone
91	58*	57	115	208	117	59	42	339	Dextropropoxyphene
91	58*	59	42	129	0	0	0	265	Dextropropoxyphene Metabolite
91	58*	191	129	128	205	178	174	265	Hydrolysed Dextropropoxyphene
91	58*	56	65	51	77	134	0	149	Methamphetamine
91	58*	56	65	42	39	51	41	149	Methamphetamine
91	58*	57	56	78	65	77	42	179	Methoxyphenamine
91	58*	59	56	30	42	121	78	179	Methoxyphenamine*
91	58*	43	45	30	59	56	42	179	Methoxyphenamine (ortho-)
91	58*	59	134	65	56	42	57	149	Methylamphetamine*
91	58*	59	44	42	134	65	57	149	Phentermine
91	58*	42	41	134	65	59	40	149	Phentermine*
91	58*	72	71	65	42	79	78	255	Tripelennamine
91	58*	72	71	197	185	184	92	255	Tripelennamine*
91	71*	57	106	65	72	55	77	177	Bethanidine
91	71*	106	177	57	72	65	30	177	Bethanidine*
91	72*	73	193	165	58	115	178	311	Alphamethadol
91	72*	73	56	148	41	57	42	163	Mephentermine
91	72*	114	145	75	160	117	92	467	Oxethazaine*
91	81*	106	95	79	68	67	78	167	Ethinamate (Ethynylcyclohexyl Carbamate)
91	82*	68	42	92	159	158	65	159	Pargyline
91	84*	55	77	85	182	83	65	265	Antazoline
91	84*	77	65	104	85	83	182	265	Antazoline
91	84*	36	51	55	0	0	0	265	Antazoline
91	84*	77	55	182	85	65	104	265	Antazoline*
91	84*	56	55	121	118	117	85	233	Methylphenidate
91	84*	55	150	146	56	85	83	233	Methylphenidate
91	84*	56	90	65	89	77	115	233	Methylphenidate
91	84*	85	56	55	150	41	118	233	Methylphenidate
91	86*	99	145	115	58	87	56	289	Caramiphen
91	86*	145	84	144	85	58	87	333	Carbetapentane
91	86*	87	145	58	144	30	44	333	Carbetapentane*
91	86*	105	87	144	58	100	56	335	Oxeladin*
91	92*	42	81	41	43	109	55	152	Sabinol
91	92*	65	103	93	39	163	77	284	Tropicamide
91	93*	92	79	80	77	43	121	136	Car-3-ene
91	93*	77	119	92	136	41	39	136	Phellandrene(alpha-)
91	93*	77	136	121	39	43	27	136	Terpinene(gamma-)
91	98*	84	41	56	58	39	115	215	Fencamfamine
91	105*	104	77	133	92	51	65	136	Phenelzine
91	115*	77	116	117	129	163	0	163	Nitrostyrene
91	115*	105	116	40	39	51	77	163	Nitrostyrene
91	115*	105	116	51	39	77	63	163	1-Phenyl-2-nitropropene
91	118*	44	92	178	65	0	0	178	Phenacemide

Asterisk Indicates True Base Peak

							MWT	COMPOUND
91	119*	211	65	120	212	92	90	211	2-Methylbenzanilide
91	119*	227	65	120	228	92	63	227	2-Methyl-4'-hydroxybenzanilide
91	120*	103	121	77	65	42	39	211	Benzylphenethylemine(alpha)
91	120*	148	103	194	77	65	0	239	Formyl-1,3-diphenylisopropylamine (N-)
91	120*	44	121	42	65	51	77	121	Methylbenzylamine (N-Methyl)
91	126*	127	174	55	41	42	97	217	Prolintane
91	134*	42	119	65	135	58	86	225	Benzyl-N-methylphenethylamine (alpha-)
91	137*	107	79	65	92	138	77	137	Nitrotoluene (meta)
91	145*	74	72	56	86	277	43	467	Oxethazaine
91	160*	129	130	117	115	104	203	203	Ethyl-1-phenylcyclohexylamine (N-)
91	162*	44	119	163	41	70	65	253	Di-(1-isopropylphenyl)amine*
91	186*	150	187	143	65	115	116	277	2-Benzyl-2-methyl-5-phenyl-2,3-dihydropyrimid-4-one
91	198*	65	199	197	155	107	92	198	Ditolyl Ether(p-)
91	200*	84	242	243	115	129	117	243	Phencyclidine
91	200*	243	84	242	186	166	201	243	Phencyclidine
91	200*	84	86	186	259	0	0	259	4-Phenyl-4-piperidinocyclohexanol
91	235*	65	132	76	233	90	250	250	Methaqualone
91	246*	42	247	233	57	218	56	367	Benzethidine
91	246*	42	233	162	57	56	149	367	Benzethidine
91	246*	165	377	452	0	0	0	452	Diphenoxylate
91	269*	55	295	296	241	324	271	324	Prazepam
91	279*	280	322	247	121	201	189	322	Methylwarfarin
91	284*	375	81	42	36	285	175	375	Benzyl Morphine
91	323*	154	338	251	266	309	307	338	Methaqualone Metabolite TMS Derivative
91	323*	132	338	154	0	0	0	338	Methaqualone Metabolite TMS Derivative
91	350*	259	352	348	65	351	107	365	Metolazone*
92*	56	94	63	84	86	184	93	276	4-Hydroxycyclophosphamide
92*	63	56	94	65	57	171	219	220	Phosphoramide Mustard
92*	65	108	214	156	39	109	43	214	Sulphaguanidine*
92*	65	108	156	191	71	63	93	255	Sulphathiazole
92*	70	43	41	109	90	39	76	154	1-Terpin-4-ol
92*	91	42	81	41	43	109	55	152	Sabinol
92*	91	65	103	93	39	163	77	284	Tropicamide
92*	120	39	65	152	121	63	64	152	Methyl Salicylate
92*	120	138	64	39	63	65	53	138	Salicylic Acid
92*	128	65	39	66	122	80	93	138	3-Nitroaniline
92*	156	191	65	108	76	104	50	403	Phthalylsulphathiazole*
92*	156	65	191	108	55	45	174	355	Succinylsulphathiazole*
92*	156	65	108	191	45	39	55	255	Sulphathiazole*
92*	216	215	108	65	156	54	125	280	Sulphamethoxypyrazine*
92*	254	163	91	93	65	103	104	284	Tropicamide
92*	270	65	156	108	106	93	59	270	Sulphamethizole*
92*	284	156	108	65	106	93	220	284	Sulphaethidole*
92	84*	120	65	56	111	119	82	248	Piridocaine
92	91*	65	155	185	58	77	121	185	Methyl-p-tolylsulphonamide (N-Methyl)
92	91*	425	72	160	44	114	65	999	Oxethazaine Probe Product 2*
92	91*	65	44	117	39	90	89	178	Phenacemide
92	91*	118	44	43	135	65	178	178	Phenacemide*
92	91*	120	65	63	51	39	89	120	Phenylacetaldehyde
92	91*	65	39	63	51	93	45	92	Toluene
92	93*	91	79	77	80	121	105	136	Ocimene
92	93*	91	79	77	80	121	105	136	Ocimene-X(beta-)
92	93*	91	77	41	79	121	39	136	Pinene(alpha-)
92	107*	106	65	79	39	108	77	107	3-Ethylpyridine

								MWT	COMPOUND
92	107*	152	121	65	0	0		0	152 Phenoxyacetic Acid
92	120*	165	65	137	0	0		0	165 Benzocaine
92	120*	105	148	150	121	133		65	165 Ethenzamide*
92	120*	152	121	65	64	93		63	152 Methyl Salicylate
92	120*	91	65	121	77	93		137	137 Nitrotoluene (ortho)
92	120*	137	65	121	39	64		53	137 Salicylamide*
92	120*	138	64	39	63	121		65	138 Salicylic acid*
92	121*	120	65	162	63	93		64	336 Dicoumarol
92	135*	65	108	138	80	39		106	272 Vacor
92	137*	120	65	39	52	66		138	137 3-Aminobenzoic Acid
92	137*	80	77	123	109	122		107	137 3-Methyl-4-nitrosophenol
92	156*	108	65	140	43	157		42	267 Sulphafurazole*
92	156*	108	65	140	253	157		43	253 Sulphamethoxazole*
92	156*	108	65	269	39	45		270	269 Sulphasomizole*
92	169*	289	65	290	39	333		184	398 Sulphasalazine*
92	172*	156	65	108	39	44		41	214 Sulphacetamide*
92	172*	156	65	108	173	39		174	172 Sulphanilamide*
92	184*	185	65	39	108	66		186	249 Sulphapyridine*
92	199*	135	76	324	119	93		134	324 Oxyphenbutazone
92	214*	65	213	108	42	215		39	278 Sulphasomidine*
92	215*	216	65	108	53	69		39	280 Sulphamethoxypyridazine*
92	246*	65	245	108	247	39		260	310 Sulphadimethoxine*
93*	41	69	77	91	53	67		94	136 Myrcene
93*	41	27	39	79	80	77		43	136 Ocimene
93*	41	69	79	91	77	94		80	136 Pinene(beta-)
93*	41	77	91	79	27	39		69	136 Sabinene
93*	43	41	109	94	91	39		55	152 Verbenol(trans-(-))
93*	63	27	95	65	31	94		36	142 Bis(2-chloroethyl) Ether
93*	66	92	78	65	39	51		94	93 2-Methylpyridine
93*	66	92	65	39	94	67		59	93 3-Methylpyridine
93*	66	92	65	39	94	54		67	93 4-Methylpyridine
93*	66	92	39	65	67	51		54	93 4-Methylpyridine(4-Picoline)
93*	66	92	65	78	39	51		67	93 2-Picoline(2-Methylpyridine)
93*	66	92	65	39	67	40		94	93 3-Picoline(3-Methylpyridine)
93*	71	41	69	55	43	80		121	154 Linalool
93*	77	79	91	80	41	92		136	136 Car-3-ene
93*	77	119	51	211	0	0		0	211 1,3-Diphenylguanidine
93*	77	91	41	79	39	136		94	136 Sabinene
93*	79	80	121	94	107	81		136	136 Fenchene(alpha-)
93*	80	92	91	79	121	105		77	136 Ocimene-Y(beta-)
93*	91	92	79	80	77	43		121	136 Car-3-ene
93*	91	77	119	92	136	41		39	136 Phellandrene(alpha-)
93*	91	77	136	121	39	43		27	136 Terpinene(gamma-)
93*	92	91	79	77	80	121		105	136 Ocimene
93*	92	91	79	80	121	105		77	136 Ocimene-X(beta-)
93*	92	91	77	41	79	121		39	136 Pinene(alpha-)
93*	109	199	77	119	162	55		324	324 Oxyphenbutazone
93*	121	79	67	107	95	94		68	136 Camphene
93*	121	136	81	43	55	41		107	154 Caran-trans-3-ol(cis-)
93*	121	136	111	43	41	55		79	154 Caran-trans-2-ol((-)cis-)
93*	121	136	95	43	81	110		107	154 Neoisothujyl Alcohol
93*	121	136	39	41	79	91		27	136 Terpinolene
93*	135	94	66	43	65	39		0	135 Acetanilide
93*	135	65	43	39	66	63		92	135 Acetanilide

Asterisk Indicates True Base Peak

	PEAKS							MWT	COMPOUND
93*	136	111	78	79	40	94	80	136	Fenchene(alpha-)
93*	136	77	94	79	91	80	92	136	Phellandrene(beta-)
93*	136	121	92	91	43	77	79	136	Terpinene(gamma-)
93*	174	95	59	172	176	74	43	172	Dibromomethane
93*	213	79	77	106	121	119	104	213	2-Methylsulphonylacetanilide
93	41*	69	79	91	77	67	53	136	Myrcene
93	43*	41	179	39	137	65	32	179	Isopropyl-N-phenylcarbamate (IPC)
93	44*	65	43	166	137	111	139	183	Adrenaline
93	59*	121	43	136	28	81	41	154	2-Terpineol
93	59*	121	81	43	136	68	92	154	Terpineol(alpha-)
93	68*	67	136	41	121	79	39	136	Dipentene
93	68*	67	39	79	53	27	94	136	Limonene
93	68*	67	136	94	41	121	53	136	Limonene
93	68*	67	39	41	53	136	79	136	1-Methyl-4-isopropenylcyclohexene
93	71*	47	43	55	69	80	154	154	Linalool
93	71*	111	43	86	69	55	68	154	Terpineol
93	87*	125	58	47	79	63	229	229	Dimethoate (Cygon)
93	95*	43	121	81	55	110	70	154	Thujyl Alcohol
93	119*	41	69	105	161	55	91	204	Cedrene(alpha-)
93	121*	136	119	106	91	79	77	136	Terpinene(alpha-)
93	125*	127	173	158	99	55	79	330	Malathion
93	156*	281	125	43	157	55	63	281	Menazon
93	174*	95	172	176	91	81	79	172	Methylene Bromide
93	274*	125	121	91	107	135	246	320	Phenthoate
94*	41	39	42	77	151	95	93	151	Amantadine
94*	45	79	49	46	47	48	61	94	Dimethyldisulphide
94*	65	66	95	55	51	63	47	94	Phenol
94*	66	39	65	40	95	38	55	94	Phenol
94*	67	26	39	40	53	38	42	94	Methylpyrazine
94*	77	138	95	51	66	172	45	472	5-Aceto-8-hydroxyquinoline Sulphate
94*	95	80	39	41	27	67	53	95	2,4-Dimethylpyrrole
94*	95	141	64	47	46	79	45	141	Methamidophos
94*	96	44	47	59	35	24	98	94	Dichloroacetylene
94*	138	42	108	154	303	136	137	303	Hyoscine
94*	138	42	108	136	41	96	97	303	Hyoscine*
94*	151	57	95	40	41	58	108	151	Amantadine*
94	42*	77	250	249	291	198	51	291	Cruformate Metabolite
94	53*	57	30	82	99	67	111	252	Cimetidine
94	84*	177	148	42	93	134	41	177	Modaline
94	124*	82	77	79	123	96	67	275	Homatropine
94	138*	73	108	42	154	137	136	375	Hyoscine TMS Derivative
94	138*	108	136	301	154	77	42	303	Hyoscine
94	151*	168	106	150	122	123	0	168	Pridoxamine
94	151*	122	106	51	53	149	150	169	Pyridoxine
95*	41	81	69	83	108	109	67	152	Camphor
95*	41	27	43	39	93	55	29	154	Isoborneol
95*	41	43	110	93	55	136	67	154	Isoborneol
95*	81	108	152	41	83	109	69	152	Bornanone-2
95*	81	41	69	55	83	67	137	152	Camphor
95*	81	69	41	55	67	108	109	152	Camphor
95*	81	108	55	109	152	67	83	152	Camphor
95*	81	108	152	69	109	83	110	152	Camphor
95*	93	43	121	81	55	110	70	154	Thujyl Alcohol
95*	96	124	42	82	83	94	41	208	Pilocarpine

PEAKS							MWT	COMPOUND
95*	96	109	208	42	41	54	83	208 Pilocarpine
95*	110	43	39	96	68	0	0	110 2-Acetylfuran
95*	110	41	55	96	69	139	136	154 Isoborneol
95*	115	196	41	67	43	96	209	252 2,4-Dimethyltalbutal (or 4,6-Dimethyl-)
95*	130	132	60	97	35	134	47	130 Trichloroethylene
95*	327	39	329	96	122	244	67	327 Diloxanide Furoate*
95	68*	180	53	41	40	67	42	180 Theophylline
95	71*	81	41	55	43	69	82	156 Neomenthol
95	81*	71	41	67	55	138	123	156 Menthol
95	82*	137	54	109	152	41	110	152 Piperitone
95	94*	80	39	41	27	67	53	95 2,4-Dimethylpyrrole
95	94*	141	64	47	46	79	45	141 Methamidophos
95	96*	43	53	81	27	51	50	96 2,5-Dimethylfuran
95	96*	39	38	29	37	97	67	96 Furfural
95	123*	164	69	0	0	0	0	164 Droperidol Hofmann Reaction Product
95	123*	224	42	237	206	56	84	375 Haloperidol
95	123*	164	69	0	0	0	0	164 Haloperidol Hofmann Reaction Product 1
95	130*	132	97	134	60	99	62	130 Trichloroethylene
95	139*	109	41	121	43	55	136	154 Isofenchol(endo-)
95	180*	181	68	41	123	53	96	420 Aminophylline*
95	180*	68	41	58	53	123	96	269 Bufylline*
95	180*	224	123	68	109	193	194	224 Oxyethyltheophylline
95	180*	68	41	43	53	57	55	180 Theophylline
95	180*	68	41	53	181	96	40	180 Theophylline*
95	208*	193	67	180	123	73	43	208 7-Ethyltheophylline
95	222*	194	166	207	179	123	67	222 1,7-Diethyl-3-methylxanthine
95	250*	39	235	66	207	41	193	250 3'-Ketohexobarbitone
96*	44	36	173	172	56	70	129	247 3-Desmethylprodine
96*	44	43	31	45	27	78	264	280 Hydroxystilbamidine*
96*	73	94	42	57	227	142	212	227 Scopoline TMS Ether
96*	82	83	42	97	55	68	199	199 Methylecgonine
96*	83	82	182	93	36	42	131	329 Cinnamoylcocaine
96*	95	43	53	81	27	51	50	96 2,5-Dimethylfuran
96*	95	39	38	29	37	97	67	96 Furfural
96*	105	77	97	216	42	218	51	337 Lobeline*
96	39*	95	29	0	0	0	0	96 Furfural
96	43*	95	53	81	91	27	41	96 2,5-Dimethylfuran
96	61*	98	63	26	60	62	100	96 1,2-Dichloroethylene
96	77*	188	105	38	82	93	106	188 Antipyrine
96	79*	346	81	345	70	78	113	446 Thioproperazine
96	82*	83	124	138	168	185	186	185 Ecgonine
96	82*	83	97	42	199	94	168	199 Methylecgonine
96	94*	44	47	59	35	24	98	94 Dichloroacetylene
96	95*	124	42	82	83	94	41	208 Pilocarpine
96	95*	109	208	42	41	54	83	208 Pilocarpine
96	107*	106	97	143	248	140	79	248 Thionazin
96	109*	15	79	185	47	45	95	220 Dichlorvos
96	111*	175	128	157	316	262	331	331 Ethyl-3-piperidylcyclopentyl Glycolate (N-)
96	111*	167	165	152	105	128	139	323 Ethyl-3-piperidyldiphenyl Acetate (N-)
96	111*	165	42	167	43	41	56	323 Piperidolate
96	111*	167	112	165	71	43	42	323 Piperidolate*
96	155*	140	43	42	81	94	53	155 Arecoline
96	166*	209	55	167	71	42	69	209 Ethirimol
96	188*	77	56	105	189	55	51	188 Phenazone*

```
        PEAKS                        MWT    COMPOUND


96  285* 229  228   70  214  115  200  285 Hydromorphone
96  287* 215  286  229  213  228  243  287 Cyproheptadine
96  287* 286  215   70   44   58   42  287 Cyproheptadine*
96  299* 242  243  185  228  214  300  299 Metopon
96  303* 302  231  202  245  259  288  303 Cyproheptadine Metabolite
96  303* 302  231  202  245  259  304  303 3-Hydroxycyproheptadine
97*  41  126   39   69   53   51  125  126 5-(Hydroxymethyl)-2-furfuraldehyde
97*  55   69   72   71   98   43   62  232 Mebutamate*
97*  56   32   91   68   98   65   57  202 Fendroprox
97*  58   72  191   71   78   79  190  261 Methapyrilene
97*  72   55   71   62  110   69  158  232 Mebutamate
97*  83   99   85   61   26   96   63  132 1,1,2-Trichloroethane
97*  98   55   82  199  198  180  296  296 Methdilazine
97*  99   61   63  117  119  101   62  132 1,1,1-Trichloroethane
97* 105   77   96   42  183   98   82  419 Mepenzolate Bromide
97* 105   77  183   84   36   42   51  325 Methyl-3-piperidylbenzilate (N-)
97* 105   77   96   84   98  183   70  325 Methyl-3-piperidylbenzilate(N-)
97* 105   77  183  165  114  167  152  325 Methyl-3-piperidylbenzilate (N-Methyl)
97* 107  105  189  128  249  248  331  331 Methyl-3-piperidylphenylcyclohexyl Glycolate (N-Methyl)
97* 109  291  139  125  137  155  123  291 Parathion
97* 111   83  112   55   41   69   56  154 Menthone(o-)
97* 125   65  153   93   45  121   73  456 Dioxathion
97* 165  164  206   84  249  135  136  249 Thiophencyclidine (Thio-PCP)
97* 172   91  105   57   42   77  201  275 Alphameprodine
97* 172   91  105   57   42   77  201  275 Betameprodine
97* 197  199  314  316  258  286  125  349 Chlorpyrifos
97* 291  109  137  139  155  125   65  291 Parathion
97* 359  242  303  357  331  240  301  392 Bromophos Ethyl
97   43*  61   99   45   27  119  117  132 1,1,1-Trichloroethane
97   56* 231   42  111   77   71  112  231 Amidopyrine
97   56*  77  231   42   91   51   55  231 Aminopyrine
97   58*  72   84   71  191  192  261  261 Methapyrilene
97   58*  72   71  121  224  149  191  261 Methapyrilene
97   58*  72   71   42  191   79   78  261 Methapyrilene*
97   58*  72   71   79   42   78   40  261 Thenyldiamine
97   58*  71   72   91   79  203   78  261 Thenyldiamine
97   58*  72   71  203  191  190   42  261 Thenyldiamine*
97   62*  99   64   61   36  107   45  142 Dalapon
97   77* 129  157  125   31  103   51  298 Phoxim
97   82*  42   83   96   57   94   55  185 Ecgonine
97   82* 103   77   72   42   91  115  273 1-Phenyl-1-carbethoxy-2-dimethylaminocyclohex-3-ene
97  109* 291  139  125  137  155  123  291 Parathion
97  115*  73   43   65  121  125   93  285 Prothoate
97  131*  58  159  125  160  329   65  329 Mecarbam
97  157* 121  342  153  125  159  199  342 Trithion
97  231* 153  121  125   29   65   93  384 Ethion
97  248* 219  218  111  249  217  263  263 Dimethylthiambutene*
97  279* 223   88  251  162  281  164  314 Dichlofenthion
98*  41   42   99   55   69   77  105  287 Cycrimine
98*  41   81   42   39   97   53   69   98 Furfuryl Alcohol
98*  41   99   42   55   98  233   69  233 Piperoxan*
98*  42   65   78  176    0    0    0  176 Cotinine
98*  55  218   99   41   96  219   42  301 Trihexyphenidyl
98*  58   79  129   91   77   52  172  287 Cycrimine
```

98*	70	41	77	91	99	96	120	264	Carbocaine
98*	70	99	36	42	96	99	38	246	Mepivacaine
98*	70	42	94	77	51	99	126	386	Mesoridazine
98*	70	99	42	386	126	55	41	386	Mesoridazine*
98*	70	214	111	134	341	199	326	341	Sulpyrid
98*	70	42	185	126	99	96	55	370	Thioridazine
98*	70	370	126	99	185	244	125	370	Thioridazine (Melleril)
98*	71	84	56	99	124	41	167	167	Mecamylamine
98*	77	94	99	126	0	0	0	386	Mesoridazine
98*	84	71	56	41	42	99	124	167	Mecamylamine*
98*	91	84	41	56	58	39	115	215	Fencamfamine
98*	99	218	55	41	77	96	219	301	Benzhexol
98*	99	218	85	131	219	84	69	287	Cycrimine
98*	99	105	77	55	41	127	111	309	Diphenidol*
98*	99	70	42	96	55	41	40	246	Mepivacaine*
98*	105	55	41	99	77	218	84	301	Benzhexol*
98*	105	77	41	42	55	99	84	309	Diphenidol
98*	105	183	77	96	55	325	114	325	Methyl-4-piperidylbenzilate (N-Methyl)
98*	111	55	41	112	99	96	97	321	Dihexyverine
98*	111	99	147	55	41	42	96	391	Flavoxate*
98*	111	99	42	55	41	29	112	335	Norpipanone*
98*	111	99	199	41	288	200	55	399	Pipazethate*
98*	112	99	55	42	41	211	84	322	Fenpipramide*
98*	119	91	99	124	64	41	55	397	Diperodon*
98*	119	160	84	85	91	245	230	245	Mydocalm (Tolperisone)
98*	125	42	41	29	99	120	190	274	Phenampromide
98*	125	99	120	41	42	57	77	274	Phenampromide
98*	126	370	99	70	258	125	42	370	Thioridazine
98*	134	70	111	198	242	383	382	383	Sulpyrid Trimethyl Derivative
98*	147	111	42	55	115	70	96	391	Flavoxate
98*	152	154	42	69	174	208	153	273	Chlormezanone*
98*	176	42	118	41	119	51	39	176	Cotinine
98*	215	58	84	91	56	71	186	215	Fencamfamin*
98*	218	99	219	55	41	42	85	301	Benzhexol
98*	218	99	55	130	41	42	85	311	Biperiden
98*	218	99	55	41	42	77	84	311	Biperiden*
98*	218	97	55	41	219	105	77	301	Trihexyphenidyl
98*	231	232	197	0	0	0	0	232	Chlorprothixene Metabolite
98*	321	83	55	140	41	155	182	999	Methyprylone Probe Product (Dimer)*
98*	370	126	99	40	70	371	258	370	Thioridazine*
98	58*	202	231	189	215	218	205	319	Oxetorone
98	70*	69	77	204	89	83	105	204	Thozalinone
98	83*	55	166	153	41	84	155	199	Methyprylon Metabolite (6-Hydroxy)
98	97*	55	82	199	198	180	296	296	Methdilazine
98	124*	105	84	216	96	259	200	259	3-Phenyl-3-piperidinocyclohexanol
98	172*	201	91	202	275	96	200	275	Alphameprodine
99*	41	39	72	45	98	71	38	99	Allyl Isothiocyanate
99*	41	56	57	39	55	29	155	266	Tributyl Phosphate
99*	42	98	70	114	96	44	43	281	Diphenylpyraline
99*	56	72	165	300	228	229	242	300	Chlorcyclizine
99*	56	165	43	228	229	241	242	300	Chlorcyclizine
99*	56	167	207	194	164	195	208	266	Cyclizine
99*	56	42	43	167	70	44	207	266	Cyclizine
99*	56	167	207	194	266	195	165	266	Cyclizine*

Asterisk Indicates True Base Peak

	PEAKS						MWT	COMPOUND
99*	81	55	44	41	43	120	79	181 Hexapropymate
99*	83	101	85	61	70	31	0	132 Trichloroethane
99*	114	98	167	70	165	57	43	281 Diphenylpyraline*
99*	155	127	81	109	0	0	0	224 Diethylamyl Phosphate
99*	155	109	127	45	82	81	27	182 Triethyl Phosphate
99*	165	56	194	167	207	152	208	266 Cyclizine
99*	197	44	58	112	309	41	42	309 Methixene*
99	30*	28	42	43	41	27	39	99 Piperidone
99	41*	39	72	45	38	27	40	99 Allyl Isothiocyanate
99	43*	71	57	114	27	42	72	114 2,5-Hexanedione(Acetonyl Acetone)
99	56*	165	167	42	0	0	0	266 Cyclizine
99	86*	30	58	87	120	162	206	278 Acetylprocaine Hydrochloride
99	86*	71	58	87	140	84	100	308 Benoxinate
99	86*	30	58	71	41	136	56	308 Benoxinate
99	86*	91	191	87	119	58	248	263 Butethamate*
99	86*	91	144	58	56	41	87	289 Caramiphen
99	86*	154	30	87	58	29	156	270 Chloroprocaine*
99	86*	154	58	56	87	42	84	266 Chlorprocaine
99	86*	55	100	87	41	44	165	309 Dicyclomine
99	86*	58	87	56	85	183	387	387 Flurazepam
99	86*	184	58	30	87	201	186	299 Metoclopramide*
99	86*	29	30	100	71	87	192	308 Oxybuprocaine*
99	86*	120	30	92	87	58	65	235 Procainamide*
99	86*	120	58	65	42	92	56	236 Procaine
99	86*	120	58	87	30	92	71	236 Procaine*
99	86*	58	87	56	100	84	70	294 Proparcaine
99	86*	58	71	136	56	41	80	294 Proxymetacaine
99	97*	61	63	117	119	101	62	132 1,1,1-Trichloroethane
99	98*	218	55	41	77	96	219	301 Benzhexol
99	98*	218	85	131	219	84	69	287 Cycrimine
99	98*	105	77	55	41	127	111	309 Diphenidol*
99	98*	70	42	96	55	41	40	246 Mepivacaine*
99	120*	30	86	56	44	57	193	236 Butethamine*
99	124*	142	141	154	98	129	125	228 5-Butyl-5-(2-hydroxyethyl)barbituric Acid
99	139*	124	153	78	79	95	107	153 2-Methoxy-4-nitrosophenol
100*	36	56	101	42	38	165	91	339 Moramide Intermediate
100*	42	56	101	114	353	165	91	353 Dioxaphetyl butyrate
100*	42	28	41	70	45	40	30	100 3-Methylsydnone
100*	44	72	101	77	56	42	105	205 Diethylpropion*
100*	44	101	72	42	56	180	198	312 Ethopropazine
100*	55	45	74	119	73	56	41	118 Succinic Acid
100*	56	101	42	185	184	41	128	324 Furaltadone*
100*	56	42	176	98	120	70	189	276 Molindone
100*	56	101	42	165	91	115	70	339 Moramide Intermediate
100*	56	42	101	55	54	41	30	226 Nimorazole*
100*	58	41	65	43	30	192	56	239 Isoetharine
100*	58	41	101	43	56	65	30	239 Isoetharine*
100*	58	105	77	56	101	41	70	235 Meprylcaine
100*	72	240	340	44	197	254	43	442 Acepromazine Maleate*
100*	72	39	71	55	45	46	101	100 2,5-Dihydrothiophene(2-oxy-)
100*	72	56	165	42	101	91	115	353 Dioxaphetyl Butyrate
100*	77	44	42	72	105	101	51	205 Diethylpropion
100*	83	39	55	82	85	29	27	100 Senecioic Acid
100*	101	72	197	312	84	179	212	312 Ethopropazine

								MWT	COMPOUND
100*	101	44	72	198	180	42	29	312	Ethopropazine*
100*	113	91	101	87	165	115	56	378	Doxapram
100*	113	56	101	87	378	194	91	378	Doxapram*
100*	114	353	91	165	70	161	178	353	Dioxaphetyl butyrate
100*	114	42	56	115	70	101	55	398	Pholcodeine
100*	120	70	41	56	42	101	293	293	Pramoxine
100*	128	265	55	56	41	42	40	392	Dextromoramide
100*	128	265	266	129	246	236	306	392	Moramide
100*	128	70	41	42	293	56	101	293	Pramoxine*
100*	131	311	218	0	0	0	0	311	Fomocaine
100*	133	77	52	148	50	76	0	148	Ethyl Phenylmethyl Ketone (ortho)
100*	265	128	266	44	98	56	101	392	Dextromoramide*
100*	265	128	266	56	129	55	101	392	Levomoramide
100*	272	274	270	237	102	65	276	370	Heptachlor
100*	378	113	56	101	87	379	194	378	Doxapram
100	41*	42	43	30	69	39	68	100	Nitrosopyrrolidine
100	42*	71	41	70	55	39	45	100	Tetrahydropyran-3-one
100	42*	43	29	72	41	30	69	100	Tetronic Acid
100	44*	34	55	115	147	41	0	349	Ampicillin
100	44*	234	88	57	91	105	129	325	Norpropoxyphene Amide
100	44*	41	60	29	27	30	86	232	Piperazine Adipate*
100	44*	42	41	30	55	39	45	115	Tuaminoheptane
100	55*	27	29	39	54	85	53	100	Tiglic Acid(2-Methyl-but-2-enoic Acid)
100	56*	98	58	43	123	41	192	239	Isoetharine Mesylate
100	56*	138	110	57	237	70	41	237	Viloxazine*
100	85*	125	101	199	227	99	153	398	Tributoxyethyl Phosphate
100	114*	56	42	115	101	70	398	398	Pholcodeine*
100	128*	45	73	43	143	0	0	143	2-Hydroxy-4-methyl-5-ethylthiazole
100	132*	91	77	65	89	79	0	250	Amphetaminil
100	142*	99	125	143	98	124	156	186	5-(2-Hydroxyethyl)-5-methylbarbituric Acid
100	171*	173	75	74	136	99	50	171	Dichlobenil
100	246*	42	82	91	56	232	41	346	Morpheridine
101*	43	72	41	27	42	39	45	101	Propyl Isothiocyanate
101*	59	69	41	142	107	77	102	242	MCPB Methyl Ester
101*	103	147	145	105	31	149	66	179	Fluorodichlorobromomethane
101*	103	66	105	47	35	31	82	136	Trichlorofluoromethane(Freon 11)
101*	151	103	153	85	87	31	105	147	1,1,2-Trichlorotrifluoroethane
101	43*	546	428	331	402	275	487	673	Iopronic Acid
101	44*	58	30	28	41	42	43	101	Methyl Amylamine(n-)
101	44*	28	43	30	41	43	45	101	Methyl Isoamylamine
101	58*	44	42	59	29	30	0	101	Dimethylbutylamine
101	88*	73	60	81	93	109	111	258	Demeton PO
101	88*	43	61	60	41	73	70	200	Ethyl Caprate
101	88*	43	41	73	61	70	55	228	Ethyl Laurate
101	88*	43	41	73	55	57	70	256	Ethyl Myristate
101	88*	43	57	55	41	89	73	284	Ethyl Palmitate
101	88*	43	57	89	55	41	73	312	Ethyl Stearate
101	100*	72	197	312	84	179	212	312	Ethopropazine
101	100*	44	72	198	180	42	29	312	Ethopropazine*
101	128*	50	75	129	76	64	51	128	Isophthalonitrile
101	128*	50	75	129	76	51	64	128	Phthalodinitrile(o-)
102*	30	72	44	116	55	173	71	204	Ethambutol*
102*	43	42	57	56	41	40	39	102	Methyl Isopropyl Nitrosamine
102*	44	42	29	56	57	27	30	102	Diethyl Nitrosamine

			PEAKS					MWT	COMPOUND

102*	84	58	56	42	30	103	91	385	Bamifylline*
102*	87	69	41	59	0	0	0	101	3,4-Dihydroxyproline
102*	104	75	48	77	50	38	47	102	4-Chloropyrazole
102*	245	247	304	305	306	58	126	335	Hydroxychloroquine*
102	43*	42	28	29	59	44	41	102	Methyl Pyruvate
102	60*	41	45	74	87	56	39	102	3-Methyltetrahydrothiophene
102	73*	57	41	43	55	101	115	144	Valproic Acid
102	87*	45	41	59	39	60	74	102	2-Methyltetrahydrothiophene
102	129*	128	51	130	76	75	103	129	Isoquinoline
102	129*	128	51	130	76	50	103	129	Quinoline
102	170*	169	115	77	51	60	61	170	4-Methyl-5-phenylpyrimidine
103*	28	160	77	207	41	150	0	207	1-(3,4-Methylenedioxyphenyl)-2-nitropropene
103*	76	50	104	51	75	77	52	103	Benzonitrile
103*	77	82	29	97	42	51	104	273	Tilidine
103*	131	43	77	51	145	146	50	308	Warfarin
103*	148	94	96	108	41	77	91	397	Methscopolamine Bromide
103*	160	77	207	41	39	150	51	207	1,2-Methylenedioxy-4-(2-nitropropenyl)benzene
103	44*	91	65	42	104	45	102	135	Amphetamine
103	101*	147	145	105	31	149	66	179	Fluorodichlorobromomethane
103	101*	66	105	47	35	31	82	136	Trichlorofluoromethane(Freon 11)
103	104*	78	51	77	50	105	52	104	Styrene
103	160*	89	131	115	76	161	104	160	Hydrallazine*
103	189*	190	104	188	102	77	174	189	Phensuximide(ii)*
104*	43	91	39	51	103	78	27	164	1-Phenylethyl Acetate
104*	43	91	105	65	39	51	77	164	Phenylethyl Acetate
104*	55	43	56	44	58	41	62	260	Carisoprodol
104*	71	147	42	44	70	56	102	237	Cartap
104*	71	147	42	44	70	56	102	171	Crimidine
104*	76	50	67	54	82	41	148	320	Dicyclohexyl Phthalate
104*	103	78	51	77	50	105	52	104	Styrene
104*	105	176	133	0	0	0	0	176	Dealkylated Ethotoin
104*	105	204	77	78	133	51	132	204	Ethotoin
104*	132	218	51	103	77	78	52	218	Phenylmethylbarbituric Acid*
104*	164	91	105	133	103	165	78	164	Methyl beta-Phenylpropionate
104*	175	77	119	51	190	42	161	190	5-Phenylhydantoin(5-Methyl-)
104*	189	77	51	105	103	190	56	218	Mephenytoin
104*	189	103	78	51	77	105	52	189	Phensuximide(i)*
104*	189	146	105	0	0	0	0	189	Phenylglutarimide(alpha)
104*	189	103	117	78	91	51	146	189	3-Phenylpiperidin-2,6-dione(Glutethimide Metabolite)
104	77*	75	51	76	111	106	138	300	Temazepam
104	91*	150	105	78	77	65	51	150	Phenylpropionic Acid
104	102*	75	48	77	50	38	47	102	4-Chloropyrazole
104	105*	78	52	51	77	50	79	105	4-Vinylpyridine
104	122*	78	106	51	0	0	0	122	Nicotinamide
104	132*	78	130	77	103	43	48	175	Debrisoquine
104	132*	44	175	130	117	103	43	175	Debrisoquine*
104	132*	246	103	78	77	51	133	246	Permethylated Phenylmethylbarbituric Acid*
104	133*	76	105	197	0	0	0	197	Methylsaccharin
104	149*	76	50	150	41	43	42	250	Diisopropyl Phthalate
104	175*	77	176	105	132	51	204	204	Methoin Metabolite (Phenylethylhydantoin)
104	180*	77	51	209	223	165	181	252	Diphenylhydantoin
104	180*	266	77	237	57	209	71	266	3-Methyldilantin
104	180*	223	77	209	252	51	165	252	Phenytoin*
104	189*	77	190	51	105	103	132	218	Mephenytoin

Peak Index   *Asterisk Indicates True Base Peak*

		PEAKS					MWT	COMPOUND
104	189*	77	190	51	0	0	0	218 Methetoin
104	189*	190	77	44	105	132	103	218 Methoin*
105*	77	182	44	86	165	51	30	999 Benactyzine Probe Product*
105*	77	122	51	50	44	78	52	289 Benzoylecgonine
105*	77	96	183	51	42	182	94	431 Clidinium Bromide
105*	77	51	135	134	50	106	117	179 Hippuric Acid
105*	77	120	43	51	96	106	97	337 Lobeline
105*	77	134	51	106	193	161	50	193 Methyl Hippurate
105*	77	235	96	111	97	183	250	323 Piperidolate
105*	77	30	45	227	51	165	183	227 Prenylamine Metabolite
105*	91	104	77	133	92	51	65	136 Phenelzine
105*	104	78	52	51	77	50	79	105 4-Vinylpyridine
105*	111	77	96	97	183	42	51	433 Pipenzolate Bromide
105*	112	77	41	55	44	42	56	261 Piperocaine
105*	120	40	29	44	77	79	51	120 Isopropylbenzene
105*	120	119	77	91	39	106	79	120 Mesitylene(1,3,5-Trimethylbenzene)
105*	122	77	51	50	39	74	78	122 Benzoic Acid
105*	129	112	77	42	313	41	55	344 Oxyphencyclimine*
105*	134	77	106	51	135	235	160	235 Butyl Hippurate (n-)
105*	135	51	134	77	106	50	78	179 Hippuric Acid
105*	177	77	209	254	210	103	181	254 Ketoprofen*
105*	189	165	42	166	106	79	77	390 Meclizine
105*	189	36	38	201	165	285	166	390 Meclozine
105*	254	177	209	77	210	181	255	254 Ketoprofen
105	44*	91	77	178	208	65	130	269 Norpropoxyphene Carbinol
105	58*	77	42	59	113	51	30	271 Amylocaine
105	58*	77	98	122	42	113	51	271 Amylocaine*
105	58*	57	43	71	55	41	167	327 Dimenoxadole
105	58*	77	42	106	56	57	51	165 Ephedrine
105	79*	104	51	78	52	50	77	136 Betahistine
105	82*	122	77	51	42	94	81	289 Benzoylecgonine
105	83*	77	183	165	152	139	128	311 3-Piperidylbenzilate
105	84*	85	77	55	56	183	42	267 Azacyclonol
105	84*	77	56	85	248	249	36	267 Pipradrol
105	86*	77	87	182	99	183	58	327 Benactyzine*
105	96*	77	97	216	42	218	51	337 Lobeline*
105	97*	77	96	42	183	98	82	419 Mepenzolate Bromide
105	97*	77	183	84	36	42	51	325 Methyl-3-piperidylbenzilate (N-)
105	97*	77	96	84	98	183	70	325 Methyl-3-piperidylbenzilate(N-)
105	97*	77	183	165	114	167	152	325 Methyl-3-piperidylbenzilate (N-Methyl)
105	98*	55	41	99	77	218	84	301 Benzhexol*
105	98*	77	41	42	55	99	84	309 Diphenidol
105	98*	183	77	96	55	325	114	325 Methyl-4-piperidylbenzilate (N-Methyl)
105	104*	176	133	0	0	0	0	176 Dealkylated Ethotoin
105	104*	204	77	78	133	51	132	204 Ethotoin
105	111*	96	77	183	165	128	324	339 Ethyl-3-piperidyl Benzilate (N-)
105	112*	246	77	55	41	98	97	261 Piperocaine
105	123*	77	96	183	110	165	351	351 Allyl-3-piperidylbenzilate (N-Allyl)
105	123*	78	51	106	77	124	50	123 Nicotinic Acid*
105	127*	193	104	121	43	179	166	314 Crotoxyphos
105	148*	91	44	77	79	65	56	239 Methyldiphenethylamine (N-)
105	151*	77	79	103	121	78	39	151 4-Nitroxylene(o-)
105	162*	44	163	57	190	106	58	324 Diampromide
105	162*	190	163	106	29	134	77	324 Diampromide*

Asterisk Indicates True Base Peak   Peak Inde

			PEAKS					MWT	COMPOUND
105	162*	57	163	114	190	44	77	351	Phenadoxone
105	183*	77	84	82	165	152	129	311	4-Piperidylbenzilate
105	189*	201	285	165	166	190	134	390	Meclozine*
105	189*	107	77	55	91	79	190	344	Oxyphencyclimine
105	204*	104	133	77	72	205	78	204	Ethotoin (Peganone)
105	209*	77	268	191	103	210	51	268	Ketoprofen Methyl Ester
106*	30	77	185	105	104	89	141	186	Mafenide*
106*	42	108	44	63	43	49	41	156	Mechlorethamine
106*	78	51	137	50	79	107	31	137	Isoniazid*
106*	78	70	59	42	284	300	44	406	Nicodicodeine
106*	78	176	51	148	162	56	0	178	Nikethamide
106*	78	177	51	178	107	149	163	178	Nikethamide
106*	107	79	78	52	51	65	80	107	2-Ethylpyridine
106*	107	77	108	79	78	53	39	107	Toluidine(p-)
106*	121	77	107	79	53	120	78	121	2-Ethylaniline
106*	121	77	107	120	104	51	79	121	Ethylaniline(N-)
106*	137	78	136	79	105	107	138	137	Methyl Pyridine-3-carboxylate
106*	162	78	51	79	123	43	41	179	Iproniazid
106*	177	78	178	51	105	107	149	178	Nikethamide
106	42*	77	57	246	43	56	91	366	Piminodine
106	77*	105	51	50	78	52	39	106	Benzaldehyde
106	78*	51	137	79	0	0	0	137	Iproniazid Metabolite
106	78*	51	137	50	79	107	52	137	Isoniazid
106	78*	51	104	77	137	50	0	137	Isoniazid
106	78*	135	136	51	50	79	52	137	Methylnicotinamide (N-)
106	78*	51	137	50	136	107	79	137	Methylnicotinate*
106	91*	197	162	107	148	27	63	197	Beclamide*
106	91*	51	65	77	92	39	78	106	Ethylbenzene
106	91*	127	110	57	92	65	104	231	Isocarboxazid
106	91*	44	78	51	177	79	107	298	Nialamide
106	91*	78	51	58	44	79	149	298	Nialamide
106	91*	105	77	51	92	39	79	106	Xylene (meta)
106	91*	105	77	51	92	39	78	106	Xylene (ortho)
106	91*	105	77	51	92	79	103	106	Xylene (para)
106	107*	66	39	65	92	79	108	107	2,3-Dimethylpyridine
106	107*	79	77	92	39	65	108	107	2,5-Dimethylpyridine
106	107*	66	92	65	39	108	79	107	2,6-Dimethylpyridine
106	107*	79	92	77	39	108	65	107	3,4-Dimethylpyridine
106	107*	79	92	77	39	108	80	107	3,5-Dimethylpyridine
106	107*	79	92	39	65	80	77	107	2,4-Dimethylpyridine(2,4-Lutidine)
106	107*	92	65	79	39	51	108	107	4-Ethylpyridine
106	107*	77	79	108	39	65	78	107	Toluidine(m-)
106	120*	78	92	51	226	41	39	226	Metyrapone*
106	123*	43	58	78	79	51	164	179	Iproniazid
106	123*	78	51	105	50	77	52	600	Nicofuranose*
106	152*	77	79	122	59	0	0	152	2-Methyl-4-nitroaniline
106	152*	77	79	91	104	78	105	152	4-Methyl-2-nitroaniline
106	177*	0	0	0	0	0	0	177	Bethanidine
106	282*	229	267	78	42	124	81	404	Nicocodeine
107*	36	108	77	38	39	51	79	137	Tyramine
107*	44	77	40	108	39	51	43	182	Hydrophenyllactic Acid(para)
107*	69	125	83	79	55	41	77	276	Cyclandelate
107*	77	51	39	53	78	50	52	152	Hydroxyphenylacetic Acid(para)
107*	79	77	51	152	105	50	78	152	Pemoline Metabolite 2

								MWT	COMPOUND
107*	79	77	91	51	105	108	78	138	1-Phenyl-1,2-dihydroxyethane
107*	79	120	77	75	44	43	51	181	Styramate
107*	92	106	65	79	39	108	77	107	3-Ethylpyridine
107*	92	152	121	65	0	0	0	152	Phenoxyacetic Acid
107*	96	106	97	143	248	140	79	248	Thionazin
107*	106	66	39	65	92	79	108	107	2,3-Dimethylpyridine
107*	106	79	77	92	39	65	108	107	2,5-Dimethylpyridine
107*	106	66	92	65	39	108	79	107	2,6-Dimethylpyridine
107*	106	79	92	77	39	108	65	107	3,4-Dimethylpyridine
107*	106	79	92	77	39	108	80	107	3,5-Dimethylpyridine
107*	106	79	92	39	65	80	77	107	2,4-Dimethylpyridine(2,4-Lutidine)
107*	106	92	65	79	39	51	108	107	4-Ethylpyridine
107*	106	77	79	108	39	65	78	107	Toluidine(m-)
107*	108	77	51	79	39	53	50	108	Cresol(p-)
107*	108	77	138	51	53	79	78	138	Hydroxyphenylethanol (beta-m-Hydroxy)
107*	108	78	79	80	77	51	52	214	Phenyramidol*
107*	120	79	75	77	44	91	45	181	Styramate*
107*	122	121	77	91	79	51	39	122	2,4-Dimethylphenol
107*	145	268	238	121	133	159	224	268	Diethylstilbestrol
107*	149	57	78	108	79	77	72	261	Ethoheptazine
107*	152	77	39	108	51	79	53	152	Hydroxyphenylacetic Acid(meta)
107*	163	223	108	29	164	41	57	223	Bufexamac*
107*	165	123	95	121	0	0	0	327	Azaperone
107*	166	77	39	108	45	65	120	166	3-(4-Hydroxyphenyl)propionic Acid
107*	173	77	175	79	253	251	105	330	Triarimol
107*	176	90	77	70	105	42	79	176	Pemoline*
107	27*'	109	26	45	108	105	95	186	1,2-Dibromoethane
107	44*	77	43	78	51	42	108	151	Hydroxyamphetamine
107	77*	51	89	42	105	90	79	176	Pemoline
107	81*	105	91	95	79	93	67	386	Cholesterol
107	97*	105	189	128	249	248	331	331	Methyl-3-piperidylphenylcyclohexyl Glycolate (N-Methyl)
107	106*	79	78	52	51	65	80	107	2-Ethylpyridine
107	106*	77	108	79	78	53	39	107	Toluidine(p-)
107	108*	79	77	39	90	109	51	108	Cresol(m-)
107	108*	77	79	90	39	51	109	108	Cresol(o-)
107	108*	77	79	109	90	39	53	108	Cresol(p-)
107	108*	91	182	109	77	79	31	182	Mephenesin*
107	109*	73	108	106	123	43	121	186	Ethylene dibromide
107	121*	135	163	177	298	77	136	298	Benzestrol
107	122*	121	77	91	79	39	78	122	2,3-Xylenol
107	122*	121	77	91	123	39	79	122	2,4-Xylenol
107	122*	121	77	91	79	123	39	122	2,5-Xylenol
107	122*	121	77	91	79	78	103	122	2,6-Xylenol
107	122*	121	77	91	123	39	108	122	3,4-Xylenol
107	122*	121	77	79	91	123	39	122	3,5-Xylenol
107	123*	80	78	95	79	93	81	123	4-Nitrosophenol
107	124*	82	83	42	77	79	94	275	Homatropine*
107	135*	41	150	95	136	77	91	150	Butylphenol(p-t-)
107	142*	144	87	43	77	45	108	228	MCPB
107	149*	30	74	73	60	0	0	149	Propylmethyldithiocarbamate (S-n-propyl-)
107	176*	90	89	77	105	79	70	176	Pemoline
108*	42	39	40	81	38	52	41	108	2,5-Dimethylpyrazine
108*	67	42	40	41	26	39	109	108	2,3-Dimethylpyrazine
108*	78	65	39	77	93	79	51	108	Anisole(Methyl Phenyl Ether)

```
              PEAKS                        MWT      COMPOUND


108*  80   81  107   28    0    0    0   108  Phenylenediamine(ortho-)
108*  81  213   54   77  136   51   43   213  Phenazopyridine
108* 107   79   77   39   90  109   51   108  Cresol(m-)
108* 107   77   79   90   39   51  109   108  Cresol(o-)
108* 107   77   79  109   90   39   53   108  Cresol(p-)
108* 107   91  182  109   77   79   31   182  Mephenesin*
108* 109  179  137   43   81   80  110   179  Phenacetin*
108* 118   91  107   43   57  182   75   182  Mephenesin
108* 118   91  107   57   43   44   75   225  Mephenesin Carbamate
108* 123  165   43   52   80   53  122   165  Paracetamol Methyl Ester
108* 199  163  201   92  149  132  170   248  Dimethylcyclophosphoramide Mustard
108* 246  151  229   95  152  153   92   246  Cruformate Metabolite
108* 248  140   65   92  141  109   80   248  Dapsone
108   30* 107   77  137   43   41   55   137  Tyramine
108   30* 107   77  137   51  109   78   137  Tyramine
108   42*  39   40   38   27   37   41   108  2,6-Dimethylpyrazine
108   79* 107   77   51   91   50   78   108  Benzyl Alcohol
108   79* 263  277  279  345  378    0   378  Dieldrin
108  107*  77   51   79   39   53   50   108  Cresol(p-)
108  107*  77  138   51   53   79   78   138  Hydroxyphenylethanol (beta-m-Hydroxy)
108  107*  78   79   80   77   51   52   214  Phenyramidol*
108  109*  80   57   43   55   53   71   109  Nicotinyl Alcohol
108  109*  80   53   51   39   91   27   109  Nicotinyl Alcohol*
108  118* 182   91  225  107   57   75   225  Mephenesin Carbamate*
108  123*  80   53   65  124  109   52   123  Anisidine(o-)
108  135*  54   53   81   43  136   66   135  Adenine
108  173*  65   92   80   39   63  156   173  Sulphanilic Acid
108  202* 174  137  109   80  203  145   217  Ethoxyquin
108  213*  81   36   44  136   54   77   213  Azopyridine
108  243* 123  122  164  136   95   79   243  4'-Methoxy-2-(methylsulphonyl)acetanilide
109*  44   77  125   58  105   51    0   247  Betacaine
109*  76  104   92  239   65  108   43   362  Phthalylsulphacetamide*
109*  79   47  195  197  145   83  220   220  Vapona (DDVP)
109*  80  151  108   43   81   53   52   151  Acetaminophen
109*  80   53   52   39   64   63   81   109  2-Aminophenol
109*  80   81  110   41   39   53   74   109  3-Aminophenol
109*  80   53   81  108   52   54  110   109  4-Aminophenol
109*  80   44   81   53   39   52   54   153  Aminosalicylic Acid*
109*  81  149   81  275  139   99   65   275  Diethyl-p-nitrophenyl Phosphate
109*  81  275  149   99  139  127  247   275  Paraoxone
109*  96   15   79  185   47   45   95   220  Dichlorvos
109*  97  291  139  125  137  155  123   291  Parathion
109* 107   73  108  106  123   43  121   186  Ethylene dibromide
109* 108   80   57   43   55   53   71   109  Nicotinyl Alcohol
109* 108   80   53   51   39   91   27   109  Nicotinyl Alcohol*
109* 125   79  297  128   47   63   93   297  Chlorthion
109* 125  263   79   93   63   47  264   263  Methyl Parathion
109* 137  246  290  110  305  276   81   305  Pirimiphos Methyl
109* 145   79  185  147   35  187   47   378  Naled
109* 145   15   79   96  185   73   47   378  Naled
109* 148  110   81   88   58   57   39   205  2-(2'Propynyloxy)phenyl N-Methylcarbamate
109* 150   78   41  110   81   58   39   207  2-(Allyloxyphenyl) N-Methylcarbamate
109* 151   43   80   81    0    0    0   151  Hydroxyacetanilide(p-)
109* 151   43   79   80   53  108  110   151  Paracetamol
```

			PEAKS					MWT	COMPOUND
109*	151	43	80	108	81	53	52	151	Paracetamol*
109*	185	79	202	145	187	200	199	220	Dichlorvos
109*	185	79	145	187	47	220	110	220	Dichlorvos
109*	219	181	183	111	193	288	0	288	Lindane(beta)
109*	263	125	79	63	93	264	64	263	Parathion-methyl
109	79*	110	139	145	80	112	95	256	Trichlorfon
109	86*	30	151	87	81	99	58	279	Etamphylline*
109	87*	145	58	79	142	112	88	287	Vamidithion
109	93*	199	77	119	162	55	324	324	Oxyphenbutazone
109	97*	291	139	125	137	155	123	291	Parathion
109	108*	179	137	43	81	80	110	179	Phenacetin*
109	110*	79	156	80	47	125	126	230	Demeton Methyl
109	110*	79	95	80	140	47	29	140	Trimethyl Phosphate
109	118*	124	77	81	62	95	75	241	Methocarbamol
109	121*	151	65	93	43	39	271	271	Acetaminosalol (Phenetsal, Salophen)
109	124*	81	52	77	51	65	95	198	Guaiacol Glyceryl Ether
109	124*	198	77	81	43	180	122	198	Guaiphenesin
109	124*	198	31	81	77	95	0	198	Guaiphenesin
109	124*	77	52	223	95	81	122	223	Mephenoxalone
109	124*	118	43	198	122	123	125	241	Methocarbamol
109	127*	125	277	260	79	192	93	277	Fenitrothion
109	169*	125	168	79	110	142	170	262	Demeton S-Methyl Sulphone
109	169*	125	79	59	0	0	0	230	Metasystox
109	194*	55	67	82	42	137	41	194	Caffeine
109	194*	67	82	55	193	195	81	194	Caffeine
109	194*	55	67	82	193	42	40	194	Caffeine
109	194*	55	67	82	195	42	110	194	Caffeine*
109	230*	82	187	243	363	42	123	381	Benperidol*
109	262*	166	264	293	95	123	75	293	Flurazepam Benzophenone 2
109	262*	166	264	275	311	313	123	311	Flurazepam Chloroethyl Artifact Benzophenone
109	262*	166	264	293	123	168	95	293	2-Hydroxyethylamino-5-Chloro-2'-Fluorobenzophenone
109	291*	97	137	29	139	78	292	291	Parathion
109	291*	97	137	139	155	125	123	291	Parathion-ethyl
109	336*	227	249	115	338	0	0	336	Cruformate Metabolite
109	362*	97	226	210	125	364	29	362	Coumaphos
110*	63	64	81	53	55	92	111	110	Catechol
110*	64	63	81	39	92	55	111	110	Catechol
110*	82	39	81	53	69	55	111	110	Resorcinol
110*	109	79	156	80	47	125	126	230	Demeton Methyl
110*	109	79	95	80	140	47	29	140	Trimethyl Phosphate
110*	126	44	77	41	58	42	105	247	Benzamine*
110*	136	44	108	154	81	69	80	154	1,4-Dihydroxybenzoic Acid
110*	152	111	81	58	43	41	52	209	Isopropoxyphenyl N-methylcarbamate (o-Isopropoxy-)
110	91*	79	231	272	298	41	77	298	Norethindrone
110	95*	43	39	96	68	0	0	110	2-Acetylfuran
110	95*	41	55	96	69	139	136	154	Isoborneol
110	156*	58	109	79	80	126	47	213	Omethoate
110	274*	91	79	105	147	215	256	274	19-Nortesterone
111*	36	77	105	96	112	183	42	339	Ethyl-3-piperidyl Benzilate (N-)
111*	41	43	93	55	121	69	95	154	Caran-trans-2-ol(trans-)
111*	75	157	113	50	127	99	159	157	1-Chloro-4-nitrobenzene
111*	75	141	205	171	112	50	113	205	Methyl-p-chlorobenzenesulphonamide (N-Methyl)
111*	96	175	128	157	316	262	331	331	Ethyl-3-piperidylcyclopentyl Glycolate (N-)
111*	96	167	165	152	105	128	139	323	Ethyl-3-piperidyldiphenyl Acetate (N-)

Asterisk Indicates True Base Peak   Peak Inde

							MWT	COMPOUND
111*	96	165	42	167	43	41	56	323 Piperidolate
111*	96	167	112	165	71	43	42	323 Piperidolate*
111*	105	96	77	183	165	128	324	339 Ethyl-3-piperidyl Benzilate (N-)
111*	113	192	41	43	190	194	79	190 Dibromofluoromethane
111*	127	55	83	59	41	112	42	345 Clopamide*
111*	138	81	109	82	93	97	68	294 Cyanthoate
111*	175	75	99	113	177	73	63	302 Chlorfenson
111*	175	75	85	30	276	127	113	276 Chlorpropamide*
111*	175	205	141	113	75	112	56	290 Methylchlorpropamide (N-Methyl)
111*	175	205	75	113	207	84	177	205 Methyl-p-chlorobenzenesulphonamide (N-Methyl)
111	58*	71	42	75	59	141	113	257 Meclofenoxate*
111	69*	41	68	93	123	67	81	154 Lavandulol
111	97*	83	112	55	41	69	56	154 Menthone(o-)
111	98*	55	41	112	99	96	97	321 Dihexyverine
111	98*	99	147	55	41	42	96	391 Flavoxate*
111	98*	99	42	55	41	29	112	335 Norpipanone*
111	98*	99	199	41	288	200	55	399 Pipazethate*
111	105*	77	96	97	183	42	51	433 Pipenzolate Bromide
111	139*	141	75	170	50	113	172	170 4-Chlorobenzoic Acid Methyl Ester
111	139*	141	138	140	75	158	113	357 Indomethacin
111	238*	240	75	113	50	74	127	238 1-Chloro-2-iodobenzene
111	262*	219	97	263	86	42	264	277 Ethylmethylthiambutene*
111	276*	219	277	42	97	100	135	291 Diethylthiambutene
111	310*	112	58	199	212	96	41	310 Pecazine (Mepazine)
112*	41	113	334	56	44	55	69	349 Dipipanone
112*	45	85	161	163	0	0	0	161 Chlormethiazole
112*	69	42	40	41	68	39	70	112 Uracil
112*	77	114	51	50	113	75	76	112 Chlorobenzene
112*	77	105	139	55	41	96	56	261 Hexylcaine
112*	82	55	54	45	84	41	53	112 Maleinehydrazide
112*	85	45	141	113	59	71	69	141 Chlormethiazole Metabolite
112*	105	246	77	55	41	98	97	261 Piperocaine
112*	113	56	57	110	91	55	42	349 Dipipanone
112*	113	41	56	69	334	55	91	349 Dipipanone
112*	114	33	93	95	91	81	79	112 Bromofluoromethane
112*	121	344	55	110	41	179	96	359 Cyclomethycaine
112*	143	113	85	45	0	0	0	143 4-Methyl-5-b-hydroxyethiazole
112*	157	85	45	158	0	0	0	157 4-Methyl-5-thiazoleacetic Acid
112*	161	85	45	163	113	114	59	161 Chlormethiazole*
112*	246	105	77	55	41	44	247	261 Piperocaine*
112*	264	113	91	179	110	178	115	349 Dipipanone*
112*	344	121	41	67	55	54	345	359 Cyclomethycaine*
112	58*	105	96	42	111	77	51	278 Amydricaine
112	70*	293	264	44	42	305	41	374 Bromhexine*
112	98*	99	55	42	41	211	84	322 Fenpipramide*
112	105*	77	41	55	44	42	56	261 Piperocaine
112	129*	55	41	42	44	262	105	344 Oxyphencyclimine
112	165*	71	185	113	164	182	149	406 Bis(2-ethylhexyl)-3-hydroxyphthalate
112	184*	42	55	170	183	212	58	240 1,3-Dimethylbutethal
113*	55	70	42	41	39	85	69	141 Ethosuximide
113*	70	55	42	41	39	85	69	141 Ethosuximide(ii)*
113*	70	373	141	43	72	42	127	373 Prochlorperazine*
113*	70	42	221	56	43	222	114	443 Thiothixene
113*	70	407	43	141	42	127	71	407 Trifluoperazine*

								MWT	COMPOUND
113*	339	44	70	141	43	60	340	339	Perazine*
113	70*	43	42	71	141	56	72	409	Butaperazine
113	70*	42	141	43	56	399	71	399	Thiethylperazine
113	70*	43	42	127	71	44	56	446	Thioproperazine
113	70*	127	446	198	71	445	43	446	Thioproperazine
113	100*	91	101	87	165	115	56	378	Doxapram
113	100*	56	101	87	378	194	91	378	Doxapram*
113	111*	192	41	43	190	194	79	190	Dibromofluoromethane
113	112*	56	57	110	91	55	42	349	Dipipanone
113	112*	41	56	69	334	55	91	349	Dipipanone
113	169*	78	171	115	76	63	51	169	Chlorzoxazone
113	399*	70	141	43	72	400	71	399	Thiethylperazine Maleate (Torecane)
114*	42	72	113	69	81	54	115	114	Methimazole*
114*	44	42	41	142	43	223	56	365	Pericyazine
114*	44	142	365	42	223	115	205	365	Pericyazine*
114*	72	30	115	160	43	174	145	274	5-Methoxydiisopropyltryptamine
114*	72	337	100	43	237	115	85	337	Propanidid
114*	100	56	42	115	101	70	398	398	Pholcodeine*
114*	115	70	57	56	336	91	223	351	Phenadoxone
114	43*	85	30	101	86	44	72	157	Buformin*
114	86*	100	44	238	115	56	72	480	Isopropamide Iodide
114	91*	43	85	139	104	72	243	999	Phenformin Probe Product*
114	99*	98	167	70	165	57	43	281	Diphenylpyraline*
114	100*	353	91	165	70	161	178	353	Dioxaphetyl butyrate
114	100*	42	56	115	70	101	55	398	Pholcodeine
114	112*	33	93	95	91	81	79	112	Bromofluoromethane
114	115*	88	62	116	63	89	51	171	Tetrazolophthalazine
114	170*	156	155	113	197	83	42	538	Morfamquat
114	186*	29	72	42	113	27	109	186	Carbimazole*
115*	91	77	116	117	129	163	0	163	Nitrostyrene
115*	91	105	116	40	39	51	77	163	Nitrostyrene
115*	91	105	116	51	39	77	63	163	1-Phenyl-2-nitropropene
115*	97	73	43	65	121	125	93	285	Prothoate
115*	114	88	62	116	63	89	51	171	Tetrazolophthalazine
115*	117	89	53	109	51	91	39	144	Ethchlorvynol (Placidyl)
115*	175	111	58	177	290	113	75	290	Methyl Enol Ether of Chlorpropamide
115*	285	131	128	162	127	77	152	285	Morphine
115	72*	144	259	116	215	127	254	259	Propranolol
115	91*	128	117	129	219	157	0	275	Dihydroxyphencyclidine
115	95*	196	41	67	43	96	209	252	2,4-Dimethyltalbutal (or 4,6-Dimethyl-)
115	114*	70	57	56	336	91	223	351	Phenadoxone
115	116*	58	63	89	117	39	62	116	Indene
115	143*	142	116	144	89	72	59	143	3-Methylisoquinoline
115	143*	142	144	116	89	77	39	143	4-Methylquinoline
115	144*	116	58	55	63	201	40	201	1-Naphthyl-N-methylcarbamate (Carbaryl, Sevin)
115	144*	116	145	72	89	63	58	144	1-Napthanol
115	153*	129	91	143	0	0	0	158	Phenylcyclohexene
115	170*	114	88	171	62	116	89	170	Triazolophthalazine
115	184*	88	62	185	114	51	50	184	3-Methyl-s-triazolopthalazine (Hydralazine Metab)
115	184*	185	114	183	88	129	155	184	3-Methyltriazolophthalazine
115	185*	141	230	170	45	153	142	230	Naproxen
115	185*	200	171	128	44	91	199	200	Tetrahydrozoline
115	201*	285	173	143	84	202	174	285	Piperine
115	208*	117	91	193	0	0	0	208	Dextropropoxyphene Metabolite

       Asterisk Indicates True Base Peak

115	209*	141	210	152	153	139	208	210	Naphazoline
115	397*	242	398	88	143	271	62	397	Diiodohydroxyquinoline*
116*	31	66	85	118	97	47	81	116	Chlorotrifluoroethylene
116*	88	43	148	60	56	72	117	296	Disulfiram
116*	88	29	44	60	148	56	27	296	Disulfiram*
116*	115	58	63	89	117	39	62	116	Indene
116*	197	196	119	67	129	69	98	197	Halothane
116	58*	198	199	72	59	42	44	358	Dimethoxanate*
116	86*	41	58	73	56	57	228	343	Cinchocaine
116	87*	55	113	57	145	59	143	174	Methyl 3-Hydroxydipropylacetate
116	91*	131	65	149	92	150	90	149	Benzyl Methyl Ketoxime
116	117*	90	89	51	77	118	63	117	Phenylacetonitrile
117*	73	327	75	143	256	118	69	342	1,3-Dimethylhydroxypentobarbitone TMS Ether
117*	90	116	89	63	51	50	39	117	Benzyl Cyanide
117*	90	89	118	116	63	59	39	117	Indole
117*	90	116	89	51	63	39	118	117	Phenylacetonitrile
117*	116	90	89	51	77	118	63	117	Phenylacetonitrile
117*	118	115	91	58	39	103	65	118	Methylstyrene(p-)
117*	118	162	91	77	78	103	105	162	Phenylbutyrolactone (Primidone, Phenobarb, Glutethimide Metab)
117*	119	163	82	161	121	47	165	196	Bromotrichloromethane
117*	119	121	82	47	84	35	49	152	Carbon Tetrachloride
117*	119	35	47	121	82	84	49	152	Carbon Tetrachloride
117*	119	201	166	164	203	199	168	234	Hexachloroethane
117*	119	47	121	82	84	49	35	152	Tetrachloromethane
117*	146	43	91	32	115	39	116	218	Primidone
117*	189	132	115	91	160	77	39	217	Glutethimide (Doriden)
117*	199	105	77	183	427	91	336	427	Cinnamyl-3-piperidylbenzilate (N-Cinnamyl)
117	58*	208	115	193	91	179	130	339	Dextropropoxyphene*
117	73*	75	103	205	147	145	129	596	Chloralose (alpha) TMS Ether
117	91*	118	42	92	65	77	136	179	Phenprobamate
117	115*	89	53	109	51	91	39	144	Ethchlorvynol (Placidyl)
117	118*	203	103	77	78	119	91	203	Methsuximide
117	118*	103	78	77	115	51	91	118	Methylstyrene(alpha-)
117	118*	91	92	119	65	77	103	179	Phenprobamate*
117	118*	43	91	119	105	65	77	178	3-Phenyl-1-propanol Acetate
117	132*	133	92	91	115	118	59	132	1-Methyl-4-isopropenylbenzene
117	145*	122	89	105	90	63	146	145	Hydroxyquinoline*
117	146*	190	118	115	103	91	161	218	Primidone
117	146*	103	115	91	0	0	0	216	Primidone Metabolite
117	161*	177	91	119	118	220	121	220	Ibuprofen Methyl Ester
117	189*	132	160	91	115	77	103	217	Glutethimide
117	201*	167	202	251	165	118	115	368	Cinnarizine*
117	201*	119	203	199	166	94	47	234	Hexachloroethane
117	204*	77	115	51	91	103	118	232	Phenobarbitone
117	204*	146	161	77	103	115	118	232	Phenobarbitone*
117	218*	146	118	219	161	39	51	246	Mephobarbitone (Mebaral, Prominal)
117	218*	118	146	103	77	91	115	246	Methylphenobarbitone*
118*	91	44	92	178	65	0	0	178	Phenacemide
118*	108	182	91	225	107	57	75	225	Mephenesin Carbamate*
118*	109	124	77	81	62	95	75	241	Methocarbamol
118*	117	203	103	77	78	119	91	203	Methsuximide
118*	117	103	78	77	115	51	91	118	Methylstyrene(alpha-)
118*	117	91	92	119	65	77	103	179	Phenprobamate*
118*	117	43	91	119	105	65	77	178	3-Phenyl-1-propanol Acetate

```
                    PEAKS                        MWT     COMPOUND


118* 124  109   44    57    77    81    45   241  Methocarbamol
118* 128   75  130    44    57    62    61   245  Chlorphenesin Carbamate
118* 133   93   88    91    42    27     0   152  Carveol(cis-)
118* 203   56   77   218   141   132   204   218  5-Phenylhydantoin(5-methyl-) Dimethyl Derivative
118* 243  272  104    77   129   128    89   272  Alphenal Dimethyl Derivative
118* 243  104   77   129   231   130   128   272  1,3-Dimethylalphenal
118* 280  203  194    77   165   223   222   280  Phenytoin Dimethyl Derivative
118  108*  91  107    43    57   182    75   182  Mephenesin
118  108*  91  107    57    43    44    75   225  Mephenesin Carbamate
118  117* 115   91    58    39   103    65   118  Methylstyrene(p-)
118  117* 162   91    77    78   103   105   162  Phenylbutyrolactone (Primidone, Phenobarb, Glutethimide Metab)
118  124* 198  109    43    57   125    77   241  Methocarbamol*
118  128* 130   75    57    43    44    61   245  Chlorphenesin Carbamate
118  140*  91   29    46    69    65    39   231  Amphetamine TFA Derivative
118  140*  91   69    45     0     0     0   231  Amphetamine Trifluoroacetate
118  153* 138  102   130     0     0     0   221  Ketamine Metabolite
118  154* 110   69    91    45     0     0   245  Methylamphetamine TFA
118  218* 146  117   103    77   219    91   246  Mephobarbitone
118  232* 117  146   175   233   188   260   260  Dimethylphenobarbitone (N,N-)
118  232* 146  117   103    77   175   115   260  Dimethylphenobarbitone (N,N-Dimethyl)
118  232* 117  146   175   103   233    77   260  Permethylated Methylphenobarbitone*
118  232* 117  146   175   103    77    91   260  Permethylated Phenobarbitone*
118  330*  91  319   219    64    92   421   421  Bendrofluazide*
119*  41   93  204    69   105    91    55   204  Cedrene(alpha-)
119*  91  211   65   120   212    92    90   211  2-Methylbenzanilide
119*  91  227   65   120   228    92    63   227  2-Methyl-4'-hydroxybenzanilide
119*  93   41   69   105   161    55    91   204  Cedrene(alpha-)
119* 120  136   94    93   138   121   246   333  Seneciphylline
119* 137   92   65    39    64   120    91   137  2-Aminobenzoic Acid
119* 151  120   92   152    91    93    65   151  Methylaminobenzoate (ortho)
119   85*  69   31    87   135    29    50   154  Chloropentafluoroethane
119   98*  91   99   124    64    41    55   397  Diperodon*
119   98* 160   84    85    91   245   230   245  Mydocalm (Tolperisone)
119  117* 163   82   161   121    47   165   196  Bromotrichloromethane
119  117* 121   82    47    84    35    49   152  Carbon Tetrachloride
119  117*  35   47   121    82    84    49   152  Carbon Tetrachloride
119  117* 201  166   164   203   199   168   234  Hexachloroethane
119  117*  47  121    82    84    49    35   152  Tetrachloromethane
119  165*  30   78    90    92    91    79   182  2,4-Dinitrotoluene
119  165*  30  118    92    91    90    78   182  2,6-Dinitrotoluene
120*  41  100  142   178    42    92    44   306  Butacaine
120*  43  138   92    42   121    39    45   180  Acetylsalicylic Acid
120*  43  138   92   121    39    64    63   180  Acetylsalicylic Acid*
120*  43  121  138    39    65    41    29   208  Isoamyl Salicylate
120*  57  164  165    70    69    84    98   294  2-(2-Ethylhexyl)-4-hydroxyphthalate
120*  65   86   92    56    99    41    44   236  Butethamine
120*  65   92   39   165   137    63    41   165  Ethyl-4-aminobenzoate
120*  65   91   92    39    77   121    89   137  2-Nitrotoluene
120*  77  176   93    43    57   169   211   211  Propachlor (tech)
120*  85   58   99    56    44    57   149   236  Butethamine
120*  86   58   99    56    44    57    36   236  Butethamine
120*  91  103  121    77    65    42    39   211  Benzylphenethylemine(alpha)
120*  91  148  103   194    77    65     0   239  Formyl-1,3-diphenylisopropylamine (N-)
120*  91   44  121    42    65    51    77   121  Methylbenzylamine (N-Methyl)
```

```
        PEAKS                       MWT     COMPOUND


120*  92  165   65  137    0    0    0   165 Benzocaine
120*  92  105  148  150  121  133   65   165 Ethenzamide*
120*  92  152  121   65   64   93   63   152 Methyl Salicylate
120*  92   91   65  121   77   93  137   137 Nitrotoluene (ortho)
120*  92  137   65  121   39   64   53   137 Salicylamide*
120*  92  138   64   39   63  121   65   138 Salicylic acid*
120*  99   30   86   56   44   57  193   236 Butethamine*
120* 106   78   92   51  226   41   39   226 Metyrapone*
120* 121   91   65   77  119  122  118   121 Methyltoluidine(N-Me,p-)
120* 122  281   91  160   65  280  162   281 Phentolamine
120* 137  193   92   65  121  138   41   193 Butylaminobenzoate
120* 137  193   65   92   39   41  121   193 Isobutylaminobenzoate(p-Amino-)*
120* 138   43   92   64   65   63   42   180 Acetylsalicylic Acid
120* 138   43   92  121   39   63   64   180 Acetylsalicylic Acid
120* 149  106  121   77   91  132  150   149 2,4-Dimethylformanilide
120* 151   92  121  152   65   93  122   151 Methylaminobenzoate (para)
120* 152   43  121   92   63   65   64   194 Methyl Acetylsalicylate
120* 165   92   65  137   39  121   93   165 Benzocaine*
120* 263  142  178  100  264   41   29   306 Butacaine*
120   58*  72   70  136   18   42    0   154 Terpineol(alpha-)
120   66*  39   65  269  205  118   77   389 Cyclothiazide*
120   70*  43   91   41   65   77   98   161 Aletamine
120   71*  43   91   41    0    0    0   161 Aletamine
120   72*  55   91   65   39   77   31   163 1-Isopropyl-2-phenylethylamine
120   86*  30   87   58   92   84   56   278 Dimethocaine*
120   86*  65   92   42   43   41   56   236 Procaine
120   91* 121   92   65  106   30  135   240 Benzathine
120   91* 121   92   65  167  119  106   999 Oxethazaine Probe Product 1*
120   92*  39   65  152  121   63   64   152 Methyl Salicylate
120   92* 138   64   39   63   65   53   138 Salicylic Acid
120  100*  70   41   56   42  101  293   293 Pramoxine
120  105*  40   29   44   77   79   51   120 Isopropylbenzene
120  105* 119   77   91   39  106   79   120 Mesitylene(1,3,5-Trimethylbenzene)
120  107*  79   75   77   44   91   45   181 Styramate*
120  119* 136   94   93  138  121  246   333 Seneciphylline
120  121* 106   77   91  122   93  118   121 2,4-Dimethylaniline
120  121*  92   65  195   39   93  149   195 Hydroxyhippuric Acid(ortho)
120  121* 220  148    0    0    0    0   220 Hydroxylated Ethotoin
120  121*  69   92  195   39   93   45   195 Salicyluric Acid
120  121*  79  106  122   77  107   39   121 2,4,6-Trimethylpyridine
120  121* 354  122   43  234  106   64   354 Xipamide*
120  130*  91  119  129  115  147  148   148 1-Hydroxy-1,2,3,4,-tetrahydronaphthalene
120  136* 220  119  121  138  335  246   335 Senecionine
120  137*  92   65   39  138  121   63   137 4-Aminobenzoic Acid*
120  142* 143   86  142   92   72   56   292 Leucinocaine*
120  164*  92   83   70  112  221  165   294 1-(2-Ethylhexyl)-3-hydroxyphthalate
120  195*  43  210  135   75   73   92   252 Acetylsalicylic Acid TMS Ester
120  200* 185  121   40   41   69   44   270 Permethylated Thiopentone*
120  222*  92  194   64   63  221   97   222 Flavone
120  265* 249   43  103  121  162  131   308 Warfarin
121*  43  122  223  147   91   41  135   392 Betamethasone
121*  58   72   71  214  122  215   78   285 Mepyramine
121*  58   72   79   71   78   42  122   285 Pyrilamine
121*  58   72   71   78  215  122   77   286 Thonzylamine
```

	PEAKS							MWT	COMPOUND
121*	65	39	93	63	92	64	122	214	Phenylsalicylate
121*	92	120	65	162	63	93	64	336	Dicoumarol
121*	93	136	119	106	91	79	77	136	Terpinene(alpha-)
121*	107	135	163	177	298	77	136	298	Benzestrol
121*	109	151	65	93	43	39	271	271	Acetaminosalol (Phenetsal, Salophen)
121*	120	106	77	91	122	93	118	121	2,4-Dimethylaniline
121*	120	92	65	195	39	93	149	195	Hydroxvhippuric Acid(ortho)
121*	120	220	148	0	0	0	0	220	Hydroxylated Ethotoin
121*	120	69	92	195	39	93	45	195	Salicyluric Acid
121*	120	79	106	122	77	107	39	121	2,4,6-Trimethylpyridine
121*	120	354	122	43	234	106	64	354	Xipamide*
121*	122	315	43	147	223	135	41	392	Dexamethason
121*	122	123	65	39	93	63	66	122	4-Hydroxybenzaldehyde
121*	122	39	56	27	42	94	54	122	2-Methyl-5-ethylpyrazine
121*	122	91	147	225	43	135	120	360	Prednisolone
121*	136	91	77	58	39	41	193	193	3-Isopropylphenyl-N-methylcarbamate (UC-10854, H 8757)
121*	136	105	93	79	91	122	107	136	Ocimene(allo-)
121*	138	93	65	39	63	127	53	138	4-Hydroxybenzoic Acid
121*	138	65	93	39	122	41	63	180	Propylparaben
121*	150	65	93	168	0	0	0	168	2-Hydroxymethyl-4-hydroxybenzoic Acid
121*	150	93	65	168	0	0	0	168	2-Hydroxymethyl-5-hydroxybenzoic Acid
121*	150	65	93	0	0	0	0	150	2-Hydroxymethyl-4-hydroxybenzolactone
121*	150	93	65	0	0	0	0	150	2-Hydroxymethyl-5-hydroxybenzolactone
121*	152	93	65	39	122	63	153	336	Fentanyl
121*	152	40	93	65	0	0	0	152	Methyl Hydroxybenzoate
121*	152	93	65	39	0	0	0	152	Methylparaben
121*	163	151	109	43	122	108	65	313	Benorylate*
121*	180	91	181	122	148	107	93	180	Methyl Methoxyphenyl-(ortho)-acetate
121*	180	122	181	0	0	0	0	180	Methyl Methoxyphenyl-(para)-acetate
121*	189	292	310	263	265	131	251	310	Warfarin Alcohol
121*	194	134	163	122	195	135	108	194	Methyl Methoxyphenyl-(para)-propionate
121*	232	139	273	234	332	141	275	332	Procetofene Metabolite Methyl Ester
121*	252	120	132	106	122	105	383	383	BTS 29101
121*	252	105	77	126	79	237	251	252	2,4-Dimethylphenylformamidine (N,N'-bis-2,4-Dimethyl-)
121*	318	317	173	120	362	44	31	408	Ethyl Biscoumacetate*
121	41*	123	39	202	27	204	200	200	1,3-Dibromopropane
121	41*	123	39	27	42	40	38	200	1,2-Dibromopropane
121	43*	77	105	51	103	78	48	166	2-Phenyllactic Acid
121	45*	186	97	65	93	214	153	376	Phenkapton
121	58*	134	127	148	238	270	284	369	Methoxypropoxyphene
121	58*	134	59	57	127	105	77	369	Methoxypropoxyphene
121	58*	72	78	42	77	122	91	286	Thonzylamine
121	58*	216	72	215	71	122	59	286	Thonzylamine
121	58*	72	71	216	215	122	78	286	Thonzylamine*
121	59*	93	136	43	81	41	55	154	Terpineol
121	75*	97	260	47	65	93	69	260	Phorate
121	86*	99	30	58	65	42	56	265	Paraethoxycaine
121	91*	65	122	309	64	230	123	431	Benzthiazide*
121	93*	79	67	107	95	94	68	136	Camphene
121	93*	136	81	43	55	41	107	154	Caran-trans-3-ol(cis-)
121	93*	136	111	43	41	55	79	154	Caran-trans-2-ol((-)cis-)
121	93*	136	95	43	81	110	107	154	Neoisothujyl Alcohol
121	93*	136	39	41	79	91	27	136	Terpinolene
121	106*	77	107	79	53	120	78	121	2-Ethylaniline

Asterisk Indicates True Base Peak

```
                    PEAKS                        MWT    COMPOUND


121  106*  77  107  120  104   51   79   121  Ethylaniline(N-)
121  112* 344   55  110   41  179   96   359  Cyclomethycaine
121  120*  91   65   77  119  122  118   121  Methyltoluidine(N-Me,p-)
121  122*  91  312  342    0    0    0   392  Betamethasone
121  122*  91  123  147  108   77  286   286  Boldenone
121  122*  56   58  194   55   45   73   251  C 10015
121  122*  91  147  161   43  107  120   300  Ethisterone
121  137* 120   94   77   80  107   92   137  2-Methyl-4-nitrosophenol
121  150*  56   93   39  151   66   94   445  Glipizide*
121  156* 158   93   58   65  141   51   213  Banol
121  156* 158   91   77   65  157  122   156  4-Chloro-3,5-xylenol
121  157* 342   97  153   45  159  125   342  Trithion
121  162* 132  147  293  344  120  106   293  Amitraz
121  180* 181  122   59   91  148  182   180  Methyl Methoxyphenyl-(meta)-acetate
121  182* 184  154  367   97  241   58   367  Phosalone
121  182*  97  184  111  154   65  138   367  Phosalone
121  191*  77  104  122   43   51  192   191  Amiphenazole*
121  256* 185   43  257   65   42  214   285  Probenecid*
121  310*  92   43  120  311   63   65   353  Acenocoumarin
121  310* 353  311   43  120   92  296   353  Nicoumalone*
121  336* 120  215  162   92  337  187   336  Dicoumarol*
122*  77  107  105   91   79  123   30   221  Metaxalone
122*  78  106   51   50   52   44  123   122  Nicotinamide*
122* 104   78  106   51    0    0    0   122  Nicotinamide
122* 107  121   77   91   79   39   78   122  2,3-Xylenol
122* 107  121   77   91  123   39   79   122  2,4-Xylenol
122* 107  121   77   91   79  123   39   122  2,5-Xylenol
122* 107  121   77   91   79   78  103   122  2,6-Xylenol
122* 107  121   77   91  123   39  108   122  3,4-Xylenol
122* 107  121   77   79   91  123   39   122  3,5-Xylenol
122* 121   91  312  342    0    0    0   392  Betamethasone
122* 121   91  123  147  108   77  286   286  Boldenone
122* 121   56   58  194   55   45   73   251  C 10015
122* 121   91  147  161   43  107  120   300  Ethisterone
122* 161   94  160   57  120  123  162   218  2-(Methyl-2'-propynylamino)phenyl N-Methylcarbamate
122   42*  39   27   81   54   53   52   122  2,3,5-Trimethylpyrazine
122   44* 121   78   77   42   51   52   165  Methoxyamphetamine
122   44* 121   42   78   77   91  107   165  Methoxyamphetamine (para)
122  105*  77   51   50   39   74   78   122  Benzoic Acid
122  107* 121   77   91   79   51   39   122  2,4-Dimethylphenol
122  120* 281   91  160   65  280  162   281  Phentolamine
122  121* 315   43  147  223  135   41   392  Dexamethason
122  121* 123   65   39   93   63   66   122  4-Hydroxybenzaldehyde
122  121*  39   56   27   42   94   54   122  2-Methyl-5-ethylpyrazine
122  121*  91  147  225   43  135  120   360  Prednisolone
122  123*  94  108   81   29   60   43   123  Methylaminophenol (para)
123*  31   58  106   79   43   78   51   179  Iproniazid*
123*  42  222  207  179  149  150  166   222  1,3-Diethyl-9-methylxanthine
123*  43   44   51   78   77   50  105   167  Isocinchomeronic Acid
123*  51   78  106   50   52  105   39   123  Isonicotinic Acid
123*  51  105   78   50   77   44   52   167  Quinolinic Acid
123*  57   42  103   44   85  124   51   226  Carbidopa*
123*  77  167   92   44   91  224   42   332  Dichlofluanid
123*  77   51  124   39   74   53   78   197  Levodopa
```

```
           PEAKS                      MWT    COMPOUND


123*  83   55   41   151   81   69   39   230  Bromocamphor
123*  95  164   69     0    0    0    0   164  Droperidol Hofmann Reaction Product
123*  95  224   42   237  206   56   84   375  Haloperidol
123*  95  164   69     0    0    0    0   164  Haloperidol Hofmann Reaction Product 1
123* 105   77   96   183  110  165  351   351  Allyl-3-piperidylbenzilate (N-Allyl)
123* 105   78   51   106   77  124   50   123  Nicotinic Acid*
123* 106   43   58    78   79   51  164   179  Iproniazid
123* 106   78   51   105   50   77   52   600  Nicofuranose*
123* 107   80   78    95   79   93   81   123  4-Nitrosophenol
123* 108   80   53    65  124  109   52   123  Anisidine(o-)
123* 122   94  108    81   29   60   43   123  Methylaminophenol (para)
123* 124   77   74   197  152  105  179   197  Levodopa
123* 124   77   44    51   74   53   39   197  Levodopa*
123* 139  221  122    43  140   81  124   221  Cyclofuramid
123* 141   43  151    41   42   57  231   401  Pipamazine
123* 171  143   81   128   91  172   41   338  Bioresmethrin
123* 215   43   81    53  124  216   94   215  Furcarbanil
123* 312   59   58    70  115  171  270   312  6-Deoxy-6-azidodihydroisomorphine
123* 326   59  298    70  198  185  115   326  6-Deoxy-6-azidodihydroisocodeine
123* 340   59  283   115  322  199   70   340  3-Ethoxy-6-deoxy-6-azidodihydroisomorphine
123   56* 217   83    57   77  215  216   217  Methylaminophenazone
123   56* 215  217    83  119   77   57   217  Noraminophenazone
123   77*  51   65    78   93   50   30   123  Nitrobenzene
123   88*  44   42   124   41  122   77   211  Methyldopa
123  108* 165   43    52   80   53  122   165  Paracetamol Methyl Ester
123  124*  78   77    51   39  107  110   124  4-Methylcatechol
123  124*  39   95    51   67   55   69   124  Orcinol
123  141*  42   41   151   55  169   96   401  Pipamazine
123  165* 246  199    95   42  108  214   379  Droperidol
123  245* 258  165   186   95  246    0   396  Aceperone
123  279* 412  165   206  281  292   95   430  Amiperone
124*  30  123  153    77   51  125   78   153  4-(2-Aminoethyl)pyrocatechol
124*  42   82   94   289   67   83   96   289  Atropine
124*  42   82   83    94   67   96   77   275  Homatropine
124*  43  314   79    91  229  272  105   314  Progesterone
124*  55  139   99   142  141  125  154   346  5-(2-Ethoxythiocarbonylthioethyl)-5-isopentylbarbituric Acid
124*  82   83   57    41   94   42   43   267  Anisotropine
124*  82   94   83    42   96  103   67   289  Atropine*
124*  82   42   83    94   67   96   39   289  Atropine (dl-Hyoscyamine)
124*  82  168   77   105   42   94   83   289  Benzoylecgonine*
124*  82   83  275    94   96   67   80   275  Homatropine
124*  82   94   77    42   44  105  106   275  Homatropine
124*  82   83  245    94   77  105   67   245  Tropocaine
124*  83   82   94   289   42   96  125   289  Atropine
124*  83   82   73    94   96  125   42   361  Atropine TMS Ether
124*  94   82   77    79  123   96   67   275  Homatropine
124*  98  105   84   216   96  259  200   259  3-Phenyl-3-piperidinocyclohexanol
124*  99  142  141   154   98  129  125   228  5-Butyl-5-(2-hydroxyethyl)barbituric Acid
124* 107   82   83    42   77   79   94   275  Homatropine*
124* 109   81   52    77   51   65   95   198  Guaiacol Glyceryl Ether
124* 109  198   77    81   43  180  122   198  Guaiphenesin
124* 109  198   31    81   77   95    0   198  Guaiphenesin
124* 109   77   52   223   95   81  122   223  Mephenoxalone
124* 109  118   43   198  122  123  125   241  Methocarbamol
```

Asterisk Indicates True Base Peak

124*	118	198	109	43	57	125	77	241 Methocarbamol*
124*	123	78	77	51	39	107	110	124 4-Methylcatechol
124*	123	39	95	51	67	55	69	124 Orcinol
124*	141	154	167	155	184	142	98	291 5-(2-Bromoethyl)-5-butylbarbituric Acid
124*	141	167	154	125	112	98	190	246 5-Butyl-5-(2-chloroethyl)-barbituric Acid
124*	141	156	139	98	99	101	114	200 5-Ethyl-5-(2-hydroxyethyl)barbituric Acid
124*	141	154	153	170	98	125	155	255 5-Propyl-5-(2-thiocyanatoethyl)-barbituric Acid
124*	142	154	141	99	172	155	98	242 5-(2-Hydroxyethyl)-5-isopentylbarbituric Acid
124*	154	155	141	142	181	125	198	305 5-(2-Bromoethyl)-5-isopentylbarbituric Acid
124*	154	141	181	198	142	153	155	269 5-Butyl-5-(2-thiocyanatoethyl)-barbituric Acid
124*	154	181	125	141	153	142	112	283 5-Pentyl-5-(2-thiocyanatoethyl)-barbituric Acid
124*	155	154	141	142	181	153	98	305 5-(2-Bromoethyl)-5-pentylbarbituric Acid
124*	198	109	77	81	110	125	167	198 Glycerol guiacolate
124*	223	109	77	122	123	52	95	223 Mephenoxalone (Trepidone)
124*	225	154	155	128	127	125	141	352 5-(2-Iodoethyl)-5-isopentylbarbituric Acid
124*	276	58	140	56	72	125	41	291 Eucatropine*
124	58*	71	199	125	167	154	141	255 5-Butyl-5-(2-dimethylaminoethyl)-barbituric Acid
124	81*	54	53	125	171	45	42	171 Metronidazole*
124	82*	105	77	168	122	42	83	289 Benzoylecgonine
124	118*	109	44	57	77	81	45	241 Methocarbamol
124	123*	77	74	197	152	105	179	197 Levodopa
124	123*	77	44	51	74	53	39	197 Levodopa*
124	138*	141	169	168	247	249	98	318 5-(3-Bromopropyl)-5-isopropylbarbituric Acid
124	159*	260	161	262	126	163	88	260 Methazole
124	180*	152	91	44	40	42	105	271 Phosphorylated Amphetamine
124	206*	176	208	160	178	162	126	206 2,6-Dichloro-4-nitroaniline (DCNA)
124	221*	64	206	150	151	163	178	221 3,4,5-Trimethoxybenzoyl Cyanide
124	302*	43	91	79	121	105	122	302 Methyltestosterone
125*	44	70	91	41	40	244	153	581 Ergotamine*
125*	70	91	153	41	244	43	71	611 Dihydroergocristine*
125*	70	91	153	41	43	244	349	611 Dihydroergocristine Mesilate
125*	70	91	153	244	40	43	314	581 Ergotamine Tartrate
125*	93	127	173	158	99	55	79	330 Malathion
125*	189	230	127	191	63	232	90	230 Diazoxide*
125	54*	168	69	97	53	43	0	168 Uric Acid
125	58*	42	168	60	41	89	127	183 Chlorphentermine
125	70*	91	153	43	41	44	244	583 Dihydroergotamine*
125	70*	71	91	153	267	154	221	609 Ergocristine*
125	97*	65	153	93	45	121	73	456 Dioxathion
125	98*	42	41	29	99	120	190	274 Phenampromide
125	98*	99	120	41	42	57	77	274 Phenampromide
125	109*	79	297	128	47	63	93	297 Chlorthion
125	109*	263	79	93	63	47	264	263 Methyl Parathion
125	157*	159	314	93	45	316	171	314 Methyl Trithion
125	173*	127	93	158	99	143	0	330 Malathion
125	190*	134	162	91	41	39	135	278 Absisic Acid
125	261*	109	108	97	80	233	205	261 Aminoparathion
125	263*	109	79	93	47	63	264	263 Parathion-methyl
125	278*	109	153	168	169	93	79	278 Fenthion
125	285*	287	109	93	79	47	289	320 Fenchlorvos
125	287*	285	79	109	93	289	47	320 Fenchlorvos
126*	44	55	58	139	42	43	96	198 Guanethidine
126*	52	80	51	108	39	79	53	126 Pyrogallol
126*	69	85	52	43	97	80	51	126 Phloroglucinol

	PEAKS							MWT	COMPOUND
126*	69	85	52	43	63	42	41	126	Phloroglucinol
126*	80	52	108	51	63	79	39	126	Pyrogallol
126*	91	127	174	55	41	42	97	217	Prolintane
126*	127	174	91	70	69	55	42	217	Prolintane*
126*	127	42	91	55	41	65	119	245	Pyrovalerone
126*	139	44	58	43	55	42	41	198	Guanethidine
126*	141	43	45	71	98	69	72	141	Chlormethiazole Metab (5-Acetyl-4-methylthiazole)
126	55*	54	127	83	52	82	56	126	Thymine
126	86*	259	58	112	99	400	74	399	Quinacrine
126	91*	128	125	65	127	92	89	126	3-Chlorotoluene
126	91*	128	89	125	90	65	63	126	2-Chlorotoluene
126	91*	125	128	65	127	92	89	126	4-Chlorotoluene
126	98*	370	99	70	258	125	42	370	Thioridazine
126	110*	44	77	41	58	42	105	247	Benzamine*
126	151*	166	31	51	43	58	223	223	Bendiocarb
126	168*	405	43	368	326	446	0	560	Sotalol, tri TFA Derivative
126	440*	69	427	0	0	0	0	553	Noradrenaline TFA Derivative
127*	15	43	67	109	79	192	39	224	Phosdrin
127*	43	229	44	161	231	85	186	287	Chlorproguanil*
127*	55	70	41	42	128	69	112	155	Methylethosuximide (N-)
127*	55	70	41	42	0	0	0	155	Methylsuximide(N-)
127*	67	72	193	237	44	109	111	237	Dicrotophos
127*	72	264	138	109	67	193	42	299	Phosphamidon
127*	105	193	104	121	43	179	166	314	Crotoxyphos
127*	109	125	277	260	79	192	93	277	Fenitrothion
127*	129	65	92	128	100	99	91	127	4-Chloroaniline
127*	129	65	92	64	39	91	63	127	2-Chloroaniline
127*	161	163	153	129	187	90	189	314	Triclocarban*
127*	173	115	126	77	101	128	143	173	1-Nitronaphthalene
127*	192	109	67	43	193	39	79	224	Mevinphos
127*	192	109	67	43	193	79	39	224	Mevinphos GC peak 2
127*	192	109	67	43	164	193	224	224	Phosdrin
127*	296	169	281	183	154	141	128	296	MK-251 Metabolite
127	43*	213	171	129	41	154	153	213	Chlorpropham
127	57*	41	43	55	88	243	42	371	Ioxynil
127	91*	106	110	43	92	65	120	231	Isocarboxazid*
127	111*	55	83	59	41	112	42	345	Clopamide*
127	126*	174	91	70	69	55	42	217	Prolintane*
127	126*	42	91	55	41	65	119	245	Pyrovalerone
127	129*	131	48	47	91	93	81	206	Chlorodibromoethane
127	129*	131	48	47	93	91	79	206	Dibromochloromethane
127	167*	168	294	295	285	0	0	294	MK-251 Metabolite
127	198*	183	95	123	181	196	199	199	3,4,5-Trimethoxybenzyl Alcohol (Mescaline Precursor)
127	295*	168	153	167	296	276	311	341	MK-251 Metabolite
127	297*	185	170	298	312	311	293	312	MK-251 Metabolite
127	311*	169	312	295	153	199	184	326	MK-251 Metabolite
128*	45	100	73	43	71	97	125	143	Chlormethiazole Metabolite (5-(1-hydroxyethyl)-4-methylthiazole)
128*	45	73	100	177	0	0	0	177	2-Hydroxy-4-methyl-5-b-chloroethiazole
128*	64	130	63	65	92	39	129	128	2-Chlorophenol
128*	69	143	73	169	75	159	41	324	Clofibrate Derivative
128*	100	45	73	43	143	0	0	143	2-Hydroxy-4-methyl-5-ethylthiazole
128*	101	50	75	129	76	64	51	128	Isophthalonitrile
128*	101	50	75	129	76	51	64	128	Phthalodinitrile(o-)
128*	118	130	75	57	43	44	61	245	Chlorphenesin Carbamate

Asterisk Indicates True Base Peak   Peak Ind

```
        PEAKS                    MWT     COMPOUND


128* 129  115   41  202   43   70   55   297 Tridemorph
128* 130   65   64   39   26   63  129   128 3-Chlorophenol
128* 130   65   39   64   63  129   99   128 4-Chlorophenol
128* 130  202   31   29   65   43  111   202 Chlorphenesin*
128* 130  202   43  129  111   75  204   245 Chlorphenesin Carbamate*
128* 130  169   87   41  129  242  171   242 Clofibrate*
128* 130  169   41  129  228   75   69   228 Clofibrate Derivative
128* 141  268  130  270   77  143  233   268 Dichlorophen*
128* 217  216  173  103  146  119  130   217 5-(3'-Pyridylmethylene)barbituric Acid
128   42*  85   43   44   41   69   70   128 Barbituric Acid
128   58* 127   59  193  179  178  115   309 Isomethadone
128   84*  42   43   77  139   82  179   343 Clemastine
128   84* 179   42   85  178  214   98   343 Clemastine*
128   92*  65   39   66  122   80   93   138 3-Nitroaniline
128  100* 265   55   56   41   42   40   392 Dextromoramide
128  100* 265  266  129  246  236  306   392 Moramide
128  100*  70   41   42  293   56  101   293 Pramoxine*
128  118*  75  130   44   57   62   61   245 Chlorphenesin Carbamate
128  142* 121  120  171   83   87  143   202 Methyl 2-Propylglutarate
128  143* 115  144  142  101   51   77   143 2-Methylquinoline
128  174* 101  116   77  102   75   51   174 6-Nitroquinoline
128  174* 116  101   77   89  102   75   174 8-Nitroquinoline
128  206* 107  119   92  135  178  106   206 5-Furfurylidenebarbituric Acid
128  217* 118  103  146  119  145  174   217 5-(4'-Pyridylmethylene)barbituric Acid
128  242* 127  155  171  143  154  156   242 5-Cinnamylidenebarbituric Acid
129*  57   70   71   55  112   43   41   370 Dioctyl Adipate*
129* 102  128   51  130   76   75  103   129 Isoquinoline
129* 102  128   51  130   76   50  103   129 Quinoline
129* 112   55   41   42   44  262  105   344 Oxyphencyclimine
129* 127  131   48   47   91   93   81   206 Chlorodibromoethane
129* 127  131   48   47   93   91   79   206 Dibromochloromethane
129* 131  191   81   79   50   31  189   208 Difluorodibromomethane
129* 157  128  130  187  158  156  188   187 2-Carbomethoxyquinoline
129* 158  130  143  115  128   91   77   158 1-Phenylcyclohexene
129   43*  69   41   86   97   55   44   278 Acetylcarbromal*
129   43*  57   56   41   72   39   58   157 Paramethadione
129   73* 305  128  304  276  103  232   305 Norpethidinic Acid Acid Ethyl Ester N-TMS Derivative
129   73* 128  114  334  349  115  130   349 Norpethidinic Acid TMS Ester N-TMS Derivative
129  105* 112   77   42  313   41   55   344 Oxyphencyclimine*
129  127*  65   92  128  100   99   91   127 4-Chloroaniline
129  127*  65   92   64   39   91   63   127 2-Chloroaniline
129  128* 115   41  202   43   70   55   297 Tridemorph
129  207*  72   91  208  105  174  128   325 Methadone-N-Oxide
130*  44  205   77  129  131  103  102   205 3-Indolelactic Acid
130*  45  131  159   77   62   43  103   175 3-(3-Indolyl)acetic Acid
130*  76  103   50  102   75  129   52   130 Phthalazine
130*  77   51  132  115  103  133   78   133 Tranylcypromine
130*  95  132   97  134   60   99   62   130 Trichloroethylene
130* 120   91  119  129  115  147  148   148 1-Hydroxy-1,2,3,4,-tetrahydronaphthalene
130* 131  205  129   77   43   41   45   205 3-Indolelactic Acid
130* 131   36   77  108   65  160   38   160 Tryptamine
130* 132   79   83   85  134  109   81   374 S 421
130* 148  243  299  209  102  194   76   299 Dowco 199
130* 159   77   64  131  103   48   51   159 3-Indoleacetaldehyde
```

130*	161	143	131	103	77	115	102	161	Tryptophol
130*	174	77	131	103	102	129	51	174	3-Indoleacetamide
130*	175	77	131	103	102	129	176	175	Indolyl-3-acetic Acid
130*	189	131	79	190	52	103	51	189	Methyl 3-Indoleacetate
130*	217	143	131	186	218	144	117	217	Methyl-3-indolebutyrate
130	45*	44	71	42	61	115	55	158	Ethylbutyrylurea(Carbromal Metabolite)
130	49*	128	51	93	132	95	47	128	Bromochloromethane
130	49*	128	51	93	81	79	95	128	Methylenechlorobromide
130	50*	76	74	75	63	51	38	160	Hydrallazine
130	95*	132	60	97	35	134	47	130	Trichloroethylene
130	128*	65	64	39	26	63	129	128	3-Chlorophenol
130	128*	65	39	64	63	129	99	128	4-Chlorophenol
130	128*	202	31	29	65	43	111	202	Chlorphenesin*
130	128*	202	43	129	111	75	204	245	Chlorphenesin Carbamate*
130	128*	169	87	41	129	242	171	242	Clofibrate*
130	128*	169	41	129	228	75	69	228	Clofibrate Derivative
130	186*	77	53	120	93	221	51	221	Prynachlor
130	334*	120	333	143	107	162	144	334	Strychnine
131*	97	58	159	125	160	329	65	329	Mecarbam
131*	162	79	103	161	52	132	163	162	Methyl Cinnamate
131*	327	73	143	75	132	328	169	342	1,3-Dimethylhydroxyamobarbitone TMS Ether
131	58*	72	71	79	42	30	78	295	Chloropyrilene (Chlorothen, Tagathen)
131	58*	71	72	42	79	133	78	295	Chlorothen
131	58*	130	30	77	132	103	59	188	Ethyltryptamine (N-)
131	69*	100	502	119	414	464	614	614	Heptacosafluorobutylamine
131	100*	311	218	0	0	0	0	311	Fomocaine
131	103*	43	77	51	145	146	50	308	Warfarin
131	129*	191	81	79	50	31	189	208	Difluorodibromomethane
131	130*	205	129	77	43	41	45	205	3-Indolelactic Acid
131	130*	36	77	108	65	160	38	160	Tryptamine
131	147*	130	132	119	148	365	218	365	Indapamide*
131	166*	102	138	77	91	195	0	223	Ketamine Metabolite
131	219*	100	218	0	0	0	0	327	Fomocaine Metabolite
131	255*	254	125	257	256	57	58	325	Clemizole
131	255*	60	73	255	43	125	258	325	Clemizole
131	265*	43	103	121	120	145	146	308	Warfarin
132*	77	160	104	105	76	97	129	345	Azinphos Ethyl
132*	100	91	77	65	89	79	0	250	Amphetaminil
132*	104	78	130	77	103	43	48	175	Debrisoquine
132*	104	44	175	130	117	103	43	175	Debrisoquine*
132*	104	246	103	78	77	51	133	246	Permethylated Phenylmethylbarbituric Acid*
132*	117	133	92	91	115	118	59	132	1-Methyl-4-isopropenylbenzene
132*	133	280	262	235	104	77	89	280	7-Hydroxydiftalone
132*	147	262	133	104	77	148	190	322	7-Ureidodiftalone
132*	264	90	133	89	105	118	265	264	Diftalone
132*	280	133	104	235	262	77	89	280	Oxo-2-phthalazinylmethyl Benzoic Acid (2-(1-(2H))) (Diftalone Metab)
132	91*	42	105	92	43	41	77	133	Methylphenylaziridine
132	104*	218	51	103	77	78	52	218	Phenylmethylbarbituric Acid*
132	130*	79	83	85	134	109	81	374	S 421
132	133*	56	115	30	117	91	77	133	Tranylcypromine*
132	160*	77	93	125	104	147	105	317	Azinphos Methyl
132	160*	77	104	105	76	159	109	301	Dimethyl-s-(4-oxo-1,2,3-benzotriazin-3(4h)-ylmethyl) Phosphorothiolate
132	160*	77	93	105	76	104	51	317	Guthion
132	162*	121	120	106	118	147	77	162	2,4-Dimethylphenyl-N'-methylformamidine (N-2,4-Dimethyl-)

								MWT	COMPOUND
132	189*	117	160	91	115	103	77	217	Glutethimide*
132	203*	117	174	115	91	231	103	231	Glutethimide Methyl Derivative
133*	104	76	105	197	0	0	0	197	Methylsaccharin
133*	132	56	115	30	117	91	77	133	Tranylcypromine*
133*	146	89	278	105	77	132	118	296	7,14-Dihydroxydiftalone
133*	147	280	105	77	132	134	90	280	7-Hydroxydiftalone
133	53*	135	54	39	27	89	51	212	1,4-Dibromo-2-butene
133	72*	30	116	248	134	56	41	248	Pindolol*
133	84*	42	162	161	39	51	41	162	Nicotine
133	84*	76	42	162	161	58	119	162	Nicotine
133	84*	42	162	161	105	77	119	162	Nicotine*
133	91*	176	174	44	107	92	177	299	Buphenine*
133	100*	77	52	148	50	76	0	148	Ethyl Phenylmethyl Ketone (ortho)
133	118*	93	88	91	42	27	0	152	Carveol(cis-)
133	132*	280	262	235	104	77	89	280	7-Hydroxydiftalone
133	204*	176	233	148	131	188	119	233	2-Ethyl-2-(4-hydroxyphenyl)glutarimide (Glutethimide Metab)
133	206*	150	370	160	134	178	151	370	Thiophanate
133	263*	132	308	264	105	77	279	308	7-Ethoxydiftalone
134*	79	77	106	105	103	151	78	151	2-Nitroxylene(m-)
134*	91	42	119	65	135	58	86	225	Benzyl-N-methylphenethylamine (alpha-)
134	72*	91	42	57	31	65	44	163	1-Ethyl-N-methyl-2-phenylethylamine
134	78*	106	51	77	39	40	107	152	Hydroxyphenylacetic Acid(ortho)
134	91*	92	43	65	77	89	51	134	Benzyl Methyl Ketone
134	98*	70	111	198	242	383	382	383	Sulpyrid Trimethyl Derivative
134	105*	77	106	51	135	235	160	235	Butyl Hippurate (n-)
134	150*	164	123	136	79	122	285	285	4'-Methoxy-N-isopropyl-2-(methylsulphonyl)acetanilide
134	170*	199	243	172	198	201	200	243	Benazolin
135*	57	91	77	43	119	105	180	209	Hydroxyphenamate
135*	66	77	107	92	136	63	64	166	Methylparaben Dimethyl Ether
135*	77	90	92	51	64	136	63	223	Salicyluric Acid Methyl Ester Methyl Ether
135*	81	173	39	41	57	150	201	350	Propargite
135*	90	77	92	136	134	51	64	223	Salicyluric Acid Methyl Ester Methyl Ether
135*	92	65	108	138	80	39	106	272	Vacor
135*	107	41	150	95	136	77	91	150	Butylphenol(p-t-)
135*	108	54	53	81	43	136	66	135	Adenine
135*	136	45	97	149	164	91	77	249	1-(1(2-Trienyl)cyclohexl)piperidine
135*	150	91	107	117	0	0	0	150	Thymol
135*	152	77	92	64	63	136	107	194	Propyl Hydroxybenzoate Methyl Ester
135*	166	133	77	105	137	134	136	166	Methyl methoxybenzoate(ortho)
135*	166	136	167	77	107	92	137	166	Methyl methoxybenzoate(para)
135*	194	179	136	91	137	180	40	194	Methyl Acetylsalicylate
135	43*	170	212	219	69	172	15	256	3,5-Diacetoxybenzoyl Chloride
135	79*	153	107	52	53	80	51	153	Aminosalicylic Acid(p-)
135	93*	94	66	43	65	39	0	135	Acetanilide
135	93*	65	43	39	66	63	92	135	Acetanilide
135	105*	51	134	77	106	50	78	179	Hippuric Acid
135	136*	52	28	137	109	29	18	136	Allopurinol
135	136*	310	149	122	225	155	186	310	Ibogaine
135	160*	44	294	104	92	161	77	294	Triamifos
135	164*	136	149	97	84	163	91	164	1-Thiophenylcyclohexene
135	166*	107	167	136	77	108	59	166	Methyl methoxybenzoate(meta)
135	178*	176	107	70	77	84	0	301	Isoxsuprine (Duvadilan)
135	262*	111	219	97	263	56	0	277	Ethylmethylthiambutene
136*	41	42	55	158	173	81	79	324	Quinidine

```
                PEAKS                      MWT    COMPOUND


136*  42  94  43  95  96  47  79   183  Acephate
136*  42  41  81  55  79 130  77   294  Cinchonidine
136*  42  58  55  81 294  41 159   294  Cinchonine
136*  42  41  55  81  79 130  77   294  Cinchonine
136*  81 137  42  41  55 130 128   294  Cinchonidine*
136* 120 220 119 121 138 335 246   335  Senecionine
136* 135  52  28 137 109  29  18   136  Allopurinol
136* 135 310 149 122 225 155 186   310  Ibogaine
136* 137  81  79  42  41  55 159   294  Cinchonidine
136* 137  83 117 189 158  85 160   324  Quinine
136* 137  81  42  41 189  55 117   324  Quinine*
136* 154 108  80  52  95 137  69   154  2,4-Dihydroxybenzoic Acid*
136* 158 159 186  81   0   0   0   324  Quinine
136* 167 108  80  53  52 137  51   167  Orthocaine*
136* 294  81 159  55  42  41 143   294  Cinchonine*
136* 310 135 225 149 122 155 311   310  Ibogaine*
136* 324 189  81 173 138  55   0   324  Quinidine
136   44* 135  77  51 179  45  78   179  3,4-Methylenedioxyamphetamine
136   91*  92  65  39  63  45  51   136  Phenylacetic Acid
136   93* 111  78  79  40  94  80   136  Fenchene(alpha-)
136   93*  77  94  79  91  80  92   136  Phellandrene(beta-)
136   93* 121  92  91  43  77  79   136  Terpinene(gamma-)
136  110*  44 108 154  81  69  80   154  1,4-Dihydroxybenzoic Acid
136  121*  91  77  58  39  41 193   193  3-Isopropylphenyl-N-methylcarbamate (UC-10854, H 8757)
136  121* 105  93  79  91 122 107   136  Ocimene(allo-)
136  135*  45  97 149 164  91  77   249  1-(1(2-Trienyl)cyclohexl)piperidine
136  152*  96 124 180  51  39  52   180  4-Oxo-4,5,6,7-tetrahydrocoumarone-3-carboxylic Acid
136  164* 135 163  96 108 165  57   221  Methabenzthiazuron
136  194* 193  81 137 109  39 197   236  Ibomal Methylation Artifact*
136  222* 166 194 150 123 207  67   222  1,3-Diethyl-7-methylxanthine
137*  81 138 109 166   0   0   0   166  Vanillal
137*  91 107  79  65  92 138  77   137  Nitrotoluene (meta)
137*  92 120  65  39  52  66 138   137  3-Aminobenzoic Acid
137*  92  80  77 123 109 122 107   137  3-Methyl-4-nitrosophenol
137* 120  92  65  39 138 121  63   137  4-Aminobenzoic Acid*
137* 121 120  94  77  80 107  92   137  2-Methyl-4-nitrosophenol
137* 152 109  77  78  39 138  15   152  2,5-Dimethoxytoluene
137* 180 147 151 162  58 134  65   237  3-Hydroxycarbofuran
137* 238 181  40  37 240  44 138   346  Tolylfluanide
137* 256  44  45  43  41 120  55   256  2',4,4'-Trihydroxychalcone
137   44* 136 163 179   0   0   0   179  3,4-Methylenedioxyamphetamine
137   67*  91 138  55  79  41  95   340  Dimethisterone
137   77* 101 177 176 204 100 203   320  Lorazepam
137   91*  65  39 107  63  89  77   137  Nitrotoluene(p-)
137  106*  78 136  79 105 107 138   137  Methyl Pyridine-3-carboxylate
137  109* 246 290 110 305 276  81   305  Pirimiphos Methyl
137  119*  92  65  39  64 120  91   137  2-Aminobenzoic Acid
137  120* 193  92  65 121 138  41   193  Butylaminobenzoate
137  120* 193  65  92  39  41 121   193  Isobutylaminobenzoate(p-Amino-)*
137  136*  81  79  42  41  55 159   294  Cinchonidine
137  136*  83 117 189 158  85 160   324  Quinine
137  136*  81  42  41 189  55 117   324  Quinine*
137  154* 109  69  81  53 155  51   154  3,5-Dihydroxybenzoic Acid
137  168*  44 139  43 152 124 167   211  Methoxamine
```

PEAKS								MWT	COMPOUND
137	169*	125	95	55	41	84	43	240	Methyl-6-n-pentyl-4-hydroxy-2-oxocyclohex-3-ene-1-carboxylate
137	179*	152	304	93	153	97	135	304	Diazinon
137	313*	315	164	312	0	0	0	313	Flurazepam Metabolite
137	313*	315	314	0	0	0	0	313	Flurazepam Metabolite 1
138*	44	92	43	195	134	90	106	270	2-Nitrotartranilic Acid
138*	56	55	41	44	30	28	82	181	Dicyclohexylamine
138*	65	92	77	104	39	90	64	138	2-Nitroaniline
138*	65	92	39	108	106	66	52	138	4-Nitroaniline
138*	94	73	108	42	154	137	136	375	Hyoscine TMS Derivative
138*	94	108	136	301	154	77	42	303	Hyoscine
138*	124	141	169	168	247	249	98	318	5-(3-Bromopropyl)-5-isopropylbarbituric Acid
138*	326	55	110	189	82	160	139	326	Hydroquinidine*
138*	352	215	310	214	69	321	354	352	Griseofulvin
138	30*	195	140	103	197	77	196	213	Baclofen*
138	39*	195	136	119	53	91	120	234	Brallobarbitone Methylation Artifact*
138	58*	195	180	43	41	152	69	195	2,4-Diisopropylamino-s-triazine
138	81*	53	82	91	139	56	42	215	Furfurylmethylamphetamine
138	94*	42	108	154	303	136	137	303	Hyoscine
138	94*	42	108	136	41	96	97	303	Hyoscine*
138	111*	81	109	82	93	97	68	294	Cyanthoate
138	120*	43	92	64	65	63	42	180	Acetylsalicylic Acid
138	120*	43	92	121	39	63	64	180	Acetylsalicylic Acid
138	121*	93	65	39	63	127	53	138	4-Hydroxybenzoic Acid
138	121*	65	93	39	122	41	63	180	Propylparaben
138	195*	41	53	194	80	110	58	236	1,3-Dimethylallobarbitone
138	195*	41	80	53	194	39	110	236	Permethylated Allobarbitone*
138	313*	42	282	229	176	115	146	313	Morphine Dimethyl Derivative
138	386*	387	84	110	42	301	263	430	Piritramide*
139*	43	251	253	178	111	141	248	266	Chlorfenethol
139*	65	39	109	93	53	81	63	139	Nitrophenol(para)
139*	73	141	111	487	370	75	140	487	Indomethacin Derivative TMS Derivative
139*	95	109	41	121	43	55	136	154	Isofenchol(endo-)
139*	99	124	153	78	79	95	107	153	2-Methoxy-4-nitrosophenol
139*	111	141	75	170	50	113	172	170	4-Chlorobenzoic Acid Methyl Ester
139*	111	141	138	140	75	158	113	357	Indomethacin
139*	141	357	111	359	140	113	75	357	Indomethacin*
139*	141	429	111	73	140	431	113	429	Indomethacin Derivative TMS Derivative
139*	141	111	371	140	75	113	158	371	Indomethacin Methyl Ester
139*	141	73	111	312	429	140	75	429	Indomethacin TMS Ester
139*	156	111	44	75	141	230	50	265	2-Amino-5,2'-Dichlorobenzophenone
139*	156	111	75	141	50	158	113	156	4-Chlorobenzoic Acid*
139*	231	141	111	233	75	140	113	231	2-Chlorobenzanilide
139*	247	141	111	75	249	140	248	247	2-Chloro-4'-hydroxybenzanilide
139*	282	69	55	41	271	91	105	414	Diosgenin*
139	31*	93	153	151	77	137	123	169	Noradrenaline
139	58*	254	77	178	105	179	111	289	Chlophedianol
139	123*	221	122	43	140	81	124	221	Cyclofuramid
139	126*	44	58	43	55	42	41	198	Guanethidine
139	213*	75	169	111	215	141	77	228	Indomethacin Derivative (4-Chlorobenzoic Acid TMS Ester)
139	241*	276	195	111	44	165	119	276	2-Amino-5-Nitro-2'-Chlorobenzophenone
139	264*	266	111	141	251	152	253	292	Proclonol
140*	83	82	124	96	97	167	125	307	Benztropine
140*	118	91	29	46	69	65	39	231	Amphetamine TFA Derivative
140*	118	91	69	45	0	0	0	231	Amphetamine Trifluoroacetate

Peak Index   *Asterisk Indicates True Base Peak*   Page: 229

							MWT	COMPOUND
140*	141	84	98	56	138	41	96	288 Bupivacaine
140*	141	84	41	29	96	56	55	288 Bupivacaine*
140*	155	83	38	69	98	55	183	183 Methyprylone
140	42*	112	41	85	43	71	141	140 Hexamine*
140	42*	41	85	112	44	43	40	140 Methenamine
140	58*	67	72	83	81	155	156	155 Propylhexedrine
140	71*	70	263	262	189	57	42	263 Hydroxypethidine
140	83*	82	124	96	97	42	125	307 Benztropine*
140	83*	82	124	43	96	97	42	333 Deptropine*
140	155*	83	98	55	168	183	0	183 Methyprylone
140	155*	83	98	55	41	84	69	183 Methyprylone*
140	168*	169	64	63	167	114	141	168 Norharman
141*	43	183	44	140	108	80	52	254 Paracetamol Cysteine Conjugate
141*	70	55	41	42	69	112	126	169 Ethylethosuximide (N-)
141*	77	176	111	113	178	75	143	176 Dybenal
141*	123	42	41	151	55	169	96	401 Pipamazine
141*	143	77	51	105	0	0	0	186 Alclofenac Metabolite
141*	143	156	158	0	0	0	0	242 3'-Hydroxyamobarbitone
141*	156	157	40	69	98	142	70	226 Amylobarbitone
141*	156	41	39	55	98	112	44	184 Barbitone
141*	156	55	41	29	142	39	98	212 Butethal
141*	156	41	55	98	142	184	40	212 Butobarbitone
141*	156	41	55	98	39	142	155	212 Butobarbitone*
141*	156	43	41	157	55	39	98	226 Pentobarbitone*
141*	156	41	43	39	98	155	169	198 Probarbital*
141*	156	40	69	98	38	70	67	212 Secbutobarbitone
141*	156	41	57	39	98	157	47	212 Secbutobarbitone*
141*	169	401	42	96	41	142	70	401 Pipamazine*
141*	186	77	143	105	0	0	0	260 Alclofenac Metabolite
141*	200	77	143	155	125	142	202	200 MCPA
141*	445	169	123	155	96	42	317	445 Metopimazine*
141	41*	226	143	181	0	0	0	226 Alclofenac
141	41*	39	156	57	55	53	44	212 Butobarbitone
141	57*	167	41	181	182	83	223	238 Nealbarbitone
141	123*	43	151	41	42	57	231	401 Pipamazine
141	124*	154	167	155	184	142	98	291 5-(2-Bromoethyl)-5-butylbarbituric Acid
141	124*	167	154	125	112	98	190	246 5-Butyl-5-(2-chloroethyl)-barbituric Acid
141	124*	156	139	98	99	101	114	200 5-Ethyl-5-(2-hydroxyethyl)barbituric Acid
141	124*	154	153	170	98	125	155	255 5-Propyl-5-(2-thiocyanatoethyl)-barbituric Acid
141	126*	43	45	71	98	69	72	141 Chlormethiazole Metab (5-Acetyl-4-methylthiazole)
141	128*	268	130	270	77	143	233	268 Dichlorophen*
141	139*	357	111	359	140	113	75	357 Indomethacin*
141	139*	429	111	73	140	431	113	429 Indomethacin Derivative TMS Derivative
141	139*	111	371	140	75	113	158	371 Indomethacin Methyl Ester
141	139*	73	111	312	429	140	75	429 Indomethacin TMS Ester
141	140*	84	98	56	138	41	96	288 Bupivacaine
141	140*	84	41	29	96	56	55	288 Bupivacaine*
141	156*	157	41	43	142	197	55	226 Amobarbitone (Amytal)
141	156*	55	157	142	197	69	98	226 Amylobarbitone
141	156*	157	41	55	142	98	39	226 Amylobarbitone*
141	156*	98	155	55	112	41	83	184 Barbitone
141	156*	55	155	98	39	82	43	184 Barbitone*
141	156*	124	139	98	157	155	112	263 5-(2-Bromoethyl)-5-ethylbarbituric Acid
141	156*	41	57	157	39	98	55	212 Butabarbitone

								MWT	COMPOUND
141	156*	184	55	155	142	170	98	212	Butobarbitone
141	156*	55	41	157	43	98	39	240	Hexethal*
141	156*	157	45	211	199	181	55	228	Hydroxybutobarbitone
141	156*	157	69	45	197	195	98	242	Hydroxypentobarbitone
141	156*	157	55	69	155	98	197	226	Pentobarbitone
141	156*	98	112	197	226	0	0	226	Pentobarbitone
141	156*	41	43	155	157	98	169	198	Probarbitone
141	190*	154	124	191	155	142	181	260	5-(2-Chloroethyl)-5-isopentyl-barbituric Acid
141	200*	155	0	0	0	0	0	200	2-Methyl-4-chlorophenoxyacetic Acid (MCPA)
141	207*	40	81	79	44	77	41	236	Cyclobarbitone
141	207*	81	79	67	80	41	77	236	Cyclobarbitone*
141	214*	155	125	216	77	45	143	214	MCPA Methyl Ester
141	221*	81	79	222	41	67	93	250	Heptabarbitone
141	221*	79	81	77	40	38	67	250	Heptabarbitone
142*	42	44	170	43	41	140	96	410	Piperacetazine
142*	57	42	197	185	339	240	226	339	Hexetidine
142*	100	99	125	143	98	124	156	186	5-(2-Hydroxyethyl)-5-methylbarbituric Acid
142*	107	144	87	43	77	45	108	228	MCPB
142*	120	143	86	142	92	72	56	292	Leucinocaine*
142*	128	121	120	171	83	87	143	202	Methyl 2-Propylglutarate
142*	143	100	155	44	112	57	29	316	Butalamine*
142*	169	228	107	141	144	171	77	228	Mecoprop Methyl Ester
142*	214	141	169	107	77	0	0	214	4-Chloro-2-methylphenoxypropionic Acid (CMPP)
142*	228	141	0	0	0	0	0	228	4-Chloro-2-methylphenoxybutyric Acid (MCPB)
142	124*	154	141	99	172	155	98	242	5-(2-Hydroxyethyl)-5-isopentylbarbituric Acid
142	143*	115	144	141	89	116	63	143	6-Methylquinoline
142	143*	115	144	141	116	89	39	143	7-Methylquinoline
142	143*	115	144	141	89	116	39	143	8-Methylquinoline
143*	70	100	144	42	56	98	221	400	Clopenthixol*
143*	70	100	144	42	98	58	56	434	Flupenthixol*
143*	115	142	116	144	89	72	59	143	3-Methylisoquinoline
143*	115	142	144	116	89	77	39	143	4-Methylquinoline
143*	128	115	144	142	101	51	77	143	2-Methylquinoline
143*	142	115	144	141	89	116	63	143	6-Methylquinoline
143*	142	115	144	141	116	89	39	143	7-Methylquinoline
143*	142	115	144	141	89	116	39	143	8-Methylquinoline
143*	248	115	77	69	171	81	249	248	1-Phenylazonapth-2-ol
143	43*	58	42	41	39	128	56	143	Trimethadione
143	112*	113	85	45	0	0	0	143	4-Methyl-5-b-hydroxyethiazole
143	141*	77	51	105	0	0	0	186	Alclofenac Metabolite
143	141*	156	158	0	0	0	0	242	3'-Hydroxyamobarbitone
143	142*	100	155	44	112	57	29	316	Butalamine*
143	200*	185	169	126	201	170	155	242	Hydroxyamylobarbitone (N-Hydroxy)
143	246*	403	70	404	42	248	113	403	Perphenazine*
143	254*	411	70	255	42	380	157	411	Acetophenazine
143	268*	425	70	269	42	394	157	425	Carphenazine*
143	268*	425	55	70	41	40	269	425	Carphenazine (Proketazine)
143	280*	42	70	437	406	113	56	437	Fluphenazine*
144*	43	100	86	116	145	128	89	272	Indol-3-yl-N,N-disopropylglyoxamide
144*	43	42	116	69	59	145	41	144	Thiobarbituric Acid
144*	72	30	115	43	130	144	56	244	Diisopropyltryptamine (N,N-)
144*	72	145	116	89	29	100	244	244	Indol-3-yl-N,N-diethylglyoxamide
144*	115	116	58	55	63	201	40	201	1-Naphthyl-N-methylcarbamate (Carbaryl, Sevin)
144*	115	116	145	72	89	63	58	144	1-Napthanol

								MWT	COMPOUND
144*	145	116	89	63	146	90	58	145	3-Indolealdehyde
144*	216	116	89	72	63	0	0	216	3-Indolylglyoxyldimethylamide
144	280*	115	224	154	252	29	281	280	Naphthylethyldiethyl Phosphate (beta-)
145*	85	93	125	146	69	58	302	302	Methidathion (Technical)
145*	91	74	72	56	86	277	43	467	Oxethazaine
145*	117	122	89	105	90	63	146	145	Hydroxyquinoline*
145	107*	268	238	121	133	159	224	268	Diethylstilbestrol
145	109*	79	185	147	35	187	47	378	Naled
145	109*	15	79	96	185	73	47	378	Naled
145	144*	116	89	63	146	90	58	145	3-Indolealdehyde
146*	57	245	189	42	91	77	132	336	Fentanyl
146*	73	232	334	247	117	246	147	362	Primidone 1,3-di-TMS Derivative
146*	117	190	118	115	103	91	161	218	Primidone
146*	117	103	115	91	0	0	0	216	Primidone Metabolite
146*	148	111	75	113	74	50	150	146	1,2-Dichlorobenzene
146*	148	111	75	50	113	150	147	146	1,3-Dichlorobenzene
146*	148	111	75	50	150	113	147	146	1,4-Dichlorobenzene
146*	190	117	118	161	189	103	91	218	Primidone*
146*	190	117	118	189	161	103	115	218	Primidone (5-Ethylhexahydro-5-phenylpyrimidine-4,6-dione)
146*	191	147	91	117	63	65	39	191	5-Hydroxyindole-3-acetic Acid
146*	191	147	130	57	145	117	89	191	5-Hydroxyindole-3-acetic Acid
146*	218	117	118	44	217	42	103	246	Primidone Dimethyl Derivative
146*	233	103	133	91	117	115	77	233	4-Hydroxyglutethimide
146*	260	118	117	103	261	91	232	288	1,3-Diethylphenobarbitone
146*	298	157	156	129	118	90	97	298	Quinalphos
146	58*	56	105	77	42	106	40	165	Ephedrine*
146	91*	65	219	92	41	39	190	219	Encyprate
146	117*	43	91	32	115	39	116	218	Primidone
146	133*	89	278	105	77	132	118	296	7,14-Dihydroxydiftalone
146	159*	218	160	147	219	0	0	218	Acetylserotonin (N-)
146	232*	117	175	118	233	103	188	260	1,3-Dimethylphenobarbitone
146	232*	118	175	117	120	121	188	260	Dimethylphenobarbitone
146	233*	188	104	103	0	0	0	233	Glutethimide Metabolite
146	234*	179	206	131	91	117	118	234	Methisazone*
146	246*	117	118	247	175	103	77	274	3-Ethylmephobarbitone
147*	45	91	64	48	105	65	77	999	Phenelzine Probe Product 1*
147*	131	130	132	119	148	365	218	365	Indapamide*
147*	148	66	73	149	45	59	131	162	Hexamethyldisiloxane*
147*	148	88	91	58	110	81	145	205	3-(2'-Propynyloxy)phenyl N-Methylcarbamate
147*	165	201	167	166	105	203	117	432	Buclizine
147*	176	148	91	39	65	51	63	176	6-Hydroxy-4-methylcoumarin
147*	189	73	74	148	75	190	146	204	Urea di-TMS Derivative
147*	231	165	285	132	166	201	117	432	Buclizine
147	41*	79	93	55	105	67	91	329	Trilostane*
147	73*	332	117	205	45	74	133	464	Ascorbic Acid(Tetra TMS)
147	91*	45	64	105	65	77	48	136	Phenelzine
147	98*	111	42	55	115	70	96	391	Flavoxate
147	132*	262	133	104	77	148	190	322	7-Ureidodiftalone
147	133*	280	105	77	132	134	90	280	7-Hydroxydiftalone
147	216*	158	57	217	131	160	187	216	Dioxyparaquat
147	231*	285	232	201	132	165	166	432	Buclizine*
148*	76	130	75	102	104	239	50	338	Chlorthalidone
148*	105	91	44	77	79	65	56	239	Methyldiphenethylamine (N-)
148*	149	77	42	51	132	105	50	149	Dimethylaminobenzaldehyde(p-)

		PEAKS					MWT	COMPOUND
148*	163	91	103	120	115	117	77	163 2-Ethyl-2-phenylmalondiamide (Primidone Metab)
148*	176	147	91	39	51	120	63	176 7-Hydroxy-4-methylcoumarin
148*	268	267	224	269	251	120	118	268 Michler's Ketone
148	69*	150	129	131	79	81	50	148 Bromotrifluoromethane
148	91*	149	65	92	42	56	39	239 Benzphetamine*
148	103*	94	96	108	41	77	91	397 Methscopolamine Bromide
148	109*	110	81	88	58	57	39	205 2-(2'Propynyloxy)phenyl N-Methylcarbamate
148	130*	243	299	209	102	194	76	299 Dowco 199
148	146*	111	75	113	74	50	150	146 1,2-Dichlorobenzene
148	146*	111	75	50	113	150	147	146 1,3-Dichlorobenzene
148	146*	111	75	50	150	113	147	146 1,4-Dichlorobenzene
148	147*	66	73	149	45	59	131	162 Hexamethyldisiloxane*
148	147*	88	91	58	110	81	145	205 3-(2'-Propynyloxy)phenyl N-Methylcarbamate
148	163*	91	103	117	120	44	77	206 2-Phenyl-2-Ethylmalonamide(PEMA-Phenobarb./Primidone metab.)
148	163*	91	103	164	0	0	0	206 Primidone Metabolite
148	184*	58	118	186	154	149	104	255 3-Chloro-6-cyano-2-norbornanon-o-(methylcarbamoyl) Oxime
148	239*	194	164	210	193	182	147	239 3,5-Dimethylpyrrole-2,4-dicarboxylic Acid Ethyl Ester
149*	41	29	57	56	104	32	65	278 Dibutyl Phthalate
149*	43	150	55	41	42	237	71	306 Dipentyl Phthalate
149*	43	167	57	104	112	279	113	278 2-Ethylhexyl Phthalate
149*	57	41	150	56	223	104	76	278 Diisobutyl Phthalate
149*	57	167	71	70	43	113	41	390 Dioctyl Phthalate(n-)
149*	57	167	43	71	70	41	55	390 Isooctyl Phthalate
149*	104	76	50	150	41	43	42	250 Diisopropyl Phthalate
149*	107	30	74	73	60	0	0	149 Propylmethyldithiocarbamate (S-n-propyl-)
149*	150	41	57	56	76	104	205	336 Butylcarbobutoxymethyl Phthalate
149*	150	41	57	223	205	56	104	278 Dibutyl Phthalate(n-)
149*	150	63	121	65	62	91	39	150 Piperonal
149*	167	57	71	70	43	'150	41	390 Di-2-ethylhexyl Phthalate
149*	177	150	65	76	105	176	104	222 Diethyl Phthalate
149*	178	65	177	150	76	121	105	222 Diethyl Phthalate
149*	312	270	354	297	269	167	256	414 Acetoxytetrahydrocannabinol (8-alpha-Acetoxy-delta 9-)
149	57*	71	43	41	69	55	113	390 Di-iso-octyl Phthalate
149	57*	71	70	293	69	111	43	418 Dinonyl Phthalate
149	72*	269	255	197	73	254	340	340 Propiomazine
149	107*	57	78	108	79	77	72	261 Ethoheptazine
149	120*	106	121	77	91	132	150	149 2,4-Dimethylformanilide
149	148*	77	42	51	132	105	50	149 Dimethylaminobenzaldehyde(p-)
149	164*	57	122	123	131	165	121	221 Carbofuran
149	164*	131	137	103	77	133	165	164 Eugenol
149	177*	166	65	194	178	104	121	222 Diethyl Terephthalate
149	225*	185	240	103	131	89	129	240 Budralazine
150*	121	56	93	39	151	66	94	445 Glipizide*
150*	134	164	123	136	79	122	285	285 4'-Methoxy-N-isopropyl-2-(methylsulphonyl)acetanilide
150*	151	121	65	93	77	135	0	151 3,4-Methylenedioxybenzylamine
150*	152	165	41	167	43	44	130	209 2-Bromo-2-ethyl-3-hydroxybutyramide(Carbromal Metabolite)
150*	189	246	57	151	174	188	190	246 4-(Methyl-2'-propynylamino)-3,5-dimethylphenyl N-Methylcarbamate
150	59*	271	270	214	171	128	212	271 Dextromethorphan
150	70*	151	153	77	51	118	60	322 Chloramphenicol
150	91*	65	92	89	59	0	0	150 Methyl Phenyl Acetate (Methyl Phenidate Decomposition Prod)
150	109*	78	41	110	81	58	39	207 2-(Allyloxyphenyl) N-Methylcarbamate
150	121*	65	93	168	0	0	0	168 2-Hydroxymethyl-4-hydroxybenzoic Acid
150	121*	93	65	168	0	0	0	168 2-Hydroxymethyl-5-hydroxybenzoic Acid
150	121*	65	93	0	0	0	0	150 2-Hydroxymethyl-4-hydroxybenzolactone

			PEAKS					MWT	COMPOUND
150	121*	93	65	0	0	0	0	150	2-Hydroxymethyl-5-hydroxybenzolactone
150	135*	91	107	117	0	0	0	150	Thymol
150	149*	41	57	56	76	104	205	336	Butylcarbobutoxymethyl Phthalate
150	149*	41	57	223	205	56	104	278	Dibutyl Phthalate(n-)
150	149*	63	121	65	62	91	39	150	Piperonal
150	151*	136	208	58	77	45	40	208	4-Dimethylamino-3-methylphenyl n-methylcarbamate (Bayer 44646)
150	222*	194	179	166	207	109	43	222	3,7-Diethyl-1-methylxanthine
150	235*	165	236	79	137	250	164	250	2-Methylhexobarbitone (or 4-Methyl-)
150	305*	307	115	152	114	306	123	305	Cliquinol*
151*	52	123	108	42	65	51	72	223	Ethamivan
151*	72	223	123	222	152	52	29	223	Ethamivan*
151*	94	168	106	150	122	123	0	168	Pridoxamine
151*	94	122	106	51	53	149	150	169	Pyridoxine
151*	105	77	79	103	121	78	39	151	4-Nitroxylene(o-)
151*	126	166	31	51	43	58	223	223	Bendiocarb
151*	150	136	208	58	77	45	40	208	4-Dimethylamino-3-methylphenyl n-methylcarbamate (Bayer 44646)
151*	153	117	119	70	101	121	29	186	1,1,1-Trifluoro-2,2,2-trifluoroethane(Freon 113)
151*	210	152	211	107	195	59	153	210	Methyl 3,4-dimethoxyphenylacetate
151*	223	222	72	123	152	108	224	223	Ethamivan
151*	224	164	152	225	149	165	193	224	Methyl 3,4-Dimethoxyphenylpropionate
151	94*	57	95	40	41	58	108	151	Amantadine*
151	101*	103	153	85	87	31	105	147	1,1,2-Trichlorotrifluoroethane
151	109*	43	80	81	0	0	0	151	Hydroxyacetanilide(p-)
151	109*	43	79	80	53	108	110	151	Paracetamol
151	109*	43	80	108	81	53	52	151	Paracetamol*
151	119*	120	92	152	91	93	65	151	Methylaminobenzoate (ortho)
151	120*	92	121	152	65	93	122	151	Methylaminobenzoate (para)
151	150*	121	65	93	77	135	0	151	3,4-Methylenedioxybenzylamine
151	152*	81	109	51	0	0	0	152	Vanillin
151	166*	30	135	195	165	44	167	195	2,5-Dimethoxy-4-methylphenethylamine
152*	42	98	154	153	174	69	56	273	Chlormezanone (Trancopal)
152*	54	109	53	81	0	0	0	152	Xanthine
152*	83	42	82	85	181	57	122	329	Cinnamoylcocaine
152*	106	77	79	122	59	0	0	152	2-Methyl-4-nitroaniline
152*	106	77	79	91	104	78	105	152	4-Methyl-2-nitroaniline
152*	136	96	124	180	51	39	52	180	4-Oxo-4,5,6,7-tetrahydrocoumarone-3-carboxylic Acid
152*	151	81	109	51	0	0	0	152	Vanillin
152*	153	154	42	155	125	111	56	273	Chlormezanone
152*	180	138	102	154	182	209	0	237	Ketamine Impurity
152	44*	137	77	65	91	78	121	195	2,5-Dimethoxyamphetamine
152	44*	137	195	179	0	0	0	195	3,5-Dimethoxyamphetamine
152	44*	195	151	77	78	65	51	195	3,5-Dimethoxyamphetamine
152	44*	91	77	65	153	151	195	195	2,6-Dimethoxyamphetamine
152	44*	151	65	51	137	78	77	195	3,4-Dimethoxyamphetamine
152	44*	151	121	153	77	91	78	195	2,4-Dimethoxyamphetamine
152	44*	137	121	153	65	77	91	195	2,5-Dimethoxyamphetamine
152	44*	40	121	151	91	77	153	195	2,4-Dimethoxyamphetamine
152	44*	40	137	91	77	65	121	195	2,5-Dimethoxyamphetamine
152	44*	91	77	65	151	121	78	195	2,6-Dimethoxyamphetamine
152	44*	151	137	153	107	65	43	195	3,4-Dimethoxyamphetamine
152	44*	40	77	42	91	151	78	195	3,5-Dimethoxyamphetamine
152	77*	79	105	78	104	118	51	152	Methylnitroaniline(N-Me,o-)
152	98*	154	42	69	174	208	153	273	Chlormezanone*
152	107*	77	39	108	51	79	53	152	Hydroxyphenylacetic Acid(meta)

Asterisk Indicates True Base Peak

PEAKS							MWT	COMPOUND
152	110*	111	81	58	43	41	52	209 Isopropoxyphenyl N-methylcarbamate (o-Isopropoxy-)
152	120*	43	121	92	63	65	64	194 Methyl Acetylsalicylate
152	121*	93	65	39	122	63	153	336 Fentanyl
152	121*	40	93	65	0	0	0	152 Methyl Hydroxybenzoate
152	121*	93	65	39	0	0	0	152 Methylparaben
152	135*	77	92	64	63	136	107	194 Propyl Hydroxybenzoate Methyl Ester
152	137*	109	77	78	39	138	15	152 2,5-Dimethoxytoluene
152	150*	165	41	167	43	44	130	209 2-Bromo-2-ethyl-3-hydroxybutyramide(Carbromal Metabolite)
152	188*	190	153	76	189	151	150	188 2-Chlorobiphenyl
152	222*	224	151	223	150	75	153	222 4,4'-Dichlorobiphenyl
153*	55	183	138	69	0	0	0	199 Methaprylon Metabolite
153*	115	129	91	143	0	0	0	158 Phenylcyclohexene
153*	118	138	102	130	0	0	0	221 Ketamine Metabolite
153*	154	77	94	110	53	40	51	234 Lenacil
153*	170	155	152	172	136	118	60	322 Chloramphenicol
153*	188	126	190	161	154	137	189	188 Chlorobenzylidenemalononitrile (o-Chloro - CS gas)
153	70*	36	118	117	60	155	150	322 Chloramphenicol
153	151*	117	119	70	101	121	29	186 1,1,1-Trifluoro-2,2,2-trifluoroethane(Freon 113)
153	152*	154	42	155	125	111	56	273 Chlormezanone
153	168*	109	65	58	139	91	39	225 Methiocarb
153	168*	225	91	58	45	39	77	225 (4-(Methylthio)-3,5-dimethylphenyl N-methylcarbamate (Bayer 37344)
153	168*	109	91	225	45	169	154	225 (4-(Methylthio)-3,5-dimethylphenyl N-methylcarbamate (Methiocarb)
153	170*	39	51	79	125	53	126	170 Gallic Acid
154*	43	196	137	155	69	238	42	238 3,5-Diacetoxybenzoic Acid
154*	55	41	141	42	155	281	70	281 Dodemorph
154*	59	45	91	69	132	114	117	245 Phentermine TFA
154*	70	155	167	223	225	349	153	577 Dihydroergocryptine*
154*	83	71	55	155	43	70	67	224 Butylvinal
154*	118	110	69	91	45	0	0	245 Methylamphetamine TFA
154*	137	109	69	81	53	155	51	154 3,5-Dihydroxybenzoic Acid
154*	169	83	55	41	42	98	69	197 Methyprylone Methyl Derivative
154*	246	70	349	43	223	225	167	577 Dihydroergocryptine
154*	309	197	77	155	195	307	353	864 Decafentin
154*	324	89	338	77	51	339	102	339 Papaverine
154	70*	69	224	210	55	196	209	547 Ergosinine*
154	124*	155	141	142	181	125	198	305 5-(2-Bromoethyl)-5-isopentylbarbituric Acid
154	124*	141	181	198	142	153	155	269 5-Butyl-5-(2-thiocyanatoethyl)-barbituric Acid
154	124*	181	125	141	153	142	112	283 5-Pentyl-5-(2-thiocyanatoethyl)-barbituric Acid
154	136*	108	80	52	95	137	69	154 2,4-Dihydroxybenzoic Acid*
154	153*	77	94	110	53	40	51	234 Lenacil
154	155*	156	77	127	128	51	78	155 2-Phenylpyridine
154	268*	224	180	207	192	221	223	268 Lysergic Acid (d)
155*	96	140	43	42	81	94	53	155 Arecoline
155*	140	83	98	55	168	183	0	183 Methyprylone
155*	140	83	98	55	41	84	69	183 Methyprylone*
155*	154	156	77	127	128	51	78	155 2-Phenylpyridine
155*	157	227	154	185	44	30	156	227 Buclosamide*
155*	157	170	172	156	171	158	173	240 Methylamobarbitone (N-methyl)
155*	170	112	169	55	82	41	39	198 Metharbitone*
155*	170	156	98	169	112	197	55	226 Methylbutobarbitone (N-Methyl)
155	91*	197	65	84	39	95	41	366 Glibornuride*
155	91*	129	41	284	229	163	184	284 Methyl Enol Ether of Tolbutamide
155	91*	185	65	92	0	0	0	185 Methyl-p-tolylsulphonamide (N-Methyl)
155	91*	185	65	41	92	121	56	284 Methyltolbutamide (N-Methyl)

Peak Index   *Asterisk Indicates True Base Peak*   Page: 235

155	91*	114	65	197	42	41	85	311 Tolazamide*
155	91*	171	65	107	39	108	63	270 Tolbutamide
155	99*	127	81	109	0	0	0	224 Diethylamyl Phosphate
155	99*	109	127	45	82	81	27	182 Triethyl Phosphate
155	124*	154	141	142	181	153	98	305 5-(2-Bromoethyl)-5-pentylbarbituric Acid
155	140*	83	38	69	98	55	183	183 Methyprylone
155	170*	41	169	55	112	39	83	198 Metharbitone (Gemonil)
155	170*	55	156	171	69	212	112	240 Methylamylobarbitone (N-Methyl)
155	170*	169	55	112	83	98	69	198 Methylbarbitone (N-Methyl)
155	197*	140	0	0	0	0	0	197 Phenacetin Metabolite
156*	50	155	52	51	128	157	76	256 Paraquat Dichloride
156*	82	80	128	81	155	78	79	342 Diquat
156*	92	108	65	140	43	157	42	267 Sulphafurazole*
156*	92	108	65	140	253	157	43	253 Sulphamethoxazole*
156*	92	108	65	269	39	45	270	269 Sulphasomizole*
156*	93	281	125	43	157	55	63	281 Menazon
156*	110	58	109	79	80	126	47	213 Omethoate
156*	121	158	93	58	65	141	51	213 Banol
156*	121	158	91	77	65	157	122	156 4-Chloro-3,5-xylenol
156*	141	157	41	43	142	197	55	226 Amobarbitone (Amytal)
156*	141	55	157	142	197	69	98	226 Amylobarbitone
156*	141	157	41	55	142	98	39	226 Amylobarbitone*
156*	141	98	155	55	112	41	83	184 Barbitone
156*	141	55	155	98	39	82	43	184 Barbitone*
156*	141	124	139	98	157	155	112	263 5-(2-Bromoethyl)-5-ethylbarbituric Acid
156*	141	41	57	157	39	98	55	212 Butabarbitone
156*	141	184	55	155	142	170	98	212 Butobarbitone
156*	141	55	41	157	43	98	39	240 Hexethal*
156*	141	157	45	211	199	181	55	228 Hydroxybutobarbitone
156*	141	157	69	45	197	195	98	242 Hydroxypentobarbitone
156*	141	157	55	69	155	98	197	226 Pentobarbitone
156*	141	98	112	197	226	0	0	226 Pentobarbitone
156*	141	41	43	155	157	98	169	198 Probarbitone
156*	158	141	143	157	159	142	0	226 Amobarbitone
156*	158	92	77	157	314	108	65	314 Sulphaphenazole*
156*	352	351	184	169	209	353	129	352 Ajmalicine
156	42*	127	228	142	140	83	72	228 Allantoin Pentamethyl Derivative
156	59*	157	69	56	43	41	55	242 3-Hydroxyamylobarbitone
156	92*	191	65	108	76	104	50	403 Phthalylsulphathiazole*
156	92*	65	191	108	55	45	174	355 Succinylsulphathiazole*
156	92*	65	108	191	45	39	55	255 Sulphathiazole*
156	139*	111	44	75	141	230	50	265 2-Amino-5,2'-Dichlorobenzophenone
156	139*	111	75	141	50	158	113	156 4-Chlorobenzoic Acid*
156	141*	157	40	69	98	142	70	226 Amylobarbitone
156	141*	41	39	55	98	112	44	184 Barbitone
156	141*	55	41	29	142	39	98	212 Butethal
156	141*	41	55	98	142	184	40	212 Butobarbitone
156	141*	41	55	98	39	142	155	212 Butobarbitone*
156	141*	43	41	157	55	39	98	226 Pentobarbitone*
156	141*	41	43	39	98	155	169	198 Probarbital*
156	141*	40	69	98	38	70	67	212 Secbutobarbitone
156	141*	41	57	39	98	157	47	212 Secbutobarbitone*
156	157*	158	115	142	128	89	154	157 2,6-Dimethylquinoline
156	157*	115	158	142	116	128	77	157 2,4-Dimethylquinoline

PEAKS							MWT	COMPOUND
156	169*	170	143	144	115	129	142	354 Yohimbine
156	181*	55	141	180	138	155	39	210 Crotylbarbitone
157*	97	121	342	153	125	159	199	342 Trithion
157*	121	342	97	153	45	159	125	342 Trithion
157*	125	159	314	93	45	316	171	314 Methyl Trithion
157*	156	158	115	142	128	89	154	157 2,6-Dimethylquinoline
157*	156	115	158	142	116	128	77	157 2,4-Dimethylquinoline
157*	172	41	173	43	69	55	39	242 Thiopentone
157	57*	75	155	39	49	77	41	234 Dibromochloropropane
157	59*	156	141	43	41	71	69	242 3'-Hydroxyamylobarbitone
157	112*	85	45	158	0	0	0	157 4-Methyl-5-thiazoleacetic Acid
157	129*	128	130	187	158	156	188	187 2-Carbomethoxyquinoline
157	155*	227	154	185	44	30	156	227 Buclosamide*
157	155*	170	172	156	171	158	173	240 Methylamobarbitone (N-methyl)
157	172*	42	173	242	69	71	97	242 Thiopentone
157	172*	173	43	41	55	69	71	242 Thiopentone*
158*	159	144	115	157	160	156	143	159 2,3,5-Trimethylindole
158*	214	57	145	124	45	70	96	214 Niridazole*
158	129*	130	143	115	128	91	77	158 1-Phenylcyclohexene
158	136*	159	186	81	0	0	0	324 Quinine
158	156*	141	143	157	159	142	0	226 Amobarbitone
158	156*	92	77	157	314	108	65	314 Sulphaphenazole*
159*	124	260	161	262	126	163	88	260 Methazole
159*	146	218	160	147	219	0	0	218 Acetylserotonin (N-)
159*	161	81	335	333	163	337	205	414 Miconazole*
159*	205	131	76	187	51	103	104	205 Xanthurenic Acid
159	91*	65	160	131	39	51	81	160 Tolazoline
159	130*	77	64	131	103	48	51	159 3-Indoleacetaldehyde
159	158*	144	115	157	160	156	143	159 2,3,5-Trimethylindole
160*	61	76	77	133	104	161	50	317 Phosmet
160*	91	129	130	117	115	104	203	203 Ethyl-1-phenylcyclohexylamine (N-)
160*	103	89	131	115	76	161	104	160 Hydrallazine*
160*	132	77	93	125	104	147	105	317 Azinphos Methyl
160*	132	77	104	105	76	159	109	301 Dimethyl-s-(4-oxo-1,2,3-benzotriazin-3(4h)-ylmethyl) Phosphorothiolate
160*	132	77	93	105	76	104	51	317 Guthion
160*	135	44	294	104	92	161	77	294 Triamifos
160*	173	232	161	174	145	158	0	232 Melatonin
160*	203	91	146	161	117	44	104	203 Ethyl-1-phenylcyclohexylamine (N-)
160*	219	161	220	145	74	69	83	219 Methyl-5-methoxyindoleacetate
160*	266	235	251	77	146	58	247	266 Hydroxymethaqualone
160*	300	189	145	188	301	161	42	300 Azapropazone*
160	45*	188	237	162	224	146	161	269 Alachlor
160	84*	167	159	165	152	55	77	266 Diphenazoline*
160	91*	65	174	175	279	146	124	299 Nylidrin
160	103*	77	207	41	39	150	51	207 1,2-Methylenedioxy-4-(2-nitropropenyl)benzene
160	174*	161	218	275	175	162	149	275 Physostigmine
160	192*	342	59	150	177	105	209	342 Thiophanate Methyl
160	213*	159	296	133	145	212	157	300 Ethinylestradiol
160	213*	296	133	159	145	146	214	296 Ethynylestradiol
160	235*	266	77	235	251	146	0	266 Methaqualone Metabolite
160	252*	206	254	125	162	217	148	252 Chlorodinitronaphthalene
161*	117	177	91	119	118	220	121	220 Ibuprofen Methyl Ester
161*	163	99	90	165	126	134	73	161 3,4-Dichloroaniline
161*	163	90	99	63	165	162	126	161 2,6-Dichloroaniline

		PEAKS					MWT	COMPOUND

```
161* 163  275  277  217  165  219  165  275  (3,4-Dichlorophenyl)-2-methyl-3'-hydroxyvaleramide (N-)
161* 163  259  261  217  165  219  162  259  (3,4-Dichlorophenyl)-2'-methylvaleramide (N-)
161* 163  179  235   29  162   99  207  274  Tetraethylphosphate
161* 216  160  218  163  162  117  118  216  Terbacil
161   56*  57   43  144  117   91  129  161  Cypenamine
161   75* 163  263  265  127  126  162  263  (3,4-Dichlorophenyl)-2'-methyl-2',3'-dihydroxypropionamide (N-)
161   91* 117  103  115  189  143   55  233  Glutethimide Metabolite
161  112*  85   45  163  113  114   59  161  Chlormethiazole*
161  122*  94  160   57  120  123  162  218  2-(Methyl-2'-propynylamino)phenyl N-Methylcarbamate
161  127* 163  153  129  187   90  189  314  Triclocarban*
161  130* 143  131  103   77  115  102  161  Tryptophol
161  159*  81  335  333  163  337  205  414  Miconazole*
161  163* 165   79   81   82   47   36  240  Dibromodichloromethane
161  163* 119   91  206  117  107  164  206  Ibuprofen*
161  176* 178  163  133  135   63  180  176  2,4-Dichlorophenyl Methyl Ether
162*  63  164   98   99   49   62   73  162  2,4-Dichlorophenol
162*  91   44  119  163   41   70   65  253  Di-(1-isopropylphenyl)amine*
162* 105   44  163   57  190  106   58  324  Diampromide
162* 105  190  163  106   29  134   77  324  Diampromide*
162* 105   57  163  114  190   44   77  351  Phenadoxone
162* 121  132  147  293  344  120  106  293  Amitraz
162* 132  121  120  106  118  147   77  162  2,4-Dimethylphenyl-N'-methylformamidine (N-2,4-Dimethyl-)
162* 164  220  175  185  111  147  222  220  2,3-D
162* 164  220  161   63  133  111  222  220  2,4-D
162* 164   63   98   73  166  100   99  162  2,4-Dichlorophenol
162* 164   63   98  126  166   73   99  162  2,6-Dichlorophenol
162* 164  189   59  191   55  248   87  248  2,4-Dichloroprop Methyl Ester
162* 220  161  175    0    0    0    0  220  2,4-Dichlorophenoxyacetic Acid
162* 234  189  161    0    0    0    0  234  2,4-Dichlorophenoxypropionic Acid  (2,4-DP)
162* 285  327  215  124   81  115  267  327  3-Acetylmorphine
162   42* 313   44  124   59  115   81  313  Ethylmorphine
162  106*  78   51   79  123   43   41  179  Iproniazid
162  131*  79  103  161   52  132  163  162  Methyl Cinnamate
162  209* 133  148   77   51  147  105  209  2,5-Dimethoxy-b-nitrostyrene
162  223* 161  147  176   91   77  119  223  2,5-Dimethoxy-b-methyl-b-nitrostyrene
162  285*  42   44   31  215   70  200  285  Morphine
162  285*  42  215  286  124   44  284  285  Morphine*
162  285* 284  215  286  124  268  226  301  Morphine N-oxide
162  299* 229  124  300  214  298   42  299  Codeine
162  299* 229  298  124  214  297  282  299  Codeine
162  299* 298  229  115   42  214  124  299  Codeine
162  299* 229  123   59   42   44  300  299  Neopine
162  313* 314  124  284   59   42  243  313  Ethylmorphine*
163*  77  164  135  194   76   92   50  194  Dimethyl Phthalate
163* 148   91  103  117  120   44   77  206  2-Phenyl-2-Ethylmalonamide(PEMA-Phenobarb./Primidone metab.)
163* 148   91  103  164    0    0    0  206  Primidone Metabolite
163* 161  165   79   81   82   47   36  240  Dibromodichloromethane
163* 161  119   91  206  117  107  164  206  Ibuprofen*
163* 165  280   97  224  128  252  174  315  Diethyl-O-(3,6-dichloro-2-pyridyl) Phosphorothioate
163* 194  135   76  164  103  120   77  194  Dimethyl Isophthalate
163   91* 161  119  107  117   41  118  206  Ibuprofen
163  107* 223  108   29  164   41   57  223  Bufexamac*
163  121* 151  109   43  122  108   65  313  Benorylate*
163  148*  91  103  120  115  117   77  163  2-Ethyl-2-phenylmalondiamide (Primidone Metab)
```

163	161*	99	90	165	126	134	73	161	3,4-Dichloroaniline
163	161*	90	99	63	165	162	126	161	2,6-Dichloroaniline
163	161*	275	277	217	165	219	165	275	(3,4-Dichlorophenyl)-2-methyl-3'-hydroxyvaleramide (N-)
163	161*	259	261	217	165	219	162	259	(3,4-Dichlorophenyl)-2'-methylvaleramide (N-)
163	161*	179	235	29	162	99	207	274	Tetraethylphosphate
163	183*	184	165	91	77	127	89	390	Permethrin
163	211*	147	117	240	77	89	205	240	Dinoseb
163	221*	164	44	149	122	42	36	221	Formetanate
163	279*	323	58	277	308	280	322	323	Dimefline*
163	298*	43	41	55	162	57	175	680	Tubocurarine Chloride
164*	120	92	83	70	112	221	165	294	1-(2-Ethylhexyl)-3-hydroxyphthalate
164*	135	136	149	97	84	163	91	164	1-Thiophenylcyclohexene
164*	136	135	163	96	108	165	57	221	Methabenzthiazuron
164*	149	57	122	123	131	165	121	221	Carbofuran
164*	149	131	137	103	77	133	165	164	Eugenol
164	104*	91	105	133	103	165	78	164	Methyl beta-Phenylpropionate
164	162*	220	175	185	111	147	222	220	2,3-D
164	162*	220	161	63	133	111	222	220	2,4-D
164	162*	63	98	73	166	100	99	162	2,4-Dichlorophenol
164	162*	63	98	126	166	73	99	162	2,6-Dichlorophenol
164	162*	189	59	191	55	248	87	248	2,4-Dichloroprop Methyl Ester
164	165*	150	222	58	134	77	39	222	Zectran
164	166*	129	131	168	94	96	47	164	Tetrachloroethylene
164	250*	222	194	235	251	207	150	250	1,3,7-Triethyl-8-methylxanthine
165*	41	234	193	179	42	39	178	235	Azapetine
165*	58	163	42	166	164	139	63	660	Hexafluorenium Bromide
165*	89	63	30	39	119	182	78	182	2,4-Dinitrotoluene
165*	112	71	185	113	164	182	149	406	Bis(2-ethylhexyl)-3-hydroxyphthalate
165*	119	30	78	90	92	91	79	182	2,4-Dinitrotoluene
165*	119	30	118	92	91	90	78	182	2,6-Dinitrotoluene
165*	123	246	199	95	42	108	214	379	Droperidol
165*	164	150	222	58	134	77	39	222	Zectran
165*	183	70	59	71	43	295	112	406	Bis-(2-ethylhexyl)-4-hydroxyphthalate
165*	194	147	77	121	176	118	51	194	Meconin
165*	234	194	235	193	196	166	179	235	Azapetine*
165*	251	135	0	0	0	0	0	251	STP-NCS Derivative
165	57*	70	120	43	41	92	83	294	1-(2-Ethylhexyl)-4-hydroxyphthalate
165	58*	41	73	57	166	44	59	333	Bromodiphenhydramine
165	58*	255	359	166	73	199	45	359	Captodiamine
165	72*	42	180	178	179	91	73	309	Methadone
165	86*	167	99	152	166	58	56	311	Adiphenine
165	97*	164	206	84	249	135	136	249	Thiophencyclidine (Thio-PCP)
165	99*	56	194	167	207	152	208	266	Cyclizine
165	107*	123	95	121	0	0	0	327	Azaperone
165	120*	92	65	137	39	121	93	165	Benzocaine*
165	147*	201	167	166	105	203	117	432	Buclizine
165	163*	280	97	224	128	252	174	315	Diethyl-O-(3,6-dichloro-2-pyridyl) Phosphorothioate
165	166*	167	138	133	105	60	106	166	Ethionamide*
165	167*	117	119	83	169	130	132	200	Pentachloroethane
165	180*	137	179	134	77	39	109	180	4-Methyl-2,5-dimethoxybenzaldehyde
165	192*	251	193	190	77	51	166	251	3-Cyano-3,3-Diphenylpropionic Acid(Difenoxin metab.)
165	193*	121	120	194	93	65	92	358	Carbethyl Salicylate
165	196*	197	181	166	59	94	121	196	Methyl 3,4-dimethoxybenzoate
165	201*	45	166	56	203	42	58	374	Hydroxyzine

								MWT	COMPOUND
165	222*	76	223	166	105	104	90	222	Phenindione*
165	222*	223	221	194	166	164	76	222	Phenylindanione
165	246*	42	123	199	247	214	108	379	Droperidol*
165	256*	258	193	257	76	89	104	256	Clorindione
165	308*	309	121	55	154	98	56	429	Mebeverine*
165	368*	91	179	181	90	180	107	368	Tricresyl Phosphate(o-)
166*	54	109	53	81	136	137	0	166	1-Methylxanthine
166*	68	95	41	53	123	0	0	166	3-Methylxanthine
166*	68	123	53	42	41	95	0	166	7-Methylxanthine
166*	72	238	167	138	123	152	109	238	Pirimicarb
166*	96	209	55	167	71	42	69	209	Ethirimol
166*	131	102	138	77	91	195	0	223	Ketamine Metabolite
166*	135	107	167	136	77	108	59	166	Methyl methoxybenzoate(meta)
166*	151	30	135	195	165	44	167	195	2,5-Dimethoxy-4-methylphenethylamine
166*	164	129	131	168	94	96	47	164	Tetrachloroethylene
166*	165	167	138	133	105	60	106	166	Ethionamide*
166*	209	96	167	71	180	93	55	209	Dimethirimol (tech)
166	44*	151	91	77	79	39	42	209	2,5-Dimethoxy-4-methylamphetamine(STP)
166	44*	168	137	151	63	53	95	211	Methoxamine
166	44*	165	42	77	39	64	194	209	3-Methoxy-4,5-methylenedioxyamphetamine
166	44*	151	57	43	91	135	209	209	STP
166	72*	238	167	42	44	138	109	238	Pirimicarb
166	107*	77	39	108	45	65	120	166	3-(4-Hydroxyphenyl)propionic Acid
166	135*	133	77	105	137	134	136	166	Methyl methoxybenzoate(ortho)
166	135*	136	167	77	107	92	137	166	Methyl methoxybenzoate(para)
166	181*	223	73	43	45	208	75	223	4-Acetylaminophenol(Mono-TMS-)
167*	41	124	80	39	32	166	53	208	Allobarbitone
167*	41	168	124	43	97	169	96	210	Aprobarbitone
167*	41	124	168	97	39	169	45	210	Aprobarbitone*
167*	41	39	124	168	29	43	169	210	Aprobarbitone (Allypropymal, Aprozal, Alurate)
167*	41	168	124	39	97	141	67	224	Idobutal*
167*	41	168	39	97	57	124	53	224	Talbutal
167*	43	41	39	78	55	247	122	316	Sigmodal*
167*	71	166	139	41	78	140	168	290	Carbinoxamine
167*	127	168	294	295	285	0	0	294	MK-251 Metabolite
167*	165	117	119	83	169	130	132	200	Pentachloroethane
167*	168	41	43	97	124	39	55	238	Quinalbarbitone*
167*	168	41	97	124	39	57	53	224	Talbutal*
167*	209	43	124	39	41	53	140	288	Ibomal*
167	41*	124	39	80	53	68	141	208	Allobarbitone*
167	41*	168	39	124	97	141	181	224	Butalbital*
167	41*	168	39	125	97	44	53	224	Talbutal (5-Allyl-5-s-butylbarbituric Acid)
167	43*	237	41	39	124	168	55	316	Rectidon
167	72*	73	165	152	253	166	168	253	Emepromium Bromide GC Breakdown Product
167	86*	58	87	165	152	263	42	267	Emepromium Bromide GC Breakdown Product
167	136*	108	80	53	52	137	51	167	Orthocaine*
167	149*	57	71	70	43	150	41	390	Di-2-ethylhexyl Phthalate
167	168*	43	97	169	41	124	153	246	5-(2-Bromoallyl)-barbituric Acid
167	168*	41	181	124	97	169	141	224	Butalbital
167	168*	41	39	124	97	43	141	224	Butalbital
167	168*	41	43	124	97	169	195	238	Quinalbarbitone
167	168*	41	43	97	195	124	169	238	Secobarbitone
167	199*	198	166	99	154	69	77	199	Phenothiazine
167	239*	165	59	194	181	152	193	239	Prenylamine Metabolite

PEAKS								MWT	COMPOUND
168*	41	124	181	167	141	97	98	224	Butalbital
168*	126	405	43	368	326	446	0	560	Sotalol, tri TFA Derivative
168*	137	44	139	43	152	124	167	211	Methoxamine
168*	140	169	64	63	167	114	141	168	Norharman
168*	153	109	65	58	139	91	39	225	Methiocarb
168*	153	225	91	58	45	39	77	225	(4-(Methylthio)-3,5-dimethylphenyl N-methylcarbamate (Bayer 37344)
168*	153	109	91	225	45	169	154	225	(4-(Methylthio)-3,5-dimethylphenyl N-methylcarbamate (Methiocarb)
168*	167	43	97	169	41	124	153	246	5-(2-Bromoallyl)-barbituric Acid
168*	167	41	181	124	97	169	141	224	Butalbital
168*	167	41	39	124	97	43	141	224	Butalbital
168*	167	41	43	124	97	169	195	238	Quinalbarbitone
168*	167	41	43	97	195	124	169	238	Secobarbitone
168*	170	113	78	63	169	43	76	168	Zoxazolamine
168	41*	167	43	39	55	97	124	238	Secobarbitone
168	167*	41	43	97	124	39	55	238	Quinalbarbitone*
168	167*	41	97	124	39	57	53	224	Talbutal*
168	198*	180	140	197	169	179	196	198	Hydroxyharman
169*	58	149	168	170	57	167	43	240	Pheniramine
169*	58	168	170	72	167	44	42	240	Pheniramine*
169*	69	131	133	197	227	95	97	344	2,2,2-Trichloroethyl Heptafluorobutyrate
169*	73	170	75	249	233	79	171	338	3-Methylhydroxyhexobarbitone TMS Ether
169*	78	113	171	51	63	44	50	169	Chlorzoxazone
169*	78	171	113	115	63	51	170	169	Chlorzoxazone*
169*	92	289	65	290	39	333	184	398	Sulphasalazine*
169*	109	125	168	79	110	142	170	262	Demeton S-Methyl Sulphone
169*	109	125	79	59	0	0	0	230	Metasystox
169*	113	78	171	115	76	63	51	169	Chlorzoxazone
169*	137	125	95	55	41	84	43	240	Methyl-6-n-pentyl-4-hydroxy-2-oxocyclohex-3-ene-1-carboxylate
169*	156	170	143	144	115	129	142	354	Yohimbine
169*	170	91	115	142	65	116	143	170	4-Benzylpyrimidine
169*	184	140	185	126	226	55	41	254	1,2-Dimethylamobarbital (or 1,4-Dimethyl-)
169*	184	185	170	55	112	183	69	254	1,3-Dimethylamobarbitone
169*	184	120	112	121	126	225	226	254	Dimethylamobarbitone
169*	184	170	207	57	55	69	225	254	Dimethylamylobarbitone (N,N-Dimethyl)
169*	184	183	126	112	83	41	40	212	1,3-Dimethylbarbitone
169*	184	59	55	112	183	126	69	212	Dimethylbarbitone (N,N-Dimethyl-)
169*	184	41	112	185	69	183	55	240	1,3-Dimethylbutabarbitone
169*	184	183	170	112	212	55	185	240	Dimethylbutobarbitone (N,N-Dimethyl)
169*	184	41	43	183	69	112	55	254	1,3-Dimethylpentobarbitone
169*	184	112	183	69	55	185	58	254	Dimethylpentobarbitone (N,N-Dimethyl)
169*	184	170	41	185	112	55	183	254	Permethylated Amylobarbitone*
169*	184	126	183	41	112	83	55	212	Permethylated Barbitone*
169*	184	55	112	41	170	58	183	240	Permethylated Butobarbitone*
169*	184	55	112	41	185	170	183	268	Permethylated Hexethal*
169*	184	126	183	112	83	55	170	212	Permethylated Metharbitone*
169*	184	41	43	112	183	58	55	254	Permethylated Pentobarbitone*
169*	184	183	112	197	41	126	170	226	Permethylated Probarbital*
169*	184	41	58	112	57	128	183	240	Permethylated Secbutobarbitone*
169*	184	41	112	69	43	185	55	270	Thiopentone Dimethyl Derivative
169*	195	57	209	251	112	210	196	266	Dimethylnealbarbitone (N,N-Dimethyl)
169*	195	209	57	41	112	138	251	266	Nealbarbitone Dimethyl Derivative
169*	195	57	209	41	112	250	138	264	Permethylated Nealbarbitone*
169	55*	83	70	127	140	41	97	169	Bemegride Methyl Derivative
169	58*	168	42	167	170	72	51	240	Pheniramine

169	141*	401	42	96	41	142	70	401 Pipamazine*
169	142*	228	107	141	144	171	77	228 Mecoprop Methyl Ester
169	154*	83	55	41	42	98	69	197 Methyprylone Methyl Derivative
169	170*	141	115	171	139	142	39	170 2-Hydroxybiphenyl
169	170*	102	115	116	171	51	76	170 4-Methyl-5-phenylpyrimidine
169	184*	0	0	0	0	0	0	270 Dimethyl-3-hydroxyamylobarbitone (N,N'-)
169	193*	67	66	57	233	65	205	234 Cyclopal
169	195*	351	184	156	221	170	365	578 Deserpidine
169	196*	225	210	43	94	197	44	225 2-Butylamino-4-ethylamino-6-methoxy-s-triazine (s-Butyl)
169	212*	197	213	106	211	170	168	212 Harmine*
169	235*	79	236	77	91	81	112	264 Cyclobarbitone Dimethyl Derivative
169	249*	250	41	133	67	79	93	278 Permethylated Heptabarbitone*
169	296*	297	127	310	311	0	0	311 MK-251 (Dimethyl-4-(a,a,b,b-tetrafluorophenethyl)benzylamine)
170*	77	51	141	169	171	142	115	170 Diphenyl Ether
170*	102	169	115	77	51	60	61	170 4-Methyl-5-phenylpyrimidine
170*	114	156	155	113	197	83	42	538 Morfamquat
170*	115	114	88	171	62	116	89	170 Triazolophthalazine
170*	134	199	243	172	198	201	200	243 Benazolin
170*	153	39	51	79	125	53	126	170 Gallic Acid
170*	155	41	169	55	112	39	83	198 Metharbitone (Gemonil)
170*	155	55	156	171	69	212	112	240 Methylamylobarbitone (N-Methyl)
170*	155	169	55	112	83	98	69	198 Methylbarbitone (N-Methyl)
170*	169	141	115	171	139	142	39	170 2-Hydroxybiphenyl
170*	169	102	115	116	171	51	76	170 4-Methyl-5-phenylpyrimidine
170*	171	167	142	172	129	341	323	341 Lumi-LSD (10-hydroxy-9,10-dihydro-LSD)
170*	205	172	171	207	173	60	136	205 Chlorthiamid
170	153*	155	152	172	136	118	60	322 Chloramphenicol
170	155*	112	169	55	82	41	39	198 Metharbitone*
170	155*	156	98	169	112	197	55	226 Methylbutobarbitone (N-Methyl)
170	168*	113	78	63	169	43	76	168 Zoxazolamine
170	169*	91	115	142	65	116	143	170 4-Benzylpyrimidine
170	201*	203	188	165	166	202	142	374 Hydroxyzine
170	235*	40	236	180	81	73	121	264 Permethylated Cyclobarbitone*
171*	43	143	41	128	55	141	159	270 2,3-Dihydroxyquinalbarbitone
171*	100	173	75	74	136	99	50	171 Dichlobenil
171*	173	169	188	76	160	170	129	204 2-Isopropylquinoxaline-1,4-dioxide
171*	186	156	155	185	130	116	144	186 1-Methyl-1,2,3,4-tetrahydro-beta-carboline
171	123*	143	81	128	91	172	41	338 Bioresmethrin
171	170*	167	142	172	129	341	323	341 Lumi-LSD (10-hydroxy-9,10-dihydro-LSD)
171	173*	175	91	93	92	94	79	256 Bromoform
171	200*	129	117	115	145	183	199	200 Hydrallazine Metabolite
172*	42	98	57	44	201	91	36	275 Alphameprodine
172*	92	156	65	108	39	44	41	214 Sulphacetamide*
172*	92	156	65	108	173	39	174	172 Sulphanilamide*
172*	98	201	91	202	275	96	200	275 Alphameprodine
172*	157	42	173	242	69	71	97	242 Thiopentone
172*	157	173	43	41	55	69	71	242 Thiopentone*
172*	187	84	57	42	188	44	43	261 Alphaprodine
172*	187	144	84	42	57	188	44	261 Alphaprodine
172*	187	84	42	57	171	144	186	261 Alphaprodine
172*	187	42	84	129	144	44	91	261 Alphaprodine (alpha-Prodine)
172*	187	144	84	42	57	188	44	261 Betaprodine
172*	187	84	42	44	57	29	43	261 Betaprodine*
172*	214	42	110	57	173	91	44	287 Allylprodine*

PEAKS								MWT	COMPOUND
172	36*	106	42	57	78	44	214	495	Nicomorphine
172	97*	91	105	57	42	77	201	275	Alphameprodine
172	97*	91	105	57	42	77	201	275	Betameprodine
172	157*	41	173	43	69	55	39	242	Thiopentone
173*	108	65	92	80	39	63	156	173	Sulphanilic Acid
173*	125	127	93	158	99	143	0	330	Malathion
173*	171	175	91	93	92	94	79	256	Bromoform
173*	340	168	167	165	341	174	322	340	Diphenadione*
173	76*	104	111	148	50	169	130	258	Thalidomide
173	107*	77	175	79	253	251	105	330	Triarimol
173	127*	115	126	77	101	128	143	173	1-Nitronaphthalene
173	160*	232	161	174	145	158	0	232	Melatonin
173	171*	169	188	76	160	170	129	204	2-Isopropylquinoxaline-1,4-dioxide
173	229*	44	41	115	128	91	129	244	Xylometazoline
173	246*	41	91	135	0	0	0	246	Santonin (alpha)
173	260*	259	244	201	216	128	145	262	5-(4'-Hydroxy-2'-methoxybenzylidene)-barbituric Acid
173	518*	264	519	373	376	520	249	518	Benziodarone*
174*	63	64	201	90	52	39	65	201	Thiabendazole
174*	77	91	105	132	0	0	0	174	Norantipyrine
174*	93	95	172	176	91	81	79	172	Methylene Bromide
174*	128	101	116	77	102	75	51	174	6-Nitroquinoline
174*	128	116	101	77	89	102	75	174	8-Nitroquinoline
174*	160	161	218	275	175	162	149	275	Physostigmine
174*	233	175	159	131	130	234	158	233	2-Methyl-5-methoxyindole-3-acetic Acid Methyl Ester
174*	247	98	175	173	248	112	245	330	Dipiproverine
174*	291	73	175	159	131	75	158	291	Indomethacin Derivative TMS Ester
174*	382	161	187	159	175	383	41	382	3,5-Cholestadiene-7-one
174	77*	107	135	176	65	281	146	301	Isoxsuprine
174	93*	95	59	172	176	74	43	172	Dibromomethane
174	130*	77	131	103	102	129	51	174	3-Indoleacetamide
174	201*	63	64	202	65	129	90	201	Thiabendazole
174	218*	160	161	275	219	175	162	275	Physostigmine*
175*	70	176	132	379	204	56	217	379	Oxypertine*
175*	104	77	176	105	132	51	204	204	Methoin Metabolite (Phenylethylhydantoin)
175*	188	132	123	326	95	165	311	326	Butropipazone
175*	234	45	199	177	236	73	161	234	2,4-Dichlorophenoxymethyl Acetate
175*	248	177	250	185	145	69	57	248	Ethyl 2,4-dichlorophenoxyacetate
175	104*	77	119	51	190	42	161	190	5-Phenylhydantoin(5-Methyl-)
175	111*	75	99	113	177	73	63	302	Chlorfenson
175	111*	75	85	30	276	127	113	276	Chlorpropamide*
175	111*	205	141	113	75	112	56	290	Methylchlorpropamide (N-Methyl)
175	111*	205	75	113	207	84	177	205	Methyl-p-chlorobenzenesulphonamide (N-Methyl)
175	115*	111	58	177	290	113	75	290	Methyl Enol Ether of Chlorpropamide
175	130*	77	131	103	102	129	176	175	Indolyl-3-acetic Acid
175	210*	212	177	111	75	214	73	210	Trichlorocresol(m-)
175	220*	162	222	177	145	147	164	220	3,4-D
175	306*	291	115	91	202	187	129	306	Methylated Coumatetralyl*
176*	107	90	89	77	105	79	70	176	Pemoline
176*	161	178	163	133	135	63	180	176	2,4-Dichlorophenyl Methyl Ether
176	72*	91	58	177	65	281	42	281	Alverine
176	72*	91	58	177	41	42	30	281	Alverine*
176	91*	148	133	44	174	65	107	299	Buphenine
176	98*	42	118	41	119	51	39	176	Cotinine
176	107*	90	77	70	105	42	79	176	Pemoline*

								MWT	COMPOUND
176	147*	148	91	39	65	51	63	176	6-Hydroxy-4-methylcoumarin
176	148*	147	91	39	51	120	63	176	7-Hydroxy-4-methylcoumarin
176	178*	97	128	31	180	126	47	176	Bromochloro-1,1-difluoroethylene
176	178*	97	61	128	180	31	177	176	Bromochloro-1,2-difluoroethylene
176	201*	202	56	258	70	71	42	377	Morazone*
176	363*	286	255	365	288	192	220	394	Chlorthalidone Tetramethyl Derivative
177*	43	176	219	147	302	69	111	302	Cannabielsoin (C3)
177*	43	41	135	91	178	93	95	192	Ionone(beta-)
177*	106	0	0	0	0	0	0	177	Bethanidine
177*	149	166	65	194	178	104	121	222	Diethyl Terephthalate
177	90*	105	77	106	51	89	50	177	5-Phenyloxazilidindione-(2,4)(Pemoline Metabolite 1)
177	91*	44	106	45	78	123	51	298	Nialamide*
177	105*	77	209	254	210	103	181	254	Ketoprofen*
177	106*	78	178	51	105	107	149	178	Nikethamide
177	149*	150	65	76	105	176	104	222	Diethyl Phthalate
177	225*	131	41	77	240	38	103	240	Dinoterb
177	239*	192	149	63	77	92	134	239	Mescaline Precursor
178*	44	135	179	77	84	107	41	301	Isoxsuprine*
178*	135	176	107	70	77	84	0	301	Isoxsuprine (Duvadilan)
178*	176	97	128	31	180	126	47	176	Bromochloro-1,1-difluoroethylene
178*	176	97	61	128	180	31	177	176	Bromochloro-1,2-difluoroethylene
178*	192	272	466	244	288	191	273	466	Cephaeline*
178	84*	161	133	118	0	0	0	178	Nicotine-N-oxide
178	149*	65	177	150	76	121	105	222	Diethyl Phthalate
178	179*	151	180	152	177	150	90	179	5,6-Benzoquinoline
178	179*	180	208	152	177	176	206	208	Dibenzo(a,d)cycloheptene-10,11-epoxide (5H-Dibenzo)
178	220*	221	205	163	177	179	42	221	Hydrocotarnine
179*	137	152	304	93	153	97	135	304	Diazinon
179*	178	151	180	152	177	150	90	179	5,6-Benzoquinoline
179*	178	180	208	152	177	176	206	208	Dibenzo(a,d)cycloheptene-10,11-epoxide (5H-Dibenzo)
179*	180	178	165	225	89	195	210	253	Nefopam
179*	194	164	151	91	77	121	149	194	Butylmethoxyphenol(3-t-) Methyl Ether
179	58*	180	225	178	165	42	210	253	Nefopam*
179	180*	223	252	0	0	0	0	252	Carbamazepine Metabolite
179	209*	210	135	59	89	161	193	224	Salicylic Acid Methyl Ester TMS Ether
179	289*	43	150	231	304	219	55	304	Cannabichromanon (C3)
180*	36	102	115	152	138	146	125	237	Ketamine
180*	55	109	67	82	137	181	42	180	Theobromine
180*	67	82	109	55	181	66	81	180	Theobromine
180*	68	123	53	42	95	150	151	180	Paraxanthine
180*	77	104	209	223	252	51	181	252	Diphenylhydantoin
180*	77	104	266	237	209	165	0	266	Methyldilantin
180*	77	118	91	104	65	89	51	266	5-Methylphenyl-5-phenylhydantoin(p-5-phenylhydantoin)(Metabolite)
180*	95	181	68	41	123	53	96	420	Aminophylline*
180*	95	68	41	58	53	123	96	269	Bufylline*
180*	95	224	123	68	109	193	194	224	Oxyethyltheophylline
180*	95	68	41	43	53	57	55	180	Theophylline
180*	95	68	41	53	181	96	40	180	Theophylline*
180*	104	77	51	209	223	165	181	252	Diphenylhydantoin
180*	104	266	77	237	57	209	71	266	3-Methyldilantin
180*	104	223	77	209	252	51	165	252	Phenytoin*
180*	121	181	122	59	91	148	182	180	Methyl Methoxyphenyl-(meta)-acetate
180*	124	152	91	44	40	42	105	271	Phosphorylated Amphetamine
180*	165	137	179	134	77	39	109	180	4-Methyl-2,5-dimethoxybenzaldehyde

```
        PEAKS                      MWT    COMPOUND


180* 179  223  252    0    0    0    0   252 Carbamazepine Metabolite
180* 182  209  152  138  211  102  154   237 Ketamine
180* 182  184  145  109  147   75  181   180 1,2,4-Trichlorobenzene
180* 194  223  109   95   42  254  193   254 Diphylline
180* 194  223  109   95  193  166   42   254 Dyphylline
180* 194   77  222  241  179  242  224   241 Mefenamic Acid
180* 209  104  223  252   77  181  165   252 Diphenylhydantoin
180* 209  280   77  181  104  251  165   280 3-Ethyl-5,5-diphenylhydantoin
180* 209  182  152  181   30  211  138   237 Ketamine*
180* 209  223  252  104   73    0    0   252 Phenytoin
180* 223  194  254  109   95  193  166   254 Diprophylline*
180* 223  252  179  152    0    0    0   252 10,11-Epoxycarbamazepine
180* 264  193  109  194  181   67   42   264 Pentifylline*
180* 266  237  208  118  223  194   77   266 5-Phenylhydantoin(5-(p-methylphenyl)-)
180* 308  279  208   77  251  165  104   308 1,3-Diethyl-5,5-diphenylhydantoin
180   43*  42   45   44   64  100  222   222 Acetazolamide*
180   72*  84  135   44   57   45   41   279 Karbutilate
180   72* 198   73   42   56   70   71   284 Promethazine
180   86*  58   87  198  298   41  212   298 Diethazine
180  121*  91  181  122  148  107   93   180 Methyl Methoxyphenyl-(ortho)-acetate
180  121* 122  181    0    0    0    0   180 Methyl Methoxyphenyl-(para)-acetate
180  137* 147  151  162   58  134   65   237 3-Hydroxycarbofuran
180  152* 138  102  154  182  209    0   237 Ketamine Impurity
180  179* 178  165  225   89  195  210   253 Nefopam
180  208* 205  207  206  209  104   77   208 9,10-Diaminophenanthrene
180  208* 109   67   42  137   55  179   208 1-Ethyltheobromine
180  296* 267  210  104   77  134  297   296 3-Methyl-3-hydroxydilantin Methyl Ether
181*  41  182   39  124   53  138   97   224 Enallylpropymal*
181*  86   72  182   85  152  108   99   419 Methantheline Bromide (Banthine)
181* 156   55  141  180  138  155   39   210 Crotylbarbitone
181* 166  223   73   43   45  208   75   223 4-Acetylaminophenol(Mono-TMS-)
181* 183  219  217  109  111  221   51   288 BHC(A-)
181* 183  109  219  111  217   51  221   288 Lindane(G-BHC)
181* 223  166   73   43   45  208   75   223 Paracetamol TMS Ester
181* 253   72    0    0    0    0    0   253 Mescaline-NCS Derivative
181* 395  198  251  397  396  199  666   666 Syrosingopine
181   72*  85  152   42   44   58   43   419 Methantheline Bromide
181   86*  43   44   41  114   42  152   353 Propantheline
181  182* 149   44  167  211   40   41   211 Mescaline
181  182* 167  211  151  148  139  136   211 Mescaline
181  182*  97  209  167   53   55  124   252 Methylquinalbarbitone (N-Methyl)
181  183* 109  111  219  217   51  221   288 Lindane
181  196* 125   39  110   95   93   51   196 3,4,5-Trimethoxybenzaldehyde (Mescaline Precursor)
181  221* 207  196  223  325  222  180   325 Ergonovine
181  241*  91   66   30   75  224   90   287 Tetryl
182*  30  167  181   51  107  151  211   211 Mescaline
182*  30  181  167  211  183  151  148   211 Mescaline*
182*  57   43   55   40   69   41  181   182 Harman
182*  82   83  105   94   77   96  303   303 Cocaine(beta)
182* 121  184  154  367   97  241   58   367 Phosalone
182* 121   97  184  111  154   65  138   367 Phosalone
182* 181  149   44  167  211   40   41   211 Mescaline
182* 181  167  211  151  148  139  136   211 Mescaline
182* 181   97  209  167   53   55  124   252 Methylquinalbarbitone (N-Methyl)
```

182*	276	108	169	184	278	171	305	305	Cruformate Metabolite
182	30*	94	66	65	78	92	53	182	3,4-Dinitrotoluene
182	43*	167	225	181	183	151	142	225	3,4,5-Trimethoxyamphetamine
182	44*	167	151	181	139	183	136	225	2,4,5-Trimethoxyamphetamine
182	82*	83	96	103	147	148	131	329	Cinnamoylcocaine
182	82*	83	105	303	77	94	96	303	Cocaine*
182	180*	209	152	138	211	102	154	237	Ketamine
182	180*	184	145	109	147	75	181	180	1,2,4-Trichlorobenzene
182	228*	363	88	229	76	276	257	398	Clefamide*
182	256*	276	108	169	184	291	278	291	Cruformate
183*	58	94	42	182	40	196	73	240	GS 29696
183*	76	90	50	91	120	106	92	183	Saccharin*
183*	77	252	184	105	91	51	64	252	1,2-Diphenyl-3,5-pyrazolidinedione
183*	77	308	184	105	93	309	91	308	Phenylbutazone
183*	105	77	84	82	165	152	129	311	4-Piperidylbenzilate
183*	163	184	165	91	77	127	89	390	Permethrin
183*	181	109	111	219	217	51	221	288	Lindane
183*	185	155	143	77	89	242	148	242	Cruformate Metabolite
183	76*	50	119	120	0	0	0	183	Saccharin
183	77*	266	105	51	91	78	65	266	4-Methyl-1,2-diphenyl-3,5-pyrazolinedione
183	77*	308	184	105	55	51	41	308	Phenylbutazone
183	77*	105	184	55	93	91	51	308	Phenylbutazone
183	77*	308	184	105	55	51	41	308	Phenylbutazone*
183	86*	182	105	77	165	116	312	327	Benactyzine
183	165*	70	59	71	43	295	112	406	Bis-(2-ethylhexyl)-4-hydroxyphthalate
183	181*	219	217	109	111	221	51	288	BHC(A-)
183	181*	109	219	111	217	51	221	288	Lindane(G-BHC)
183	245*	89	247	0	0	0	0	245	Flurazepam Metabolite
184*	40	169	140	126	183	44	55	212	1,2-Dimethylbarbitone (or 1,4-Dimethyl-)
184*	44	91	63	53	107	92	51	184	2,4-Dinitrophenol
184*	92	185	65	39	108	66	186	249	Sulphapyridine*
184*	112	42	55	170	183	212	58	240	1,3-Dimethylbutethal
184*	115	88	62	185	114	51	50	184	3-Methyl-s-triazolopthalazine (Hydralazine Metab)
184*	115	185	114	183	88	129	155	184	3-Methyltriazolophthalazine
184*	148	58	118	186	154	149	104	255	3-Chloro-6-cyano-2-norbornanon-o-(methylcarbamoyl) Oxime
184*	169	0	0	0	0	0	0	270	Dimethyl-3-hydroxyamylobarbitone (N,N'-)
184*	185	92	65	108	186	66	183	249	Sulphapyridine
184	169*	140	185	126	226	55	41	254	1,2-Dimethylamobarbital (or 1,4-Dimethyl-)
184	169*	185	170	55	112	183	69	254	1,3-Dimethylamobarbitone
184	169*	120	112	121	126	225	226	254	Dimethylamobarbitone
184	169*	170	207	57	55	69	225	254	Dimethylamylobarbitone (N,N-Dimethyl)
184	169*	183	126	112	83	41	40	212	1,3-Dimethylbarbitone
184	169*	59	55	112	183	126	69	212	Dimethylbarbitone (N,N-Dimethyl-)
184	169*	41	112	185	69	183	55	240	1,3-Dimethylbutabarbitone
184	169*	183	170	112	212	55	185	240	Dimethylbutobarbitone (N,N-Dimethyl)
184	169*	41	43	183	69	112	55	254	1,3-Dimethylpentobarbitone
184	169*	112	183	69	55	185	58	254	Dimethylpentobarbitone (N,N-Dimethyl)
184	169*	170	41	185	112	55	183	254	Permethylated Amylobarbitone*
184	169*	126	183	41	112	83	55	212	Permethylated Barbitone*
184	169*	55	112	41	170	58	183	240	Permethylated Butobarbitone*
184	169*	55	112	41	185	170	183	268	Permethylated Hexethal*
184	169*	126	183	112	83	55	170	212	Permethylated Metharbitone*
184	169*	41	43	112	183	58	55	254	Permethylated Pentobarbitone*
184	169*	183	112	197	41	126	170	226	Permethylated Probarbital*

	PEAKS							MWT	COMPOUND
184	169*	41	58	112	57	128	183	240	Permethylated Secbutobarbitone*
184	169*	41	112	69	43	185	55	270	Thiopentone Dimethyl Derivative
184	240*	241	212	92	138	63	223	240	Danthron
184	277*	88	279	275	63	278	53	459	Bromophenoxim
184	290*	185	104	77	168	291	198	290	Sulthiame*
184	343*	300	70	43	59	334	226	343	Acetyldihydrocodeine
185*	43	129	41	57	259	157	112	402	Acetyl Tri-n-butyl Citrate
185*	57	43	129	29	41	259	157	402	Acetyl Tri-n-butylcitrate
185*	58	141	111	128	185	113	75	185	Chlorophenoxyacetamide (p-Chloro) (Iproclozide Metab)
185*	115	141	230	170	45	153	142	230	Naproxen
185*	115	200	171	128	44	91	199	200	Tetrahydrozoline
185*	200	186	89	57	42	103	131	200	Acetonide
185*	230	141	186	184	115	170	153	230	Naproxen*
185*	244	141	115	170	186	153	154	244	Naproxen Methyl Ester
185	109*	79	202	145	187	200	199	220	Dichlorvos
185	109*	79	145	187	47	220	110	220	Dichlorvos
185	183*	155	143	77	89	242	148	242	Cruformate Metabolite
185	184*	92	65	108	186	66	183	249	Sulphapyridine
185	186*	92	65	108	39	93	187	250	Sulphadiazine*
185	226*	241	170	43	68	41	71	241	Terbutryn
185	246*	445	70	125	154	213	87	445	Thiopropazate
185	299*	242	115	42	214	128	243	299	Dihydrocodienone
186*	41	201	188	43	152	273	203	273	Dichlozoline
186*	70	83	42	85	71	57	56	275	Dimethyldemerol
186*	91	150	187	143	65	115	116	277	2-Benzyl-2-methyl-5-phenyl-2,3-dihydropyrimid-4-one
186*	114	29	72	42	113	27	109	186	Carbimazole*
186*	130	77	53	120	93	221	51	221	Prynachlor
186*	185	92	65	108	39	93	187	250	Sulphadiazine*
186*	201	42	202	187	56	57	71	275	Trimeperidine
186*	218	77	295	263	105	51	158	295	Mebendazole*
186	141*	77	143	105	0	0	0	260	Alclofenac Metabolite
186	171*	156	155	185	130	116	144	186	1-Methyl-1,2,3,4-tetrahydro-beta-carboline
187*	241	242	174	30	126	175	301	301	Pentaquine*
187	43*	42	171	229	86	144	170	229	Amiloride*
187	172*	84	57	42	188	44	43	261	Alphaprodine
187	172*	144	84	42	57	188	44	261	Alphaprodine
187	172*	84	42	57	171	144	186	261	Alphaprodine
187	172*	42	84	129	144	44	91	261	Alphaprodine (alpha-Prodine)
187	172*	144	84	42	57	188	44	261	Betaprodine
187	172*	84	42	44	57	29	43	261	Betaprodine*
187	230*	134	217	96	83	461	109	461	Pimozide
187	281*	324	121	43	0	0	0	324	4'-Hydroxywarfarin
188*	47	82	96	29	77	84	56	516	Dichloralphenazone*
188*	96	77	56	105	189	55	51	188	Phenazone*
188*	152	190	153	76	189	151	150	188	2-Chlorobiphenyl
188*	189	159	132	53	173	131	145	232	Nalidixic Acid*
188	58*	130	59	42	143	129	115	188	Dimethyltryptamine(N,N-)*
188	153*	126	190	161	154	137	189	188	Chlorobenzylidenemalononitrile (o-Chloro - CS gas)
188	175*	132	123	326	95	165	311	326	Butropipazone
188	203*	232	117	204	40	115	70	260	1,2-Dimethylphenobarbital (or 1,4-Dimethyl-)
188	247*	160	63	215	43	106	216	247	1,2-Dimethyl-5-methoxyindole-3-acetic Acid Me Ester (Indomethacin Der)
188	292*	121	130	115	293	129	128	292	Racumin (4-Hydroxy-3-(1,2,3,4-tetrahydro-1-naphthyl)coumarin)
189*	41	190	104	149	103	39	115	246	Diallyl Phthalate
189*	103	190	104	188	102	77	174	189	Phensuximide(ii)*

Peak Index   *Asterisk Indicates True Base Peak*   Page: 247

							MWT	COMPOUND
189*	104	77	190	51	105	103	132	218 Mephenytoin
189*	104	77	190	51	0	0	0	218 Methetoin
189*	104	190	77	44	105	132	103	218 Methoin*
189*	105	201	285	165	166	190	134	390 Meclozine*
189*	105	107	77	55	91	79	190	344 Oxyphencyclimine
189*	117	132	160	91	115	77	103	217 Glutethimide
189*	132	117	160	91	115	103	77	217 Glutethimide*
189	104*	77	51	105	103	190	56	218 Mephenytoin
189	104*	103	78	51	77	105	52	189 Phensuximide(i)*
189	104*	146	105	0	0	0	0	189 Phenylglutarimide(alpha)
189	104*	103	117	78	91	51	146	189 3-Phenylpiperidin-2,6-dione(Glutethimide Metabolite)
189	105*	165	42	166	106	79	77	390 Meclizine
189	105*	36	38	201	165	285	166	390 Meclozine
189	117*	132	115	91	160	77	39	217 Glutethimide (Doriden)
189	121*	292	310	263	265	131	251	310 Warfarin Alcohol
189	125*	230	127	191	63	232	90	230 Diazoxide*
189	130*	131	79	190	52	103	51	189 Methyl 3-Indoleacetate
189	147*	73	74	148	75	190	146	204 Urea di-TMS Derivative
189	150*	246	57	151	174	188	190	246 4-(Methyl-2'-propynylamino)-3,5-dimethylphenyl N-Methylcarbamate
189	188*	159	132	53	173	131	145	232 Nalidixic Acid*
189	247*	249	191	243	55	29	245	302 Ethacrynic Acid*
190*	125	134	162	91	41	39	135	278 Absisic Acid
190*	141	154	124	191	155	142	181	260 5-(2-Chloroethyl)-5-isopentyl-barbituric Acid
190	91*	119	72	191	41	44	162	281 Formyldi-(1-isopropylphenyl)amine* (N-)
190	91*	119	70	71	55	77	72	369 Tri-(1-isopropylphenyl)amine
190	146*	117	118	161	189	103	91	218 Primidone*
190	146*	117	118	189	161	103	115	218 Primidone (5-Ethylhexahydro-5-phenylpyrimidine-4,6-dione)
190	192*	111	75	194	50	113	74	190 1-Bromo-2-chlorobenzene
191*	70	44	58	189	84	165	192	263 Protriptyline
191*	121	77	104	122	43	51	192	191 Amiphenazole*
191*	206	193	208	141	195	163	143	206 1,4-Dichloro-2,5-dimethoxybenzene (Chloroneb)
191*	226	228	163	193	192	165	99	226 2,3-Dichloro-1,4-naphthoquinone (Dichlone)
191	70*	44	189	43	165	69	192	263 Protriptyline
191	73*	75	45	74	104	147	167	516 5-(3,4-Dihydroxycyclohexa-1,5-dienyl)-3-me-5-phenylhydantoin TMS Deriv
191	146*	147	91	117	63	65	39	191 5-Hydroxyindole-3-acetic Acid
191	146*	147	130	57	145	117	89	191 5-Hydroxyindole-3-acetic Acid
191	192*	189	165	152	190	193	139	192 Dibenzo(a,d)cycloheptene (5H-Dibenzo)
191	193*	121	195	123	158	156	228	226 Antimony Trichloride
191	222*	223	207	79	147	164	190	222 Methyl 3,4-Dimethoxycinnamate
192*	160	342	59	150	177	105	209	342 Thiophanate Methyl
192*	165	251	193	190	77	51	166	251 3-Cyano-3,3-Diphenylpropionic Acid(Difenoxin metab.)
192*	190	111	75	194	50	113	74	190 1-Bromo-2-chlorobenzene
192*	191	189	165	152	190	193	139	192 Dibenzo(a,d)cycloheptene (5H-Dibenzo)
192*	193	287	176	191	286	270	285	478 Dehydroemetine*
192*	206	272	480	288	246	205	191	480 Emetine*
192	127*	109	67	43	193	39	79	224 Mevinphos
192	127*	109	67	43	193	79	39	224 Mevinphos GC peak 2
192	127*	109	67	43	164	193	224	224 Phosdrin
192	178*	272	466	244	288	191	273	466 Cephaeline*
192	193*	165	191	194	167	190	63	236 Carbamazepine
192	193*	236	191	194	165	190	237	236 Carbamazepine*
192	207*	164	74	149	134	124	0	207 3,4,5-Trimethoxyphenylacetonitrile
192	254*	193	154	223	255	180	221	254 Lysergol*
193*	165	121	120	194	93	65	92	358 Carbethyl Salicylate

								MWT	COMPOUND
193*	169	67	66	57	233	65	205	234	Cyclopal
193*	191	121	195	123	158	156	228	226	Antimony Trichloride
193*	192	165	191	194	167	190	63	236	Carbamazepine
193*	192	236	191	194	165	190	237	236	Carbamazepine*
193*	228	89	255	77	165	51	110	256	Desmethylmedazepam (N-)
193*	231	247	123	194	233	136	316	316	Cannabigerol
193*	255	228	256	257	165	230	258	256	Desmethylmedazepam
193*	264	43	72	71	220	192	194	264	Mianserin*
193*	264	194	43	71	72	220	109	264	Mianserin (Bolvidon)
193	44*	195	194	208	234	71	235	266	Desipramine
193	58*	194	42	234	235	39	179	294	Trimipramine
193	67*	66	41	169	39	65	77	234	Cyclopentobarbitone*
193	73*	43	195	210	237	75	252	252	Propyl Hydroxybenzoate (Propylparaben) TMS Ether
193	192*	287	176	191	286	270	285	478	Dehydroemetine*
193	194*	136	43	41	109	39	137	264	Sigmodal Methylation Artifact*
193	208*	130	115	91	165	179	207	265	2-Ethyl-5-Methyl-3,3-Diphenyl-1-Pyrroline
193	235*	236	136	194	39	138	121	314	Permethylated Brallobarbitone*
193	265*	115	130	42	208	56	264	265	3,3-Diphenyl-1,5-dimethyl-2-pyrrolidone(Methadone metab.)
193	272*	209	78	192	273	194	181	272	Zolimidine
193	324*	206	73	325	75	194	45	339	Salicyluric Acid TMS Ester TMS Ether
193	335*	192	207	234	233	167	180	335	6-Cyano-6-desmethyl-LSD
194*	109	55	67	82	42	137	41	194	Caffeine
194*	109	67	82	55	193	195	81	194	Caffeine
194*	109	55	67	82	193	42	40	194	Caffeine
194*	109	55	67	82	195	42	110	194	Caffeine*
194*	136	193	81	137	109	39	197	236	Ibomal Methylation Artifact*
194*	193	136	43	41	109	39	137	264	Sigmodal Methylation Artifact*
194*	195	180	193	89	90	77	63	195	Impramine Degradation Product
194*	195	238	193	72	178	45	196	238	Nomifensine*
194	30*	167	165	193	116	77	179	211	Prenylamine Metabolite
194	44*	109	67	86	238	56	85	562	Acepifylline*
194	84*	55	85	56	41	30	99	277	Perhexiline*
194	121*	134	163	122	195	135	108	194	Methyl Methoxyphenyl-(para)-propionate
194	135*	179	136	91	137	180	40	194	Methyl Acetylsalicylate
194	163*	135	76	164	103	120	77	194	Dimethyl Isophthalate
194	165*	147	77	121	176	118	51	194	Meconin
194	179*	164	151	91	77	121	149	194	Butylmethoxyphenol(3-t-) Methyl Ether
194	180*	223	109	95	42	254	193	254	Diphylline
194	180*	223	109	95	193	166	42	254	Dyphylline
194	180*	77	222	241	179	242	224	241	Mefenamic Acid
194	195*	180	96	193	196	83	167	195	Iminodibenzyl(Imipramine Metabolite)
195*	36	237	41	241	75	239	170	404	Endosulphan
195*	41	194	53	138	70	137	79	236	1,2-Dimethylallobarbital (or 1,4-Dimethyl-)
195*	41	39	69	67	53	152	135	224	Vinbarbital
195*	41	141	69	39	152	135	196	224	Vinbarbitone*
195*	120	43	210	135	75	73	92	252	Acetylsalicylic Acid TMS Ester
195*	138	41	53	194	80	110	58	236	1,3-Dimethylallobarbitone
195*	138	41	80	53	194	39	110	236	Permethylated Allobarbitone*
195*	169	351	184	156	221	170	365	578	Deserpidine
195*	194	180	96	193	196	83	167	195	Iminodibenzyl(Imipramine Metabolite)
195*	196	41	138	53	111	181	58	238	1,3-Dimethylaprobarbitone
195*	196	138	41	111	181	110	53	238	Permethylated Aprobarbitone*
195*	196	41	138	111	58	181	53	252	Permethylated Enallylpropymal*
195*	196	41	138	181	169	111	39	252	Permethylated Idobutal*

							MWT	COMPOUND
195*	196	41	138	181	111	58	110	252 Permethylated Talbutal*
195*	196	41	138	181	55	237	111	310 Trimethylquinalbarbitone Carboxylic Acid Metabolite
195*	197	199	123	133	135	160	167	195 2,4,5-Trichloroaniline
195*	199	200	186	214	152	174	251	608 Reserpine
195*	212	114	30	194	72	44	43	339 Disopyramide*
195*	235	234	193	208	266	194	84	266 Desipramine
195*	237	138	238	196	43	39	110	316 Permethylated Ibomal*
195*	265	43	196	41	44	39	110	344 Permethylated Sigmodal*
195	58*	59	72	388	89	315	42	388 Trimethobenzamide
195	58*	42	388	59	57	43	56	388 Trimethobenzamide
195	169*	57	209	251	112	210	196	266 Dimethylnealbarbitone (N,N-Dimethyl)
195	169*	209	57	41	112	138	251	266 Nealbarbitone Dimethyl Derivative
195	169*	57	209	41	112	250	138	264 Permethylated Nealbarbitone*
195	194*	180	193	89	90	77	63	195 Impramine Degradation Product
195	194*	238	193	72	178	45	196	238 Nomifensine*
195	196*	41	138	181	111	209	169	252 1,3-Dimethylbutalbital
195	196*	237	181	138	45	41	69	268 Dimethyl-3'-hydroxyquinalbarbitone
195	196*	43	138	181	237	197	69	280 Dimethyl-3'-ketoquinalbarbitone
195	196*	181	111	138	55	197	266	266 Dimethylquinalbarbitone (N,N-Dimethyl)
195	196*	41	138	169	111	112	181	252 Permethylated Butalbital*
195	196*	41	138	181	111	40	43	266 Permethylated Quinalbarbitone*
195	196*	181	41	138	111	53	110	266 Secobarbitone Dimethyl Derivative
195	235*	208	44	234	193	194	71	266 Desipramine*
195	235*	234	208	193	266	194	45	266 ((3-Methylaminopropyl)-iminodibenzyl)(N-)(Imipramine Metabolite)
195	296*	58	297	253	196	212	84	592 Hexobendine*
195	365*	221	366	31	29	212	197	578 Deserpidine (Canescine, Harmonyl)
195	578*	577	367	351	579	366	365	578 Deserpidine*
196*	41	195	209	138	181	67	43	252 1,2-Dimethylbutalbital (or 1,4-Dimethyl-)
196*	165	197	181	166	59	94	121	196 Methyl 3,4-dimethoxybenzoate
196*	169	225	210	43	94	197	44	225 2-Butylamino-4-ethylamino-6-methoxy-s-triazine (s-Butyl)
196*	181	125	39	110	95	93	51	196 3,4,5-Trimethoxybenzaldehyde (Mescaline Precursor)
196*	195	41	138	181	111	209	169	252 1,3-Dimethylbutalbital
196*	195	237	181	138	45	41	69	268 Dimethyl-3'-hydroxyquinalbarbitone
196*	195	43	138	181	237	197	69	280 Dimethyl-3'-ketoquinalbarbitone
196*	195	181	111	138	55	197	266	266 Dimethylquinalbarbitone (N,N-Dimethyl)
196*	195	41	138	169	111	112	181	252 Permethylated Butalbital*
196*	195	41	138	181	111	40	43	266 Permethylated Quinalbarbitone*
196*	195	181	41	138	111	53	110	266 Secobarbitone Dimethyl Derivative
196*	198	161	200	163	86	242	245	240 Picloram
196*	198	197	195	200	225	223	256	254 Picloram Methyl Ester
196*	198	97	132	200	134	99	62	196 2,4,6-Trichlorophenol
196*	198	254	256	200	197	209	211	254 2,4,5-Trichlorophenoxyacetic Acid
196*	254	209	195	0	0	0	0	254 2,4,5-Trichlorophenoxyacetic Acid (2,4,5-T)
196	195*	41	138	53	111	181	58	238 1,3-Dimethylaprobarbitone
196	195*	138	41	111	181	110	53	238 Permethylated Aprobarbitone*
196	195*	41	138	111	58	181	53	252 Permethylated Enallylpropymal*
196	195*	41	138	181	169	111	39	252 Permethylated Idobutal*
196	195*	41	138	181	111	58	110	252 Permethylated Talbutal*
196	195*	41	138	181	55	237	111	310 Trimethylquinalbarbitone Carboxylic Acid Metabolite
196	198*	97	200	132	134	99	133	196 2,4,5-Trichlorophenol
196	221*	222	307	112	181	205	154	325 Ergometrine Maleate
196	239*	268	120	197	77	225	104	268 5-(4-Hydroxyphenyl)-5-Phenylhydantoin(Para HPPH)
197*	58	99	165	309	112	152	198	309 Methixene
197*	155	140	0	0	0	0	0	197 Phenacetin Metabolite

```
              PEAKS                        MWT      COMPOUND


197* 199   97  314  248  316  286  250  349  Chlorpyriphos
197* 241  198   77  242  104   91  103  242  Fenoprofen*
197* 256  240  198  152  106  257  241  256  Phenylacetylsalicylate
197   58*  86   42   43   44   85  196  326  Acepromazine
197   97* 199  314  316  258  286  125  349  Chlorpyrifos
197   99*  44   58  112  309   41   42  309  Methixene*
197  116* 196  119   67  129   69   98  197  Halothane
197  195* 199  123  133  135  160  167  195  2,4,5-Trichloroaniline
197  212* 169  213  211  170  106  168  212  Harmine
197  212* 141  160  154  111   93    0  212  3,4,5-Trimethoxybenzoic Acid
198*  70  115  199   71  269   72   41  327  Aminopromazine
198*  70   58  115   71   56   72  269  327  Proquamazine
198*  91   65  199  197  155  107   92  198  Ditolyl Ether(p-)
198* 127  183   95  123  181  196  199  199  3,4,5-Trimethoxybenzyl Alcohol (Mescaline Precursor)
198* 168  180  140  197  169  179  196  198  Hydroxyharman
198* 196   97  200  132  134   99  133  196  2,4,5-Trichlorophenol
198   43* 134   65  108  295   92  135  295  Acetylsulphamethoxazole
198   58*  42  180   30  154   57  199  298  Trimeprazine
198   72*  44  200   73   42   99  199  198  Monuron
198   72* 197  196   73   56   70  255  340  Propiomazine
198  124* 109   77   81  110  125  167  198  Glycerol guiacolate
198  196* 161  200  163   86  242  245  240  Picloram
198  196* 197  195  200  225  223  256  254  Picloram Methyl Ester
198  196*  97  132  200  134   99   62  196  2,4,6-Trichlorophenol
198  196* 254  256  200  197  209  211  254  2,4,5-Trichlorophenoxyacetic Acid
198  233* 201  199  116  166  171  197  233  2-Chlorophenothiazine
199*  92  135   76  324  119   93  134  324  Oxyphenbutazone
199* 167  198  166   99  154   69   77  199  Phenothiazine
199* 200  198  170  172  171   63   42  200  Harmalol
199* 200   92   65  108   39   66  201  264  Sulphamerazine*
199* 244  200  178  179  184 '183  245  244  Flurbiprofen*
199* 324   93   77   65   55  121  135  324  Oxyphenbutazone*
199   72* 167  198   71   42   56  200  284  Promethazine
199  108* 163  201   92  149  132  170  248  Dimethylcyclophosphoramide Mustard
199  117* 105   77  183  427   91  336  427  Cinnamyl-3-piperidylbenzilate (N-Cinnamyl)
199  195* 200  186  214  152  174  251  608  Reserpine
199  197*  97  314  248  316  286  250  349  Chlorpyriphos
199  221* 200  186  395  251   77  214  634  Rescinnamine
200*  42   58   61   43   47   75  104  214  Metribuzin
200*  91   84  242  243  115  129  117  243  Phencyclidine
200*  91  243   84  242  186  166  201  243  Phencyclidine
200*  91   84   86  186  259    0    0  259  4-Phenyl-4-piperidinocyclohexanol
200* 120  185  121   40   41   69   44  270  Permethylated Thiopentone*
200* 141  155    0    0    0    0  200  2-Methyl-4-chlorophenoxyacetic Acid (MCPA)
200* 143  185  169  126  201  170  155  242  Hydroxyamylobarbitone (N-Hydroxy)
200* 171  129  117  115  145  183  199  200  Hydrallazine Metabolite
200* 215  149  173  202   58  217   92  215  Atrazine
200* 215   58  202   43   68  173   69  215  Atrazine
200  141*  77  143  155  125  142  202  200  MCPA
200  185* 186   89   57   42  103  131  200  Acetonide
200  199* 198  170  172  171   63   42  200  Harmalol
200  199*  92   65  108   39   66  201  264  Sulphamerazine*
201*  44  186  173   68   71   55  158  201  Simazine
201*  81   98  175  259  176  202  242  259  Primaquine*
```

201*	115	285	173	143	84	202	174	285	Piperine
201*	117	167	202	251	165	118	115	368	Cinnarizine*
201*	117	119	203	199	166	94	47	234	Hexachloroethane
201*	165	45	166	56	203	42	58	374	Hydroxyzine
201*	170	203	188	165	166	202	142	374	Hydroxyzine
201*	174	63	64	202	65	129	90	201	Thiabendazole
201*	176	202	56	258	70	71	42	377	Morazone*
201*	203	165	45	299	166	202	56	374	Hydroxyzine*
201	44*	43	186	68	173	71	96	201	Simazine
201	56*	70	176	77	202	71	55	377	Morazone
201	186*	42	202	187	56	57	71	275	Trimeperidine
202*	108	174	137	109	80	203	145	217	Ethoxyquin
202	44*	259	203	218	215	217	42	263	Nortriptyline
202	44*	45	215	203	42	220	204	263	Nortriptyline
202	44*	45	220	218	215	91	0	263	Nortriptyline*
202	58*	116	198	59	44	72	42	358	Dimethoxanate
202	58*	231	215	289	0	0	0	291	Hydroxycyclobenzaprine
202	58*	57	201	42	84	44	186	275	Proheptazine
202	91*	115	129	158	57	77	143	245	1-(1-Phenylcyclohexyl)morpholine
202	232*	129	56	42	158	217	305	305	Normeperidine Ethylcarbamate
203*	58	205	204	72	167	202	168	274	Chlorpheniramine
203*	58	44	205	54	204	72	202	274	Chlorpheniramine*
203*	58	43	57	205	71	72	32	274	Chlorpheniramine(dex)
203*	76	50	330	74	75	127	165	330	1,2-Diiodobenzene
203*	132	117	174	115	91	231	103	231	Glutethimide Methyl Derivative
203*	188	232	117	204	40	115	70	260	1,2-Dimethylphenobarbital (or 1,4-Dimethyl-)
203*	205	234	236	188	201	204	190	234	Dicamba Methyl Ester
203*	232	132	175	204	233	160	118	232	Aminoglutethimide*
203	31*	93	121	123	95	44	201	280	Tribromethyl Alcohol*
203	44*	202	45	220	219	215	204	263	Nortriptyline
203	58*	167	72	205	202	168	204	274	Chlorpheniramine
203	58*	202	232	84	85	101	215	291	10,11-Epoxycyclobenzaprine
203	118*	56	77	218	141	132	204	218	5-Phenylhydantoin(5-methyl-) Dimethyl Derivative
203	160*	91	146	161	117	44	104	203	Ethyl-1-phenylcyclohexylamine (N-)
203	201*	165	45	299	166	202	56	374	Hydroxyzine*
204*	31	117	51	77	161	146	118	232	Phenobarbitone
204*	63	146	232	117	143	174	89	232	Phenobarbitone
204*	105	104	133	77	72	205	78	204	Ethotoin (Peganone)
204*	117	77	115	51	91	103	118	232	Phenobarbitone
204*	117	146	161	77	103	115	118	232	Phenobarbitone*
204*	133	176	233	148	131	188	119	233	2-Ethyl-2-(4-hydroxyphenyl)glutarimide (Glutethimide Metab)
204*	217	123	186	205	0	0	0	355	Moperone
204*	232	117	161	146	77	118	115	232	Phenobarbitone
204	58*	146	59	42	160	43	159	204	Bufotenine
204	58*	59	42	146	160	130	0	204	Psilocin
204	58*	59	146	159	205	160	60	204	Psilocin (4-Hydroxy-N-dimethyltryptamine)
204	58*	59	146	159	205	160	57	284	Psilocybin
204	58*	59	0	0	0	0	0	204	Psilocybin Metabolite
204	84*	205	85	42	55	105	77	287	Procyclidine*
205*	42	135	173	206	123	145	45	206	Pyrantel
205*	42	125	115	186	41	91	127	311	Pyrrobutamine
205*	57	220	41	206	145	29	81	220	Ionol
205*	70	231	78	135	136	42	166	371	Trazodone*
205*	77	41	39	220	130	145	131	220	Xylazine

PEAKS							MWT	COMPOUND
205*	207	161	163	110	112	260	262	260 Bromacil
205*	207	40	39	70	206	190	162	260 Bromacil
205*	207	42	162	70	164	231	188	260 Bromacil
205*	218	123	356	219	162	95	190	356 Fluanisone
205*	219	231	318	0	0	0	0	318 Cambendazole Metabolite
205*	220	57	206	41	145	219	55	220 2,6-Di-t-butyl-4-methylphenol (Antioxidant)
205*	220	42	77	147	221	119	118	413 Noscapine
205*	240	91	84	125	242	206	186	311 Pyrrobutamine*
205*	244	191	345	206	72	272	246	404 Benzquinamide*
205*	263	204	75	102	101	51	88	263 Cinchophen Methyl Ester
205	56*	223	167	149	41	57	65	278 Dibutyl Terephthalate(n-)
205	58*	42	191	178	128	91	129	339 Dextropropoxyphene
205	77*	104	51	177	75	229	151	286 Oxazepam
205	84*	42	240	81	55	91	125	311 Pyrrobutamine
205	159*	131	76	187	51	103	104	205 Xanthurenic Acid
205	170*	172	171	207	173	60	136	205 Chlorthiamid
205	203*	234	236	188	201	204	190	234 Dicamba Methyl Ester
205	220*	221	28	147	77	178	42	413 Noscapine
205	239*	221	211	139	166	195	387	405 Trazodone Metabolite
205	269*	221	297	271	62	285	124	297 Hydrochlorothiazide*
206*	124	176	208	160	178	162	126	206 2,6-Dichloro-4-nitroaniline (DCNA)
206*	128	107	119	92	135	178	106	206 5-Furfurylidenebarbituric Acid
206*	133	150	370	160	134	178	151	370 Thiophanate
206*	221	57	41	207	150	222	68	221 2-Amino-4,6-di-t-butylphenol
206	191*	193	208	141	195	163	143	206 1,4-Dichloro-2,5-dimethoxybenzene (Chloroneb)
206	192*	272	480	288	246	205	191	480 Emetine*
206	207*	208	204	103	205	177	102	207 1-Methyl-2-phenylindole
206	207*	91	179	180	178	89	208	294 Noxiptilin
206	310*	42	64	312	299	45	48	439 Polythiazide*
206	363*	143	42	70	207	218	113	363 Opipramol*
207*	41	39	124	91	165	122	44	286 Brallobarbitone*
207*	67	79	81	141	77	55	91	236 Cyclobarbitone
207*	129	72	91	208	105	174	128	325 Methadone-N-Oxide
207*	141	40	81	79	44	77	41	236 Cyclobarbitone
207*	141	81	79	67	80	41	77	236 Cyclobarbitone*
207*	192	164	74	149	134	124	0	207 3,4,5-Trimethoxyphenylacetonitrile
207*	206	208	204	103	205	177	102	207 1-Methyl-2-phenylindole
207*	206	91	179	180	178	89	208	294 Noxiptilin
207*	209	211	238	240	179	181	109	238 2,3,6-Trichlorobenzoate
207*	225	96	111	0	0	0	0	225 Haloperidol Hofmann Reaction Product
207*	242	244	165	270	243	208	269	270 Medazepam
207*	309	208	182	181	209	180	167	309 Nor LSD (6-desmethyl-LSD)
207*	371	312	342	208	372	42	328	371 Demecoline*
207	81*	80	79	141	169	41	77	248 Thialbarbitone oxygenated analogue
207	205*	161	163	110	112	260	262	260 Bromacil
207	205*	40	39	70	206	190	162	260 Bromacil
207	205*	42	162	70	164	231	188	260 Bromacil
207	222*	149	177	223	221	195	191	222 Apiol
207	242*	244	270	243	271	269	165	270 Medazepam*
207	250*	70	91	251	119	148	65	341 Fenethylline
208*	82	123	67	209	42	55	207	208 1,3,7,8-Tetramethylxanthine (or 1,3,8,9-Tetramethyl-)
208*	89	179	76	165	77	209	88	208 Dibenzosuberone
208*	95	193	67	180	123	73	43	208 7-Ethyltheophylline
208*	115	117	91	193	0	0	0	208 Dextropropoxyphene Metabolite

								MWT	COMPOUND
208*	180	205	207	206	209	104	77	208	9,10-Diaminophenanthrene
208*	180	109	67	42	137	55	179	208	1-Ethyltheobromine
208*	193	130	115	91	165	179	207	265	2-Ethyl-5-Methyl-3,3-Diphenyl-1-Pyrroline
208*	209	42	41	207	193	39	206	278	Triprolidine
208*	209	31	207	278	84	193	42	278	Triprolidine
208*	209	278	207	193	200	194	84	278	Triprolidine*
208*	210	94	40	77	44	76	209	393	Dialifor
208*	249	193	44	195	194	234	209	280	Trimipramine Metabolite
208	44*	58	117	57	193	130	115	325	Norpropoxyphene
208	209*	278	207	193	194	84	200	278	Triprolidine
208	265*	121	43	266	187	213	251	308	Warfarin*
208	273*	166	193	150	316	108	96	316	Bupirimate
208	279*	77	308	280	104	149	180	308	2,3-Diethyl-5,5-diphenylhydantoin
209*	105	77	268	191	103	210	51	268	Ketoprofen Methyl Ester
209*	115	141	210	152	153	139	208	210	Naphazoline
209*	162	133	148	77	51	147	105	209	2,5-Dimethoxy-b-nitrostyrene
209*	179	210	135	59	89	161	193	224	Salicylic Acid Methyl Ester TMS Ether
209*	208	278	207	193	194	84	200	278	Triprolidine
209*	210	141	115	153	208	46	181	210	Naphazoline*
209	166*	96	167	71	180	93	55	209	Dimethirimol (tech)
209	167*	43	124	39	41	53	140	288	Ibomal*
209	180*	104	223	252	77	181	165	252	Diphenylhydantoin
209	180*	280	77	181	104	251	165	280	3-Ethyl-5,5-diphenylhydantoin
209	180*	182	152	181	30	211	138	237	Ketamine*
209	180*	223	252	104	73	0	0	252	Phenytoin
209	207*	211	238	240	179	181	109	238	2,3,6-Trichlorobenzoate
209	208*	42	41	207	193	39	206	278	Triprolidine
209	208*	31	207	278	84	193	42	278	Triprolidine
209	208*	278	207	193	200	194	84	278	Triprolidine*
209	210*	30	183	165	115	181	133	227	Prenylamine Metabolite
210*	30	149	164	180	193	134	0	227	2,4,6-Trinitrotoluene
210*	56	43	184	211	75	99	76	324	Acetohexamide*
210*	82	67	153	125	42	55	0	210	1,3,7-Trimethyluric Acid
210*	89	30	63	39	51	76	134	227	2,4,6-Trinitrotoluene
210*	175	212	177	111	75	214	73	210	Trichlorocresol(m-)
210*	209	30	183	165	115	181	133	227	Prenylamine Metabolite
210	151*	152	211	107	195	59	153	210	Methyl 3,4-dimethoxyphenylacetate
210	208*	94	40	77	44	76	209	393	Dialifor
210	209*	141	115	153	208	46	181	210	Naphazoline*
210	211*	196	212	105	180	167	77	211	2-Hydroxyiminodibenzyl(Imipramine Metabolite)
210	235*	195	221	0	0	0	0	353	Methysergide
210	353*	235	336	72	54	236	195	353	Methysergide*
211	163	147	117	240	77	89	205	240	Dinoseb
211*	210	196	212	105	180	167	77	211	2-Hydroxyiminodibenzyl(Imipramine Metabolite)
211*	212	107	77	43	57	106	139	212	2,5-Diaminobenzophenone
211	226*	59	155	66	53	195	183	226	Methyl 3,4,5-trimethoxybenzoate
211	246*	107	43	245	57	248	80	246	2,5-Diamino-2'-Chlorobenzophenone
212*	42	187	45	70	180	56	98	427	Dixyrazine
212*	169	197	213	106	211	170	168	212	Harmine*
212*	197	169	213	211	170	106	168	212	Harmine
212*	197	141	160	154	111	93	0	212	3,4,5-Trimethoxybenzoic Acid
212*	213	122	198	44	53	91	65	257	Tolmetin*
212*	214	227	229	184	186	105	118	227	Bromodiethylaniline(p-Br,N,N-diEt)
212	195*	114	30	194	72	44	43	339	Disopyramide*

								MWT	COMPOUND
212	211*	107	77	43	57	106	139	212	2,5-Diaminobenzophenone
212	240*	241	184	138	92	128	63	240	Danthron*
213*	43	68	44	71	155	170	96	213	Simetryne
213*	108	81	36	44	136	54	77	213	Azopyridine
213*	139	75	169	111	215	141	77	228	Indomethacin Derivative (4-Chlorobenzoic Acid TMS Ester)
213*	160	159	296	133	145	212	157	300	Ethinylestradiol
213*	160	296	133	159	145	146	214	296	Ethynylestradiol
213*	214	170	198	169	115	63	143	214	Harmaline
213*	214	198	170	199	215	169	172	214	Harmaline
213	58*	198	180	214	57	212	270	270	Desmethylpromethazine
213	93*	79	77	106	121	119	104	213	2-Methylsulphonylacetanilide
213	212*	122	198	44	53	91	65	257	Tolmetin*
213	214*	92	65	215	108	39	42	278	Sulphadimidine*
214*	58	229	172	43	187	216	41	229	Propazine
214*	92	65	213	108	42	215	39	278	Sulphasomidine*
214*	141	155	125	216	77	45	143	214	MCPA Methyl Ester
214*	213	92	65	215	108	39	42	278	Sulphadimidine*
214*	242	216	309	215	311	179	151	309	Diclofenac Methyl Ester
214	61*	185	126	153	140	93	127	214	3-(4-Chlorophenyl)-1-methyl-1-methoxyurea
214	142*	141	169	107	77	0	0	214	4-Chloro-2-methylphenoxypropionic Acid (CMPP)
214	158*	57	145	124	45	70	96	214	Niridazole*
214	172*	42	110	57	173	91	44	287	Allylprodine*
214	212*	227	229	184	186	105	118	227	Bromodiethylaniline(p-Br,N,N-diEt)
214	213*	170	198	169	115	63	143	214	Harmaline
214	213*	198	170	199	215	169	172	214	Harmaline
214	216*	218	179	181	143	109	74	214	1,2,3,4-Tetrachlorobenzene
214	216*	218	181	179	109	108	143	214	1,2,3,5-Tetrachlorobenzene
214	216*	218	179	181	74	108	109	214	1,2,4,5-Tetrachlorobenzene
214	229*	186	114	215	154	230	199	229	1-Methoxyphenothiazine
214	229*	186	114	215	230	154	199	229	3-Methoxyphenothiazine
214	271*	44	270	32	228	272	42	271	Desomorphine
215*	41	104	77	132	39	128	244	244	Alphenal
215*	92	216	65	108	53	69	39	280	Sulphamethoxypyridazine*
215*	216	172	102	129	118	128	117	216	5-Benzylidenebarbituric Acid
215*	230	56	77	96	122	81	201	230	Propyphenazone
215*	230	56	77	216	96	41	39	230	Propyphenazone*
215*	287	42	96	229	189	286	216	287	Cyproheptadine
215	44*	411	324	45	164	42	216	411	Etorphine*
215	44*	202	216	45	213	91	189	261	Nortriptyline Metabolite
215	58*	59	229	228	227	226	218	275	Amitriptyline Metabolite
215	74*	115	42	39	127	0	0	215	1-Naphthyl-N,N-dimethyl carbamate
215	91*	79	105	77	55	41	298	298	Norethynodrel
215	98*	58	84	91	56	71	186	215	Fencamfamin*
215	123*	43	81	53	124	216	94	215	Furcarbanil
215	200*	149	173	202	58	217	92	215	Atrazine
215	200*	58	202	43	68	173	69	215	Atrazine
215	216*	92	65	108	54	125	39	280	Sulphamethoxydiazine*
215	242*	243	214	111	121	216	0	242	Cambendazole GC Decomposition Product
215	261*	259	213	203	201	263	217	259	Tetrachloronitrobenzene
215	285*	162	286	124	284	268	174	285	Morphine
215	285*	81	148	110	164	286	132	285	Norcodeine
216*	147	158	57	217	131	160	187	216	Dioxyparaquat
216*	214	218	179	181	143	109	74	214	1,2,3,4-Tetrachlorobenzene
216*	214	218	181	179	109	108	143	214	1,2,3,5-Tetrachlorobenzene

PEAKS							MWT	COMPOUND
216*	214	218	179	181	74	108	109	214	1,2,4,5-Tetrachlorobenzene
216*	215	92	65	108	54	125	39	280	Sulphamethoxydiazine*
216*	217	128	173	118	103	130	146	217	5-(2'-Pyridylmethylene)barbituric Acid
216	91*	77	259	0	0	0	0	259	1-(1-Phenylcyclohexyl)-4-hydroxypiperidine
216	92*	215	108	65	156	54	125	280	Sulphamethoxypyrazine*
216	144*	116	89	72	63	0	0	216	3-Indolylglyoxyldimethylamide
216	161*	160	218	163	162	117	118	216	Terbacil
216	215*	172	102	129	118	128	117	216	5-Benzylidenebarbituric Acid
216	231*	84	124	59	42	72	174	231	Metazocine
216	260*	302	215	242	0	0	0	302	Cambendazole
216	301*	44	42	70	302	203	57	301	Oxymorphone*
217*	45	41	70	69	110	202	72	285	Pentazocine
217*	70	69	110	202	270	285	284	285	Pentazocine
217*	128	118	103	146	119	145	174	217	5-(4'-Pyridylmethylene)barbituric Acid
217*	218	215	235	108	36	363	247	999	Dichloralphenazone Probe Artifact*
217*	218	132	146	117	103	246	118	274	4-Ethylmephobarbitone
217*	269	375	241	204	341	342	329	822	Oxazepam Glucuronide TMS Derivative
217*	294	132	194	118	208	222	77	294	5-Phenylhydantoin(5-(p-methylphenyl))- Dimethyl Derivative
217	45*	70	41	69	110	285	202	285	Pentazocine*
217	56*	83	42	57	98	77	218	217	4-Methylaminoantipyrine
217	73*	204	75	147	185	253	45	676	Hydroxyamobarbitone Glucuronide Me TMS Derivative (peak 2)
217	128*	216	173	103	146	119	130	217	5-(3'-Pyridylmethylene)barbituric Acid
217	130*	143	131	186	218	144	117	217	Methyl-3-indolebutyrate
217	204*	123	186	205	0	0	0	355	Moperone
217	216*	128	173	118	103	130	146	217	5-(2'-Pyridylmethylene)barbituric Acid
217	218*	117	118	146	103	115	91	246	Mephobarbitone
218*	42	219	91	155	165	115	56	424	Difenoxin
218*	44	191	221	219	178	42	180	249	Benzoctamine*
218*	117	146	118	219	161	39	51	246	Mephobarbitone (Mebaral, Prominal)
218*	117	118	146	103	77	91	115	246	Methylphenobarbitone*
218*	118	146	117	103	77	219	91	246	Mephobarbitone
218*	174	160	161	275	219	175	162	275	Physostigmine*
218*	217	117	118	146	103	115	91	246	Mephobarbitone
218*	275	174	160	161	0	0	0	275	Physostigmine
218	44*	217	58	57	261	216	219	261	Desmethylcyclobenzaprine
218	44*	203	73	217	202	215	219	351	Nortriptyline TMS Derivative
218	98*	99	219	55	41	42	85	301	Benzhexol
218	98*	99	55	130	41	42	85	311	Biperiden
218	98*	99	55	41	42	77	84	311	Biperiden*
218	98*	97	55	41	219	105	77	301	Trihexyphenidyl
218	146*	117	118	44	217	42	103	246	Primidone Dimethyl Derivative
218	186*	77	295	263	105	51	158	295	Mebendazole*
218	205*	123	356	219	162	95	190	356	Fluanisone
218	217*	215	235	108	36	363	247	999	Dichloralphenazone Probe Artifact*
218	217*	132	146	117	103	246	118	274	4-Ethylmephobarbitone
219*	131	100	218	0	0	0	0	327	Fomocaine Metabolite
219*	248	148	220	120	218	133	65	248	4-Hydroxyphenobarbitone
219	69*	131	100	264	502	119	414	614	Heptacosafluorotributylamine
219	109*	181	183	111	193	288	0	288	Lindane(beta)
219	160*	161	220	145	74	69	83	219	Methyl-5-methoxyindoleacetate
219	205*	231	318	0	0	0	0	318	Cambendazole Metabolite
219	232*	217	204	233	203	91	202	359	Nortriptyline N-Trifluoroacetate
219	262*	111	97	263	56	218	42	277	Ethylmethylthiambutene
219	276*	111	277	278	42	100	97	291	Diethylthiambutene

         Asterisk Indicates True Base Peak   Peak Inde

			PEAKS				MWT	COMPOUND
220*	42	77	205	51	147	44	53	413 Narcotine
220*	57	41	221	206	58	29	55	277 Azar
220*	175	162	222	177	145	147	164	220 3,4-D
220*	178	221	205	163	177	179	42	221 Hydrocotarnine
220*	205	221	28	147	77	178	42	413 Noscapine
220*	221	205	206	218	222	118	148	413 Narcotine
220*	221	205	147	42	193	77	118	413 Noscapine*
220*	292	290	222	150	74	294	255	290 Tetrachlorobiphenyl
220	44*	86	57	74	91	43	115	367 Acetylnorpropoxyphene (N-)
220	44*	86	74	58	91	43	115	367 Acetylnorpropoxyphene (N-)
220	44*	100	57	205	129	91	307	307 Dehydrated Norpropoxyphene Amide
220	44*	100	57	205	91	129	221	325 Dextropropoxyphene Metabolite 1
220	44*	100	205	57	0	0	0	307 Dextropropoxyphene Metabolite
220	44*	100	59	205	57	91	129	251 Dextropropoxyphene Metabolite 1
220	58*	59	219	277	179	191	193	279 Doxepin
220	58*	219	59	191	189	42	205	279 Doxepin*
220	162*	161	175	0	0	0	0	220 2,4-Dichlorophenoxyacetic Acid
220	205*	57	206	41	145	219	55	220 2,6-Di-t-butyl-4-methylphenol (Antioxidant)
220	205*	42	77	147	221	119	118	413 Noscapine
220	235*	73	75	204	145	236	130	350 Ethyl Phenylmalondiamide (Primidone Derivative) di TMS Derivative
220	249*	192	149	164	131	135	147	249 2-Ethyl-2-(3,4-dihydroxyphenyl)glutarimide (Glutethimide Metab)
221*	43	78	93	80	41	141	39	250 Heptabarbitone*
221*	54	339	72	55	196	207	181	339 Methylergonovine Hydrogen Maleate
221*	67	196	181	41	164	111	107	262 Allylcyclopentenylbarbitone Dimethyl Derivative
221*	67	196	41	164	181	222	0	248 Methylcyclopal
221*	72	325	54	196	55	207	181	325 Ergometrine*
221*	72	207	181	43	44	323	222	323 Iso-lysergide*
221*	72	207	181	42	44	222	58	323 LSD
221*	81	157	80	79	155	·41	77	236 Hexobarbitone*
221*	82	80	98	81	97	319	417	417 Cinepazide
221*	124	64	206	150	151	163	178	221 3,4,5-Trimethoxybenzoyl Cyanide
221*	141	81	79	222	41	67	93	250 Heptabarbitone
221*	141	79	81	77	40	38	67	250 Heptabarbitone
221*	163	164	44	149	122	42	36	221 Formetanate
221*	181	207	196	223	325	222	180	325 Ergonovine
221*	196	222	307	112	181	205	154	325 Ergometrine Maleate
221*	199	200	186	395	251	77	214	634 Rescinnamine
221*	222	196	321	126	205	207	181	339 Methylergometrine Maleate
221*	314	248	261	193	236	222	315	314 Tetrahydrocannabinol (delta 8)
221*	323	181	222	207	72	223	180	323 LSD (d) Tartrate
221	43*	83	236	56	223	222	55	236 Methazolamide*
221	67*	196	41	164	111	39	181	262 Permethylated Cyclopentobarbitone*
221	77*	105	220	88	223	222	51	221 Pyrazon
221	81*	79	80	39	77	157	41	236 Hexobarbitone
221	206*	57	41	207	150	222	68	221 2-Amino-4,6-di-t-butylphenol
221	220*	205	206	218	222	118	148	413 Narcotine
221	220*	205	147	42	193	77	118	413 Noscapine*
221	267*	207	180	223	154	196	268	267 Lysergamide*
221	295*	181	207	196	180	223	167	295 Lysergic Acid Monoethylamide
221	323*	44	199	222	223	207	76	323 LSD
221	323*	181	222	207	72	223	324	323 LSD*
221	323*	207	181	222	196	223	72	323 LSD (iso)
221	325*	207	196	181	44	223	42	325 Ergometrine
221	339*	196	181	207	223	222	72	339 Methylergometrine*

								MWT	COMPOUND
221	339*	223	207	196	181	222	340	339	Methylergometrine Maleate
222*	51	87	143	153	104	224	69	257	Barban
222*	95	194	166	207	179	123	67	222	1,7-Diethyl-3-methylxanthine
222*	120	92	194	64	63	221	97	222	Flavone
222*	136	166	194	150	123	207	67	222	1,3-Diethyl-7-methylxanthine
222*	150	194	179	166	207	109	43	222	3,7-Diethyl-1-methylxanthine
222*	152	224	151	223	150	75	153	222	4,4'-Dichlorobiphenyl
222*	165	76	223	166	105	104	90	222	Phenindione*
222*	165	223	221	194	166	164	76	222	Phenylindanione
222*	191	223	207	79	147	164	190	222	Methyl 3,4-Dimethoxycinnamate
222*	207	149	177	223	221	195	191	222	Apiol
222	221*	196	321	126	205	207	181	339	Methylergometrine Maleate
222	251*	223	250	252	195	110	97	251	7-Aminonitrazepam
223*	41	166	224	169	67	138	39	252	Permethylated Vinbarbitone*
223*	162	161	147	176	91	77	119	223	2,5-Dimethoxy-b-methyl-b-nitrostyrene
223*	224	117	167	179	178	165	193	306	Perthane
223*	241	208	222	194	180	77	224	241	Mefenamic Acid*
223*	255	208	77	222	180	194	96	255	Mefenamic Acid Methyl Ester
223*	279	97	162	251	225	164	281	314	Diethyl-O-(2,5-dichlorophenyl) Phosphorothioate (O,O-Diethyl)
223*	313	208	224	222	180	194	298	313	Mefenamic Acid TMS Derivative
223	53*	153	127	171	181	225	155	223	3-Butynyl N-(3-chlorophenyl)carbamate
223	81*	79	41	80	157	185	77	264	Thialbarbitone*
223	91*	294	57	42	56	165	44	309	Methadone (6-Dimethylamino-4,4-diphenyl-3-heptanone)
223	124*	109	77	122	123	52	95	223	Mephenoxalone (Trepidone)
223	151*	222	72	123	152	108	224	223	Ethamivan
223	180*	194	254	109	95	193	166	254	Diprophylline*
223	180*	252	179	152	0	0	0	252	10,11-Epoxycarbamazepine
223	181*	166	73	43	45	208	75	223	Paracetamol TMS Ester
223	224*	41	225	125	209	43	109	294	1,3-Diethylsecobarbitone
223	258*	286	257	259	75	122	251	286	Desalkyl-3-hydroxyflurazepam Dehydration Product (N1-Desalkyl)
223	317*	75	147	217	43	318	73	557	Paracetamol Glucuronide Methyl Ester TMS Ether
223	325*	225	224	326	226	182	167	325	2,3-Dihydro-LSD
223	380*	382	238	152	345	215	113	380	Chlorotrianisene
224*	42	237	123	95	57	226	206	375	Haloperidol
224*	42	237	226	123	206	239	56	375	Haloperidol*
224*	223	41	225	125	209	43	109	294	1,3-Diethylsecobarbitone
224*	238	226	123	340	0	0	0	375	Haloperidol
224	58*	209	71	225	72	210	180	295	Dibenzepin*
224	151*	164	152	225	149	165	193	224	Methyl 3,4-Dimethoxyphenylpropionate
224	223*	117	167	179	178	165	193	306	Perthane
224	268*	154	180	207	223	192	179	268	Lysergic Acid*
225*	68	44	173	240	198	172	43	240	Cyanazine
225*	77	226	104	76	149	153	43	318	Diphenyl Phthalate
225*	149	185	240	103	131	89	129	240	Budralazine
225*	177	131	41	77	240	38	103	240	Dinoterb
225*	227	223	190	260	141	118	188	258	Hexachloro-1,3-butadiene
225*	242	244	364	362	208	451	453	466	Chloramphenicol TMS Ether
225	124*	154	155	128	127	125	141	352	5-(2-Iodoethyl)-5-isopentylbarbituric Acid
225	207*	96	111	0	0	0	0	225	Haloperidol Hofmann Reaction Product
225	274*	318	273	275	226	257	121	318	Phenolphthalein*
226*	185	241	170	43	68	41	71	241	Terbutryn
226*	211	59	155	66	53	195	183	226	Methyl 3,4,5-trimethoxybenzoate
226	41*	77	143	181	141	39	145	226	Alclofenac*
226	191*	228	163	193	192	165	99	226	2,3-Dichloro-1,4-naphthoquinone (Dichlone)

```
226  285*  72   58   45   42   44  162  427  Morphine Methoiodide
227* 228  274  238  152   36  308  153  344  Methoxychlor
227* 310  174  284  147  160  173  199  310  Mestranol
227  225* 223  190  260  141  118  188  258  Hexachloro-1,3-butadiene
227  244*  55   57   95  243   69   43  244  2-Methylamino-5-amino-2'-Fluorobenzophenone
228* 182  363   88  229   76  276  257  398  Clefamide*
228   72* 183  230   44  229  168   45  228  Metoxuron
228  142* 141    0    0    0    0    0  228  4-Chloro-2-methylphenoxybutyric Acid (MCPB)
228  193*  89  255   77  165   51  110  256  Desmethylmedazepam (N-)
228  227* 274  238  152   36  308  153  344  Methoxychlor
228  229* 231  194  230  193  214  232  229  Desalkylclomipramine(N-)
229*  30  231  172  194  174  200  230  229  Clonidine*
229*  30   62   91   53   63   50  230  229  Picric Acid
229* 173   44   41  115  128   91  129  244  Xylometazoline
229* 214  186  114  215  154  230  199  229  1-Methoxyphenothiazine
229* 214  186  114  215  230  154  199  229  3-Methoxyphenothiazine
229* 228  231  194  230  193  214  232  229  Desalkylclomipramine(N-)
229* 230  215  228  101  202  226  227  291  Cyclobenzaprine N-oxide
229* 231   30  172  194  174  200  230  229  Clonidine
229* 231  172  194  174  230   36  228  229  Clonidine
229* 244  173  243   44  230   81  245  244  Xylometazoline
229   30* 172   42  231  194  174   43  229  Clonidine
229   69* 231  161  230  133  162  233  229  (3,4-Dichlorophenyl)methylacrylamide (N-)
229  230* 107   86   43  211   95  123  230  2,5-Diamino-2'-Fluorobenzophenone
229  231* 233  158  167  169  235  131  229  2,3,4,5-Tetrachloroaniline
229  269* 268  232   71   44  242  211  300  Desmethylclomipramine
229  299* 162  297  240  241  242  298  315  Codeine N-oxide
229  313* 138  162   42    0    0    0  313  Dimethylmorphine
229  314* 242   44  269   86  283   71  314  Desmethyltrimeprazine
229  314* 242  269   44   57   71   70  314  Methotrimeprazine Metabolite
230*  58  231   44  173  105   42  159  321  Phenazocine
230* 109   82  187  243  363   42  123  381  Benperidol*
230* 187  134  217   96   83  461  109  461  Pimozide
230* 229  107   86   43  211   95  123  230  2,5-Diamino-2'-Fluorobenzophenone
230* 231   77  232  105  154  233  126  231  2-Amino-5-chlorobenzophenone*
230* 231   77  232  105  154  233  126  231  Chlordiazepoxide Benzophenone
230* 231   77  232  105  154   51  233  231  Oxazepam Benzophenone
230* 231   58   44   42  105  173  159  321  Phenazocine*
230* 265  139  267  111  232  264  266  265  Lorazepam Benzophenone
230* 271   55  256  164  270   71  124  271  Cyclazocine*
230   30* 232   44  215  217   77  156  259  4-Bromo-2,5-dimethoxyphenethylamine
230   44* 232  273  275   77   42   45  273  Bromo STP
230   77* 231  105  154  232  126   51  231  2-Amino-5-chlorobenzophenone
230  185* 141  186  184  115  170  153  230  Naproxen*
230  215*  56   77   96  122   81  201  230  Propyphenazone
230  215*  56   77  216   96   41   39  230  Propyphenazone*
230  229* 215  228  101  202  226  227  291  Cyclobenzaprine N-oxide
230  232* 131  234  170   84  133  166  230  Tetrachlorophenol
230  315* 314  316  258  201   70  229  315  Oxycodone
230  315* 316   70   44   42  258  140  315  Oxycodone*
231*  41   43  232   69   55  174   57  314  Cannabichromene
231*  41   67   43  232   68   91   55  314  3-Pentylcannabidiol
231*  42  119  232   92   65   67   74  277  Azathioprine*
231*  97  153  121  125   29   65   93  384  Ethion
```

								MWT	COMPOUND
231*	147	285	232	201	132	165	166	432	Buclizine*
231*	216	84	124	59	42	72	174	231	Metazocine
231*	229	233	158	167	169	235	131	229	2,3,4,5-Tetrachloroaniline
231*	246	314	232	121	193	74	174	314	Cannabidiol
231*	314	271	258	43	193	41	246	314	Tetrahydrocannabinol (delta-8)
231	36*	147	38	285	132	201	165	432	Buclizine
231	42*	119	152	247	0	0	0	277	Azathioprine
231	56*	97	204	77	111	112	0	231	Amidopyrine
231	56*	97	111	112	42	77	71	231	Amidopyrine*
231	75*	45	97	153	47	46	125	260	Phorate
231	98*	232	197	0	0	0	0	232	Chlorprothixene Metabolite
231	139*	141	111	233	75	140	113	231	2-Chlorobenzanilide
231	147*	165	285	132	166	201	117	432	Buclizine
231	193*	247	123	194	233	136	316	316	Cannabigerol
231	229*	30	172	194	174	200	230	229	Clonidine
231	229*	172	194	174	230	36	228	229	Clonidine
231	230*	77	232	105	154	233	126	231	2-Amino-5-chlorobenzophenone*
231	230*	77	232	105	154	233	126	231	Chlordiazepoxide Benzophenone
231	230*	77	232	105	154	51	233	231	Oxazepam Benzophenone
231	230*	58	44	42	105	173	159	321	Phenazocine*
231	299*	314	43	41	295	55	271	314	Tetrahydrocannabinol(Delta-9)
232*	118	117	146	175	233	188	260	260	Dimethylphenobarbitone (N,N-)
232*	118	146	117	103	77	175	115	260	Dimethylphenobarbitone (N,N-Dimethyl)
232*	118	117	146	175	103	233	77	260	Permethylated Methylphenobarbitone*
232*	118	117	146	175	103	77	91	260	Permethylated Phenobarbitone*
232*	146	117	175	118	233	103	188	260	1,3-Dimethylphenobarbitone
232*	146	118	175	117	120	121	188	260	Dimethylphenobarbitone
232*	202	129	56	42	158	217	305	305	Normeperidine Ethylcarbamate
232*	219	217	204	233	203	91	202	359	Nortriptyline N-Trifluoroacetate
232*	230	131	234	170	84	133	166	230	Tetrachlorophenol
232*	250	175	204	176	233	102	78	250	Diflunisal*
232*	411	472	91	413	474	0	0	472	4-Chlorodiphenoxylic Acid
232*	438	378	380	246	91	407	423	438	Diphenoxylic Acid
232	72*	234	161	73	45	163	124	232	Diuron
232	121*	139	273	234	332	141	275	332	Procetofene Metabolite Methyl Ester
232	203*	132	175	204	233	160	118	232	Aminoglutethimide*
232	204*	117	161	146	77	118	115	232	Phenobarbitone
232	298*	204	252	251	279	233	280	298	4'-Hydroxyniflumic Acid
233*	146	188	104	103	0	0	0	233	Glutethimide Metabolite
233*	198	201	199	116	166	171	197	233	2-Chlorophenothiazine
233*	318	193	234	262	319	273	136	318	Tetrahydrocannabidiol
233*	383	259	43	245	31	95	205	383	Prazosin
233	57*	42	56	43	158	131	160	233	Ethyl-N-demethyl-4-phenylpiperidine-4-carboxylate
233	57*	56	158	103	91	160	77	233	Norpethidine
233	57*	42	56	158	0	0	0	233	Pethidine Metabolite
233	91*	30	232	31	276	275	65	276	Mebhydroline*
233	146*	103	133	91	117	115	77	233	4-Hydroxyglutethimide
233	174*	175	159	131	130	234	158	233	2-Methyl-5-methoxyindole-3-acetic Acid Methyl Ester
234*	86	221	219	233	235	178	217	307	Desmethyldoxepin N-Acetyl
234*	146	179	206	131	91	117	118	234	Methisazone*
234*	263	206	163	178	131	149	161	263	2-Ethyl-2-(4-hydroxy-3-methoxyphenyl)glutarimide (Glutethimide Metab)
234	44*	233	58	231	202	215	218	277	Hydroxydesmethylcyclobenzaprine
234	58*	235	85	193	194	195	192	280	Imipramine
234	58*	44	59	41	42	36	427	445	Narceine

		PEAKS					MWT	COMPOUND
234	162*	189	161	0	0	0	234	2,4-Dichlorophenoxypropionic Acid (2,4-DP)
234	165*	194	235	193	196	166	179	235	Azapetine*
234	175*	45	199	177	236	73	161	234	2,4-Dichlorophenoxymethyl Acetate
234	236*	238	155	157	75	50	76	236	1,2-Dibromobenzene
235*	58	234	85	280	195	193	35	280	Imipramine
235*	81	169	171	79	236	170	91	250	3-Methylhexobarbitone
235*	81	79	169	171	170	41	80	250	Permethylated Hexobarbitone*
235*	81	169	41	79	54	197	196	276	Permethylated Thialbarbitone Oxygen Analogue*
235*	91	65	132	76	233	90	250	250	Methaqualone
235*	150	165	236	79	137	250	164	250	2-Methylhexobarbitone (or 4-Methyl-)
235*	160	266	77	235	251	146	0	266	Methaqualone Metabolite
235*	169	79	236	77	91	81	112	264	Cyclobarbitone Dimethyl Derivative
235*	170	40	236	180	81	73	121	264	Permethylated Cyclobarbitone*
235*	193	236	136	194	39	138	121	314	Permethylated Brallobarbitone*
235*	195	208	44	234	193	194	71	266	Desipramine*
235*	195	234	208	193	266	194	45	266	((3-Methylaminopropyl)-iminodibenzyl)(N-)(Imipramine Metabolite)
235*	210	195	221	0	0	0	0	353	Methysergide
235*	220	73	75	204	145	236	130	350	Ethyl Phenylmalondiamide (Primidone Derivative) di TMS Derivative
235*	237	337	236	221	338	209	238	337	Aromatised 2-oxo-LSD
235*	237	165	212	246	75	176	36	352	DDT
235*	237	165	75	199	246	352	0	352	DDT (o,p-)
235*	237	165	75	50	51	352	0	352	DDT (p,p'-)
235*	237	165	212	176	282	236	284	352	DDT(p,p'-)
235*	237	165	236	199	239	238	75	352	Dicophane*
235*	237	165	236	199	239	82	238	318	TDE (p,p'-)
235*	250	91	233	236	65	76	132	250	Methaqualone*
235	58*	85	234	195	280	193	194	280	Imipramine
235	58*	85	234	236	195	193	208	280	Imipramine*
235	79*	53	178	81	41	195	261	276	Methohexital Methyl Derivative
235	195*	234	193	208	266	194	84	266	Desipramine
235	323*	338	73	176	91	75	247	338	Methaqualone Metabolite TMS Derivative
235	353*	210	72	54	45	221	195	353	Methysergide Hydrogen Maleate
236*	234	238	155	157	75	50	76	236	1,2-Dibromobenzene
236*	295	296	263	145	235	0	0	296	Niflumic Acid Methyl Ester
236*	315	317	44	77	91	287	316	315	Bromazepam
236*	315	317	286	288	316	208	179	315	Bromazepam
236*	317	315	288	316	286	208	78	315	Bromazepam*
236	53*	110	127	153	68	238	58	236	(3-Butynyl)-N'-(4-chlorophenyl) N-Methylcarbamate (N-)
236	58*	40	202	235	203	42	44	295	Dothiepin*
236	282*	237	281	263	145	44	93	282	Niflumic Acid
237*	239	235	241	332	334	404	402	470	Dienochlor
237*	239	339	238	209	240	196	167	339	2-Oxo-LSD
237	195*	138	238	196	43	39	110	316	Permethylated Ibomal*
237	235*	337	236	221	338	209	238	337	Aromatised 2-oxo-LSD
237	235*	165	212	246	75	176	36	352	DDT
237	235*	165	75	199	246	352	0	352	DDT (o,p-)
237	235*	165	75	50	51	352	0	352	DDT (p,p'-)
237	235*	165	212	176	282	236	284	352	DDT(p,p'-)
237	235*	165	236	199	239	238	75	352	Dicophane*
237	235*	165	236	199	239	82	238	318	TDE (p,p'-)
237	252*	253	221	238	177	209	149	252	Methyl 3,4,5-trimethoxycinnamate
237	339*	197	223	238	212	229	340	339	13-Hydroxy-LSD
237	339*	197	238	239	223	196	212	339	12-Hydroxy-LSD
238*	42	239	224	240	56	72	226	476	Loperamide*

								MWT	COMPOUND
238*	58	91	239	167	117	165	77	329	Prenylamine
238*	111	240	75	113	50	74	127	238	1-Chloro-2-iodobenzene
238*	240	236	203	201	242	205	182	418	Octachlorokepone
238	58*	345	91	107	167	165	252	345	Prenylamine Metabolite
238	137*	181	40	37	240	44	138	346	Tolylfluanide
238	224*	226	123	340	0	0	0	375	Haloperidol
239*	70	209	210	166	139	138	211	405	Dihydroxytrazodone (Trazodone Metabolite)
239*	76	240	104	285	241	177	102	338	Chlorthalidone
239*	148	194	164	210	193	182	147	239	3,5-Dimethylpyrrole-2,4-dicarboxylic Acid Ethyl Ester
239*	167	165	59	194	181	152	193	239	Prenylamine Metabolite
239*	177	192	149	63	77	92	134	239	Mescaline Precursor
239*	196	268	120	197	77	225	104	268	5-(4-Hydroxyphenyl)-5-Phenylhydantoin(Para HPPH)
239*	205	221	211	139	166	195	387	405	Trazodone Metabolite
239*	241	178	143	274	276	240	242	274	2,4-Dichloro-6-(o-chloroaniline)-s-triazine (Dyrene)
239*	241	178	143	75	87	99	89	274	Zinochlor
239*	274	273	275	240	276	245	211	274	Desmethyltetrazepam
239*	274	302	276	304	275	241	75	320	Lorazepam
239	76*	104	240	285	241	50	75	338	Chlorthalidone*
239	237*	235	241	332	334	404	402	470	Dienochlor
239	237*	339	238	209	240	196	167	339	2-Oxo-LSD
239	253*	359	373	252	238	237	36	407	Gentian Violet
239	254*	44	41	77	91	191	296	296	Medinoterbacetate
239	291*	274	293	75	302	276	138	320	Lorazepam*
240*	184	241	212	92	138	63	223	240	Danthron
240*	212	241	184	138	92	128	63	240	Danthron*
240	72*	42	39	40	41	44	29	240	Dimetilan
240	72*	169	44	42	73	170	56	240	Dimetilan
240	205*	91	84	125	242	206	186	311	Pyrrobutamine*
240	238*	236	203	201	242	205	182	418	Octachlorokepone
240	269*	241	268	270	107	121	213	269	7-Aminodesmethylflunitrazepam
241*	58	43	184	69	68	41	199	241	Prometryne
241*	77	242	105	44	43	195	57	242	2-Amino-5-nitrobenzophenone
241*	139	276	195	111	44	165	119	276	2-Amino-5-Nitro-2'-Chlorobenzophenone
241*	181	91	66	30	75	224	90	287	Tetryl
241*	242	269	77	243	103	270	51	270	Desmethyldiazepam
241*	242	77	105	195	211	212	51	242	Nitrazepam Benzophenone
241*	242	181	75	224	194	91	77	287	Tetryl
241*	276	139	165	195	111	278	242	276	2-Amino-2'-chloro-5-nitrobenzophenone*
241*	276	139	165	195	111	242	277	276	Clonazepam Benzophenone
241	187*	242	174	30	126	175	301	301	Pentaquine*
241	197*	198	77	242	104	91	103	242	Fenoprofen*
241	223*	208	222	194	180	77	224	241	Mefenamic Acid*
241	239*	178	143	274	276	240	242	274	2,4-Dichloro-6-(o-chloroaniline)-s-triazine (Dyrene)
241	239*	178	143	75	87	99	89	274	Zinochlor
242*	43	270	269	241	103	243	76	332	Chlorazepate
242*	128	127	155	171	143	154	156	242	5-Cinnamylidenebarbituric Acid
242*	207	244	270	243	271	269	165	270	Medazepam*
242*	215	243	214	111	121	216	0	242	Cambendazole GC Decomposition Product
242*	244	309	311	243	214	178	277	309	Meclofenamic Acid Methyl Ester
242*	256	228	255	115	129	141	116	256	5-Indanmethylenebarbituric Acid
242*	269	270	241	243	271	244	272	270	Desmethyldiazepam*
242*	270	269	241	243	271	244	103	270	Desmethyldiazepam
242	207*	244	165	270	243	208	269	270	Medazepam
242	214*	216	309	215	311	179	151	309	Diclofenac Methyl Ester

Asterisk Indicates True Base Peak

								MWT	COMPOUND
242	225*	244	364	362	208	451	453	466	Chloramphenicol TMS Ether
242	241*	269	77	243	103	270	51	270	Desmethyldiazepam
242	241*	77	105	195	211	212	51	242	Nitrazepam Benzophenone
242	241*	181	75	224	194	91	77	287	Tetryl
242	299*	243	59	57	214	185	300	299	Dihydrocodeinone
242	299*	256	96	243	58	60	214	299	Hydrocodone
242	299*	59	243	42	96	70	214	299	Hydrocodone*
242	327*	328	286	229	96	41	201	327	Naloxone
243*	108	123	122	164	136	95	79	243	4'-Methoxy-2-(methylsulphonyl)acetanilide
243*	256	70	245	192	227	258	326	326	Clozapine*
243	45*	136	200	159	157	198	242	243	Norlevorphanol
243	118*	272	104	77	129	128	89	272	Alphenal Dimethyl Derivative
243	118*	104	77	129	231	130	128	272	1,3-Dimethylalphenal
243	245*	247	278	262	179	244	181	278	2,3,4,5,6-Pentachlorobenzyl Alcohol (Blastin)
243	313*	162	124	59	42	112	0	313	Ethylmorphine
244*	42	72	475	109	245	85	476	475	Fluspirilene*
244*	59	77	29	43	31	45	55	309	Glymidine*
244*	227	55	57	95	243	69	43	244	2-Methylamino-5-amino-2'-Fluorobenzophenone
244	62*	279	45	229	57	61	111	279	2-Methylamino-2',5-Dichlorobenzophenone
244	185*	141	115	170	186	153	154	244	Naproxen Methyl Ester
244	199*	200	178	179	184	183	245	244	Flurbiprofen*
244	205*	191	345	206	72	272	246	404	Benzquinamide*
244	229*	173	243	44	230	81	245	244	Xylometazoline
244	242*	309	311	243	214	178	277	309	Meclofenamic Acid Methyl Ester
244	245*	77	105	193	246	228	44	245	2-Methylamino-5-chlorobenzophenone
244	246*	209	211	87	248	218	190	244	Tetrachloro-p-benzoquinone (Chloranil)
244	286*	77	218	51	217	288	215	286	Desmethylclobazam
245*	44	247	209	183	246	271	259	346	Flurazepam N1-Acetic Acid
245*	77	244	105	228	246	51	247	245	Diazepam Benzophenone
245*	77	244	228	105	246	193	247	245	Medazepam Benzophenone
245*	77	246	247	228	105	51	193	245	2-Methylamino-5-chlorobenzophenone*
245*	123	258	165	186	95	246	0	396	Aceperone
245*	183	89	247	0	0	0	0	245	Flurazepam Metabolite
245*	243	247	278	262	179	244	181	278	2,3,4,5,6-Pentachlorobenzyl Alcohol (Blastin)
245*	244	77	105	193	246	228	44	245	2-Methylamino-5-chlorobenzophenone
245*	247	183	210	89	105	122	246	245	Flurazepam N1-Acetic Acid Decomposition Product
245*	260	44	217	218	246	261	259	260	Oxymetazoline*
245	72*	44	290	45	40	75	247	290	Chloroxuron
245	77*	244	105	228	246	247	51	245	2-Methylamino-5-chlorobenzophenone
245	77*	244	105	193	228	168	246	245	2-Methylamino-5-chlorobenzophenone
245	102*	247	304	305	306	58	126	335	Hydroxychloroquine*
245	246*	231	215	202	116	217	203	307	Hydroxycyclobenzaprine N-oxide
245	246*	202	117	159	132	145	215	246	5-Methoxybenzylidene-(para)-barbituric Acid
245	341*	246	274	343	313	58	299	359	Monodesethylflurazepam
246*	30	302	36	274	211	273	248	331	Didesethylflurazepam
246*	36	100	42	38	82	91	56	346	Morpheridine
246*	42	247	120	172	80	218	106	352	Anileridine
246*	42	247	377	91	172	47	165	452	Diphenoxylate
246*	42	247	91	103	165	115	56	452	Diphenoxylate*
246*	42	43	247	71	41	56	91	361	Furethidine*
246*	42	36	247	100	218	56	232	346	Morpheridine
246*	42	70	143	403	56	248	112	403	Perphenazine
246*	42	367	247	57	56	91	77	367	Phenoperidine*
246*	45	42	366	58	57	43	106	366	Piminodine

	PEAKS							MWT	COMPOUND
246*	75	318	248	316	73	176	55	316	DDE
246*	91	42	247	233	57	218	56	367	Benzethidine
246*	91	42	233	162	57	56	149	367	Benzethidine
246*	91	165	377	452	0	0	0	452	Diphenoxylate
246*	92	65	245	108	247	39	260	310	Sulphadimethoxine*
246*	100	42	82	91	56	232	41	346	Morpheridine
246*	143	403	70	404	42	248	113	403	Perphenazine*
246*	146	117	118	247	175	103	77	274	3-Ethylmephobarbitone
246*	165	42	123	199	247	214	108	379	Droperidol*
246*	173	41	91	135	0	0	0	246	Santonin (alpha)
246*	185	445	70	125	154	213	87	445	Thiopropazate
246*	211	107	43	245	57	248	80	246	2,5-Diamino-2'-Chlorobenzophenone
246*	244	209	211	87	248	218	190	244	Tetrachloro-p-benzoquinone (Chloranil)
246*	245	231	215	202	116	217	203	307	Hydroxycyclobenzaprine N-oxide
246*	245	202	117	159	132	145	215	246	5-Methoxybenzylidene-(para)-barbituric Acid
246*	247	42	120	218	172	106	91	352	Anileridine
246*	247	42	36	45	91	56	219	321	Etoxeridine
246*	247	42	71	43	56	232	0	361	Furethidine
246*	247	36	100	42	84	232	56	346	Morpheridine
246*	247	42	100	232	56	218	172	346	Morpheridine*
246*	247	41	42	136	67	111	39	999	Zomepirac GC Decomposition Product*
246*	261	260	103	186	91	77	0	261	Ethylnorpethidine (N-)
246*	291	248	290	247	139	111	292	293	Zomepirac*
246*	302	58	71	289	341	56	274	359	Monodesethylflurazepam
246*	318	316	248	320	176	210	0	316	DDE (p,p-)
246*	366	106	247	133	260	234	367	366	Piminodine (Alvodine, Pimadin, Cimadon)
246*	366	42	106	133	247	57	260	366	Piminodine Ethane Sulphonate
246*	367	247	42	91	103	77	57	367	Phenoperidine
246*	375	42	247	91	376	156	184	452	Diphenoxylate
246	42*	232	70	56	143	214	43	403	Perphenazine
246	58*	233	318	86	272	248	232	334	Chlorpromazine Metabolite
246	58*	302	71	289	273	274	341	359	Monodesethylflurazepam
246	77*	247	105	78	45	51	248	247	2-Amino-3-Hydroxy-5-Chlorbenzophenone
246	108*	151	229	95	152	153	92	246	Cruformate Metabolite
246	112*	105	77	55	41	44	247	261	Piperocaine*
246	154*	70	349	43	223	225	167	577	Dihydroergocryptine
246	231*	314	232	121	193	74	174	314	Cannabidiol
246	318*	316	248	320	176	105	247	316	DDE (p,p'-)
247*	188	160	63	215	43	106	216	247	1,2-Dimethyl-5-methoxyindole-3-acetic Acid Me Ester (Indomethacin Der)
247*	189	249	191	243	55	29	245	302	Ethacrynic Acid*
247*	248	249	250	219	212	106	211	248	Pyrimethamine
247*	248	249	42	89	219	114	123	248	Pyrimethamine
247*	249	58	72	248	167	250	168	318	Brompheniramine*
247*	285	328	249	41	287	330	55	328	Clorexolone*
247	71*	172	218	174	103	96	91	247	Pethidine
247	71*	246	70	218	57	174	248	247	Pethidine
247	139*	141	111	75	249	140	248	247	2-Chloro-4'-hydroxybenzanilide
247	174*	98	175	173	248	112	245	330	Dipiproverine
247	245*	183	210	89	105	122	246	245	Flurazepam N1-Acetic Acid Decomposition Product
247	246*	42	120	218	172	106	91	352	Anileridine
247	246*	42	36	45	91	56	219	321	Etoxeridine
247	246*	42	71	43	56	232	0	361	Furethidine
247	246*	36	100	42	84	232	56	346	Morpheridine
247	246*	42	100	232	56	218	172	346	Morpheridine*

Asterisk Indicates True Base Peak   Peak Index

							MWT	COMPOUND
247	246*	41	42	136	67	111	39	999 Zomepirac GC Decomposition Product*
247	249*	248	250	276	278	198	200	276 Bromazepam Benzophenone
247	249*	58	72	167	248	168	250	318 Brompheniramine
248*	44	249	121	136	29	164	192	292 Ambucetamide*
248*	97	219	218	111	249	217	263	263 Dimethylthiambutene*
248*	249	123	250	251	154	95	230	249 Flurazepam Benzophenone
248	108*	140	65	92	141	109	80	248 Dapsone
248	143*	115	77	69	171	81	249	248 1-Phenylazonapth-2-ol
248	175*	177	250	185	145	69	57	248 Ethyl 2,4-dichlorophenoxyacetate
248	219*	148	220	120	218	133	65	248 4-Hydroxyphenobarbitone
248	247*	249	250	219	212	106	211	248 Pyrimethamine
248	247*	249	42	89	219	114	123	248 Pyrimethamine
248	249*	123	154	250	95	251	230	249 2-Amino-5-Chloro-2'-Fluorobenzophenone
248	250*	252	254	215	108	213	178	248 Pentachlorobenzene
248	290*	108	140	43	65	93	92	290 Acetyldapsone
249*	81	183	250	79	185	184	264	264 3-Ethylhexobarbitone
249*	169	250	41	133	67	79	93	278 Permethylated Heptabarbitone*
249*	220	192	149	164	131	135	147	249 2-Ethyl-2-(3,4-dihydroxyphenyl)glutarimide (Glutethimide Metab)
249*	247	248	250	276	278	198	200	276 Bromazepam Benzophenone
249*	247	58	72	167	248	168	250	318 Brompheniramine
249*	248	123	154	250	95	251	230	249 2-Amino-5-Chloro-2'-Fluorobenzophenone
249*	306	250	170	307	251	171	277	306 Dihydroxyzolimidine (Metab 2)
249	58*	247	72	167	248	168	42	318 Brompheniramine (dex) (d-Parabromdylamine)
249	58*	208	99	193	232	84	0	294 Trimipramine
249	58*	208	193	234	99	248	194	294 Trimipramine
249	63*	205	251	143	65	207	223	284 Tris-(2-chloroethyl) Phosphate
249	208*	193	44	195	194	234	209	280 Trimipramine Metabolite
249	247*	58	72	248	167	250	168	318 Brompheniramine*
249	248*	123	250	251	154	95	230	249 Flurazepam Benzophenone
249	295*	237	293	297	214	212	142	293 Pentachloronitrobenzene (PCNB)
250*	70	207	91	251	119	148	56	341 Fenethylline*
250*	95	39	235	66	207	41	193	250 3'-Ketohexobarbitone
250*	164	222	194	235	251	207	150	250 1,3,7-Triethyl-8-methylxanthine
250*	207	70	91	251	119	148	65	341 Fenethylline
250*	248	252	254	215	108	213	178	248 Pentachlorobenzene
250*	252	248	108	215	73	213	85	248 Pentachlorobenzene
250	232*	175	204	176	233	102	78	250 Diflunisal*
250	235*	91	233	236	65	76	132	250 Methaqualone*
251*	44	223	222	43	84	98	111	251 7-Aminonitrazepam
251*	222	223	250	252	195	110	97	251 7-Aminonitrazepam
251*	266	249	77	143	252	0	0	266 Methaqualone Metabolite
251*	266	249	77	143	76	252	39	266 2-Methyl-3-(4-hydroxy-2'-methylphenyl)-4-(3H)-quinazolinone
251*	280	72	134	77	265	208	175	280 1,3-Dimethyldilantin
251*	280	118	91	121	119	252	189	280 Phenprocoumon*
251*	286	252	78	288	111	152	271	286 7-Hydroxymecloqualone
251*	286	152	252	288	111	154	160	286 8-Hydroxymecloqualone
251*	286	252	111	288	271	152	273	286 5-Hydroxymecloqualone
251*	286	252	288	111	154	271	152	286 6-Hydroxymecloqualone
251	58*	250	296	211	85	224	209	296 ((3-Dimethylaminopropyl)-2-hydroxyiminodibenzyl)(N-)(Imipramine Metab)
251	165*	135	0	0	0	0	0	251 STP-NCS Derivative
251	353*	211	252	237	253	226	196	353 13-Methoxy-LSD
251	353*	237	211	249	338	252	351	353 14-Methoxy-LSD
251	353*	211	237	253	252	338	354	353 12-Methoxy-LSD
252*	160	206	254	125	162	217	148	252 Chlorodinitronaphthalene

252*	237	253	221	238	177	209	149	252	Methyl 3,4,5-trimethoxycinnamate
252	121*	120	132	106	122	105	383	383	BTS 29101
252	121*	105	77	126	79	237	251	252	2,4-Dimethylphenylformamidine (N,N'-bis-2,4-Dimethyl-)
252	250*	248	108	215	73	213	85	248	Pentachlorobenzene
252	253*	43	254	104	235	57	251	253	Triamterene (2,4,7-Triamino-6-phenylpyimido(4,5-b)pyrazine)
253*	239	359	373	252	238	237	36	407	Gentian Violet
253*	252	43	254	104	235	57	251	253	Triamterene (2,4,7-Triamino-6-phenylpyimido(4,5-b)pyrazine)
253*	271	91	105	213	145	107	147	304	Methandriol
253*	288	287	289	225	259	254	41	288	Tertrazepam
253	73*	217	185	317	75	204	69	676	1,3-Dimethylhydroxypentobarbitone Glucuronide Me Ester TMS Derivative
253	73*	75	217	147	185	317	45	676	Hydroxyamobarbitone Glucuronide Me TMS Derivative (peak 1)
253	73*	167	165	194	193	152	115	253	Prenylamine Metabolite
253	181*	72	0	0	0	0	0	253	Mescaline-NCS Derivative
253	280*	281	206	234	252	254	264	281	Nitrazepam*
253	395*	396	294	268	74	221	128	395	Lsd-TMS Derivative
254*	42	70	43	56	143	222	44	411	Acetophenazine
254*	143	411	70	255	42	380	157	411	Acetophenazine
254*	192	193	154	223	255	180	221	254	Lysergol*
254*	239	44	41	77	91	191	296	296	Medinoterbacetate
254*	256	118	255	303	305	63	45	303	Chlorambucil*
254	56*	256	229	231	199	201	0	285	4-Bromo-2,5-dimethoxyamphetamine Impurity
254	58*	45	44	77	42	59	72	289	Chlophedianol*
254	58*	91	56	183	115	107	343	345	Prenylamine Metabolite
254	92*	163	91	93	65	103	104	284	Tropicamide
254	105*	177	209	77	210	181	255	254	Ketoprofen
254	196*	209	195	0	0	0	0	254	2,4,5-Trichlorophenoxyacetic Acid (2,4,5-T)
255*	131	254	125	257	256	57	58	325	Clemizole
255*	131	60	73	255	43	125	258	325	Clemizole
255	58*	40	72	42	59	71	91	255	Phenyltoloxamine
255	58*	42	71	59	44	181	165	255	Phenyltoloxamine*
255	193*	228	256	257	165	230	258	256	Desmethylmedazepam
255	223*	208	77	222	180	194	96	255	Mefenamic Acid Methyl Ester
255	311*	42	44	296	310	312	174	311	Thebaine*
255	314*	302	246	316	211	315	273	331	Didesethylflurazepam
256*	43	44	283	84	69	284	257	368	Ketazolam
256*	58	257	182	199	157	105	91	347	Phenomorphan
256*	77	283	221	255	257	89	165	284	Diazepam
256*	121	185	43	257	65	42	214	285	Probenecid*
256*	165	258	193	257	76	89	104	256	Clorindione
256*	182	276	108	169	184	291	278	291	Cruformate
256*	283	284	285	257	255	258	286	284	Diazepam*
256*	284	283	285	84	257	258	255	368	Ketazolam*
256	137*	44	45	43	41	120	55	256	2',4,4'-Trihydroxychalcone
256	197*	240	198	152	106	257	241	256	Phenylacetylsalicylate
256	242*	228	255	115	129	141	116	256	5-Indanmethylenebarbituric Acid
256	243*	70	245	192	227	258	326	326	Clozapine*
256	254*	118	255	303	305	63	45	303	Chlorambucil*
256	258*	186	260	150	151	188	259	256	2,3,5-Trichlorobiphenyl
256	285*	43	110	257	84	287	111	285	7-Aminoclonazepam
257*	43	55	311	340	259	77	313	382	3-Acetoxyprazepam
257*	55	311	77	259	313	44	312	340	3-Hydroxyprazepam
257*	59	150	256	200	157	76	189	257	3-Hydroxy-N-methylmorphinan
257*	59	150	256	31	157	42	200	257	Racemorphan*
257*	77	268	239	205	267	233	259	286	Oxazepam*

```
              PEAKS                         MWT      COMPOUND


257* 259   77  228  286    0    0    0      286  Oxazepam
257   59* 150  256   44   31  200  157      257  Levorphanol*
257   59* 256  150   80   42   82  200      257  Racemorphan
257   77* 268  205  239  267  233  241      286  Oxazepam
257  268* 267  239  233   77  205  269      286  Oxazepam
257  343* 256  372  357  283  371    0      372  Methyloxazepam TMS Derivative
258* 223  286  257  259   75  122  251      286  Desalkyl-3-hydroxyflurazepam Dehydration Product (N1-Desalkyl)
258* 256  186  260  150  151  188  259      256  2,3,5-Trichlorobiphenyl
258* 259  243  260  115   91  116  108      259  4-Methyl-5-phenyl-2-(benzyl)pyridine
258* 259  244  260   91  115   65   72      259  2-Methyl-3-phenyl-6-(benzyl)pyrimidine
258  259* 260  257  216  173  217  172      259  5-Dimethylaminobenzylidene-(para)-barbituric Acid
258  300*  77  259  283  302  231  256      300  Clobazam*
259* 258  260  257  216  173  217  172      259  5-Dimethylaminobenzylidene-(para)-barbituric Acid
259* 260  288  287  261  289  262  290      288  Desalkylflurazepam(N1-)
259* 260  288  287  261  289  262  290      288  Desalkylflurazepam (N1-Desalkyl)
259* 260  115  244  116   95  108  215      259  2,6-Dimethyl-3,5-diphenylpyrimidine
259* 260  244  115  108   85   91  215      259  2,4-Dimethyl-3,5-diphenylpyrimidine
259  257*  77  228  286    0    0    0      286  Oxazepam
259  258* 243  260  115   91  116  108      259  4-Methyl-5-phenyl-2-(benzyl)pyridine
259  258* 244  260   91  115   65   72      259  2-Methyl-3-phenyl-6-(benzyl)pyrimidine
259  290* 275  291  243  123  200   43      290  Trimethoprim
260*  43  123  259   95   77   57  165      260  2-Amino-5-Nitro-2'-Fluorobenzophenone
260* 173  259  244  201  216  128  145      262  5-(4'-Hydroxy-2'-methoxybenzylidene)-barbituric Acid
260* 216  302  215  242    0    0    0      302  Cambendazole
260* 261   42  202   57  217  203  218      261  Phenindamine
260* 261  202   42  203  182  115  215      261  Phenindamine
260* 261   42   57  184  215  217  262      261  Phenindamine*
260* 262  180  145  261  179  182  181      289  Quinethazone
260* 262  180  287  261  145  286  124      289  Quinethazone*
260* 288  287  259  261  289  102  262      288  Desalkylflurazepam(N1-)
260* 302  216  215  243  189  242  188      302  Cambendazole
260  146* 118  117  103  261   91  232      288  1,3-Diethylphenobarbitone
260  245*  44  217  218  246  261  259      260  Oxymetazoline*
260  259* 288  287  261  289  262  290      288  Desalkylflurazepam(N1-)
260  259* 288  287  261  289  262  290      288  Desalkylflurazepam (N1-Desalkyl)
260  259* 115  244  116   95  108  215      259  2,6-Dimethyl-3,5-diphenylpyrimidine
260  259* 244  115  108   85   91  215      259  2,4-Dimethyl-3,5-diphenylpyrimidine
261*  45  263  243   55   73  245  316      316  Ethacrynic Acid Methyl Ester
261* 125  109  108   97   80  233  205      261  Aminoparathion
261* 215  259  213  203  201  263  217      259  Tetrachloronitrobenzene
261* 290  233  148  262  133  176  260      290  1,3-Dimethylhydroxyphenobarbitone Methyl Ether
261  246* 260  103  186   91   77    0      261  Ethylnorpethidine (N-)
261  260*  42  202   57  217  203  218      261  Phenindamine
261  260* 202   42  203  182  115  215      261  Phenindamine
261  260*  42   57  184  215  217  262      261  Phenindamine*
262* 109  166  264  293   95  123   75      293  Flurazepam Benzophenone 2
262* 109  166  264  275  311  313  123      311  Flurazepam Chloroethyl Artifact Benzophenone
262* 109  166  264  293  123  168   95      293  2-Hydroxyethylamino-5-Chloro-2'-Fluorobenzophenone
262* 111  219   97  263   86   42  264      277  Ethylmethylthiambutene*
262* 135  111  219   97  263   56    0      277  Ethylmethylthiambutene
262* 219  111   97  263   56  218   42      277  Ethylmethylthiambutene
262* 264  293  263  295  265    0    0      293  Flurazepam Hydroxyethyl Metabolite Benzophenone
262   44* 215  202  218  189  231  203      305  Desmethyloxetorone
262  260* 180  145  261  179  182  181      289  Quinethazone
```

Peak Index   *Asterisk Indicates True Base Peak*   Page: 267

262	260*	180	287	261	145	286	124	289	Quinethazone*
263*	42	265	164	278	319	304	264	319	Cruformate Metabolite
263*	125	109	79	93	47	63	264	263	Parathion-methyl
263*	133	132	308	264	105	77	279	308	7-Ethoxydiftalone
263*	281	166	92	145	167	235	139	281	Flufenamic Acid
263*	295	264	166	92	235	145	243	295	Flufenamic Acid Methyl Ester
263*	353	251	368	73	249	75	277	368	Flunixin TMS Ester
263	66*	79	91	265	101	261	65	362	Aldrin
263	109*	125	79	63	93	264	64	263	Parathion-methyl
263	120*	142	178	100	264	41	29	306	Butacaine*
263	205*	204	75	102	101	51	88	263	Cinchophen Methyl Ester
263	234*	206	163	178	131	149	161	263	2-Ethyl-2-(4-hydroxy-3-methoxyphenyl)glutarimide (Glutethimide Metab)
263	295*	310	251	294	249	277	181	310	Flunixin Methyl Ester
263	365*	221	264	223	0	0	0	365	1-Acetyllysergic Acid Diethylamide
264*	139	266	111	141	251	152	253	292	Proclonol
264	112*	113	91	179	110	178	115	349	Dipipanone*
264	132*	90	133	89	105	118	265	264	Diftalone
264	180*	193	109	194	181	67	42	264	Pentifylline*
264	193*	43	72	71	220	192	194	264	Mianserin*
264	193*	194	43	71	72	220	109	264	Mianserin (Bolvidon)
264	262*	293	263	295	265	0	0	293	Flurazepam Hydroxyethyl Metabolite Benzophenone
264	266*	268	165	167	130	202	200	264	Pentachlorophenol (PCP)
264	266*	268	124	229	231	194	159	264	2,4,5,6-Tetrachloroisophthalonitrile (Daconil)
264	293*	43	292	263	213	212	294	293	7-Acetoamidonitrazepam
265*	120	249	43	103	121	162	131	308	Warfarin
265*	131	43	103	121	120	145	146	308	Warfarin
265*	193	115	130	42	208	56	264	265	3,3-Diphenyl-1,5-dimethyl-2-pyrrolidone(Methadone metab.)
265*	208	121	43	266	187	213	251	308	Warfarin*
265*	267	26	186	263	269	105	107	342	1,1,2,2-Tetrabromoethane
265	73*	217	75	204	69	41	147	688	1,3-Dimethylhydroxysecobarbitone Me Ester TMS Ether
265	100*	128	266	44	98	56	101	392	Dextromoramide*
265	100*	128	266	56	129	55	101	392	Levomoramide
265	195*	43	196	41	44	39	110	344	Permethylated Sigmodal*
265	230*	139	267	111	232	264	266	265	Lorazepam Benzophenone
265	293*	264	292	43	222	223	294	293	7-Acetoamidonitrazepam
266*	88	102	75	115	128	131	176	284	Mazindol
266*	264	268	165	167	130	202	200	264	Pentachlorophenol (PCP)
266*	264	268	124	229	231	194	159	264	2,4,5,6-Tetrachloroisophthalonitrile (Daconil)
266*	267	42	220	152	224	165	178	267	Apomorphine
266*	267	224	220	268	44	250	248	267	Apomorphine*
266*	268	267	255	231	102	88	176	284	Mazindol*
266*	268	264	165	167	270	202	130	264	Pentachlorophenol
266	160*	235	251	77	146	58	247	266	Hydroxymethaqualone
266	180*	237	208	118	223	194	77	266	5-Phenylhydantoin(5-(p-methylphenyl)-)
266	251*	249	77	143	252	0	0	266	Methaqualone Metabolite
266	251*	249	77	143	76	252	39	266	2-Methyl-3-(4-hydroxy-2'-methylphenyl)-4-(3H)-quinazolinone
266	294*	42	43	278	58	295	207	335	Benefin
266	308*	440	43	512	152	194	126	554	Practolol, tri-TFA Derivative
267*	73	268	269	45	135	193	75	282	Salicyclic Acid TMS Ester TMS Ether
267*	77	51	91	63	117	294	248	295	Methylnitrazepam
267*	221	207	180	223	154	196	268	267	Lysergamide*
267*	268	45	238	282	43	55	41	282	Propyl Cannabinol
267*	269	323	81	325	109	295	170	358	Chlorfenvinphos (beta)
267*	323	269	325	81	109	295	170	358	Chlorfenvinphos (alpha)

267	265*	26	186	263	269	105	107	342	1,1,2,2-Tetrabromoethane
267	266*	42	220	152	224	165	178	267	Apomorphine
267	266*	224	220	268	44	250	248	267	Apomorphine*
267	296*	210	219	134	180	77	297	296	3-Methyl-p-hydroxydilantin Methyl Ether
268*	143	425	70	269	42	394	157	425	Carphenazine*
268*	143	425	55	70	41	40	269	425	Carphenazine (Proketazine)
268*	154	224	180	207	192	221	223	268	Lysergic Acid (d)
268*	224	154	180	207	223	192	179	268	Lysergic Acid*
268*	257	267	239	233	77	205	269	286	Oxazepam
268*	303	269	305	304	0	0	0	303	Chloromorphide (beta)
268*	425	143	269	197	394	157	171	425	Carphenazine
268	58*	269	85	193	229	228	192	314	Chlorimipramine
268	148*	267	224	269	251	120	118	268	Michler's Ketone
268	266*	267	255	231	102	88	176	284	Mazindol*
268	266*	264	165	167	270	202	130	264	Pentachlorophenol
268	267*	45	238	282	43	55	41	282	Propyl Cannabinol
268	285*	284	77	42	286	233	287	285	2-Desmethylchlordiazepoxide
268	295*	297	97	57	270	62	64	295	Chlorthiazide
268	327*	215	328	269	267	146	124	327	Acetylmorphine
268	327*	43	42	215	44	328	146	327	Acetylmorphine
268	327*	215	59	204	70	146	81	327	6-Acetylmorphine
268	327*	328	42	43	215	44	59	327	6-Acetylmorphine*
269*	64	205	297	271	43	44	31	297	Hydrochlorothiazide
269*	91	55	295	296	241	324	271	324	Prazepam
269*	205	221	297	271	62	285	124	297	Hydrochlorothiazide*
269*	229	268	232	71	44	242	211	300	Desmethylclomipramine
269*	240	241	268	270	107	121	213	269	7-Aminodesmethylflunitrazepam
269*	324	91	296	295	55	323	325	324	Prazepam
269	58*	85	268	270	271	'314	242	314	Clomipramine
269	91*	324	55	296	295	323	297	324	Prazepam*
269	217*	375	241	204	341	342	329	822	Oxazepam Glucuronide TMS Derivative
269	242*	270	241	243	271	244	272	270	Desmethyldiazepam*
269	267*	323	81	325	109	295	170	358	Chlorfenvinphos (beta)
270	92*	65	156	108	106	93	59	270	Sulphamethizole*
270	242*	269	241	243	271	244	103	270	Desmethyldiazepam
270	285*	77	91	105	257	166	287	285	Prazepam Benzophenone
271*	44	270	214	272	42	43	70	311	Nalorphine (N-Allylnormorphine)
271*	56	43	41	55	42	273	77	300	3-Hydroxydiazepam
271*	61	43	45	70	300	256	273	342	3-Acetoxydiazepam
271*	77	273	300	0	0	0	0	300	Diazepam Metabolite
271*	77	273	300	255	0	0	0	300	3-Hydroxydiazepam
271*	81	150	148	45	110	42	82	271	Normorphine
271*	214	44	270	32	228	272	42	271	Desomorphine
271*	273	77	272	256	300	255	257	300	3-Hydroxydiazepam(N-Methyloxazepam)
271*	273	300	272	256	77	255	257	300	Temazepam*
271	59*	150	270	31	214	42	171	271	Dextromethorphan*
271	59*	150	214	270	171	112	213	271	Methorphan
271	59*	150	31	270	214	42	171	271	Racemethorphan*
271	230*	55	256	164	270	71	124	271	Cyclazocine*
271	253*	91	105	213	145	107	147	304	Methandriol
271	298*	299	224	272	270	252	280	299	Desmethylflunitrazepam
272*	193	209	78	192	273	194	181	272	Zolimidine
272*	273	258	259	115	243	91	260	273	2,4-Dimethyl-3-phenyl-6-(benzyl)pyrimidine
272*	274	270	237	235	239	218	216	486	Kepone

								MWT	COMPOUND
272*	274	237	270	143	276	239	235	486	Kepone
272*	274	237	239	270	276	235	241	540	Mirex
272*	274	270	237	276	251	203	182	452	Nonachlorokepone
272	100*	274	270	237	102	65	276	370	Heptachlor
273*	208	166	193	150	316	108	96	316	Bupirimate
273	271*	77	272	256	300	255	257	300	3-Hydroxydiazepam(N-Methyloxazepam)
273	271*	300	272	256	77	255	257	300	Temazepam*
273	272*	258	259	115	243	91	260	273	2,4-Dimethyl-3-phenyl-6-(benzyl)pyrimidine
273	274*	211	257	123	199	275	95	274	2-Methylamino-5-nitro-2'-fluorobenzophenone
273	274*	225	318	181	257	121	197	318	Phenolphthalein
273	288*	331	287	304	290	289	275	331	Flurazepam(N1-Ethanol-)
273	288*	331	287	304	289	290	253	332	Hydroxyethylflurazepam
274*	60	273	259	181	202	218	215	335	Oxetorone N-oxide
274*	93	125	121	91	107	135	246	320	Phenthoate
274*	110	91	79	105	147	215	256	274	19-Nortesterone
274*	225	318	273	275	226	257	121	318	Phenolphthalein*
274*	273	211	257	123	199	275	95	274	2-Methylamino-5-nitro-2'-fluorobenzophenone
274*	273	225	318	181	257	121	197	318	Phenolphthalein
274	239*	273	275	240	276	245	211	274	Desmethyltetrazepam
274	239*	302	276	304	275	241	75	320	Lorazepam
274	272*	270	237	235	239	218	216	486	Kepone
274	272*	237	270	143	276	239	235	486	Kepone
274	272*	237	239	270	276	235	241	540	Mirex
274	272*	270	237	276	251	203	182	452	Nonachlorokepone
274	318*	77	226	104	44	210	105	318	Sulthiame Dimethyl Derivative
275*	277	258	257	259	75	223	122	304	Desalkyl-3-hydroxyflurazepam (N1-Desalkyl)
275*	293	96	179	0	0	0	0	293	Penfluridol Hofmann Reaction Product 1
275	218*	174	160	161	0	0	0	275	Physostigmine
276*	111	219	277	42	97	100	135	291	Diethylthiambutene
276*	219	111	277	278	42	100	97	291	Diethylthiambutene
276	58*	218	261	0	0	0	0	276	Psilocin-TMS Derivative
276	124*	58	140	56	72	125	41	291	Eucatropine*
276	182*	108	169	184	278	171	305	305	Cruformate Metabolite
276	241*	139	165	195	111	278	242	276	2-Amino-2'-chloro-5-nitrobenzophenone*
276	241*	139	165	195	111	242	277	276	Clonazepam Benzophenone
276	277*	262	42	115	105	91	56	277	Di-2-Ethyl-1,5-Dimethyl-3,3-Diphenyl-1-Pyrrolinium(Methadone metab.)
276	277*	262	105	278	220	200	91	277	Methadone Metabolite 1
276	361*	277	199	319	43	318	362	361	Bisacodyl
277*	184	88	279	275	63	278	53	459	Bromophenoxim
277*	276	262	42	115	105	91	56	277	Di-2-Ethyl-1,5-Dimethyl-3,3-Diphenyl-1-Pyrrolinium(Methadone metab.)
277*	276	262	105	278	220	200	91	277	Methadone Metabolite 1
277*	279	275	278	280	276	281	249	303	2,3,4,5,6-Pentachloromandelonitrile (Oryzon)
277*	306	288	278	209	198	222	237	306	Dihydroxyzolimidine (Metab 1)
277	40*	276	44	262	57	43	41	277	Methadone Metabolite
277	275*	258	257	259	75	223	122	304	Desalkyl-3-hydroxyflurazepam (N1-Desalkyl)
277	354*	325	104	268	73	269	355	354	5-(4-Hydroxyphenyl)-3-methyl-5-phenylhydantoin TMS Ether
277	361*	319	276	199	318	362	43	361	Bisacodyl*
278*	125	109	153	168	169	93	79	278	Fenthion
278	77*	51	78	105	279	91	39	404	Sulfinpyrazone
278	77*	105	78	51	279	252	130	404	Sulphinpyrazone*
278	321*	320	292	306	191	322	304	353	Berberine
279*	91	280	322	247	121	201	189	322	Methylwarfarin
279*	97	223	88	251	162	281	164	314	Dichlofenthion
279*	123	412	165	206	281	292	95	430	Amiperone

```
              PEAKS                    MWT     COMPOUND

279*  163  323   58  277  308  280  322   323  Dimefline*
279*  208   77  308  280  104  149  180   308  2,3-Diethyl-5,5-diphenylhydantoin
279   223*   97  162  251  225  164  281   314  Diethyl-O-(2,5-dichlorophenyl) Phosphorothioate (O,O-Diethyl)
279   277*  275  278  280  276  281  249   303  2,3,4,5,6-Pentachloromandelonitrile (Oryzon)
280*   42  143   70   56  113  281  265   437  Fluphenazine
280*   70  143   42  113  406  281  437   437  Fluphenazine
280*  143   42   70  437  406  113   56   437  Fluphenazine*
280*  144  115  224  154  252   29  281   280  Naphthylethyldiethyl Phosphate (beta-)
280*  253  281  206  234  252  254  264   281  Nitrazepam*
280*  314  315  234  289  240   75   76   315  Clonazepam
280*  314  315  286  234  288  316  240   315  Clonazepam*
280    42*  70  143   56  113   72  100   437  Fluphenazine
280   118* 203  194   77  165  223  222   280  Phenytoin Dimethyl Derivative
280   132* 133  104  235  262   77   89   280  Oxo-2-phthalazinylmethyl Benzoic Acid (2-(1-(2H))) (Diftalone Metab)
280   251*  72  134   77  265  208  175   280  1,3-Dimethyldilantin
280   251* 118   91  121  119  252  189   280  Phenprocoumon*
281*   43  282  187  107   91   55  105   384  Megesterol Acetate
281*  187  324  121   43    0    0    0   324  4'-Hydroxywarfarin
281   263* 166   92  145  167  235  139   281  Flufenamic Acid
281   296* 143  279   76  297   39   77   296  2-Methyl-3-(2'-methyl-4'-hydroxy-5'-methoxyphenyl)-quinazoline
282*  106  229  267   78   42  124   81   404  Nicocodeine
282   236  237  281  263  145   44   93   282  Niflumic Acid
282*  283  284  299  241  247   77  253   299  Chlordiazepoxide
282*  283   77  284   89   91  285  163   299  Chlordiazepoxide
282*  284  155  157   76   75   50   74   282  1-Bromo-2-iodobenzene
282*  299  218  284  283  219  241  214   299  Chlordiazepoxide
282*  299  284  283  241   77   55   91   299  Chlordiazepoxide
282*  299  284  283  241   56  301  253   299  Chlordiazepoxide*
282    43*  45   59   58  169  115  495   495  Codeine Heptafluorobutyrate Derivative
282    44* 283   77  284   56   91   57   299  Chlordiazepoxide
282    58*  30  284  355   73  283   44   355  Amodiaquine*
282    91*  79  105   77   81   67  338   392  Deoxycholic Acid
282   139*  69   55   41  271   91  105   414  Diosgenin*
282   283* 256  176  157   43   41   57   283  Levallorphan*
282   283* 160   42  284  268  110   44   283  Methyldesorphine
282   341*  43  229   42  342  204  162   341  Acetylcodeine
282   341* 229  204   81   70  149  124   341  6-Acetylcodeine
282   341*  42   43  229   59  204  342   341  Acetylcodeine
282   341* 229   42   43   59  342  204   341  6-Acetylcodeine*
282   404* 106   78  229  267   42  124   404  Nicocodeine
283*   44  255  282  254  284  264  256   283  7-Aminoflunitrazepam
283*   44  282  256   43  176   85   57   283  Levallorphan
283*  282  256  176  157   43   41   57   283  Levallorphan*
283*  282  160   42  284  268  110   44   283  Methyldesorphine
283*  285  202  139   50   75   63   76   283  Nitrofen
283   256* 284  285  257  255  258  286   284  Diazepam*
283   282* 284  299  241  247   77  253   299  Chlordiazepoxide
283   282*  77  284   89   91  285  163   299  Chlordiazepoxide
284*   91  375   81   42   36  285  175   375  Benzyl Morphine
284*  286  288  282  249  247  251  214   282  Hexachlorobenzene (HCB)
284*  465   81  285   42  175  181  162   465  Monopentafluorobenzylmorphine
284    58*  91  213  147  152  137  115   375  Prenylamine Metabolite
284    58*  86  238  198  199   85   42   284  Promazine*
284    72*  73   43   42  198  180  166   284  Promethazine
```

Peak Index   *Asterisk Indicates True Base Peak*

284	92*	156	108	65	106	93	220	284	Sulphaethidole*
284	256*	283	285	84	257	258	255	368	Ketazolam*
284	282*	155	157	76	75	50	74	282	1-Bromo-2-iodobenzene
284	329*	224	268	330	285	225	270	346	Nifedipine*
285*	36	229	42	96	228	44	286	285	Dihydromorphinone
285*	42	70	162	44	284	124	59	285	Morphine
285*	42	215	44	162	124	70	115	285	Morphine
285*	63	75	109	312	183	238	266	313	Flunitrazepam
285*	81	215	148	286	164	110	115	285	Norcodeine*
285*	96	229	228	70	214	115	200	285	Hydromorphone
285*	125	287	109	93	79	47	289	320	Fenchlorvos
285*	162	42	44	31	215	70	200	285	Morphine
285*	162	42	215	286	124	44	284	285	Morphine*
285*	162	284	215	286	124	268	226	301	Morphine N-oxide
285*	215	162	286	124	284	268	174	285	Morphine
285*	215	81	148	110	164	286	132	285	Norcodeine
285*	226	72	58	45	42	44	162	427	Morphine Methoiodide
285*	256	43	110	257	84	287	111	285	7-Aminoclonazepam
285*	268	284	77	42	286	233	287	285	2-Desmethylchlordiazepoxide
285*	270	77	91	105	257	166	287	285	Prazepam Benzophenone
285*	286	269	287	241	242	77	270	286	Demoxepam*
285*	287	125	109	289	79	47	93	320	Fenchlorvos (Ronnel)
285*	312	313	286	266	238	294	284	313	Flunitrazepam*
285	42*	115	44	96	58	228	229	285	Hydromorphone
285	55*	77	270	91	105	166	56	285	2-Cyclopropylmethylamino-5-Chlorobenzophenone
285	58*	72	42	71	59	186	44	301	Morphine-N-oxide
285	58*	214	200	86	227	85	212	285	Prothipendyl*
285	115*	131	128	162	127	77	152	285	Morphine
285	162*	327	215	124	81	115	267	327	3-Acetylmorphine
285	247*	328	249	41	287	330	55	328	Clorexolone*
285	283*	202	139	50	75	63	76	283	Nitrofen
285	313*	312	286	266	238	294	239	313	Flunitrazepam
286*	244	77	218	51	217	288	215	286	Desmethylclobazam
286*	300	96	42	244	230	301	82	492	Benzitramide*
286	43*	299	41	44	227	300	211	584	Bilirubin
286	251*	252	78	288	111	152	271	286	7-Hydroxymecloqualone
286	251*	152	252	288	111	154	160	286	8-Hydroxymecloqualone
286	251*	252	111	288	271	152	273	286	5-Hydroxymecloqualone
286	251*	252	288	111	154	271	152	286	6-Hydroxymecloqualone
286	284*	288	282	249	247	251	214	282	Hexachlorobenzene (HCB)
286	285*	269	287	241	242	77	270	286	Demoxepam*
287*	70	44	42	163	59	288	286	287	Dihydromorphine
287*	70	164	44	42	286	285	288	287	Dihydromorphine
287*	70	44	164	42	288	59	230	287	Dihydromorphine*
287*	96	215	286	229	213	228	243	287	Cyproheptadine
287*	96	286	215	70	44	58	42	287	Cyproheptadine*
287*	125	285	79	109	93	289	47	320	Fenchlorvos
287*	363	176	255	365	289	351	220	380	Chlorthalidone Trimethyl Derivative
287	215*	42	96	229	189	286	216	287	Cyproheptadine
287	285*	125	109	289	79	47	93	320	Fenchlorvos (Ronnel)
288*	273	331	287	304	290	289	275	331	Flurazepam(N1-Ethanol-)
288*	273	331	287	304	289	290	253	332	Hydroxyethylflurazepam
288	253*	287	289	225	259	254	41	288	Tertrazepam
288	260*	287	259	261	289	102	262	288	Desalkylflurazepam(N1-)

	PEAKS						MWT	COMPOUND
288	317*	359	401	43	318	289	196	401 Oxyphenisatin Acetate*
289*	179	43	150	231	304	219	55	304 Cannabichromanon (C3)
289*	318	291	320	275	290	319	317	334 Clotiazepam
290*	67	108	107	79	55	41	93	290 Androsterone
290*	184	185	104	77	168	291	198	290 Sulthiame*
290*	248	108	140	43	65	93	92	290 Acetyldapsone
290*	259	275	291	243	123	200	43	290 Trimethoprim
290	261*	233	148	262	133	176	260	290 1,3-Dimethylhydroxyphenobarbitone Methyl Ether
290	292*	220	294	222	150	293	184	290 2,2',4,4'-Tetrachlorobiphenyl
291*	109	97	137	29	139	78	292	291 Parathion
291*	109	97	137	139	155	125	123	291 Parathion-ethyl
291*	239	274	293	75	302	276	138	320 Lorazepam*
291	97*	109	137	139	155	125	65	291 Parathion
291	174*	73	175	159	131	75	158	291 Indomethacin Derivative TMS Ester
291	246*	248	290	247	139	111	292	293 Zomepirac*
291	292*	119	248	193	249	220	221	292 5-Benzylidene-1-phenylbarbituric Acid
292*	188	121	130	115	293	129	128	292 Racumin (4-Hydroxy-3-(1,2,3,4-tetrahydro-1-naphthyl)coumarin)
292*	290	220	294	222	150	293	184	290 2,2',4,4'-Tetrachlorobiphenyl
292*	291	119	248	193	249	220	221	292 5-Benzylidene-1-phenylbarbituric Acid
292	220*	290	222	150	74	294	255	290 Tetrachlorobiphenyl
293*	264	43	292	263	213	212	294	293 7-Acetoamidonitrazepam
293*	265	264	292	43	222	223	294	293 7-Acetoamidonitrazepam
293	58*	45	59	193	100	178	294	293 Butriptyline*
293	70*	264	112	305	291	262	295	374 Bromhexine
293	275*	96	179	0	0	0	0	293 Penfluridol Hofmann Reaction Product 1
294*	266	42	43	278	58	295	207	335 Benefin
294*	295	221	252	264	280	165	237	295 Apomorphine Dimethyl Ether
294	136*	81	159	55	42	41	143	294 Cinchonine*
294	217*	132	194	118	208	222	77	294 5-Phenylhydantoin(5-(p-methylphenyl))- Dimethyl Derivative
295*	127	168	153	167	296	276	311	341 MK-251 Metabolite
295*	221	181	207	196	180	223	167	295 Lysergic Acid Monoethylamide
295*	249	237	293	297	214	212	142	293 Pentachloronitrobenzene (PCNB)
295*	263	310	251	294	249	277	181	310 Flunixin Methyl Ester
295*	268	297	97	57	270	62	64	295 Chlorthiazide
295*	296	310	238	31	59	311	297	310 Cannabinol
295*	296	43	310	238	299	58	41	310 Cannabinol
295*	296	310	238	251	223	165	119	310 Cannabinol
295	236*	296	263	145	235	0	0	296 Niflumic Acid Methyl Ester
295	263*	264	166	92	235	145	243	295 Flufenamic Acid Methyl Ester
295	294*	221	252	264	280	165	237	295 Apomorphine Dimethyl Ether
296*	41	44	110	298	285	55	268	379 Cyclopenthiazide*
296*	76	104	240	50	295	268	297	296 Dithianon
296*	169	297	127	310	311	0	0	311 MK-251 (Dimethyl-4-(a,a,b,b-tetrafluorophenethyl)benzylamine)
296*	180	267	210	104	77	134	297	296 3-Methyl-3-hydroxydilantin Methyl Ether
296*	195	58	297	253	196	212	84	592 Hexobendine*
296*	267	210	219	134	180	77	297	296 3-Methyl-p-hydroxydilantin Methyl Ether
296*	281	143	279	76	297	39	77	296 2-Methyl-3-(2'-methyl-4'-hydroxy-5'-methoxyphenyl)-quinazoline
296*	298	205	221	64	63	41	125	325 Ethiazide*
296*	298	279	205	36	64	117	62	379 Trichlormethiazide
296	42*	211	239	152	165	311	139	311 Thebaine
296	84*	42	85	180	55	212	41	296 Pyrathiazine
296	127*	169	281	183	154	141	128	296 MK-251 Metabolite
296	295*	310	238	31	59	311	297	310 Cannabinol
296	295*	43	310	238	299	58	41	310 Cannabinol

Peak Index   *Asterisk Indicates True Base Peak*

								MWT	COMPOUND
296	295*	310	238	251	223	165	119	310	Cannabinol
296	311*	42	255	44	310	312	253	311	Thebaine
297*	127	185	170	298	312	311	293	312	MK-251 Metabolite
297	325*	324	43	256	306	296	326	325	7-Acetamidoflunitrazepam
298*	44	252	253	297	279	280	145	298	5-Hydroxyniflumic Acid
298*	163	43	41	55	162	57	175	680	Tubocurarine Chloride
298*	232	204	252	251	279	233	280	298	4'-Hydroxyniflumic Acid
298*	271	299	224	272	270	252	280	299	Desmethylflunitrazepam
298*	594	58	593	299	609	595	564	680	Tubocurarine Chloride
298	58*	212	198	100	299	252	199	298	Trimeprazine*
298	86*	83	58	87	30	85	180	298	Diethazine
298	86*	87	30	58	299	212	180	298	Diethazine*
298	146*	157	156	129	118	90	97	298	Quinalphos
298	296*	205	221	64	63	41	125	325	Ethiazide*
298	296*	279	205	36	64	117	62	379	Trichlormethiazide
298	341*	342	242	284	299	340	326	341	Thebacon
299*	42	162	229	44	124	59	300	299	Codeine
299*	42	124	162	300	297	44	59	299	Codeine
299*	42	162	124	229	59	300	69	299	Codeine*
299*	44	242	59	96	42	76	243	299	Hydrocodone
299*	81	42	229	162	124	300	44	299	Codeine
299*	96	242	243	185	228	214	300	299	Metopon
299*	162	229	124	300	214	298	42	299	Codeine
299*	162	229	298	124	214	297	282	299	Codeine
299*	162	298	229	115	42	214	124	299	Codeine
299*	162	229	123	59	42	44	300	299	Neopine
299*	185	242	115	42	214	128	243	299	Dihydrocodienone
299*	229	162	297	240	241	242	298	315	Codeine N-oxide
299*	231	314	43	41	295	55	271	314	Tetrahydrocannabinol(Delta-9)
299*	242	243	59	57	214	185	300	299	Dihydrocodeinone
299*	242	256	96	243	58	60	214	299	Hydrocodone
299*	242	59	243	42	96	70	214	299	Hydrocodone*
299*	300	301	283	77	43	256	302	300	Diazepam-N-Oxide
299*	330	59	43	300	0	0	0	330	11-HydroxyTHC (delta 9)
299*	343	358	231	283	290	315	326	358	Methylated THC-9-oic Acid
299*	344	329	283	276	288	273	281	344	Tetrahydrocannabinol-9-oic Acid
299	282*	218	284	283	219	241	214	299	Chlordiazepoxide
299	282*	284	283	241	77	55	91	299	Chlordiazepoxide
299	282*	284	283	241	56	301	253	299	Chlordiazepoxide*
299	301*	303	332	45	330	221	334	330	Chlorthal Methyl
299	301*	303	221	142	223	317	315	346	Glenbar
299	314*	231	271	243	258	41	43	314	Tetrahydrocannabinol
299	314*	231	271	243	258	315	300	314	Tetrahydrocannabinol (delta 9)
300*	77	258	51	255	259	256	283	300	Clobazam
300*	258	77	259	283	302	231	256	300	Clobazam*
300	160*	189	145	188	301	161	42	300	Azapropazone*
300	286*	96	42	244	230	301	82	492	Benzitramide*
300	299*	301	283	77	43	256	302	300	Diazepam-N-Oxide
300	301*	302	164	244	242	284	286	301	Dihydrocodeine
300	343*	184	43	70	59	344	326	343	Acetyldihydrocodeine
301*	44	42	59	164	70	302	242	301	Dihydrocodeine*
301*	216	44	42	70	302	203	57	301	Oxymorphone*
301*	299	303	332	45	330	221	334	330	Chlorthal Methyl
301*	299	303	221	142	223	317	315	346	Glenbar

301*	300	302	164	244	242	284	286	301 Dihydrocodeine
301	303*	305	115	194	196	114	87	301 Broxyquinoline*
302*	43	41	135	287	91	105	121	302 Abietic Acid
302*	57	85	231	91	110	79	215	302 Norethandolone
302*	124	43	91	79	121	105	122	302 Methyltestosterone
302	246*	58	71	289	341	56	274	359 Monodesethylflurazepam
302	260*	216	215	243	189	242	188	302 Cambendazole
303*	70	58	44	57	42	216	286	303 Hydromorphinol
303*	96	302	231	202	245	259	288	303 Cyproheptadine Metabolite
303*	96	302	231	202	245	259	304	303 3-Hydroxycyproheptadine
303*	301	305	115	194	196	114	87	301 Broxyquinoline*
303*	331	239	255	30	158	64	159	331 Hydroflumethiazide*
303*	359	97	357	331	242	125	109	392 Bromophos Ethyl
303	268*	269	305	304	0	0	0	303 Chloromorphide (beta)
303	332*	275	304	190	276	134	333	332 1,3-Diethylhydroxyphenobarbitone Ethyl Ether
304*	44	75	306	51	57	50	305	334 Lormetazepam
304*	306	64	74	109	177	176	48	304 Dichlorphenamide*
304	58*	214	42	59	306	232	233	304 Chlorphenethiazine
305*	150	307	115	152	114	306	123	305 Cliquinol*
306*	175	291	115	91	202	187	129	306 Methylated Coumatetralyl*
306	249*	250	170	307	251	171	277	306 Dihydroxyzolimidine (Metab 2)
306	277*	288	278	209	198	222	237	306 Dihydroxyzolimidine (Metab 1)
306	304*	64	74	109	177	176	48	304 Dichlorphenamide*
308*	165	309	121	55	154	98	56	429 Mebeverine*
308*	266	440	43	512	152	194	126	554 Practolol, tri-TFA Derivative
308	180*	279	208	77	251	165	104	308 1,3-Diethyl-5,5-diphenylhydantoin
309	154*	197	77	155	195	307	353	864 Decafentin
309	207*	208	182	181	209	180	167	309 Nor LSD (6-desmethyl-LSD)
310*	42	312	129	45	64	'30	131	439 Polythiazide
310*	64	36	312	42	43	62	63	359 Methychlothiazide (Endurone)
310*	111	112	58	199	212	96	41	310 Pecazine (Mepazine)
310*	121	92	43	120	311	63	65	353 Acenocoumarin
310*	121	353	311	43	120	92	296	353 Nicoumalone*
310*	206	42	64	312	299	45	48	439 Polythiazide*
310	136*	135	225	149	122	155	311	310 Ibogaine*
310	227*	174	284	147	160	173	199	310 Mestranol
311*	43	60	41	312	45	241	188	311 Nalorphine
311*	44	255	296	310	312	42	253	311 Thebaine
311*	127	169	312	295	153	199	184	326 MK-251 Metabolite
311*	255	42	44	296	310	312	174	311 Thebaine*
311*	296	42	255	44	310	312	253	311 Thebaine
311*	312	41	188	80	82	81	241	311 Nalorphine*
311	41*	39	188	81	115	42	77	311 Nalorphone
312*	43	399	297	356	281	371	311	399 Colchicine*
312*	371	43	297	399	281	298	313	399 Colchicine
312	43*	297	281	254	298	152	139	399 Colchicine
312	123*	59	58	70	115	171	270	312 6-Deoxy-6-azidodihydroisomorphine
312	149*	270	354	297	269	167	256	414 Acetoxytetrahydrocannabinol (8-alpha-Acetoxy-delta 9-)
312	285*	313	286	266	238	294	284	313 Flunitrazepam*
312	311*	41	188	80	82	81	241	311 Nalorphine*
313*	42	162	36	81	124	59	44	313 Ethylmorphine
313*	137	315	164	312	0	0	0	313 Flurazepam Metabolite
313*	137	315	314	0	0	0	0	313 Flurazepam Metabolite 1
313*	138	42	282	229	176	115	146	313 Morphine Dimethyl Derivative

PEAKS							MWT	COMPOUND
313*	162	314	124	284	59	42	243	313 Ethylmorphine*
313*	229	138	162	42	0	0	0	313 Dimethylmorphine
313*	243	162	124	59	42	112	0	313 Ethylmorphine
313*	285	312	286	266	238	294	239	313 Flunitrazepam
313*	315	314	312	137	250	183	273	313 Didesethylflurazepam Dehydration Product
313	223*	208	224	222	180	194	298	313 Mefenamic Acid TMS Derivative
313	342*	238	344	315	102	75	137	342 Triazolam
314*	229	242	44	269	86	283	71	314 Desmethyltrimeprazine
314*	229	242	269	44	57	71	70	314 Methotrimeprazine Metabolite
314*	255	302	246	316	211	315	273	331 Didesethylflurazepam
314*	299	231	271	243	258	41	43	314 Tetrahydrocannabinol
314*	299	231	271	243	258	315	300	314 Tetrahydrocannabinol (delta 9)
314	221*	248	261	193	236	222	315	314 Tetrahydrocannabinol (delta 8)
314	231*	271	258	43	193	41	246	314 Tetrahydrocannabinol (delta-8)
314	280*	315	234	289	240	75	76	315 Clonazepam
314	280*	315	286	234	288	316	240	315 Clonazepam*
315*	81	43	55	368	41	107	95	386 Cholesterol
315*	230	314	316	258	201	70	229	315 Oxycodone
315*	230	316	70	44	42	258	140	315 Oxycodone*
315*	316	142	144	177	179	107	285	315 1,3,5-Trichloro-2,4,6-trinitrobenzene (Bulbosan)
315	236*	317	44	77	91	287	316	315 Bromazepam
315	236*	317	286	288	316	208	179	315 Bromazepam
315	313*	314	312	137	250	183	273	313 Didesethylflurazepam Dehydration Product
316	315*	142	144	177	179	107	285	315 1,3,5-Trichloro-2,4,6-trinitrobenzene (Bulbosan)
317*	43	271	147	289	390	95	81	390 Alphadolone Acetate*
317*	223	75	147	217	43	318	73	557 Paracetamol Glucuronide Methyl Ester TMS Ether
317*	288	359	401	43	318	289	196	401 Oxyphenisatin Acetate*
317	73*	169	171	41	318	217	75	620 Clofibrate Glucuronide Methyl Ester TMS Ether
317	73*	217	75	147	318	43	79	688 5-(4-Hydroxyphenyl)-3-methyl-5-phenylhydantoin Glucuronide me Ester TMS
317	236*	315	288	316	286	208	78	315 Bromazepam*
318*	246	316	248	320	176	105	247	316 DDE (p,p'-)
318*	274	77	226	104	44	210	105	318 Sulthiame Dimethyl Derivative
318	58*	86	272	85	320	232	42	318 Chlorpromazine
318	121*	317	173	120	362	44	31	408 Ethyl Biscoumacetate*
318	233*	193	234	262	319	273	136	318 Tetrahydrocannabidiol
318	246*	316	248	320	176	210	0	316 DDE (p,p-)
318	289*	291	320	275	290	319	317	334 Clotiazepam
318	333*	57	304	168	180	71	166	333 Pirimiphos Ethyl
318	346*	259	345	347	273	348	320	346 Flurazepam N1-Acetic Acid
319	58*	86	321	85	273	36	274	318 Chlorpromazine
319	73*	291	348	206	320	45	333	348 1,3-Dimethylhydroxyphenobarbitone TMS Ether
321*	278	320	292	306	191	322	304	353 Berberine
321*	364	304	240	168	91	322	365	364 Bumetanide*
321	98*	83	55	140	41	155	182	999 Methyprylone Probe Product (Dimer)*
323*	73	338	307	75	249	154	143	338 Methaqualone Metabolite TMS Derivative
323*	91	154	338	251	266	309	307	338 Methaqualone Metabolite TMS Derivative
323*	91	132	338	154	0	0	0	338 Methaqualone Metabolite TMS Derivative
323*	221	44	199	222	223	207	76	323 LSD
323*	221	181	222	207	72	223	324	323 LSD*
323*	221	207	181	222	196	223	72	323 LSD (iso)
323*	235	338	73	176	91	75	247	338 Methaqualone Metabolite TMS Derivative
323*	338	73	321	154	143	163	249	338 Methaqualone Metabolite TMS Derivative
323*	338	73	321	143	154	0	0	338 Methaqualone Metabolite TMS Derivative
323*	338	73	75	249	143	0	0	338 Methaqualone Metabolite TMS Derivative

PEAKS   MWT   COMPOUND

323*	338	91	73	321	154	132	149	338	Methaqualone Metabolite TMS Derivative
323*	338	91	321	75	154	73	149	338	Methaqualone Metabolite TMS Derivative
323	221*	181	222	207	72	223	180	323	LSD (d) Tartrate
323	267*	269	325	81	109	295	170	358	Chlorfenvinphos (alpha)
324*	193	206	73	325	75	194	45	339	Salicyluric Acid TMS Ester TMS Ether
324*	338	339	308	293	325	220	340	339	Papaverine
324	58*	154	114	105	96	72	77	382	Dimethylaminoethyl-3-piperidylbenzilate (N-Dimethyl)
324	136*	189	81	173	138	55	0	324	Quinidine
324	154*	89	338	77	51	339	102	339	Papaverine
324	199*	93	77	65	55	121	135	324	Oxyphenbutazone*
324	269*	91	296	295	55	323	325	324	Prazepam
324	326*	328	330	254	256	184	258	324	Pentachlorobiphenyl
324	339*	338	325	340	308	154	292	339	Papaverine*
325*	221	207	196	181	44	223	42	325	Ergometrine
325*	223	225	224	326	226	182	167	325	2,3-Dihydro-LSD
325*	297	324	43	256	306	296	326	325	7-Acetamidoflunitrazepam
326*	43	144	182	58	311	183	145	326	Rauwolfine
326*	324	328	330	254	256	184	258	324	Pentachlorobiphenyl
326	123*	59	298	70	198	185	115	326	6-Deoxy-6-azidodihydroisocodeine
326	138*	55	110	189	82	160	139	326	Hydroquinidine*
327*	43	268	42	215	81	146	59	327	6-Acetylmorphine
327*	43	369	268	310	42	215	204	369	Diamorphine*
327*	242	328	286	229	96	41	201	327	Naloxone
327*	268	215	328	269	267	146	124	327	Acetylmorphine
327*	268	43	42	215	44	328	146	327	Acetylmorphine
327*	268	215	59	204	70	146	81	327	6-Acetylmorphine
327*	268	328	42	43	215	44	59	327	6-Acetylmorphine*
327*	328	41	242	286	96	229	70	327	Naloxone*
327*	369	43	268	310	42	215	128	369	Diamorphine
327	43*	299	298	292	329	328	256	327	7-Acetoamidoclonazepam
327	95*	39	329	96	122	244	67	327	Diloxanide Furoate*
327	131*	73	143	75	132	328	169	342	1,3-Dimethylhydroxyamobarbitone TMS Ether
328*	69	329	126	441	315	0	0	441	Dopamine TFA Derivative
328*	70	216	58	187	115	286	285	328	6-Deoxy-6-azido-14-hydroxydihydroisomorphine
328	58*	100	42	135	228	229	242	328	Methotrimeprazine
328	58*	100	269	229	283	242	243	328	Methotrimeprazine
328	58*	100	228	185	329	242	229	328	Methotrimeprazine*
328	58*	100	43	71	269	229	207	328	Methotrimeprazine Metabolite
328	58*	100	229	269	207	242	283	328	Methotrimeprazine Metabolite
328	58*	229	242	100	269	283	0	344	Methotrimeprazine Sulphoxide
328	327*	41	242	286	96	229	70	327	Naloxone*
329*	284	224	268	330	285	225	270	346	Nifedipine*
329	331*	125	333	93	109	332	62	364	Bromophos
329	331*	125	333	79	109	47	93	364	Bromophos
330*	118	91	319	219	64	92	421	421	Bendrofluazide*
330*	332	62	63	141	334	328	143	328	2,4,6-Tribromophenol
330	299*	59	43	300	0	0	0	330	11-HydroxyTHC (delta 9)
331*	43	71	332	121	147	131	41	408	Clobetasone*
331*	329	125	333	93	109	332	62	364	Bromophos
331*	329	125	333	79	109	47	93	364	Bromophos
331	303*	239	255	30	158	64	159	331	Hydroflumethiazide*
332*	303	275	304	190	276	134	333	332	1,3-Diethylhydroxyphenobarbitone Ethyl Ether
332*	333	304	335	176	303	334	162	353	Chelidonine
332	330*	62	63	141	334	328	143	328	2,4,6-Tribromophenol

Peak Index   *Asterisk Indicates True Base Peak*

333*	318	57	304	168	180	71	166	333	Pirimiphos Ethyl
333	332*	304	335	176	303	334	162	353	Chelidonine
334*	44	120	77	41	107	55	144	334	Strychnine
334*	130	120	333	143	107	162	144	334	Strychnine
334*	335	162	120	107	144	143	130	334	Strychnine*
335*	193	192	207	234	233	167	180	335	6-Cyano-6-desmethyl-LSD
335	334*	162	120	107	144	143	130	334	Strychnine*
336*	109	227	249	115	338	0	0	336	Cruformate Metabolite
336*	121	120	215	162	92	337	187	336	Dicoumarol*
338*	339	77	75	296	310	78	76	398	Adriamycin Metabolite
338	73*	323	247	179	139	235	154	338	Methaqualone Metabolite TMS Derivative
338	323*	73	321	154	143	163	249	338	Methaqualone Metabolite TMS Derivative
338	323*	73	321	143	154	0	0	338	Methaqualone Metabolite TMS Derivative
338	323*	73	75	249	143	0	0	338	Methaqualone Metabolite TMS Derivative
338	323*	91	73	321	154	132	149	338	Methaqualone Metabolite TMS Derivative
338	323*	91	321	75	154	73	149	338	Methaqualone Metabolite TMS Derivative
338	324*	339	308	293	325	220	340	339	Papaverine
338	352*	367	366	353	336	322	339	367	Dimoxyline
339*	221	196	181	207	223	222	72	339	Methylergometrine*
339*	221	223	207	196	181	222	340	339	Methylergometrine Maleate
339*	237	197	223	238	212	229	340	339	13-Hydroxy-LSD
339*	237	197	238	239	223	196	212	339	12-Hydroxy-LSD
339*	324	338	325	340	308	154	292	339	Papaverine*
339*	384	45	43	323	321	322	366	384	Daunomycin Metabolite
339	113*	44	70	141	43	60	340	339	Perazine*
339	338*	77	75	296	310	78	76	398	Adriamycin Metabolite
340	43*	298	325	91	41	231	280	340	Norethisterone Acetate
340	123*	59	283	115	322	199	70	340	3-Ethoxy-6-deoxy-6-azidodihydroisomorphine
340	173*	168	167	165	341	174	322	340	Diphenadione*
340	399*	73	287	43	400	342	341	399	Acetylmorphine TMS Ether
341*	43	340	374	267	107	55	342	416	Spironolactone*
341*	55	36	300	342	110	243	256	341	Naltrexone
341*	245	246	274	343	313	58	299	359	Monodesethylflurazepam
341*	282	43	229	42	342	204	162	341	Acetylcodeine
341*	282	229	204	81	70	149	124	341	6-Acetylcodeine
341*	282	42	43	229	59	204	342	341	Acetylcodeine
341*	282	229	42	43	59	342	204	341	6-Acetylcodeine*
341*	298	342	242	284	299	340	326	341	Thebacon
341	73*	342	343	45	344	327	91	342	Desmethyldiazepam TMS Derivative
341	73*	43	271	75	147	41	342	444	1,3-Dimethyldihydroxysecobarbitone di-TMS Ether
342*	70	230	201	314	115	58	300	342	6-Deoxy-6-azido-14-hydroxydihydroisocodeine
342*	313	238	344	315	102	75	137	342	Triazolam
343*	43	70	284	59	300	344	42	343	Acetyldihydrocodeine*
343*	55	110	36	98	302	84	344	343	Hydroxynaltrexone (alpha)
343*	70	344	109	42	113	491	56	491	Lidoflazine*
343*	184	300	70	43	59	334	226	343	Acetyldihydrocodeine
343*	257	256	372	357	283	371	0	372	Methyloxazepam TMS Derivative
343*	300	184	43	70	59	344	326	343	Acetyldihydrocodeine
343*	372	345	357	257	0	0	0	372	3-Hydroxydiazepam TMS Derivative
343	73*	257	256	345	372	283	45	372	3-Hydroxydiazepam TMS Derivative
343	299*	358	231	283	290	315	326	358	Methylated THC-9-oic Acid
344*	73	372	371	346	345	373	374	372	4'-Hydroxydiazepam TMS Derivative
344	112*	121	41	67	55	54	345	359	Cyclomethycaine*
344	299*	329	283	276	288	273	281	344	Tetrahydrocannabinol-9-oic Acid

Asterisk Indicates True Base Peak

								MWT	COMPOUND
346*	318	259	345	347	273	348	320	346	Flurazepam N1-Acetic Acid
347	376*	319	348	377	361	73	192	376	1,3-Diethylhydroxyphenobarbitone TMS Ether
350*	91	259	352	348	65	351	107	365	Metolazone*
352*	338	367	366	353	336	322	339	367	Dimoxyline
352	58*	86	353	85	306	42	266	352	Fluopromazine*
352	138*	215	310	214	69	321	354	352	Griseofulvin
352	156*	351	184	169	209	353	129	352	Ajmalicine
353*	210	235	336	72	54	236	195	353	Methysergide*
353*	235	210	72	54	45	221	195	353	Methysergide Hydrogen Maleate
353*	251	211	252	237	253	226	196	353	13-Methoxy-LSD
353*	251	237	211	249	338	252	351	353	14-Methoxy-LSD
353*	251	211	237	253	252	338	354	353	12-Methoxy-LSD
353*	354	169	170	355	156	184	144	354	Yohimbine*
353	42*	310	138	44	75	288	218	353	Hydrochlorothiazide Tetramethyl Derivative
353	81*	355	351	357	237	386	0	386	Heptachlor Epoxide
353	263*	368	73	249	75	277		368	Flunixin TMS Ester
354*	73	104	268	325	282	355	77	354	5-(3-Hydroxyphenyl)-3-methyl-5-phenylhydantoin TMS Ether
354*	277	325	104	268	73	269	355	354	5-(4-Hydroxyphenyl)-3-methyl-5-phenylhydantoin TMS Ether
354*	355	340	324	281	162	297	312	355	Boldine Dimethyl Ether
354	353*	169	170	355	156	184	144	354	Yohimbine*
355	354*	340	324	281	162	297	312	355	Boldine Dimethyl Ether
357	359*	361	237	355	272	251	239	390	Bromodan
359*	357	361	237	355	272	251	239	390	Bromodan
359	58*	358	403	360	343	388	227	403	Amotriphene*
359	97*	242	303	357	331	240	301	392	Bromophos Ethyl
359	303*	97	357	331	242	125	109	392	Bromophos Ethyl
360*	362	290	358	288	364	145	292	358	2,4,5,2',3',4'-Hexachlorobiphenyl
360*	362	358	290	288	364	292	145	358	Hexachlorobiphenyl(bis-2,4,5-)
361*	276	277	199	319	43	318	362	361	Bisacodyl
361*	277	319	276	199	318	362	43	361	Bisacodyl*
362*	109	97	226	210	125	364	29	362	Coumaphos
362	360*	290	358	288	364	145	292	358	2,4,5,2',3',4'-Hexachlorobiphenyl
362	360*	358	290	288	364	292	145	358	Hexachlorobiphenyl(bis-2,4,5-)
363*	176	286	255	365	288	192	220	394	Chlorthalidone Tetramethyl Derivative
363*	206	143	42	70	207	218	113	363	Opipramol*
363	287*	176	255	365	289	351	220	380	Chlorthalidone Trimethyl Derivative
364	321*	304	240	168	91	322	365	364	Bumetanide*
365*	195	221	366	31	29	212	197	578	Deserpidine (Canescine, Harmonyl)
365*	263	221	264	223	0	0	0	365	1-Acetyllysergic Acid Diethylamide
366	246*	106	247	133	260	234	367	366	Piminodine (Alvodine, Pimadin, Cimadon)
366	246*	42	106	133	247	57	260	366	Piminodine Ethane Sulphonate
367	246*	247	42	91	103	77	57	367	Phenoperidine
367	368*	91	369	165	108	90	65	368	Tricresyl Phosphate(m-)
367	368*	107	108	369	91	198	165	368	Tricresyl Phosphate(p-)
368*	165	91	179	181	90	180	107	368	Tricresyl Phosphate(o-)
368*	367	91	369	165	108	90	65	368	Tricresyl Phosphate(m-)
368*	367	107	108	369	91	198	165	368	Tricresyl Phosphate(p-)
368*	386	275	149	353	147	145	247	386	Cholesterol
368	386*	43	55	275	81	57	95	386	Cholesterol*
368	397*	382	398	354	29	340	312	397	Octaverine*
369	68*	41	86	370	73	40	140	455	Salbutamol TMS Derivative
369	327*	43	268	310	42	215	128	369	Diamorphine
370	98*	126	99	40	70	371	258	370	Thioridazine*
371	207*	312	342	208	372	42	328	371	Demecoline*

371	312*	43	297	399	281	298	313	399	Colchicine
372	343*	345	357	257	0	0	0	372	3-Hydroxydiazepam TMS Derivative
375	246*	42	247	91	376	156	184	452	Diphenoxylate
376*	347	319	348	377	361	73	192	376	1,3-Diethylhydroxyphenobarbitone TMS Ether
378	100*	113	56	101	87	379	194	378	Doxapram
379	394*	395	197	107	120	203	55	394	Brucine
380*	223	382	238	152	345	215	113	380	Chlorotrianisene
380*	382	345	381	190	223	238	113	380	Chlorotrianisene
382	174*	161	187	159	175	383	41	382	3,5-Cholestadiene-7-one
382	380*	345	381	190	223	238	113	380	Chlorotrianisene
383	233*	259	43	245	31	95	205	383	Prazosin
384	339*	45	43	323	321	322	366	384	Daunomycin Metabolite
386*	138	387	84	110	42	301	263	430	Piritramide*
386*	368	43	55	275	81	57	95	386	Cholesterol*
386	368*	275	149	353	147	145	247	386	Cholesterol
394*	379	395	197	107	120	203	55	394	Brucine
394*	395	120	107	146	91	134	79	394	Brucine
394*	395	379	392	120	197	203	393	394	Brucine*
395*	253	396	294	268	74	221	128	395	Lsd-TMS Derivative
395	181*	198	251	397	396	199	666	666	Syrosingopine
395	394*	120	107	146	91	134	79	394	Brucine
395	394*	379	392	120	197	203	393	394	Brucine*
397*	115	242	398	88	143	271	62	397	Diiodohydroxyquinoline*
397*	368	382	398	354	29	340	312	397	Octaverine*
399*	113	70	141	43	72	400	71	399	Thiethylperazine Maleate (Torecane)
399*	340	73	287	43	400	342	341	399	Acetylmorphine TMS Ether
404*	282	106	78	229	267	42	124	404	Nicocodeine
411	232*	472	91	413	474	0	0	472	4-Chlorodiphenoxylic Acid
423*	55	364	255	84	59'	424	121	423	Cyprenorphine*
425	268*	143	269	197	394	157	171	425	Carphenazine
427	58*	234	59	50	42	428	91	445	Narceine*
429	73*	430	45	431	147	432	75	430	3-Hydroxydesmethyldiazepam
429	73*	236	196	146	414	430	287	429	Morphine bis TMS Ether
438	232*	378	380	246	91	407	423	438	Diphenoxylic Acid
440*	69	126	441	0	0	0	0	553	6-Hydroxydopamine TFA Derivative
440*	126	69	427	0	0	0	0	553	Noradrenaline TFA Derivative
445	141*	169	123	155	96	42	317	445	Metopimazine*
449	73*	359	450	147	269	75	243	654	Cortolone(Tetra TMS)
455*	457	472	474	459	456	458	473	472	Clofazimine*
457	455*	472	474	459	456	458	473	472	Clofazimine*
460*	43	401	359	487	545	402	461	587	3-Acetamido-2,4,6-triiodophenoxyacetic Acid
464*	43	81	523	204	69	70	411	523	6-Acetylmorphine (3-Heptafluorobutyrate Derivative)
465	284*	81	285	42	175	181	162	465	Monopentafluorobenzylmorphine
473	504*	429	505	221	474	84	430	504	Dipyridamole*
474*	43	401	359	432	487	418	475	601	3-Acetamido-2,4,6-triiodophenoxyacetic Acid Methyl Ester
481*	43	44	53	268	310	523	69	523	3-Acetylmorphine 6-Heptafluorobutyrate Derivative
494	585*	91	43	57	41	73	55	585	Myrophine
504*	473	429	505	221	474	84	430	504	Dipyridamole*
518*	173	264	519	373	376	520	249	518	Benziodarone*
546*	86	420	517	547	391	263	250	546	Amiodarone Probe Product*
560	59*	331	428	528	402	433	373	687	Iopronic Acid Methyl Ester
578*	195	577	367	351	579	366	365	578	Deserpidine*
585*	494	91	43	57	41	73	55	585	Myrophine
594	298*	58	593	299	609	595	564	680	Tubocurarine Chloride

```
          PEAKS                    MWT       COMPOUND


606  608* 195  609  395  397  212  396  608 Reserpine*
608* 606  195  609  395  397  212  396  608 Reserpine*
```